Lecture Notes in Mathematics

Edited by A. Dold, F. Takens and B. Teissier

Editorial Policy
for the publication of monographs

1. Lecture Notes aim to report new developments in all areas of mathematics – quickly, informally and at a high level. Monograph manuscripts should be reasonably self-contained and rounded off. Thus they may, and often will, present not only results of the author but also related work by other people. They may be based on specialized lecture courses. Furthermore, the manuscripts should provide sufficient motivation, examples and applications. This clearly distinguishes Lecture Notes from journal articles or technical reports which normally are very concise. Articles intended for a journal but too long to be accepted by most journals, usually do not have this "lecture notes" character. For similar reasons it is unusual for doctoral theses to be accepted for the Lecture Notes series.

2. Manuscripts should be submitted (preferably in duplicate) either to one of the series editors or to Springer-Verlag, Heidelberg. In general, manuscripts will be sent out to 2 external referees for evaluation. If a decision cannot yet be reached on the basis of the first 2 reports, further referees may be contacted: the author will be informed of this. A final decision to publish can be made only on the basis of the complete manuscript, however a refereeing process leading to a preliminary decision can be based on a pre-final or incomplete manuscript. The strict minimum amount of material that will be considered should include a detailed outline describing the planned contents of each chapter, a bibliography and several sample chapters.
Authors should be aware that incomplete or insufficiently close to final manuscripts almost always result in longer refereeing times and nevertheless unclear referees' recommendations, making further refereeing of a final draft necessary.
Authors should also be aware that parallel submission of their manuscript to another publisher while under consideration for LNM will in general lead to immediate rejection.

3. Manuscripts should in general be submitted in English.
Final manuscripts should contain at least 100 pages of mathematical text and should include
- a table of contents;
- an informative introduction, with adequate motivation and perhaps some historical remarks: it should be accessible to a reader not intimately familiar with the topic treated;
- a subject index: as a rule this is genuinely helpful for the reader.

Continued on back inside cover

Lecture Notes in Mathematics 1740

Springer
Berlin
Heidelberg
New York
Barcelona
Hong Kong
London
Milan
Paris
Singapore
Tokyo

B. Kawohl O. Pironneau L. Tartar
J.-P. Zolésio

Optimal Shape Design

Lectures given at the joint C.I.M./C.I.M.E.
Summer School held in Tróia, Portugal,
June 1–6, 1998

Editor: A. Cellina and A. Ornelas

Fondazione
C.I.M.E.

Springer

Author

Bernhard Kawohl
Mathematisches Institut
Universität zu Köln
Weyertal 86-90
50931 Köln, Germany
E-mail: kawohl@MI.uni-koeln.de

Olivier Pironneau
Université Pierre et Marie Curie
Département de Mathématique
4, Place Jussieu
75252 Paris, France
E-mail: Olivier.Pironneau@inria.fr

Luc Tartar
Carnegie Mellon University
Department of Mathematical Sciences
Schenley Park
Pittsburgh, PA, 15213-3890, USA
E-mail: tartar@andrew.cmu.edu

Jean-Paul Zolésio
INRIA
Centre de Mathématiques Appliquées
2004 Route de Lucioles, B.P. 93
06902 Sophia Antipolis cedex, France
E-mail: Jean-Paul.Zolesio@sophia.inria.fr

Editors

Arrigo Cellina
Universita di Milano - Bicocca
Dipartimento di Matematica
e Applicazioni
Via Bicocca degli Arcimboldi 8
20126 Milano, Italy
E-mail: cellina@ares.mat.unimi.it

António Ornelas
Centro de Investigação em
Matemática e Aplicações
Universidade de Evora
rua Romão Ramalho 59
7000-671 Évora Portugal
E-mail: ornelas@uevora.pt

Cataloging-in-Publication Data applied for

Die Deutsche Bibliothek - CIP-Einheitsaufnahme

Optimal shape design : lectures given at the joint CIM, CIME summer school, held in Troia, Portugal, June 1 - 6, 1998 / Fondazione CIME. B. Kawohl ... Ed.: A. Cellina and A. Ornelas. - Berlin ; Heidelberg ; New York ; Barcelona ; Hong Kong ; London ; Milan ; Paris ; Singapore ; Tokyo : Springer, 2000
(Lecture notes in mathematics ; Vol. 1740 : Subseries: Fondazione CIME)
ISBN 3-540-67971-5

Mathematics Subject Classification (2000): 49K20, 65K10, 65N55

ISSN 0075-8434
ISBN 3-540-67971-5 Springer-Verlag Berlin Heidelberg New York

Springer-Verlag Berlin Heidelberg New York
a member of BertelsmannSpringer Science+Business Media GmbH

Printed in Germany

Typesetting: Camera-ready TEX output by the author
SPIN: 10724274 41/3142-543210 - Printed on acid-free paper

Table of Contents

Optimal Shape Design

Introduction

These lectures were presented at the joint C.I.M./C.I.M.E. Summer School on Optimal Shape Design held in Tróia (Portugal), June 02 to June 07, 1998. The mathematical problems that can be described by the label "Optimal shape design" form a broad area: it concerns the optimization of some performance criterion where the criterion depends, besides constraints that qualify the problem, on the "shape" of some region. A classical setting is Structural Mechanics of elastic bodies such as bridges, beams, plates, shells, arches. These structures have to satisfy requirements of load and have to be designed in an optimal way; for example, should be built using the least amount of material. Alternatively, one might seek the optimal shape of a geometrical object moving in a fluid: B. Kawohl devotes most of his lectures to the classical Newton's problem of minimal resistance. This fascinating problem, studied by Newton in the interest of "Her Majesty's Navy", as reported by Kawohl, goes back about three hundred years and has been a subject of controversy and discussion from the very beginning (the functional to be minimized, although rotationally symmetric, is not convex, and this explains the difficulty of the problem). Alternatively, we may think of seeking the optimal shape of a wing. to be designed so as to reduce the drag while keeping a given value for the lift. Or we might wish to design the optimal shape of a region (a harbor), given suitable constraints on the size of the entrance to the harbor, subject to incoming waves, so as to minimize the height of the waves inside it. Or we might wish to design some electrical device consisting of a (simply connected) region (partially) coated with a conducting material, say copper (the non-covered portion of the region is considered to be a perfect insulator): the goal is to minimize the cost of the device, subject to constraints on the performance of the resulting design. Or we might try to design materials obtained layering several materials, with different characteristics: the goal in this case could become that of computing the effective properties of the limit material.

A large class of problems of this kind can be reduced to a standard formulation of the Calculus of Variations, i.e. that of minimizing an integral functional of the kind

$$\int_\Omega f(x, u(x), \nabla u(x))\, dx$$

subject to boundary conditions and possibly to additional constraints on $u(x)$ and $\nabla u(x)$ (a variant of this formulation would be that of a Control Problem; in this formulation the shape is -in general- the control). In this case one minimizes an energy or work functional with respect to the design parameters. However some interesting shape optimization problems do not fall into this formulation: a remarkable class of problems not of this form are the "opaque square" problem and its generalizations, as presented here by B. Kawohl.

In an optimal shape problem the following (not easy) questions arise: will there be an optimal solution? And, if the answer is "no", how should one relax the problem? Then: how does one proceed to derive necessary conditions?

To see the relevance of the existence part, consider that it might very well happen that it is convenient, in order to reduce the overall cost, to perforate the material (Ω) with many fine holes (with a geometry depending on the problem). If this is the case, it is likely that there will not be a limit to the scale of the perforation i.e. that the optimal design will not be realized, or that it will be realized only in a suitable generalized class of designs. The reason is the inherent lack of convexity of the problem: in fact, the natural setting is to assume, as possible solutions, the class of characteristic functions of subsets of Ω, a non-convex set. At this point one might try to overcome the lack of convexity from the beginning by using some special tools (Liapunov's Theorem on the range of non-atomic measures has been used for some problem, so far when Ω is one-dimensional or for rotationally symmetric functionals; explicit construction of solutions has been possible in some problems, in the case of affine boundary conditions; the use of Baire's Category theorem as an existence tool, seeing it as a Theorem of non-empty intersection, in the case of more general boundary conditions). The references at the end of this Introduction provide an overview of papers in the above directions and are meant to be an addition to the references presented at the end of each Chapter; largely they concern the existence of Minima in Problems of the Calculus of Variations without convexity condition. These references are not necessary relevant to the special problems of Optimal Shape Design. In fact, the reader will notice that a chapter on "General Theory of Existence of solutions on Problems of Optimal Shape" is missing from this book. In fact, such a general theory does not exist. A device that has been used by several investigators consists in replacing characteristic functions by density functions, whose values range from zero to one. In the case the final optimal solution is a characteristic function, one has a true solution; otherwise it will be some sort of generalized solution.

The remarkable set of lectures by L. Tartar deals with the Homogenization method. In a terse and strongly personal way the author revisits his own contributions to the history of the mathematical discoveries and progress in this area. A typical problem is that of considering a fixed region Ω to be covered by r symmetric materials M_i in quantities k_i. One wants to minimize a given functional subject to the state equation

$$-div(Agrad(u)) = f$$

and to the constraints $\sum_i k_i \geq meas(\Omega)$ and $\int_\Omega \chi_i dx \leq k_i$. Since, as we have noted, the set of characteristic functions is not convex, one has to identify a relaxed formulation for the problem.

Domain variations in Partial Differential Equations are the subject of the comprehensive set of lectures by J. P. Zolesio. To optimize the geometrical boundaries of some problem involving solutions to some P.D.E. in order to improve a given cost functional, the main question is that of applying "variations" to the boundary. In these lectures the moving domain is the image of a given domain by the flow of a vector field, often a non-smooth vector field.

A technique for the actual numerical implementation is essential to a field as "Optimal Shape Design" and this is exposed in the chapter by O. Pironneau. Several case studies are presented, and it is shown how the finite element method is well suited for this class of problems. The references at the end of the chapter refer to several very recent applications of numerical methods to the solution of industrial problems.

Besides the lectures presented here, Prof. P. Villaggio of the University of Pisa presented an interesting set of lectures on "Explicit Solutions in Elastic Optimization". Unfortunately, his many commitments prevented him from producing a set of Notes for this Book.

This joint C.I.M./ C.I.M.E. summer course was made possible by a generous contribution from the European Commission, DG XII, contract ERBFMMACT970273.

Arrigo Cellina
António Ornelas
December, 1999

References

P. Celada, A. Cellina, Existence and Nonexistence of Solutions to a Variational Problem on a Square, Houston J. of Math. 24 (1998), 345-375

P. Celada, S. Perrotta, G. Treu, Existence of solutions for a class of nonconvex minimum problems, Math. Z. 228 (1998), 177-199

A. Cellina, On Minima of a Functional of the Gradient: Sufficient Conditions, Nonlinear Anal. TMA, 20 (1993) 337-341

A. Cellina, Minimizing a functional depending on ∇u and u, Ann. Inst. H. Poincare', Anal. Nonlin. 14 (1997), 339-352

A. Cellina, G. Colombo, On a classical problem of the Calculus of variations without convexity assumptions, Ann. Inst. H. Poincare', Anal. Nonlin. 7 (1990), 97-106

A. Cellina, F. Flores, Radially symmetric solutions of a class of problems in the Calculus of Variations without convexity assumptions, Ann. Instit. H. Poincare', Analise nonlineaire, 9 (1992), 485-478

A. Cellina, S. Perrotta, On a Problem of Potential Wells, Journal of Convex Anaysis 2 (1995), 103-115

A. Cellina, S. Perrotta, On Minima of Radially Symmetric Functionals of the Gradient, Nonlinear Anal. TMA, 23 (1994), 239-249

A. Cellina, S. Perrotta, On the validity of the Maximum Principle and of the Euler Lagrange equation for a minimum problem depending on the gradient, SIAM J. Control Optim. 36 (1998), 1987-1998

A. Cellina, S. Zagatti, An Existence Result in a Problem of the Vectorial Case of the Calculus of Variations, SIAM J. Control Optim. 33 (1995) 960-970

L. Cesari, Optimization, Theory and Applications, Springer Verlag, New York, 1983

G. Crasta, On the minimum problem for a class of noncoercive nonconvex variational problems, SIAM J. Control Optim.

G. Crasta, An existence result for non-coercive non-convex Problems in the Calculus of Variations, Nonlinear Anal. TMA, 26 (1996), 1527-1533

G. Crasta, Existence of minimizers for non-convex variational problems with slow growth, J. Optim. Theory Appl. 99 (1998), 381-401

G. Crasta, Existence, uniqueness and qualitative properties of minima to radially symmetric noncoercive nonconvex variational problems, Math. Z.

G. Crasta, Variational Problems for a Class of Functional on Convex Domains, preprint 1999

G. Crasta, A. Malusa, Existence results for non-convex variational problems, SIAM J. Control Optim. 34 (1996), 2064-2076

G. Crasta, A. Malusa, Non-convex Minimization Problems for functionals defined on vector-valued Functions, preprint 1998

B. Dacorogna, P. Marcellini, Existence of Minimizers for non-quasiconvex integrals, Arch. Rat. Mech. Anal. 131 (1995) 359-399

B. Dacorogna, P. Marcellini, Theorems d'existence dans les cas scalaire et vectorielle pour les equations de Hamilton Jacobi, C.R. Acad. Sci. Paris 322 (1996), 237-240

B. Dacorogna, P. Marcellini, Sur le Probleme de Cauchy-Dirichlet pour les systemes d'equations non-lineaires du premier ordre, C.R. Acad. Sci. Paris, 323 (1996), 599-602

B. Dacorogna, P. Marcellini, Implicit Partial Differential Equations, Birkhauser, Basel, 1999

F.S. De Blasi, G. Pianigiani, A Baire category approach to the existence of solutions of multivalued differential equations in Banach spaces, Functialaj Ekvacioj 25 (1982), 153-162

F. S. De Blasi, G. Pianigiani On the Dirichlet problem for first order partial differential equations. A Baire category approach". NoDEA Nonlinear Diff. Eq. Appl. 6(1999) 13-34

N. Fusco, P. Marcellini , A. Ornelas, Existence of Minimizers for Some Noncovex One-dimensional Integrals, Portugaliae Mathematica 55 (1998), 167-185

F. Gazzola, Existence of Minima for nonconvex functionals depending on the distance from the boundary, Arch. Rat. Mech. Anal., to appear

P. Marcellini, Alcune osservazioni sull'esistenza del minimo di integrali del Calcolo delle Variazioni senza ipotesi di convessita', Rendiconti Matem. 13 (1980) 271-281

P. Marcellini, A relation between existence of minima for non-convex integrals and uniqueness for non strictly convex integrals of the Calculus of variations, Lecture Notes in Mathematics Vol 979, 216-232, Springer, New York 1983

A. Ornelas, Existence of Scalar Minimizers for Nonconvex Simple Integrals of Sum Type, J. Math. Anal. Appl. 221 (1998)

J.P. Raymond, Existence and Uniqueness Results for Minimization Problems with Nonconvex Functionals, J. Optim. Theory Appl. 82 (1994), 571-592

R. Taharaoui, Sur une classe de fonctionnelles non convexes et applications, SIAM J. Math. Anal. 21 (1990), 37-52

G. Treu, An existence result for a Class of Non-convex Problems of the Calculus of variations, J. Convex Anal. 5 (1998), 31-44

M. Vornicescu, A Variational Problem on subsets of $\Re^n$, Proc. Roy. Soc. Edinburg, Sect. A 127 (1997), 1089-1101

Some nonconvex shape optimization problems

Bernd Kawohl

Mathematisches Institut, Universität zu Köln, D-50923 Köln, Germany
email: kawohl@mi.uni-koeln.de

1. Minimizing paths, the opaque square

The following problem was pointed out to me by Fred Almgren. Imagine a piece of land, which is quadratic and of size one. The owner of the land is an unpleasant person. He does not want his neighbours to be able to communicate with each other by talking across his property. One way to reach his goal is to build a fence around the land which would block any line of sight across it. Moreover, the landowner is somewhat stingy and does not want to spend more money than necessary on building the fence. In fact a fence on three sides of his property would serve the same purpose and be cheaper. In that case a neighbour on the free side could look inside his property but not across it to a neighbour on any of the other three sides. What is the shortest fence that would have the desired function of blocking every line of sight across the property?

Although this shape optimization problem looks fairly simple it offers a couple of surprises even for mathematicians working on shape problems. Clearly we want the vision across each of the four corners to be blocked, so an optimal fence will contain all four corners of the square. So building two diagonal fences of total length 2 times $\sqrt{2}$, i.e. ≈ 2.82, would be preferable to building on three sides around. (Never mind the discomfort that this fence across his land may cause to the owner, he is just happy to get away cheaper.)

Finding the shortest path that connects four points is a well-understood problem in optimization. Its solution is the Steiner tree depicted in Figure 1.1, and its length is $1 + \sqrt{3}$, that is ≈ 2.73. Now this tree has only two reflection symmetries, although the square has four. The Steiner tree is not symmetric across the diagonals. There is, however, an even shorter fence, the first record of which I have found in [34]. According to [32] the same solution was found by Maurice Poirier, a secondary school teacher in Orléans, Ontario, who might have been inspired by a test for mathematically gifted school children. The test can be found in [60] and is quoted in [32]. The shortest known fence is depicted in Figure 1.2 and cosists of two components, namely a Steiner tree connecting three corners. This tree blocks vision from its convex hull, so the remainder of the fence is the shortest line from the fourth corner to the triangle generated by the Steiner tree. It has length $(2+\sqrt{3})/\sqrt{2} \approx 2.64$, and it features only one reflection symmetry.

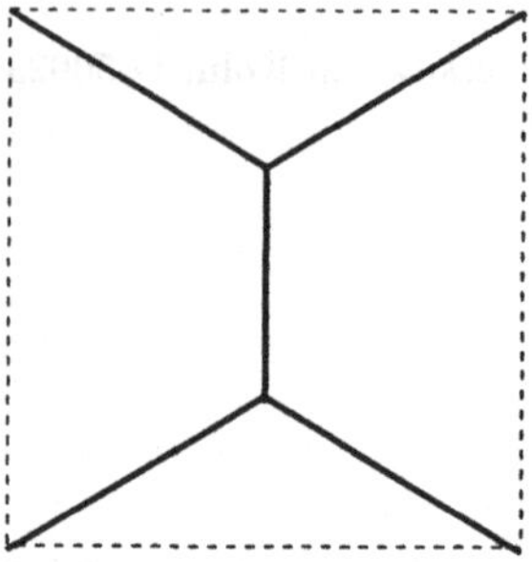

Fig.1.1. The Steiner tree

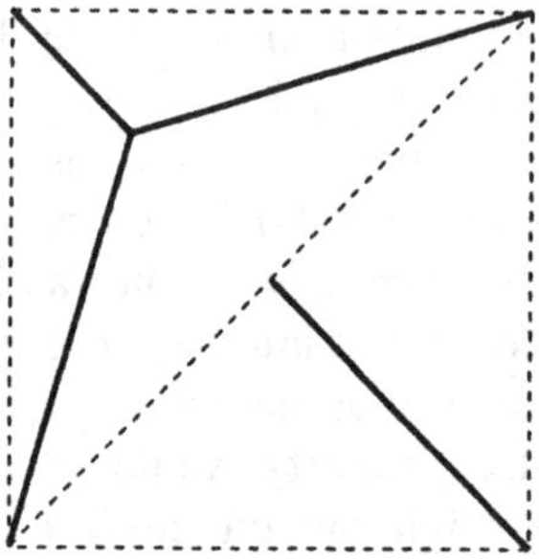

Fig.1.2. The shortest known fence

In [39] I outlined a proof that the fence depicted in Figure 1.2 is a curve which minimizes one-dimensional Hausdorff measure in the class of all doubly connected curves C which have the property that every straight line intersecting the square D also intersects C. To fix some ideas let me give two definitions.

Definition 1.1:
Let $C \subset \mathbb{R}^n$. Then the 1-dimensional Hausdorff measure of C is defined as

$$\mathcal{H}^1(C) := \lim_{\delta \to 0} \inf\{\sum_{i=1}^{\infty} \operatorname{diam}(U_i)^1 \mid U_i \text{ cover } C \text{ and have diam} \leq \delta\}$$

Definition 1.2:
Let E and F be subsets of $\mathbb{R}^n$. Then their Euclidean δ-neighbourhood is denoted by $[E]_\delta$ and $[F]_\delta$, and their Hausdorff distance is defined as

$$\delta(E, F) := \inf\{\delta \mid E \subset [F]_\delta \text{ and } F \subset [E]_\delta\}.$$

This distance gives rise to a metric, the Hausdorff metric.

Theorem 1.1 [22]:
For each maximum number $k \in \mathbb{N}$ of components and for every convex compact $D \subset \mathbb{R}^n$, there exists a curve $C_k \subset D$ minimizing one-dimensional Hausdorff-measure under the geometric side constraint, that every straight line intersecting D also intersects the curve.

This was apparently first shown in [22] but may be more accessible in [24, p.39]. Let me sketch the proof. Clearly the Hausdorff-measure $\mathcal{H}^1(C)$ is bounded below and thus there exists a minimizing sequence C^j of admissible sets, each consisting of k pairwise disjoint connected components $C_1^j, \ldots, C_k^j$. Due to Blaschke's selection theorem [24, p.37], for $m = 1, \ldots, k$ and after passing to a subsequence the sequence of connected sets C_m^j converges in Hausdorff metric to a connected limit set C_m^∞. Fortunately Hausdorff measure is lower semicontinuous along this subsequence, that is

$$\mathcal{H}^1(C_m^\infty) \leq \mathcal{H}^1(C_m^j). \tag{1.1}$$

The union over $m = 1, \ldots, k$ of these C_m^∞ is a minimizer C_k, since by the subadditivity of $\mathcal{H}^1$ we have

$$\mathcal{H}^1(C_k) \leq \sum_{m=1}^{k} \mathcal{H}^1(C_m^j) = \mathcal{H}^1(C^j). \tag{1.2}$$

Note first that the last equality follows for each fixed j from the pairwise disjointness of the sets C_m^j, second that the geometric constraint is preserved under convergence of sets in the Hausdorff metric, so that C_k blocks vision just like each of the C^j did, and third that it is conceivable that C_k has less than k components, because some of the C_m^∞ might have nonempty pairwise intersection. This proves the existence result of a minimizing C_k for every k. To verify the geometric constraint one simply has to realize that the points x_j of intersection of a given line of sight with C^j converge to a point both in C^∞ and on the given line.

Now C_1 and C_2 are shown in Figures 1.1 and 1.2, and it is conjectured that $C_k = C_2$ for any natural number $k \geq 2$, and not only for even prime numbers.

What happens as $k \to \infty$? I do not know of any existence result in the class of admissible curves with a countable number of components. The difficulty of proving this lies in the fact that the semicontinuity result (1.2) is no longer true if the sets C^j have a countable number of components. A counterexample which I learned from G.Dal Maso is provided by the following sequence of subsets C^j of the unit interval $[0, 1]$ in $\mathbb{R}$.

$$C^1 = \left[0, \frac{1}{2}\right],$$

$$C^2 = \left[0, \frac{1}{4}\right] \cup \left[\frac{1}{2}, \frac{3}{4}\right],$$

$$C^3 = \left[0, \frac{1}{8}\right] \cup \left[\frac{1}{4}, \frac{3}{8}\right] \cup \left[\frac{1}{2}, \frac{5}{8}\right] \cup \left[\frac{3}{4}, \frac{7}{8}\right] \quad \text{etc.}$$

Clearly the Hausdorff distance of C^j to $[0,1]$ is 2^{-j} so that $C^j \to C_\infty = [0,1]$ in Hausdorff distance, but $\mathcal{H}^1(C_\infty) = 1 > 0.5 = \mathcal{H}^1(C^j)$. We shall see a similar example below.

Why is C_2 from Figure 1.2 optimal in the class of doubly connected curves? The optimal C must contain all four corners, otherwise one could look across a corner. Therefore its two connected but disjoint components C^1 and C^2 must emanate from the corners of the square. Several cases are conceivable. If C^1 contains exactly two diagonal corners, then C^2 must cross C^1, a contradiction to being disjoint. If C^1 contains exactly two adjacent corners, then the convex hulls of C^1 and C^2 must have nonempty intersection. Let z be a point from this intersection. Connecting all four corners with straight line segments to z will yield a shorter, and simply connected curve; again a contradiction. Thus the case that C^1 (or equivalently C^2) contains exactly three corners remains to be treated. If C^1 remains in the triangle spanned by its three corners, it must be the Steiner tree depicted in Figure 1.2, and then C^2 is half a diagonal, so that they make up the two components in Figure 1.2. Finally the case that C^1 exceeds the triangle spanned by its three corners a, b and c was neglected in [39] and can be treated as follows. I am indebted to Marino Belloni for pointing this out to me. To be disconnected from C^1 but still block vision, C^2 must touch the convex hull of C^1 but not C^1 in some point g. So g lies on a line segment, the endpoints of which, e and f, are elements of C^1, see Figure 1.3.

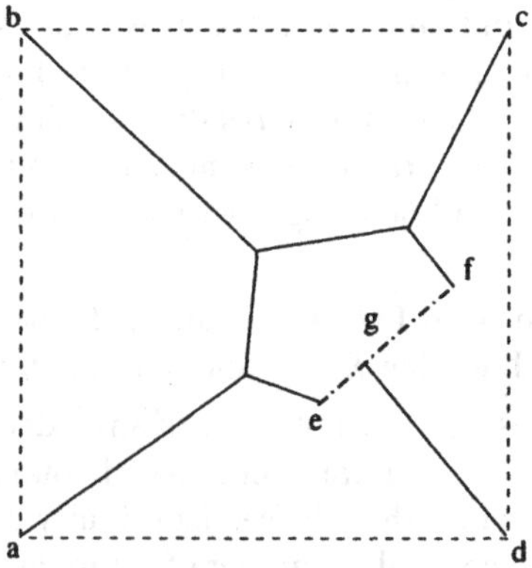

Fig.1.3. Not an optimal situation

Since C^2 minimizes distance from the fourth corner d to the convex hull of C^1, the segments ef and dg must be orthogonal to each other. Now I move e and f away from the fourth corner d in the direction of dg. This will shorten the path connecting a, b, c, e and f more than it will lengthen the one from d to g. This completes the proof that the solution depicted in Figure 1.2 is optimal in the class of curves with at most two components.

There are several ways to generalize this problem. One can look at the three-dimensional cube instead of a twodimensional square and try to build a twodimensional (minimal) surface which blocks vision through the cube. A rather small surface consists of a rounded square in the middle with surfaces extending from the edges of the small rounded square to the large original cube. It shows up, when you dip a cubic wire frame into soap. An even better surface was calculated by Brakke, see [9], and yields an interesting surface which forms a tunnel within it. Of course to pass through the tunnel, you cannot go straight.

Another way is to pass from a square to other polygons or even to a circle in two dimensions. The circle has an interesting "application". In [61] Sherlock Holmes has to find an underground tunnel very fast. It is a straight tunnel that runs under a lawn within a given radius 1 of a given point, and the only way to find it is by digging. More realistic seems to be the application of locating a straight telephone line in the ground, provided it is within radius 1 of a given marker [23]. In fact, when I presented this problem in Oberwolfach in 1995, M.Plum asked me what could be said about the circle. From the experience with the square we expect the optimal fence, trench or simply curve to contain all the boundary points of the circle. Since those form are a set of length 2π, the optimal curve should have at least (one-dimensional Hausdorff) measure 2π? The answer is again not intuitive even for the convex analyst.

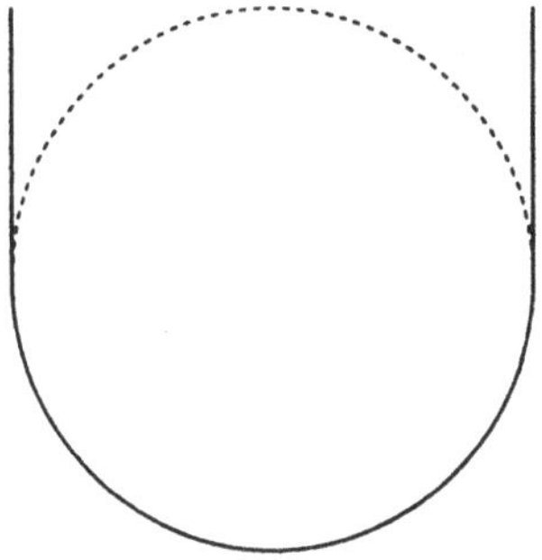

Fig.1.4. How to find a telephone line

Figure 1.4 shows a solution that, according to [23], has been known to M.Magidor in 1974. It leaves the unit disk, has total length $2 + \pi =\approx 5.14$, and is the shortest simply connected curve which intersects all the lines that intersect the circle [22]. If one admits multiply connected curves which may leave the disk, one can do even better and come up with the nowadays apparently best known doubly connected one of length ≈ 4.8189 [22 p.264, St. p.113], a triply connected one of length $1 + \sqrt{3} + 2\pi/3 \approx 4.7998$ and so on.

When I saw those solutions, I wondered if one could reach similar results under the additional constraint that the (possibly multiply connected) admissible curves C had to stay inside the unit disk. One way to approach this problem consists of approximating the disk by a regular polygon with many corners. Clearly, for a general polygon, with many corners we cannot expect to prove that a certain curve is optimal. But we can make good and maybe numerically supported guesses. Incidentally, computer scientists are studying the complexity of such a problem, but they call the admissible curves "opaque forests"of a polygon, see [1,19,57]. Maybe the term "forest" comes from the fact that it tends to be made up of Steiner "trees". To investigate a regular polygon with many corners, it is instructive to look first at polygons with few corners and to draw conclusions from them. Clearly, an equilateral triangle has the Steiner tree in Figure 1.5 as an optimal solution.

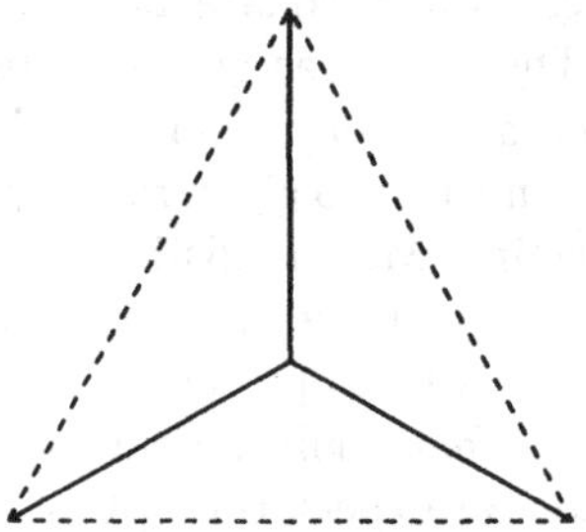

Fig. 1.5. Steiner tree of length $\sqrt{3} \approx 1.732$

Incidentally, do you know how to construct the so-called Torricelli point, the meeting point of the three segments, in a general triangle and how to prove that the lines meet in general under the angle $2\pi/3$? This is an old (1640) elementary geometric problem, the solution of which can be traced back to Torricelli, a student of Galileo Galilei and to Cavalieri [14, p.2], but its solution is not so easy to find in textbooks on the calculus of variations. A reference is [14, p.103, 16, Ch. VII] or the one written up for high school students 33].

The best known curve for a regular square was given above in Figure 1.2, the one for a pentagon looks like the one on the left in Figure 1.6, while the one on the right had been suggested by one of the participants of this summer school.

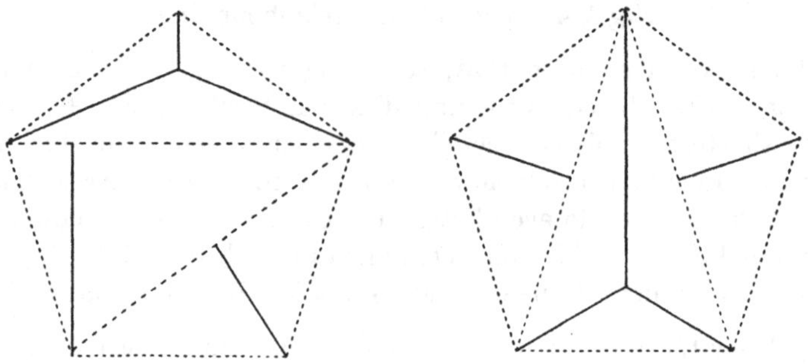

Fig. 1.6. Good curves of length ≈ 3.528 [23] and ≈ 3.580 [51]

When we reach the hexagon, the Toricelli point creeps into a corner and the best known even at most doubly connected curve becomes simply connected and connects four boundary points, see Figure 1.7 left. Again the one on the right was brought up during the summer school but is not optimal.

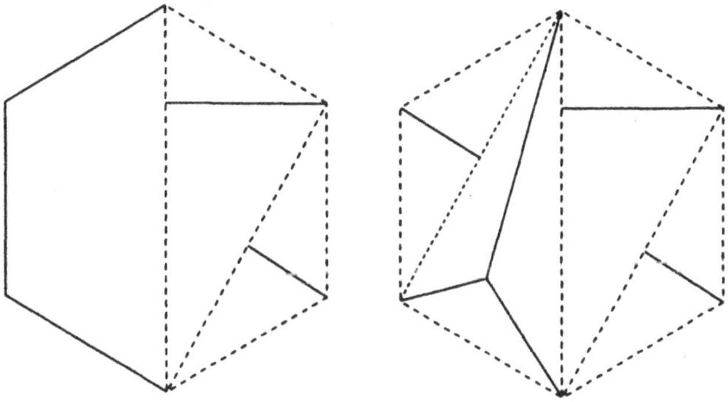

Fig.1.7. Good curves of length $(7+\sqrt{3})/2 \approx 4.366$ [23] and ≈ 4.512 [51]

Now we can extrapolate to a regular $2N$-gon, whose corners lie on the unit circle, and do the following construction: We set C_0 = half of the boundary of the polygon, with P_0 and P_N as endpoints, and suppose that the remaining $(N-1)$ corners are numbered $P_1, P_2, \dots, P_{N-1}$ in consecutive order. For $i = 1, \dots, N-1$ let C_i be the straight line segment that connects P_i to the closest point Q_i on the convex hull of $\left(\bigcup_{j=1}^{i} C_j\right)$, see Figure 1.8.

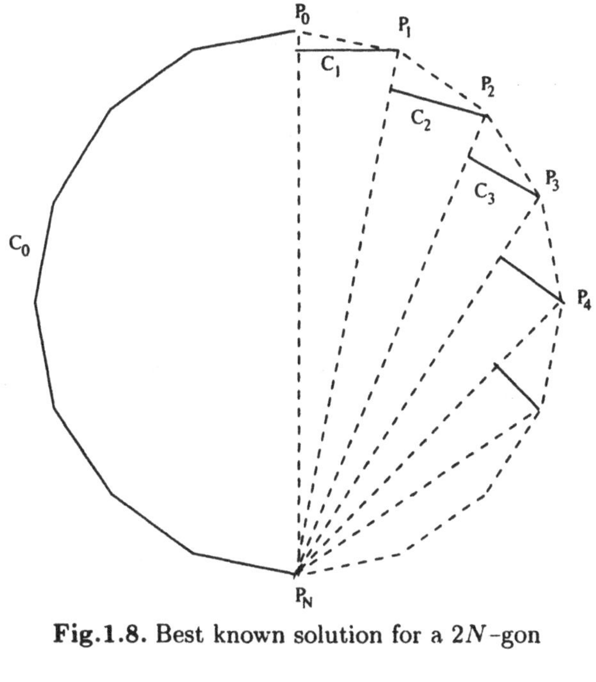

Fig.1.8. Best known solution for a $2N$-gon

Finally set $C_N := \left(\bigcup_{j=0}^{N-1} C_j\right)$. C_N intersects every straight line that crosses the polygon and a fortiori the disk of radius $1-\sin(\pi/2N)$, because the latter is contained in the polygon. Moreover C_N stays inside the closed unit disk. As the number of corners N tends to ∞, we may expect a limit set C_∞ with a countable number of components, which intersects every straight line that intersects the unit disk. So let us calculate the length ℓ_N of C_N by doing some trigonometry.

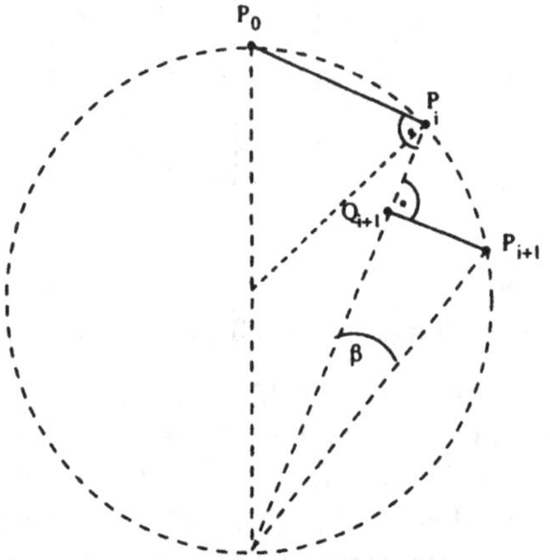

Fig.1.9. Computation of ℓ_N

Let us denote the left endpoint of C_{i+1} by Q_{i+1}. The angle β between P_iP_N and $P_{i+1}P_N$ is $\beta = \pi/2N$, since the angle between P_0P_N and P_iP_N is $i\pi/2N$. The length of the segment $\overline{P_iP_N}$ can be calculated using Thales' Theorem as $\overline{P_iP_N} = 2\cos(i\pi/2N)$, so $\sin\beta = \overline{Q_{i+1}P_{i+1}}\ /\ \overline{P_{i+1}P_N}$ or length of $C_{i+1} = \sin(\pi/2N)\cdot 2\cos((i+1)\pi/2N)$ for $i = 1,\dots,N-1$.
To get the length of all the arcs we have to add them up

$$\ell_N = \text{length of } C_0 + \sum_{i=1}^{N-1} \text{length of } C_i \quad .$$

Now the length of C_0 is $N\cdot\overline{P_0P_1} = N\cdot 2\sin(\pi/2N)$, and so

$$\ell_N = 2N\ \sin\left(\frac{\pi}{2N}\right) + \sum_{j=1}^{N-1} \sin\left(\frac{\pi}{2N}\right)\cdot 2\cos\left(\frac{j\pi}{2N}\right)$$

or, observing that $\cos\pi/2 = 0$,

$$l_N = 2N\ \sin\left(\frac{\pi}{2N}\right)\left(1+\frac{1}{N}\sum_{j=1}^{N}\cos\left(\frac{j\pi}{2N}\right)\right) \quad .$$

But $\frac{1}{N}\sum_{j=1}^{N}\cos\left(\frac{j}{N}\cdot\frac{\pi}{2}\right)$ is the Riemann sum for

$$\frac{2}{\pi}\int_0^{\frac{\pi}{2}} \cos x \, dx = \frac{2}{\pi} \quad ,$$

and for large N we can approximate $\sin \pi/2N$ by $\pi/2N$, so that

$$\lim_{N\to\infty} \ell_N = \pi\left(1+\frac{2}{\pi}\right) = \pi + 2 \quad .$$

Therefore $\lim_{N\to\infty} \mathcal{H}^1(C_N) = 2+\pi$, but unfortunately the limit object C_∞ in the Hausdorff topology is the unit circle with $\mathcal{H}^1(C_\infty) = 2\pi$. This provides another example that demonstrates the lack of weak lower semicontinuity of $\mathcal{H}^1$ along sequences of curves whose number of components becomes unbounded. However, if you want to find a water pipe (of positive) thickness under a circular garden bed, without digging outside the bed, the above construction of C_N offers a pretty good strategy.

I tried to trace the origins of the opaque square problem and discovered that the opaque disk seemed to be much younger. It is described for instance in [18, p.39-41], where a related problem is also described, namely that of finding one's way out of a forest. More on the forest problem can be found in 35] and [41, 42].

2. Newton's problem of minimal resistance

More than 300 years ago Isaac Newton solved a puzzling variational problem. He tried to minimize the resistance of a rotational body travelling through a liquid, and one of his motivations was that his ivestigations would be of interest to "Her Majesty's Navy". In a place like Troia, where this summer school is held, the importance of building fuel efficient ships is quite obvious and one way to optimize fuel efficiency is undoubtedly to minimize drag. But what exactly is drag? Newton placed himself, so to speak, in the boat and thought of a liquid as a (rarefied) particle stream that flows past the boat or more generally, past a body in $\mathbb{R}^3$. He stated for instance in his *Principia Mathematica* (1686):

"If in a rare medium consisting of equal particles freely disposed at equal distances from each other, a globe and a cylinder described on equal diameters move with equal velocities in the direction of the axis of the cylinder, the resistance of the globe will be but half as great as that of the cylinder."

Today we would briefly say: "The drag coefficient of a ball is 0.5."

Incidentally, the drag coefficient of a TGV or a sportscar is somewhere below 0.3. Drag coefficient is the ratio of the drag of a body B over the drag of a flat plate whose shape is the projection of B in direction of the flow.

Drag itself is measured in absolute terms by the total momentum received from fluid particles to which the body is exposed at any given time. This momentum is caused by collisions of the fluid particle with the body B, and for lack of a more sophisticated model at his time, Newton used Newtonian mechanics. He assumed that the particles do not interact with each other, and that upon impact on the obstacle they are reflected in a perfectly elastic way and disappear to infinity. These assumptions are violated in real gases around us, but they are valid in rarefied gases such as the upper layers of our atmosphere. In fact, in the 1960s, NASA tested the validity of the model when predicting the drag of orbiting rockets, see [47] and the references therein. In order to express drag in terms of the geometry of a body, let us imagine that the direction of flow is vertically downward. Only the upper part of a body (at rest) is then exposed to collisions with the particle stream, and each particle can lose up to 100 percent of its kinetic energy when striking the body.

To fix ideas, let us assume a rotational body B, whose axis of rotation coincides with the direction of the flow, and call its maximal cross section Ω. Any shape of the body below the plane of maximal cross section, does no longer influence the flow (according to this model). The surface of B that lies above Ω is supposed to be parametrized by a function u of r, where r measures the distance of a point x in Ω to its center, i.e. without loss of generality to the origin. Moreover we assume that u is nonnegative in Ω. If Ω is the unit disk D, and B a cylinder in the direction of the flow, then $u \equiv const.$ in D. If B is a ball of radius 1, however, then $u(r) = \sqrt{1-r^2}$, if B is a circular cone with base D and height h, then $u(r) = (1-r)h$ and so on.

Now the path of a particle approaching the body B and its reflection span a vertical plane and are depicted in Figure 2.1. The tangent to u in the point of reflection and the vertical form an angle θ. If we split the momentum vector of the particle into a component tangential and one normal to the tangent of u at the point of impact, then the tangential momentum is conserved, but the normal one is reversed. Trigonometry shows that the proportion of the normal component is $\sin(\theta)$ times the original momentum. Thus the normal component exerts drag, but only its vertical component in the direction of the flow slows the body down. The horizontal component might cause a drift to the side, however, there is a balance for this drift coming from the opposite side of the rotational body, so that they annihilate each other.

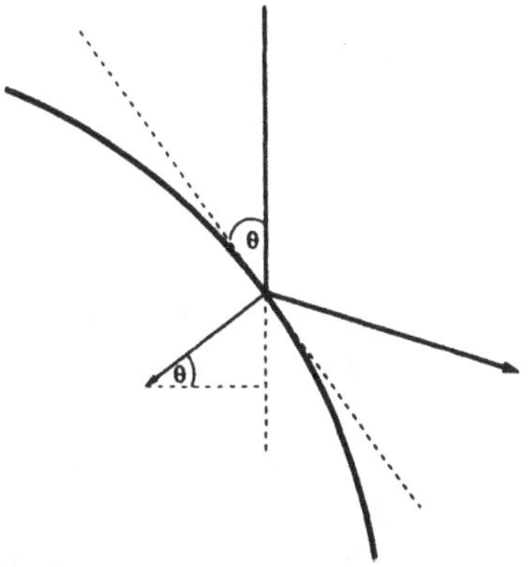

Fig.2.1. The $\sin^2(\theta)$-law of Newton

Thus the drag exerted in vertical direction at a point x is proportional to $\sin^2(\theta)$, that is to $(1+u_r^2)^{-1}$, and the total drag on the body is proprtional to

$$R(u) = \int_D \frac{1}{1+u_r^2}\,dx = 2\pi \int_0^1 \frac{1}{1+(u_r(\rho))^2}\,\rho\,d\rho. \tag{2.1}$$

Using (2.1) as a definition for (absolute) resistance, it is now a simple exercise in integration to verify that the drag coefficient of a ball is 0.5, and the one of a circular cone of height h on the unit disk is $1/(1+h^2)$. The higher and more slender the cone, the smaller is its resistance according to Newton. What is the body of minimal resistance? This is a shape optimization problem par excellence, and Newton offered a surprising solution for it. Recall that he was one of the founders of infinitesimal calculus as well as of the calculus of variations. Thus he did not exactly specify the class of admissible functions u for the functional R. In fact, he had imagined the rotational body of length M as having a horizontal axis of rotation and being exposed to a horizontal flow. Therefore setting $v(t) = u^{-1}(M-t)$ and assuming that $u_r \neq 0$, the functional can be rewritten in the form

$$\int_0^M \frac{vv'^3}{1+v'^2}\,dt, \tag{2.2}$$

in which it shows up in classical books such as [65, Fu, 30]. Rotating v around the horizontal t-axis would generate the surface of B. This is an indication that Newton might have had radially decreasing functions u satisfying $o \le u \le M$ as admissible functions in mind when he attacked the problem of finding a body of minimal resistance.

In this class of (formerly unspecified) functions one can seek to minimize R by solving the associated Euler-Lagrange equation. The first variation of R in direction of a test function $\phi(r)$ should vanish at the minimizing u, i.e

$$\int_0^1 \frac{r}{(1+u_r^2)^2}\,2u_r\phi_r\,dr = -2\int_0^1 \frac{d}{dr}\left(\frac{ru_r}{(1+u_r^2)^2}\right)\phi\,dr = 0, \tag{2.3}$$

which implies that

$$\frac{d}{dr}\left(\frac{ru_r}{(1+u_r^2)^2}\right)=0$$

or

$$h(u_r):=\frac{u_r}{(1+u_r^2)^2}=\frac{const.}{r}. \tag{2.4}$$

Note that the function $s\mapsto h(s)$ is bounded on $\mathbb{R}$, so that the lefthand side of (2.4) is bounded, while the righthand side becomes unbounded as $r\to\infty$. (The case that $const.=0$ gives only the solution $u\equiv M$ which maximizes R and which is of no interest to us.) Therefore, for $const.\neq 0$ one cannot solve (2.4) in a neighborhood of the origin. Nevertheless Newton offered a body of minimal resistance. The graph of Newton's optimal solution, which I will call u_N from now on, consists of two parts, a constant top $u\equiv M$ on $[0,r_0]$, and a concave arc on $[r_0,1]$. The arc solves (2.4) for a suitable *const.*, and both r_0 and *const.* are chosen in such a way that $u_r(r_0^+)=-1$ and $u(1)=0$. For $M=1$ the function $u_N(r)$ is sketched in Figure 2.2. Its calculation is a nice exercise in ODE and can be found in [11].

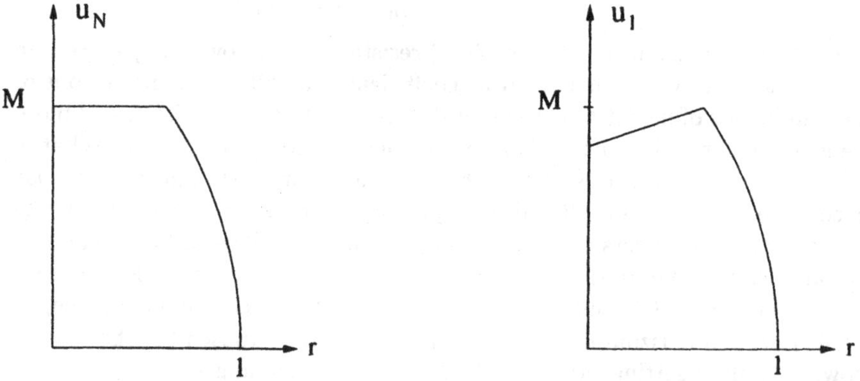

Fig.2.2. Newton's optimal solution u_N and a modification u_1

Rotating u_N about the vertical axis results in something like the shape of a bullet, and everybody who understands a little bit of fluid dynamics will immediately dismiss Newton's solution as not optimal. It has a blunt front end, a set of stagnation points, where fluid particles will be blocked, causing increased pressure and temperature and so on. This reasoning leaves the realm of Newton's physical model of "particles freely disposed at equal distances from each other".

It seemed to be a mystery how Newton had come up with the boundary condition $u_N'(r_0^+)=-1$. D.T.Whiteside has dedicated many years of his life to studying Newton's mathematical papers and has written a series of carefully researched books about Newton's papers and subsequent developments. He has reached the conclusion about the boundary condition, that "This must surely be an educated guess on Newton's part.", see [66, p.478]. Newton did in fact have deep insight, and today one can give a rigorous derivation of this condition, see [8, Thm. 2.3].

Nevertheless it is worthwile to look at Newton's original reasoning, which has been reproduced in [45]. Newton used Figure 2.3 (to which I have added a u and an r-axis) to illustrate his reasoning.

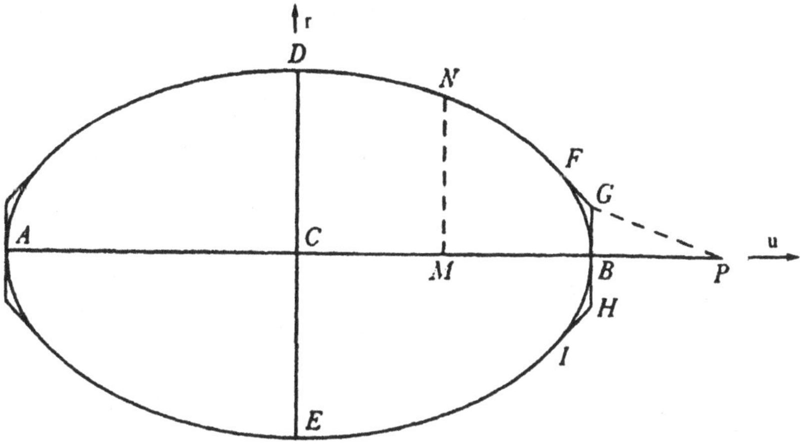

Fig.2.3. Newton's drawing, see [66], p. 462

After deriving his model of resistance, he continues as follows (calling straight lines right lines).

"Incidentally, ..., it follows from the above that, if the solid $ABDE$ be generated by the convolution of an elliptical or oval figure $ADBE$ about its axis AB, and the generating figure be touched by three right lines FG, GH, HI in the points F, B and I, so that GH shall be perpendicular to the axis in the point of contact B, and FG, HI may be inclined to GH in the abgles FGB; BHI of 135 degrees: the solid arising from the convolution of the figure $ADFGHIE$ about the same axis AB will be less resisted than the former solid, provided that both move forwards in the direction of their axis AB, and that the extremity B of each go forward. This Proposition I conceive may be of use in the building of ships."

Let me remark in passing, that the construction of w in the proof of Theorem 3.2 below was inspired by this reasoning of Newton. Newton then goes on to comment on the curved part of his object.

"If the figure $DNFG$ be such a curve, that if, from any point thereof, as N, the perpendicular NM be let fall on the axis AB, and from the given point G there be drawn the right line GP parallel to a right line touching the figure in N, and cutting the axis produced in P, MN becomes to GP as GP^3 to $4BP \cdot GB^2$, the solid desribed by the revolution of this figure about its axis AB, moving in the before.mentioned rare medium from A to B, will be less resisted than any other circular solid whatsoever, described of the same length and breadth."

To interpret this explanation in terms of formulas, I added two axes to Newton's figure, an r-axis vertically upward through D and a u-axis along AP, pointing in that direction. Then B is $(0, M)$ and D is $(1,0)$, while we denote the length of BG by r_0. Now Newton's statement transforms into current notatation in the following way. Let the optimal curve $BGFND$ joining B to D be the graph $r \mapsto u(r)$, $r \in [0,1]$. Then u is constantly M on $[0, r_0]$, at r_0 its right derivative is -1, and on $(r_0, 1)$ the function u satisfies the equation

$$MN = r = -r_0 \frac{[1 + u_r^2]^2}{4u_r} = \frac{GP^4}{4BP \cdot GB^2},$$

where we have used that $GP^2 = BG^2 + BP^2$. Notice that this is nothing but (2.4).

Newton's solution was apparently not well understood by his contemporaries, nor by subsequent generations. About a century after Newton's solution, Legendre criticized it. Legendre observed that one could diminish R to as close to zero as one would wish by choosing a function u that has a large derivate everywhere, but oscilllates in order to stay within the L^∞ bound. To find away out of this dilemma, he had to exclude such minimizing functions by restricting the class of admissible ones. Legendre chose to prescribe the one-dimensional arc length as a side constraint. This constraint has no physical meaning, and it leads to an Euler-Lagrange equation different from (2.4), which Legendre was able to solve in special cases. Belloni and I have revisited Legendre's paper in [8], and I refer to it for more details.

Other ways out of Legendre's dilemma have been suggested. Further 2 centuries later McShane argues in [45], that oscillating slopes of u would generate "troughs which fill up with stagnant fluid" so that one obtains "an effective surface giving a nonincreasing u". His reasoning ignores Newton's assumption of noninteracting "particles freely disposed" as well.

The only argument to rule out admissible functions like zig-zag functions with large slope, which is consistent with Newton's physics, is the following: Those functions generate troughs, when rotated, and fluid particles can get caught in the trough in the sense that the hit the body more than once. Their momentum at the second impact, however, is not measured by the functional R. Suppose a particle hits the body first at $(x, u(x))$, then bounces off and hits again at $(y, u(y))$, this time not vertically from above. The functional R catches only the momentum which the point $(y, u(y))$ is exposed to from a particle travelling vertically downward. So a physically correct class of admissible functions seems to be made up of those functions which satisfy the so-called single shock property, that every fluid particle interacts with the body at most once. This is a geometric condition. We want a particle path, that is reflected at $(x, u(x))$ to lie above the rest of the obstacle. In terms of formulas, if $u'(x) < 0$, this translates into (see Fig. 2.4):

$$u(y) \leq u(x) + \frac{1}{2}\left(u'(x) - \frac{1}{u'(x)}\right)(y - x) \quad \text{for } y > x. \tag{2.5}$$

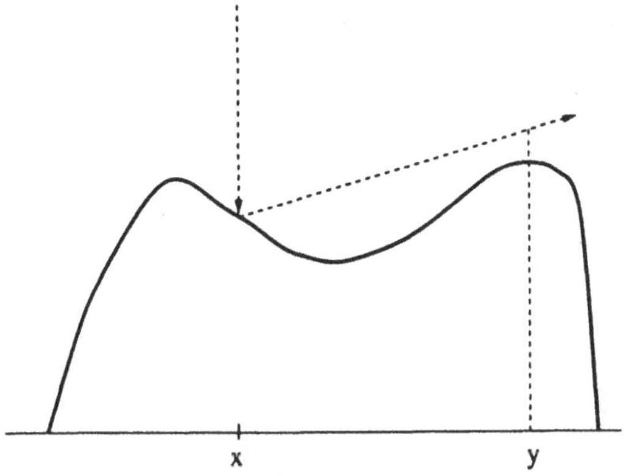

Fig.2.4. Derivation of (2.5)

Let me derive (2.5) by noting that the incoming and outgoing direction v_{in} and v_{out} of the particle can be expressed in terms of the tangential and normal vectors τ and ν to u at x as

$$v_{in} = \begin{pmatrix} 0 \\ -1 \end{pmatrix} = \left[\begin{pmatrix} 0 \\ -1 \end{pmatrix} \cdot \nu\right] \nu + \left[\begin{pmatrix} 0 \\ -1 \end{pmatrix} \cdot \tau\right] \tau$$

$$\text{and} \qquad v_{out} = -\left[\begin{pmatrix} 0 \\ -1 \end{pmatrix} \cdot \nu\right] \nu + \left[\begin{pmatrix} 0 \\ -1 \end{pmatrix} \cdot \tau\right] \tau,$$

because the normal direction of a particle is reversed upon impact. Since

$$\tau = \frac{1}{\sqrt{1+u'(x^2)}} \begin{pmatrix} 1 \\ u'(x) \end{pmatrix} \quad \text{and} \quad \nu = \frac{1}{\sqrt{1+u'(x^2)}} \begin{pmatrix} -u'(x) \\ 1 \end{pmatrix},$$

one calculates

$$v_{out} = \frac{1}{\sqrt{1+u'(x^2)}} \begin{pmatrix} 2u'(x) \\ u'^2(x) - 1 \end{pmatrix},$$

which implies that the path of the reflected particle has a slope of magnitude $\frac{1}{2}(u'(x) - \frac{1}{u'(x)})$. This proves (2.5).

Newton himself spoke of "oval bodies" when he derived his model, and he illustrated it only with convex bodies. Obviously convex bodies have the single shock property, but they form only a subclass. In fact, concave functions u would satisfy the stronger condition

$$u(y) \leq u(x) + u'(x)(y - x). \tag{2.6}$$

An example of a function satisfying (2.5) but not (2.6) is a modification of Newton's optimal solution, which is depicted in Figure 2.2. Take u_N and replace the flat top by a circular cone of opening angle $2\pi/3$. Then every particle that falls into this cone will be reflected by the body and leave the cone in a radial direction parallel to the cone without hitting the body another time. Since this modified solution, u_1 say, has a nonzero gradient on the disk with radius r_0, it has a smaller resistance than Newton's original solution u_N.

Therefore we have now two classes of admissible functions to choose from, namely

$$C_M := \{u \in W^{1,\infty}_{loc}(D), 0 \le u \le M, u \text{ convex}\} \quad \text{or}$$
$$S_M := \{u \in W^{1,\infty}_{loc}(D), 0 \le u \le M, u \text{ satisfies the single shock condition}\}.$$

As we have just seen, among radial functions the solution of Newton is not optimal in S_M, but in C_M it is. To prove any existence result via the direct method in the calculus of variations we must show that the functional R is lower semicontinuous along a minimizing sequence u_k of functions in the admissible set, be it S_M or C_M. Using (2.5) and (2.6) one can fairly easily show that functions in S_M satisfy

$$|u_r(\rho)| \le \frac{M + \sqrt{M^2 + (1-\rho)^2}}{1-\rho}, \qquad \text{for a.e. } \rho \in [0,1), \tag{2.7}$$

while those in C_M satisfy

$$|u_\rho| \le \frac{M}{1-\rho}, \qquad \text{for a.e. } \rho \in [0,1). \tag{2.8}$$

Therefore minimizing sequences in S_M or C_M are a priori uniformly bounded in $W^{1,\infty}(D')$, where D' is any disc with radius less than 1, centered at the origin, and thus they have weakly convergent subsequences in $W^{1,p}(D')$ for any $p \in [1,\infty)$. This is not enough information to obtain lower semicontinuity along a sequence in S_M. Let me introduce a sequence u_k of functions which is a modification of Newton's u_N. u_1 is defined above, and u_k (as a function of ρ) has a zig-zag shape consisting of k line segments, where u_N used to be flat. Clearly the sequence $\{u_k\}_{k\in\mathbb{N}}$ is bounded even in $W^{1,\infty}(D)$ and converges strongly in $L^\infty(D)$ and weakly in $W^{1,p}(D)$ to u_N as $k \to \infty$, but $R(u_k) = R(u_1) > R(u_N)$ for any $k \in \mathbb{N}$.

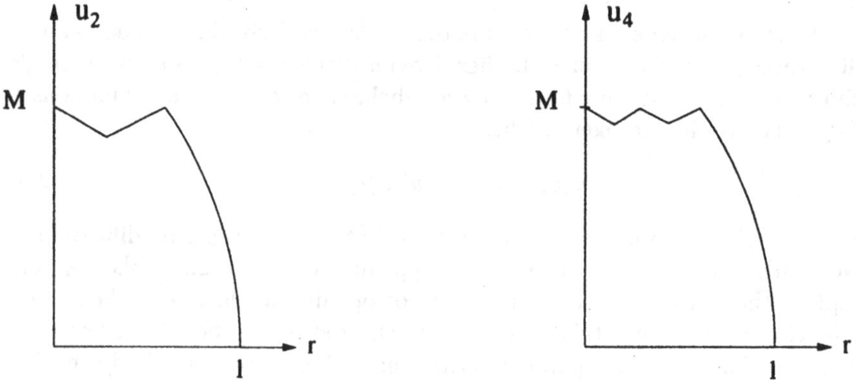

Fig.2.5. Plot of u_2 and u_4

This behaviour should not come as a surprise to us, because R is not convex in $|u_r|$ and therefore not weakly lower semicontinuous in $W^{1,p}(D)$. What is surprising, is the fact that in C_M one can get an existence proof after all, and as far as I can tell, the first rigorous existence proof for radial functions in C_M seems to have appeared only in 1990 [44], three centuries after Newton found more or less the explicit solution in the first place. His proof was later generalized in [12] and will be given below, when I will treat more general (nonrotational) convex bodies. As a matter of fact, for radial functions in C_M one can prove the existence of a minimizer of $R(v)$, and show that Newton's u_N is this uniquely determined minimizer, while in S_M it is not easy to even guess an optimal u. The picture will change, though, if nonradial functions are also admitted for variations. This will be done in Chapter III, below.

In Figure 2.5 I have sketched the qualitative behavior of a sequence of functions in S_M. This sequence oscillates on the interval $(0, r_0)$ in a regular zig-zag fashion with slope $\pm 1/\sqrt{3}$, and for $M = 1$ we had picked $r_0 = .35$ as in Newton's solution. If we do variations on r_0, then we need another curved arc reaching from (r_0, M) down to $(1, 0)$. Such arcs can be constructed from the Euler equation (2.4) or equivalently from solutions of Newton type with different M. This construction was done in great numerical detail in [62] for $M = 1$, and D.Stolz figured out the following numerical values:

- $R(u_N) = 1.163$
- $R(u_k) = 1.073$ if $r_0 = .35$ as for u_N
- $R(u_k) = 1.022$ if $r_0 = .636$
- $R(u_k) = 1.378$ if $r_0 = .765$

These values suggest that for $M = 1$, it pays to enlarge r_0 over Newton's $r_0 = .35$ and to drill a cone or a couple of concentric troughs of slope $\pm 1/\sqrt{3}$ into the blunt part, to reduce drag. Somehow such a roughening of surfaces reminds us of the dimples in golf balls or of the fact that rough surfaces are known to reduce the drag coefficients of ships and airplanes. These real world phenomena are based on other effects, though, such as the break up of turbulence (Divide et impera!).

There is another conceivable way to improve Newton's solution. Note that the particles coming out of the cone come out parallel to the cone. Can one construct a radial valley of radius r_0 such that every incoming particle just about makes it across the ridge on the opposite side? This is a question from geometric optic. To construct such a valley one has to investigate the equality case in (2.5), see also Figure 2.5.

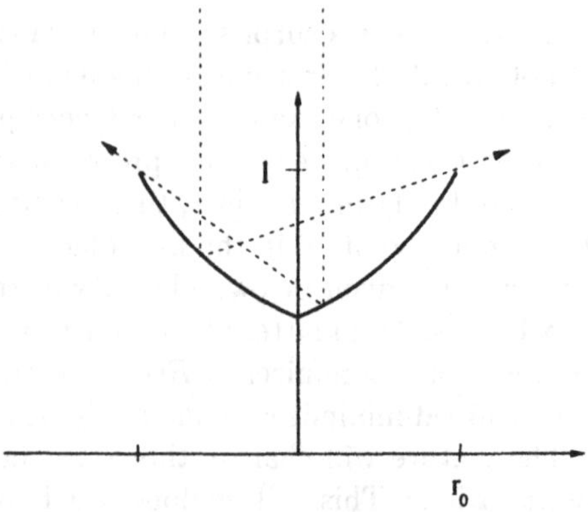

Fig.2.6. Shape of a function satisfying equality in (2.5)

The function depicted in Figure 2.6 is a piecewise polynomial of degree 2, with slope 1 at the outer edge r_0 and slope 1/2 at the origin. Note that the slope of the cone was $\sqrt{3}/2$ which is between 1 and 1/2. So which "valley" has less resistance, the conical one or the parabolic one? And does the resistance of a body change if one carves many concentric parabolic troughs into it instead of one big valley as in Figure 2.7?

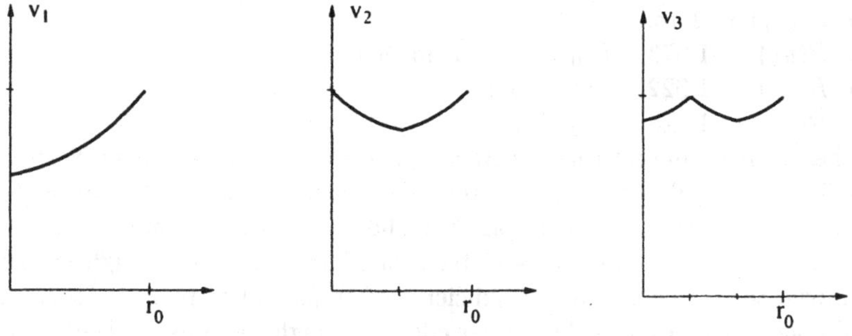

Fig.2.7. Parts of the Graphs of v_1, v_2, v_3

Only a calculation can show. This question, too, was investigated by D.Stolz. She compared the resistance of the blunt part of Newtons body with the one of the modified parts, i. e. she considered

$$\overline{R}(v) = 2\pi \int_0^{r_0} \frac{1}{1+(v_r(\rho))^2}\, \rho \, d\rho, \tag{2.9}$$

and computed that

$$\begin{aligned} \overline{R}(u_N) &= 3.142\ r_0^2, \\ \overline{R}(v_1) &= 1.854\ r_0^2, \\ \overline{R}(v_2) &= 2.020\ r_0^2 = \overline{R}(v_k) \text{ for } k \text{ even}, \end{aligned}$$

and

$$\overline{R}(v_k) = \frac{4\pi}{k^2}\left[\ln\frac{8}{5} + \left(\frac{k^2-3}{2}\right)\left(\frac{\pi}{4} + \arctan\left(-\frac{1}{2}\right)\right)\right] r_0^2$$

for k odd. Note that in this latter case $\overline{R}(v_k)$ increases in k as $k \to \infty$ to $\overline{R}(v_2) = 2\pi(\frac{\pi}{4} + \arctan(-\frac{1}{2}))r_0^2$, while $v_k \to u_N$ in L^∞ – another illustration for lack of lower semicontinuity. Therefore among the functions v_k a single valley, i.e. v_1, is optimal. Now for the cone u_1 it is easily seen that resistance is inbetween, because

$$\overline{R}(u_1) = \frac{3}{4}\pi r_0^2 = 2,355\ r_0^2.$$

Summarizing these results, it is a good strategy to carve piecewise parabolic troughs into $u_N(r)$ and to vary r_0 when one wants to minimize $R(v)$ in the class of radial functions in S_M. The minimizer there – if it exists – still remains to be found.

3. More on Newton's problem

In this lecture I shall modify Newton's original problem in several ways. First of all, I allow for nonrotational bodies, so that the admissible functions for R are no longer radial. Instead of being defined on a disk they are defined on an arbitrary domain $\Omega \subset \mathbb{R}^N$. Now the functional reads

$$R(v) := \int_\Omega \frac{1}{1+|\nabla v|^2}\, dx, \tag{3.1}$$

and we seek to minimize it in the following class of admissible functions:

$$C_M := \{\ v \in W^{1,\infty}_{loc}(\Omega), 0 \le v \le M, v\ \text{ convex }\}.$$

The following can be shown.

Theorem 3.1:
Let $\Omega \subset \mathbb{R}^N$ be convex.
a) There exists a minimizer u of R in C_M.
b) The minimizer is in general not unique, because if $\Omega = D$, the unit disk in $\mathbb{R}^2$, u is not radial.
c) The Euler-Lagrange equation associated with R is formally

$$-\text{div}\left(\frac{\nabla u}{(1+|\nabla u|^2)^2}\right) = 0; \tag{3.2}$$

it is of mixed (elliptic-hyperbolic) type, and holds in the interior of C_M, i.e. in points where u is strictly concave and where $u(x)$ lies strictly between 0 and M.

Proof of a): Since R is nonnegative, it has an infimum on C_M. Let u_k with $k \in \mathbb{N}$ be a minimizing sequence for R in C_M. What can be said about convergence of this sequence? For many coercive variational problems one can derive convergence from the fact that minimizing sequences are bounded in some norm. Our functional is not coercive, in fact a minimizer will try to have a large gradient, because this will decrease the integrand. In this problem we show that compactness of minimizing sequences comes from the suitable choice of admissible functions. They are compact in the strong topology of $W^{1,p}_{loc}(\Omega)$ for any $p \in [1, \infty)$. To realize this, one has to first observe that functions in C_M satisfy the inequality (which for radial functions degenerates to (2.8)

$$|\nabla v(x)| \leq \frac{M}{dist(x, \Omega)} \qquad \text{a.e. in } \Omega. \tag{3.3}$$

This can be seen as follows. For almost every point $x \in \Omega$ the function v is differentiable, and the normal vector to the graph of v together with the "vertical" unit vector span a plane, restricted to which we can plot the graph of v as a convex, one–dimensional function.

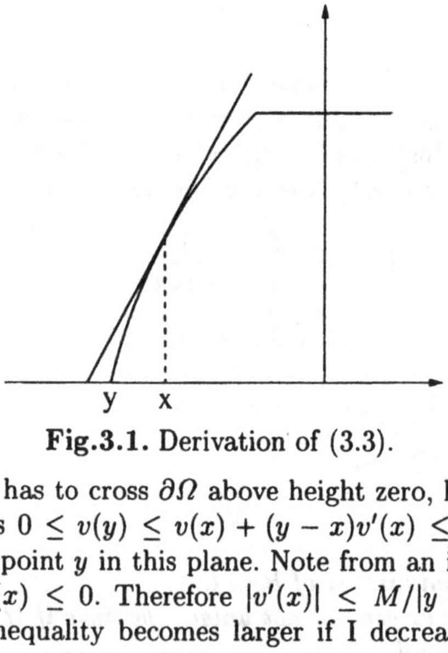

Fig.3.1. Derivation of (3.3).

Its tangent in x has to cross $\partial\Omega$ above height zero, hence we obtain the chain of inequalities $0 \leq v(y) \leq v(x) + (y - x)v'(x) \leq M - |y - x||v'(x)|$ for every boundary point y in this plane. Note from an inspection of Figure 3.1, that $(y - x)v'(x) \leq 0$. Therefore $|v'(x)| \leq M/|y - x|$, and the right hand side in this inequality becomes larger if I decrease the denominator to $dist(x, \partial\Omega)$. This establishes (3.3). Therefore any sequence u_k in C_M is locally uniformly Lipschitz-continuous, i.e. for every compact $\Omega' \subset\subset \Omega$ it is uniformly Lipschitz-continuous on Ω', and, due to the Arzela-Ascoli Theorem, after possibly passing to a subsequence, it has a locally uniform concave limit u. This shows that C_M is compact in $L^\infty_{loc}(\Omega)$. Let us now show that moreover

$$\nabla u_k(x) \to \nabla u(x) \qquad \text{a.e. in } \Omega. \tag{3.4}$$

Then (3.4), (3.3) and the Lebesgue Dominated Convergence Theorem imply compactness of C_M in every $W^{1,p}_{loc}(\Omega)$. To verify (3.4) let x be a point where all the u_k and u are differentiable, which is almost everywhere in Ω. By concavity of the one-dimensional functions $t \mapsto u(x + te_i)$ (with e_i denoting the i-th unit vector in $\mathbb{R}^N$ for $i = 1, 2, \ldots, N$) the derivative of this function at x can be bounded below and above by the difference quotient from the right and from the left, i.e.

$$\frac{u_k(x + \varepsilon e_i) - u_k(x)}{\varepsilon} \leq \frac{\partial}{\partial x_i} u_k(x) \leq \frac{u_k(x - \varepsilon e_i) - u_k(x)}{-\varepsilon} \tag{3.5}$$

holds for a.e. $x \in \Omega$, for every $k \in \mathbb{N}$ and every $i = 1, 2, \ldots, N$. There are two limits $k \to \infty$ and $\varepsilon \to 0$ that we can apply to (3.5). Passing with k to infinity first, gives

$$\begin{aligned}\frac{u(x + \varepsilon e_i) - u(x)}{\varepsilon} &\leq \liminf_{k\to\infty} \frac{\partial}{\partial x_i} u_k(x) \\ &\leq \limsup_{k\to\infty} \frac{\partial}{\partial x_i} u_k(x) \leq \frac{u(x - \varepsilon e_i) - u(x)}{-\varepsilon} ,\end{aligned} \tag{3.6}$$

and sending ε to zero in (3.6) yields the proof of (3.4), namely

$$\frac{\partial}{\partial x_i} u(x) \leq \liminf_{k\to\infty} \frac{\partial}{\partial x_i} u_k(x) \leq \limsup_{k\to\infty} \frac{\partial}{\partial x_i} u_k(x) \leq \frac{\partial}{\partial x_i} u(x) \quad \text{a.e. in } \Omega.$$

Now (3.4) and Lebesgue's Dominated Convergence Theorem imply that $R(u) \leq \liminf_{k\to\infty} R(u_k)$ and prove a).

Before proceeding with the proof of b) let me *remark*, that for v in the class S_M one can also find a locally uniform gradient bound which reminds us of (2.7)

$$|\nabla v(x)| \leq \frac{M + \sqrt{M^2 + dist^2(x, \partial\Omega)}}{dist(x, \partial\Omega)} \qquad \text{a.e. in } \Omega,$$

a circumstance, which might raise hopes of being able to prove existence of minimizers in S_M as well. But as the examples of sequences in my second lecture have shown, in general (3.4) is violated for sequences in S_M. Therefore I do not know how to get a handle on the existence problem in S_M, not to mention uniqueness and other properties that I can investigate in the C_M case.

Proof of b): One can give several proofs for the nonsymmetry. The third one (chronologically) that was found consists in showing that for a peculiar nonradial function $v \in C_M$ we have $R(v) < R(u_N)$ so that the minimum over radial functions in C_M, which is attained at Newton's solution u_N, lies above the minimum in general functions in C_M. This special function v looks like a lipstick or screwdriver and was found by P.Guasoni [31]. A picture of it can also be found in [25]. For a long time I had tried to prove the converse, that a minimizer in C_M had to be radial if $\Omega = D$ was a circular disk, see [12, p.85-86]. Then my coauthors and I followed a suggestion of H.Brezis and looked at the second variation of R at u_N. If a test function $\phi \in C^\infty(D)$ has the property that for every ε in a sufficiently small neighborhood of zero the function $u_N + \varepsilon\phi)$ is an element of C_M, and if u_N is a global minimizer of R in C_M, then $R(u_N + \varepsilon\phi)$ can be interpreted as a function of ε, which has a local minimum at $\varepsilon = 0$. Consequently the first variation $\delta R(u_N, \phi)$ in direction of this testfunction vanishes, i.e.

$$\delta R(u_N, \phi) = -2 \int_D \frac{\nabla u_N \cdot \nabla \phi}{(1 + |\nabla u_N|^2)^2}\, dx = 0, \tag{3.7}$$

and its second variation $\delta^2 R(u_N, \phi)$ is nonnegative, i.e.

$$\delta^2 R(u_N, \phi) = \int_D \frac{2\left[-(1 + |\nabla u_N|^2)|\nabla\phi|^2 + 4(\nabla u_N \cdot \nabla\phi)^2\right]}{(1 + |\nabla u_N|^2)^3}\, dx \geq 0. \tag{3.8}$$

We intend to show that (3.8) can be falsified. This implies that u_N is not a minimizer in C_M. but first let us note that as long as we stay with ϕ in the class of radial functions, the terms $(\nabla u_N \cdot \nabla\phi)^2$ and $|\nabla u_N|^2 |\nabla\phi|^2$ coincide and are equal to $u_r^2\ \phi_r^2$. Therefore in the radial case the square bracket in (3.8) becomes $(3u_r^2 - 1)\phi_r^2$, and this is positive, because any ϕ such that $u_N \pm \varepsilon\phi \in C_M$ must have its support outside the blunt part of u_N, i.e. in a region where $|\nabla u_N| \geq 1$. Incidentally, this together with c) proves optimality of u_N in the class of radial functions in C_M.

So in order to violate (3.8) we have to choose a function ϕ which is not radial. This can be achieved if the angle between the vectors ∇u_N and $\nabla\phi$ is mostly close to a right angle. So let $\phi(r, \theta) = \eta(r) sin(k\theta)$, where η has its support in $(r_0, 1)$, then the negative term in the square bracket of (3.8) wins over the positive term. This shows that the Newton's radial solution u_N can be infinitesimally perturbed into a function with smaller resistance. In fact, the second proof of nonsymmetry was based on constructing a function close to u_N, and it can be found in [10]. The (chronologically) first proof is the one that falsifies (3.8).

Proof of c): The Euler–Lagrange equation (3.2) follows from (3.7) and the fundamental lemma in the calculus of variations, provided $u \pm \varepsilon\phi \in C_M$ for any sufficiently small nonnegative ε. A sufficient criterion for this is that u is piecewise of class C^2, say $u \in C^2(\Omega')$ with $\Omega' \subset \Omega \setminus (u^{-1}(M) \cup u^{-1}(0))$, and that u is strictly concave in Ω', in the sense that its Hessian is strictly negative, bounded away from zero. In that case any $\phi \in C_0^\infty(\Omega)$ is an admissible variation, because $u \pm \varepsilon\phi$ will be in C_M for sufficiently small ε. It remains to show that (3.2) is of mixed type. To this end we differentiate out and obtain

$$(1 + |\nabla u|^2)^{-3}\ \left[\ (1 + |\nabla u|^2)\Delta u\ - 4\ u_i u_{ij} u_j\ \right] = 0. \tag{3.9}$$

Now recall that the Laplace-operator can be rewritten in any orthogonal coordinate system, for instance in the one spanned by $\nu := -\nabla u/|\nabla u|$ and

τ_i, where τ_i, $i = 1, \ldots, N-1$, denote an orthogonal system of vectors tangent to the level surface at u through x. Then

$$\Delta u = u_{\nu\nu} + \sum_{i=1}^{N-1} u_{\tau_i \tau_i} =: u_{\nu\nu} + u_{\tau\tau} \ ,$$

and the square bracket in (3.9) becomes

$$\begin{aligned} &(1 + |\nabla u|^2)\,(u_{\nu\nu} + u_{\tau\tau}) \ - 4\, u_i u_{ij} u_j \\ &= (1 - 3|u_\nu|^2)\ u_{\nu\nu} + (1 + |u_v|^2)\ u_{\tau\tau} = 0. \end{aligned} \tag{3.10}$$

Note that the coefficient in front of $u_{\nu\nu}$ can change sign. If $|\nabla u| < 1/\sqrt{3}$, then it is positive and the equation is elliptic, while $|\nabla u| > 1/\sqrt{3}$ yields a negative coefficient and a hyperbolic equation. This completes the proof of Theorem 3.1.

Remark 3.1:
Newton's solution, and as we shall see in the next Theorem, minimizers in C_M as well, have $|\nabla u| \geq 1$ in the interesting parts of their graph. Does this mean that they solve a hyperbolic equation there? Yes, but only if the assumptions for the derivation of (3.2) are satisfied. The latter seems unlikely and one could also argue that a minimizer u should satisfy the Monge-Ampère equation

$$\det D^2 u = 0 \quad \text{a.e. in } \Omega' \tag{3.11}$$

instead of (3.2). This was pointed out to me by H. Berestycki. The (heuristic) argument goes as follows. Suppose u minimizes R in C_M and is strictly concave and of class $C^2(\Omega')$ where Ω' is a subset of Ω in which u is neither M nor 0. Then, as in the proof of the nonsymmetry part of Theorem 3.1 above, we could construct an admissible testfunction which is highly oscillatory in directions orthogonal to ∇u, and for which the second variation of R at u becomes negative. This contradiction implies that u is not smooth or not strictly concave. In the latter case, because it is still concave, one of its principle curvatures must vanish, i.e. the Hessian has vanishing determinant.

So which equation, (3.2) or (3.11) is the right one for minimizers? I do not know. There is numerical evidence which supports (3.11), but maybe neither one is correct. It would be instructive to know what the minimizer in C_M looks like if D is a disk, but even this special question is still open.

Other things are known, however. In the following Theorem 3.2 I want to give a mathematical justification for Newton's "educated guess"(see Chap. II above) that a minimizer has always slope outside the interval $(0,1)$.

Theorem 3.2:
Let $\Omega \subset \mathbb{R}^N$ be convex and u be a minimizer of R in C_M. Then $|\nabla u| \notin (0,1)$ a.e. in Ω.

For the proof we argue by contradiction and show that u cannot be minimizer, if $|\nabla u(x)| \in (0,1)$ on a subset, N say, of Ω with positive measure. In this case we replace u by a function $w \in C_M$ which lies above u, and has slope outside $(0,1)$. Following the idea of Newton in Figure 2.3 above, such a function can be constructed by

$$w(x) := \inf\{\ t(x)\ :\ t(x) \text{ is a tangent plane to } u \text{ with slope } \notin (0,1)\ \},$$

and because of the concavity of u it touches u in those points where $|\nabla u| = 0$ or $|\nabla u| > 0$. To realize why $R(w) < R(u)$, it helps to "relax" R and to define its relaxation $\tilde{R} = \int_\Omega \tilde{f}(|\nabla u|)\, dx$, by replacing the map $f(s) = 1/(1+s^2)$ with something more convex, namely $\tilde{f}$. The largest convex function below f in $\mathbb{R}$ is identically zero. Therefore this function is not suitable for our purposes, but if we conceive of f as a function on $[0,\infty)$ and replace this function by its convex envelope, then

$$\tilde{f}s) = \begin{cases} 1 - s/2 & \text{if } s \in [0,1] \\ 1/(1+s^2) & \text{if } s \in [1, 1\infty) \end{cases} \tag{3.12}$$

will do the job. Notice that $f < \tilde{f}$ only in $(0,1)$, so that $\tilde{R}(w) = R(w)$ by construction of w, while $\tilde{R}(u) < R(u)$ by the assumption that N is not a nullset. If we manage to show that

$$\tilde{R}(u) \geq \tilde{R}(w), \tag{3.13}$$

then we have finished the proof. Therefore let us subtract both sides in (3.13) from each other and note that the integrands coincide when $|\nabla u| \notin (0,1)$ to arrive at

$$2\tilde{R}(u) - 2\tilde{R}(v) = \int_{\{u \neq v\}} (|\nabla v| - |\nabla u|)\, dx = \int_\Omega (|\nabla v| - |\nabla u|)\, dx. \tag{3.14}$$

By the coarea formula we have

$$\int_\Omega (|\nabla v| - |\nabla u|)\, dx = \int_0^M \left[\mathcal{H}^{N-1}(\{v = t\}) - \mathcal{H}^{N-1}(\{u = t\})\right] dt, \tag{3.15}$$

and the integrand in (3.14) is nonnegative for every $t \in [0, M]$, because the level sets of u are convex subsets of the convex level sets of v, for details of this last statement see [12, p.75]. Therefore the right hand sides of (3.15) and (3.14) are nonnegative, and the proof of Theorem 3.2 is complete.

Remark 3.2:
Other results can be shown for minimizers. I list them here without proof. The natural boundary condition asociated with the functional R is the vanishing flux condition

$$\frac{1}{(1+|\nabla u|^2)^2}\frac{\partial u}{\partial \nu} = 0 \qquad \text{on } \partial\Omega, \tag{3.16}$$

but like the Euler equation, it can only be derived under strong assumptions on u. It is also conceivable that $u = 0$ or $u = M$ on the boundary or that u is not strictly concave on $\partial\Omega$. As a matter of fact, u vanishes in at least one boundary point, because otherwise one could stretch the graph of u in vertical direction and obtain a new function in C_M with smaller resistance than u. Even (3.16) can be realized in more than one way, because it is implied by $u_\nu = 0$ or $|\nabla u| = \infty$.

This leads in a natural way to the question if derivatives of u are a priori bounded. In a way they are, see (3.3), but (3.3) breaks down at the boundary. Once we know that $|\nabla u|$ is globally bounded, we also know that $u = 0$ on $\partial\Omega$. To see this, it suffices to do a small variation in the one-dimensional situation depicted in Figure 3.2.

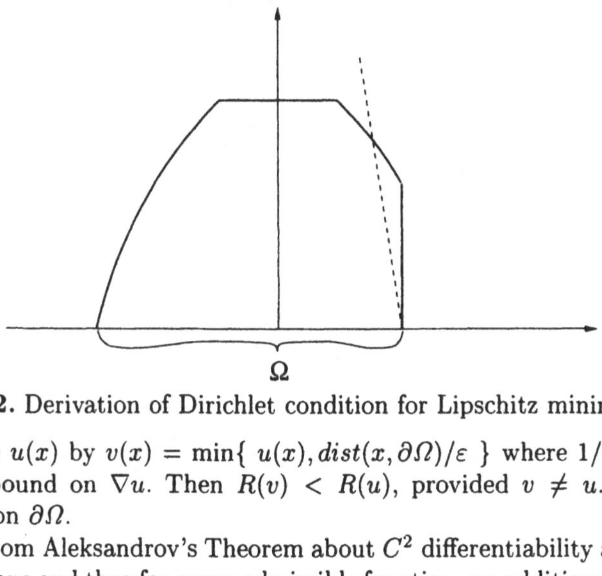

Fig.3.2. Derivation of Dirichlet condition for Lipschitz minimizers.

Replace $u(x)$ by $v(x) = \min\{\ u(x), dist(x, \partial\Omega)/\varepsilon\ \}$ where $1/\varepsilon$ is greater than the bound on ∇u. Then $R(v) < R(u)$, provided $v \neq u$. Otherwise $v = u = 0$ on $\partial\Omega$.

Aside from Aleksandrov's Theorem about C^2 differentiability a.e. for concave functions and thus for every admissible function, no additional regularity result seems to be known for minimizers of R in C_M. Incidentally, under moderate regularity assumptions, we can improve on Theorem 3.1b) and show that $|\nabla u|$ jumps precisely from 0 to 1 on the boundary of the blunt part $\partial\{u = M\}$, see [12, p.87].

What about numerical calculations of the minimizer? The functional and the admissible set of functions are somewhat nonstandard. R is neither convex nor coercive, and concavity is a pointwise constraint on second order derivatives, or a nonlocal constraint, namely

$$u(tx + (1-t)y) \geq tu(x) + (1-t)u(y) \qquad \text{for any } x, y \in \omega,\ t \in [0,1] \quad (3.17)$$

on the admissible functions themselves. If one wants to do a finite element approximation of a minimizer with piecewise linear finite elements, these are not of class C^2 across their "edges". Therefore we have to incorporate

(3.17) into a numerical formulation of the problem. I report now on joint work with C.Schwab [40]. To simplify matters let Ω be the unit square. We triangulate it in a regular way with triangles of gridsize h. Instead of looking for a function of $x \in \Omega$ we look now for a vector in $\mathbb{R}^N$ with $N = O(h^{-2})$, whose components should be u evaluated at each gridpoint. The functional R can be discretized by splitting the integral into a sum of integrals over each triangle. On a given triangle, the gradient of u can be approximated by the affine function that takes the values of u in the gridpoints. This way we can define a discretized version R_h of R, and minimizing R_h constitutes an optimization problem in a finite dimensional space $\mathbb{R}^N$. For the latter ones there are standard optimization algorithms, for instance gradient methods, in which one starts at some point v, evaluates the gradient of R_h at v and modifies v into the direction of steepest descent of R_h. With these concepts in mind it looks fairly easy to do a finite element analysis.

However, the following issues have to be addressed.

a) The discretized functional is not convex either, and therefore any descent method may end in a local, but not global minimizer of the functional. This shortcoming can realistically only be dealt with, if one starts with a good guess in a neighborhood of the optimal solution. Schwab and I took various parabolas with varying eccentricity as initial guesses, and always ended at the "same" minimizer. This gave us some confidence that we were on the right track.

b) Any decent descent algorithm applied to the discretized functional should endeavour to increase the gradient and decrease the functional. It will therefore suggest an increase of values of v in individual gridpoints above M or a decrease below 0, in which case the piecewise linear interpolation would violate the L^∞ bound in C_M. This problem can be dealt with by a simple cut-off at M resp. 0 of the new iterate suggested by the descent algorithm.

c) Finally we may start out with a polygon satisfying the concavity constraint in C_M and obtain a next iteration that is no longer concave. In this case doing local adjustments of the new iterate is not of much help, because (3.17) is a global constraint in Ω. Concavity of minimizers had to be enforced in a global way, and we found a penalty method to be the appropriate tool.

Let me elaborate on this aspect in more detail. As long as a piecewise linear function is concave, it bends the right way, or its gradient jumps in the right direction along the edges of the triangulation. Suppose we measure, how much a (nonconcave) function violates the concavity constraint with some functional C_h, which is zero for concave functions and positive for nonconcave ones, then instead of minimizing R_h we might consider minimizing

$$R_h(v) + \frac{1}{\varepsilon} C_h(v) \tag{3.18}$$

on the set S_h of all those continuous functions, which are affine on each triangle and take values in $[0, M]$. If $\varepsilon > 0$ is then chosen small enough, the minimizers of (3.18) should be almost concave. This idea works. Let T_k and

T_l be two adjacent triangles with a normal e_{kl} to their common edge pointing from T_k into T_l. Then a piecewise linear u is concave across this edge, provided the scalar product $(\nabla u|_{T_l} - \nabla u|_{T_l}) \cdot e_{kl}$ is nonpositive. Therefore a good way to define C_h is to measure the positive part of $(\nabla u|_{T_l} - \nabla u|_{T_l}) \cdot e_{kl}$ for every edge in the triangulation and to add it up in some l^s-norm, as in

$$C_h(v) := \sum_{kl} \left|[(\nabla v|_{T_l} - \nabla v|_{T_l}) \cdot e_{kl}]^+\right|^s . \tag{3.19}$$

With these ideas fixed let us denote a minimizer of (3.18) by u_ε^h and let us call it a finite element solution. The the following facts can be shown.

Theorem 3.3:
The function R_h is in $C^\infty([0, M]^N)$ and the function C_h is in $C^{\bar{s}, s-\bar{s}}([0, M]^N)$, where the functions R_h and C_h are interpreted as functions of the nodal variables, and where $\bar{s}$ is the largest integer strictly less than s.
For every $\varepsilon > 0$ and every triangulation of width h there exists a finite element solution u_ε^h.
The finite element solution is almost concave, in the sense that

$$\begin{aligned} u_\varepsilon^h(y) \leq & u_\varepsilon^h(x) + (y - x) \cdot \nabla u_\varepsilon^h(x) \\ & + c\, \varepsilon^{1/s}\, h^{(1-s)/s}\, |y - x| \end{aligned} \tag{3.18}$$

with c independent of $x, y \in \Omega$, h and ε.
If ε satisfies

$$\varepsilon \leq \begin{cases} ch^{1/\alpha} \text{ with } \alpha \in (0, 1/(s-1)) & \text{for } s > 1, \\ o(1) & \text{for } s = 1, \end{cases}$$

then, as $h \to 0$, there is a sequence of finite element solutions, which converges to a limit $u \in C_M$. Moreover, this limit u is a minimizer for the continuous problem.

A proof of Theorem 3.3 is given in [40]. Let me just stress the fact that concave functions satisfy (3.18) with $c = 0$, and that a balance between h and ε is needed so that the last term in (3.18) becomes small as ε and h go to zero.

You may wonder what kind of numerical solution this method delivers. Our numerical caculations led to an optimal u (on a square), the graph of which was a piecewise developable surface, consisting of a square-shaped blunt end and four pieces that seemed to depend only on the distance of x to the boundary. For figures I refer to [40]. Incidentally, these calcuations seem to support the validity of the Monge-Ampère equation (3.11) above. However, for $M = 2$ Paolo Guasoni was able to give an ad hoc example of a polygonal, concave, roof-shaped function, whose resistance was lower than the one of our numerical example. This suggests that we started the descent algorithm with the wrong initial guesses in the first place. So no matter what we try, Newton's problem of minimal resistance seems to resist attack after attack.

To conclude this lecture let me mention a few other aspects of the problem. Presently an existence proof in S_M seems out of reach, but one can restrict S_M a bit an try to look for bodies of fairly small resistance there. In his Diplomathesis M.Mester took cones and carved radial troughs into them, obtaining shapes such as the one depicted in Figure 3.3.

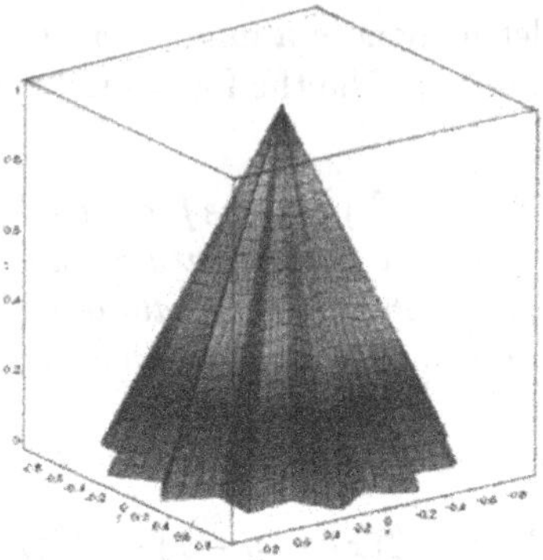

Fig.3.3. Modification of a circular cone.

As the number of creases goes to ∞, they can be less and less deep without violating their membership in S_M. This way one can construct a sequence $v_n \in S_M$ with $\lim_{n\to\infty} v_n(x) = v_\infty(x) = 1 - |x|$ and $R(v_\infty) = 0.5\pi > \liminf_{n\to\infty} R(v_n) = 3\pi/7$, providing yet another case of lacking lower semicontinuity.

What about existence proofs in other classes of admissible functions for slightly modified functionals, which at the same time account for frictional effects? Instead of being concave and bounded in L^∞, it was suggested in 20, p.259] (for the case of radial functions) to limit the surface area of the admissible bodies. In this case Eggers was able to conclude that a minimizer of R had to have conical shape. In [25] Ferone and I looked at this optimal shape problem for a base domain $\Omega \subset \mathbb{R}^N$ which is not necessarily a disk or a ball, but simply connected and bounded.

We set

$$A(S) := \{v \in BV(\Omega) \mid v \geq 0 \text{ a.e. in } \Omega \text{ and } \int_\Omega \sqrt{1 + |\nabla v(x)|^2}\, dx = S\}. \tag{3.19}$$

and look for a minimizer of R on $A(S)$.

Theorem 3.4:
For every $S \geq |\Omega|$ there exists a minimizer u of R on $A(S)$. Moreover the function $u(x) = c\, dist(x, \partial\Omega)$ with $c = \sqrt{(s/|\Omega|)^2 - 1}$ is a solution.

To prove this theorem, observe that $S \geq |\Omega|$ is needed to make A nonempty. Now let $v(x)$ be any function from A and define

$$f(x) := \sqrt{1 + |\nabla v(x)|^2}\ .$$

If we replace f by its average

$$\underline{f} = \frac{1}{|\Omega|} \int_\Omega f(x)\, dx = \frac{s}{|\Omega|}, \tag{3.20}$$

we can construct a new function $\tilde{v}$ by solving the following Hamilton Jacobi equation

$$|\nabla \tilde{v}(x)| - \sqrt{\underline{f}^2 - 1} = 0 \quad \text{in } \Omega. \tag{3.21}$$

This will not affect the surface area constraint, because $\int f\, dx = \int \underline{f}\, dx$. Certainly, there exists a nonnegative solution of (3.21), namely a positive multiple of the distance function $dist(x, \partial\Omega)$. Let $\tilde{v}$ be this function, then $\tilde{v} \in A$ as well.

Now we observe that $R(v) = \int f(x)^{-2}\, dx$, that $f \geq 1$ and that the mapping $s \mapsto s^{-2}$ is convex for $s > 0$. Therefore

$$\begin{aligned} R(v) = \int_\Omega \frac{1}{1 + |\nabla v(x)|^2}\, dx &= \int_\Omega \frac{1}{f(x)^2}\, dx \\ &\geq \int_\Omega \frac{1}{\overline{f}^2}\, dx = \int_\Omega \frac{1}{1 + |\nabla \tilde{v}(x)|^2}\, dx = R(\tilde{v}). \end{aligned} \tag{3.22}$$

Thus replacing any function $v \in A$ by a concave function of same surface area but constant slope will not increase R. As a matter of fact, unless v satisfies (3.21), $R(\tilde{v}) < R(v)$.

Remark 3.3:
In [50] F.Nicolai has studied the related problem of minimizing R on

$$\begin{aligned} C(S) := \{v \in W^{1,\infty}_{loc}(\Omega) \mid v \geq 0,\ \ v \text{ concave in } \Omega \\ \text{and} \int_\Omega \sqrt{1 + |\nabla v(x)|^2}\, dx + \int_{\partial\Omega} v\, d\mathcal{H}^{n-1} \leq S\ \}. \end{aligned}$$

She gave an existence proof along the lines of Theorem 3.1 above. In our proof of Theorem 3.4 we do not need the concavity of minimizers. However, for convex Ω our minimizer $c\ d(x, \partial\Omega)$ is also a minimizer of R on $C(S)$.

Note also that again the minimizers of our problem need not be unique. To see this one can look at other solutions of (3.21) under homogeneous Dirichlet conditions. In fact, homogeneous Dirichlet conditions are necessary for Lipschitz continuous minimizers, recall Figure 3.2 above. If we want to somehow identify a unique solution we can ask for a viscosity solution of (3.21) or for a concave solution of (3.21) under vanishing boundary values.

The function $dist(x, \partial\Omega)$, which we have identified as a solution is piecewise of class $C^{1,1}(\Omega)$ and its derivatives degenerate on a subset of Ω which is referred to in 26] as *ridge*. Some examples of domains Ω and their ridges are depicted in 26, p. 197].

Of course Theorem 3.4 remains true if the functional R is replaced by $J(v) = \int_\Omega j(\sqrt{1+|\nabla v(x)|^2})\, dx$ with convex $j : \mathbb{R}^+ \to \mathbb{R}^+$.

Let us finally imagine that the resistance of a body depends also on frictional effects, which were not included in Newton's original model. For simplicity we suppose that the resistance caused by friction is proportional to the surface area of its front and lateral surface. In this case a modified resistance functional would read as follows. For some constant $\alpha > 0$ and for $v \in W^{1,\infty}_{loc}(\Omega)$

$$R_\alpha(v) := \int_\Omega \frac{1}{1+|\nabla v(x)|^2}\, dx \;+\; \alpha \int_\Omega \sqrt{1+|\nabla v(x)|^2}\, dx + \int_{\partial\Omega} v\, d\mathcal{H}^{n-1}. \tag{3.23}$$

Theorem 3.5:
For every $\alpha > 0$ there exists a minimizer u of R on $B := W^{1,\infty}_{loc}(\Omega)$. Moreover, for $a \in (0,2)$ the function $u(x) = \underline{c}\; d(x, \partial\Omega)$ with $\underline{c}$ given in (3.24) below is a solution, while for $a \geq 2$ the function $u \equiv 0$ is a solution.

Proof: With the same reasoning as in the proof of Theorem 3.4 we may conclude that without loss of generality a solution satisfies the eiconal equation (3.21) under homogeneous boundary conditions, so that $|\nabla u| = const.$ in Ω. Now a one-dimensional optimization shows that for large α the surface area term is so dominant that it forces $|\nabla u|$ to vanish everywhere. In fact, for $c \geq 0$ set

$$g(c) := \frac{1}{1+c^2} + \alpha\, \sqrt{1+c^2}$$

and note that $g(0) = 1+\alpha$, $g'(0) = 0$ and $g(\infty) = \infty$. Moreover, for $\alpha < 2$ the function g has a global minimum at

$$\underline{c} = \sqrt{\left(\frac{2}{\alpha}\right)^{2/3} - 1} \tag{3.24}$$

and

$$g(\underline{c}) = \left[\left(\frac{1}{2}\right)^{2/3} + 2^{1/3}\right] \alpha^{2/3} < 1+\alpha,$$

while for $\alpha \geq 2$ the function g is monotone increasing on $\mathbb{R}^+$.

4. Extremal eigenvalue problems

Eigenvalues of differential operators on bounded domains depend on the underlying domain Ω. A well-known result is the Faber-Krahn inequality, which states that *among all membranes of given area the circular one has the lowest fundamental frequency*. A more precise way of stating this result is the following.

Theorem 4.1: Faber-Krahn inequality
Let $\lambda_1(\Omega)$ be the first eigenvalue for

$$\begin{aligned} \Delta u + \lambda u &= 0 \qquad \text{in } \Omega \subset \mathbb{R}^N, \\ u &= 0 \qquad \text{on } \partial\Omega, \end{aligned} \tag{4.1}$$

then the problem of minimizing $\lambda_1(\Omega)$ among all open sets of prescribed volume (i.e. N-dimensional Lebesgue measure) has a solution.
Moreover, the domain which minimizes λ_1 is a ball, i.e. $\lambda_1(\Omega) \geq \lambda_1(\Omega^*)$.

Here as well as later Ω^* denotes the ball of the same volume as Ω, centered at the origin. Notice that Theorem 4.1 contains two statements, namely one about the existence of a minimizing domain and the other one about its shape. There are two conceptually different ways to prove Theorem 1.

a) The first proof relies on a constructive procedure, by establishing the Faber-Krahn inequality directly. Suppose Ω is any admissible domain and u_1 is a positive first eigenfunction on Ω. Then we can rearrange u_1 into its Schwarz-symmetrization u_1^* as follows. Each level set $\Omega_c := \{x \in \Omega \mid u(x) > c\}$ is transformed into Ω_c^*, a ball of the same volume as Ω_c, and u_1^* is defined as the unique radially decreasing function whose level sets are given by Ω_c^*. Now Cavalieri's principle implies $\int_\Omega |u_1|^2 dx = \int_{\Omega^*} |u_1^*|^2 dx$, and (this is the hard part, see [36]) $\int_\Omega |\nabla u_1|^2 dx \geq \int_{\Omega^*} |\nabla u_1^*|^2 dx$. But now the validity of the Faber-Krahn inequality is apparent from the variational characterization of λ_1 and from the following chain of inequalities.

$$\begin{aligned} \lambda_1(\Omega) &= \frac{\int_\Omega |\nabla u_1|^2 dx}{\int_\Omega |u_1|^2 dx} \\ &\geq \frac{\int_{\Omega^*} |\nabla u_1^*|^2 dx}{\int_{\Omega^*} |u_1^*|^2 dx} \geq \inf_{v \in W_0^{1,2}(\Omega^*)} \frac{\int_{\Omega^*} |\nabla v|^2 dx}{\int_{\Omega^*} |v|^2 dx} = \lambda_1(\Omega^*) \end{aligned} \tag{4.2}$$

b) The second "proof" uses a necessary condition for minimizers. If there exists a minimizing domain Ω, then Ω must be a ball. In fact, if there exists a minimizing smooth domain Ω, then we can perturb its boundary in the direction of a vector field $V(x)$ in such a way, that the perturbation is volume preserving. Therefore, using $\nu(x)$ as normal vector to $\partial\Omega$,

$$\int_{\partial\Omega}(V(x)\cdot\nu(x))ds=0. \tag{4.3}$$

The first variation of λ_1 with respect to domain variations is well-known as Hadamard's formula, and since it vanishes, we have

$$\int_{\partial\Omega}|\nabla u_1|^2(V(x)\cdot\nu(x))ds=0. \tag{4.4}$$

Derivations of the Hadamard formula are given in [15, p.262] or [58, p.138]. To apply [15], set $F(u,\nabla u)=|\nabla u|^2-\lambda_1(\Omega)\,u^2$.

By the way, according to [28] Hadamard's condition was already known to Lord Rayleigh, who had conjectured the Faber-Krahn inequality more than a century ago in 1877. Relation (4.4) tells us that $|\nabla u_1|^2$ is orthogonal in $L^2(\partial\Omega)$ to the function $(V(x)\cdot\nu(x))$, but according to (4.3), $(V(x)\cdot\nu(x))$ has as its range the orthogonal complement to the constant function 1. Therefore (4.3) and (4.4) imply

$$\frac{\partial u_1}{\partial\nu}=const. \qquad \text{on } \partial\Omega \tag{4.5}$$

as additional boundary condition to (4.1). But due to a famous result of Serrin [56] the overdetermined boundary value problem (4.1) (4.5) can only have a solution if Ω is a ball.

Notice that approach b) contains two gaps. First of all it uses the assumption that there exists a minimizing domain Ω. Secondly it assumes smoothness of its boundary in the derivation of Hadamard's formula. A similar state of the art has been reached in a fourth order problem, which models a vibrating clamped plate rather than a membrane. Incidentally, both Theorems 4.1 and 4.2 had been conjectured by Lord Rayleigh in 1877, see [49, 48].

Theorem 4.2: Nadirashvili-Ashbaugh-Benguria inequality
Let $\gamma_1(\Omega)$ be the first eigenvalue in

$$\begin{aligned}\Delta\Delta u-\gamma u=0 &\qquad \text{in } \Omega\subset\mathbb{R}^2,\\ u=\frac{\partial u}{\partial n}=0 &\qquad \text{on } \partial\Omega,\end{aligned} \tag{4.6}$$

then the problem of minimizing $\gamma_1(\Omega)$ among all open sets of prescribed area (i.e. 2-dimensional Lebesgue measure) has a solution.
Moreover, the domain which minimizes γ_1 is a circular disc, i.e. $\gamma_1(\Omega)\geq\gamma_1(\Omega^*)$.

Again there are two conceptually different ways to prove the Theorem.

a) The first proof relies on a delicate rearrangement argument, rearranging functions like the restriction of Δu to the set on which u is positive and so on. After all, $\gamma_1(\Omega)$ is characterised via a Rayleigh quotient, namely

$$\gamma_1(\Omega) = \inf_{v \in W_0^{2,2}(\Omega)} \frac{\int_\Omega |\Delta v|^2 dx}{\int_\Omega |v|^2 dx}.$$

The foundation for this proof was laid by Talenti in [64], who had already shown that $\gamma_1(\Omega) \geq 0.97768\ \gamma_1(\Omega^*)$. Nadirashvili was able to improve his result for $N = 2$ and thus finished a whole series of increasingly successful attempts to prove it completely. In fact Talenti, as well as later Ashbaugh and Laugesen, had attacked this problem for any space dimension N. For $N = 3$ the result of Nadirashvili could be recovered by Ashbaugh and Benguria in [3], but for general $N \geq 4$ nowadays one has only an estimate of the type

$$\gamma_1(\Omega) \geq c_N \gamma_1(\Omega^*)$$

with $c_N < 1$.

b) The second "proof" uses a necessary condition for minimizers. If there exists a minimizing domain Ω then Ω must be a ball. This strategy had been chosen by Mohr prior to Talenti and was also suggested by me in [38, p.172], while I was still unaware of [46]. In fact, using a variant of the Hadamard formula Mohr arrives at

$$\int_{\partial\Omega} |\Delta u_1|^2 (V(x) \cdot \nu(x)) ds = 0 \tag{4.7}$$

instead of (4.4), and thus – using (4.6) – at

$$\Delta u_1 = \frac{\partial^2 u_1}{\partial \nu^2} = const. \qquad \text{on } \partial\Omega \tag{4.8}$$

as an additional boundary condition. He then proceeds to show that the overdetermined problem (4.6) (4.8) has a solution only if Ω is a circular disc.

As mentioned above, approach b) has two shortcomings, because it is based on the assumptions that there exists a minimizing domain Ω and that its boundary is smooth. There is a third eigenvalue inequality which has been considered in [55, Note F] and [63]. It is meant to describe the buckling of a clamped plate. For one space dimension this problem is known as buckling of the Euler rod. The eigenvalue problem is

$$\begin{aligned} \Delta\Delta u + \Lambda \Delta u &= 0 \qquad \text{in } \Omega \subset \mathbb{R}^2, \\ u = \frac{\partial u}{\partial n} &= 0 \qquad \text{on } \partial\Omega, \end{aligned} \tag{4.9}$$

and the variational characterization of its first eigenvalue Λ_1 reads

$$\Lambda_1(\Omega) = \inf_{v \in W_0^{2,2}(\Omega)} \frac{\int_\Omega |\Delta v|^2 dx}{\int_\Omega |\nabla v|^2 dx}.$$

After Theorems 4.1 and 4.2 it is only natural to ask if $\Lambda_1(\Omega) \geq \Lambda_1(\Omega^*)$. However, only partial answers to this question are curently available. The first one was given by Szegö himself, using Schwarz symmetrization.

Proposition 4.3: [63, 55]
If Ω is such that the first eigenfunction of (4.9) *is positive, then* $\Lambda_1(\Omega) \geq \Lambda_1(\Omega^*)$.

Unfortunately the assumption of Proposition 4.3 is known to be violated for many domains, for instance domains with corners, see [67]. Incidentally, Szegö could also prove Theorem 4.2 under this assumption, but there it is known to be violated again [67]. If one wants a lower bound for $\Lambda_1(\Omega)$ which does not depend on this positivity assumption, the best estimate up to now has been obtained by Ashbaugh and Laugesen in [5], who showed

$$\Lambda_1(\Omega) \geq c_N \; \Lambda_1(\Omega^*)$$

for general domains in $\mathbb{R}^N$, with special constants $\tilde{c}_N < 1$ and $\lim_{N\to\infty} \tilde{c}_N = 1$. In particular $\tilde{c}_2 \geq 0.7877$.

While Proposition 4.3 follows strategy a) above, there is also a (hitherto unpublished) approach of Willms following strategy b). N.B. Willms can be reached at Bishops University in Quebec and he has kindly allowed me to reproduce his proof here. The final chain of inequalities in this proof was strung together by B.Willms and H.Weinberger, and H.Weinberger as well agreed to its publication here.

Proposition 4.4: [68]
If there exists a domain $\tilde{\Omega}$ which minimizes Λ_1 among all plane domains of prescribed area and which has a smooth boundary $\partial\tilde{\Omega}$, then $\tilde{\Omega}$ must be a circular disc, in other words then $\Lambda_1(\Omega) \geq \Lambda_1(\Omega^*)$ holds for any admissible Ω.

The proof of Proposition 4.4 is tricky and was presented by Willms at an Oberwolfach meeting in 1995. Suppose Ω minimizes Λ_1 among all plane domains of given area and $\partial\Omega$ is smooth. Again we denote the associated eigenfunction by u_1. Then from a domain variation as in [48] or [15], one can conclude (4.7) or

$$\Delta u_1 = c \quad \text{on } \partial\Omega, \tag{4.10}$$

where c is a real constant. Now we want to show that the overdetermined boundary value problem (4.9) (4.10) can only have a solution if Ω is a circular disc.

To do this we use the substitution $z := \Delta u_1 + \Lambda_1 u_1$. Then the differential equation from (4.9) and the boundary condition (4.10) become

$$\Delta z = 0 \quad \text{in } \Omega, \qquad z = c \quad \text{on } \partial\Omega,$$

so that $z \equiv c$ by the maximum principle for harmonic functions. Next we use the substitution

$$v(x) = u_1(x) - \frac{c}{\Lambda_1},$$

so that (4.9) and (4.10) are transformed into the overdetermined boundary value problem

$$\left.\begin{aligned} \Delta v + \Lambda_1 v &= 0 && \text{in } \Omega, \\ v &= -\frac{c}{\Lambda_1} && \text{on } \partial\Omega, \\ \frac{\partial v}{\partial \nu} &= 0 && \text{on } \partial\Omega. \end{aligned}\right\} \tag{4.11}$$

Problem (4.11) is reminiscent of the (unsolved) Pompeiu problem (4.12),

$$\left.\begin{aligned} \Delta v + \nu v &= 0 && \text{in } \Omega, \\ v &= c && \text{on } \partial\Omega, \\ \frac{\partial v}{\partial \nu} &= 0 && \text{on } \partial\Omega, \end{aligned}\right\} \tag{4.12}$$

in which one wants to show that an eigenfunction v of the Laplace operator under Neumann boundary conditions, which in addition is constant on the boundary, can only exist if the underlying domain is a ball. For a recent partial result on this problem I refer to [29]. It was also shown in [6] that for problem (4.12) the domain Ω must be a ball, provided $\nu < \lambda_2$, the second Dirichlet eigenvalue of the Laplace operator from (4.1). The eigenvalue in (4.11) violates this constraint because an inequality due to Payne [52, 53, Thm. 6] states that

$$\Lambda_1(\Omega) \geq \lambda_2(\Omega), \tag{4.13}$$

and equality holds only if Ω is a ball, in which case there is nothing to prove. Nevertheless we will conclude from (4.11) that Ω must be a circular disc after all.

To this end let us first observe that by translation of Ω we can make sure that the origin is contained in Ω, and that v has a critical point at the origin. Then we follow an idea that has been used by Payne [52] and Aviles [6] and set $w := xv_y - yv_x$. In polar coordinates (r, θ) this can be read as $w = rv_\theta$. Now a simple calculation shows

$$\begin{aligned} \Delta w + \Lambda_1 w &= 0 && \text{in } \Omega, \\ w &= 0 && \text{on } \partial\Omega. \end{aligned} \tag{4.14}$$

Therefore either $w \equiv 0$ or $\Lambda_1 = \lambda_k(\Omega)$, a Dirichlet eigenvalue for the Laplace operator on Ω and $k \geq 3$. In the first of these cases $v_\theta \equiv 0$, i.e. v depends only on r and thus Ω is a circular disc. Thus the proof of Proposition 4.4 is complete, provided we can rule out the second case. In the second case we note that both w and ∇w vanish in the origin by construction. In fact $w_x = v_y + xv_{xy} - yv_{xx}$, and $v_y(0) = 0$, so $w_x(0) = 0$, and similarly $w_y = 0$. This means that the origin is a nodal point of w, and a point where a nodal line intersects itself transversally. But then for topological reasons this nodal line divides Ω into at least three nodal domains, at least one of which has area not exceeding $|\Omega|/3$. But now we get the following chain of inequalities

$$\begin{aligned}\Lambda_1(\Omega) = \lambda_k(\Omega) &= \lambda_1(\text{subdomain of area } \leq |\Omega|/3)\\ &\geq \lambda_1(\text{subdomain of area } = |\Omega|/3)\\ &\geq \lambda_1(\text{disc of area } = |\Omega|/3)\\ &= 3\lambda_1(\text{disc of area } = |\Omega|)\\ &> \lambda_2\ (\text{disc of area } = |\Omega|)\\ &= \Lambda_1(\text{disc of area } = |\Omega|),\end{aligned}$$

which contradicts the minimality of Λ_1 in Ω. In this chain of inequalities we have used one after the other the monotonicity of λ_1 with respect to the domain, the Faber-Krahn inequality, a scaling argument, the Payne-Polya-Weinberger inequality from [54], see also [2,4], and the discussion of the equality sign in (4.13).

So much for the proof of Proposition 4.4. The discomforting aspect of approach b), that existence of a minimizing domain is assumed, can be partly resolved in several ways. There is a general existence result by Buttazzo and Dal Maso, according to which a minimizing domain of prescribed area exists among all quasi-open sets, see [13]. A more satisfactory answer might be provided by the following Proposition, whose more geometric proof is taken from [16, Sect. 2], where it was stated for the special case $\alpha = 1$.

Proposition 4.5: [16]
Among convex plane domains of prescribed area, the eigenvalues $\lambda_1(\Omega)$ from (4.1), $\gamma_1(\Omega)$ from (4.6) and $\Gamma_1(\Omega)$ from (4.9) have a minimizer. In fact any functional $j(\Omega)$ which has the property that it is
i) bounded below,
ii) monotone in the sense $j(\omega) > j(\Omega)$ for $\omega \subset \Omega$, and
iii) coercive in the sense that it tends to ∞ on strips whose width tends to 0,
iv) homogeneous of degree -2α in the sense that there exists an $\alpha > 0$ such that $j(t\Omega) = t^{-2\alpha} j(\Omega)$ for every $t > 0$,
can be minimized among convex plane domains Ω of prescribed area.

For the proof we set $A := \{\, G \subset \mathbb{R}^2 \ |G \text{ convex}, \ |G| = c > 0 \,\}$ and choose a sequence G_j from A which minimizes λ_1 (or γ_1 or Λ_1). Notice that for λ_1 and Λ_1 we have to pick $\alpha = 1$, while for γ_1 one has to choose $\alpha = 2$ in iv). We shall only use the properties of a general j.
The sequence G_j has width $w(G_j)$ uniformly bounded below by a positive constant,

$$w(G_j) \geq \delta > 0, \tag{4.15}$$

because otherwise G_j fits into arbitrarily thin strips S_j of width ε_j with $\varepsilon_j \to 0$, but due to properties i) and ii) this would imply $j(G_j) \geq j(S_j) \to \infty$, which

contradicts the property that the sequence G_j is a minimizing sequence. The width of a convex domain is the minimal distance of two parallel supporting hyperplanes.
After possible translation, the sequence G_j, is uniformly bounded, i.e. without loss of generality we may assume that there exists an $R < \infty$ such that

$$G_j \subset B_R(0) \tag{4.16}$$

for all G_j. Otherwise, since each G_j is convex and of prescribed area, it would become too thin, which contradicts (4.15).
According to a result of Sholander [59, Thm. 20], for any plane convex set $6\,|G| \geq |\partial G|\, w(G)$. This and (4.15) imply that the perimeters $|\partial G_j|$ of G_j are uniformly bounded above.
Therefore Bonnesen's inequality, see e.g. [Ba, p.8], implies that a ball of uniform radius $\rho > 0$ fits into each G_j, because $\pi\rho^2(G_j) + |G_j| \leq |\partial G_j|\rho(G_j)$ holds for the radius $\rho(G_j)$ of the largest disc in G_j. Thus the boundary of G_j can be represented in polar coordinates by a 2π-periodic function $r_j(\theta)$, and $r_j(\theta) \in [\rho, R]$ for any natural j and any $\theta \in [0, 2\pi]$.
The fact that $B_\rho(0) \subset G_j \subset B_R(0)$ and the convexity of G_j imply that G_j is starshaped with respect to every $x \in B_\rho(0)$ and that the sequence r_j is equi-Lipschitz continuous in θ. Hence by the Arzela-Ascoli theorem $r_j \to r_\infty$ uniformly, $G_j \to G_\infty$ in a strong topology, and $2|G_\infty| = \int_0^{2\pi} [r_\infty(\theta)]^2 d\theta = c$, so that G_∞ is admissible.
Finally $j(G_\infty) \leq \liminf j(G_j)$ by a continuity of j with respect to this convergence, which was shown in [17, Lemma 2.1] for the case $\alpha = 1$. In fact, if $||r_k - r_m||_\infty = d \leq \rho$ then $r_k = d + (r_k - d) \leq r_j + d = r_j(1 + d/\rho)$ so that

$$\begin{aligned} j(\Omega_m) &\leq j(\Omega_k(1 + d/\rho)^{-1}) \\ &= (1 + d/\rho)^{2\alpha} j(\Omega_k) \\ &\leq (1 + c_\alpha\, d/\rho) j(\Omega_k) \\ &\leq j(\Omega_k) + (c_\alpha/\rho) j(B_\rho)\, d \\ &= j(\Omega_k) + (c_\alpha/\rho) j(B_\rho) ||r_k - r_m||_\infty . \end{aligned} \tag{4.17}$$

In this chain of inequalities we have one after the other used the domain monotonicity i), the homogeneity iv), the fact that on the interval $(0, 1)$ the function $(1 + s)^{2\alpha}$ can be bounded by a linear function $1 + c_\alpha s$, and again the monotonicity property i). Reversing the role of Ω_m and Ω_k gives

$$|j(\Omega_m) - j(\Omega_k)| \leq (c_\alpha/\rho) j(B_\rho) ||r_k - r_m||_\infty, \tag{4.18}$$

that is the strong continuity of j with respect to this domain convergence.

While Proposition 4.5 provides us with a convex, and thus a minimizing Lipschitz domain, I cannot prove more regularity of this minimizer. Is Ω strictly convex? Is its boundary smooth enough to derive the Hadamard condition? These appear to be open questions.

Acknowledgement: The results presented in these notes are to a great part due to others, among them notably N.B.Willms and H.Weinberger (Proposition 4.4) and my coauthors M.Belloni, G.Buttazzo, F.Brock and V.Ferone. It is a pleasure to thank various colleagues for helpful discussions that led to an improvement of these notes, namely M.Ashbaugh (Columbia, MO) and M.Belloni (Parma). Finally I thank my colleagues S.Fekete and K.Veselic for providing me with copies of [1, 19] and [32] and my students M.Mester, N.Oll and A.D.Stolz for their calculations.

References

[1] Akman, V., An algorithm for determining an opaque minimal forest of a convex polygon. Inform. Proc. Lett. **24** (1987) p.193–198.

[2] Ashbaugh, M.S. & R.D.Benguria, Proof of the Payne-Polya-Weinberger conjecture, Bull. Amer. Math. Soc. **25** (1991) p.19–29.

[3] Ashbaugh, M.S. & R.D.Benguria, On Rayleigh's conjecture for the clamped plate and its generalization to three dimensions. Duke Math. J. **78** (1995) p.1–17.

[4] Ashbaugh, M.S., R.D.Benguria & R.S.Laugesen, Inequalities for the first eigenvalues of the clamped plate and buckling problems. in: *General Inequalities 7*, Eds. C.Bandle et al., Int. Ser. Num. Math. **123** (1997) p.95–110.

[5] Ashbaugh, M.S. & R.S.Laugesen, Fundamental tones and buckling loads of clamped plates, Ann. Sc. Norm. Sup. Pisa Ser. IV **23** (1996) p.383–402.

[6] Aviles, P., Symmetry theorems related to Pompeiu's problem. Amer. J. Math. **108** (1986) p.1023–1036.

[7] Belloni, M., personal communication

[8] Belloni,M. & B.Kawohl, A paper of Legendre revisited, Forum Mathematicum **9** (1997) p.655–668.

[9] Bracke, K.A., The opaque cube problem, Amer. Math. Monthly **99** (1992) p.866–871.

[10] Brock, F., V.Ferone & B.Kawohl, A symmetry problem in the calculus of variations, Calculus of Variations **4** (1996) p.593–599.

[11] Buttazzo,G., & B.Kawohl, On Newton's problem of minimal resistance, Math. Intell. **15** (1993) p.7–12.

[12] Buttazzo, G., V.Ferone & B.Kawohl, Minimum problems over sets of concave functions and related questions, Mathematische Nachrichten **173** (1995), p.71–89.

[13] Buttazzo, G. & G.Dal Maso, An existence result for aclass of shape optimization problems, Arch. Ration. Mech. Anal. **122** (1993) p.183–195.

[14] Cieslik, D., *Steiner Minimal Trees,* Kluwer Acad. Publ., Dordrecht 1998.

[15] Courant, R. & D.Hilbert, *Methods of Mathematical Physics, Vol. I,* Interscience Publ., New York 1953

[16] Courant, R. & H.Robbins, *What is Mathematics ?*, Oxford Univ. Press, New York 1941

[17] Cox, S. & M.Ross, Extremal eigenvalue problems for starlike planar domains. J. Differ. Equations **120** (1995) p.174–197.

[18] Croft, H.T., K.J.Falconer & R.K.Guy, *Unsolved Problems in Geometry.* Springer-Verlag, Heidelberg 1991

[19] Dublish, P., An $O(n^3)$ algorithm for finding the minimal opaque forest of a convex polygon. Inform. Proc. Lett. **29** (1988) p.275–276.

20] Eggers, A.J.Jr., M.M.Resnikoff & D.H.Dennis, Bodies of revolution having minimum drag at high supersonic air speeds. Technical Note 3666, National Advisory Committee for Aeronautics (1957)

[21] Faber, G., Beweis daß unter allen homogenen Membranen von gleicher Fläche und gleicher Spannung die kreisförmige den tiefsten Grundton gibt. Sitzungsber. Bayer. Akad. Wiss., Math.-Naturwiss. Kl. (1923) p.169–172.

[22] Faber, V., Mycielski,J. & P.Pedersen, On the shortest curve which meets all the lines which meet a circle. Ann. Polon. Math. **44** (1984) p.249–266.

[23] Faber, V. & J.Mycielski, On the shortest curve that meets all the lines that meet a convex body, Amer. Math. Monthly **93** (1986) p.796–801.

[24] Falconer, K.J., *The geometry of fractal sets.* Cambridge Univ. Press, Cambridge 1985.

[25] Ferone, B. & B.Kawohl, Bodies of minimal resistance under prescribed surface area, Zeitschrift Angew. Math. Mech. (ZAMM), Research Article, **79** (1999) p.277–280.

26] Friedman, A.: *Variational Principles and Free-Boundary Problems*, Robert Krieger Publ., Malabar, Florida 1988

[27] Funk, P., *Variationsrechnung und ihre Anwendungen in Physik und Technik.* Springer-Verlag, Heidelberg 1962

[28] Garabedian, P.R. & M.Schiffer, Convexity of domain functionals, J. d'Analyse Math. **2** (1953) p.281–368.

[29] Garofalo, N. & F.Segala, New results on the Pompeiu problem. Trans. Amer. Math. Soc. **325** (1991) p.273–286.

[30] Goldstine, H.H., *A History of the Calculus of Variations from the 17th through the 19th Century.* Springer-Verlag, Heidelberg 1980

[31] Guasoni, P., Problemi di ottimazione di forma su classi insiemi convessi. Tesi di Laurea, Pisa 1996

[32] Honsberger,R., *Mathematical Morsels*, The Dulciane Mathematical Expositions **3** The Math. Assoc. of America 1978

33] Isenberg, C., Minimum-Wege-Strukturen, alpha Berlin **20** (1986) p.121–123.

[34] Jones, R.E.D., Opaque sets of degree α. Amer. Math. Monthly **71** (1964) p.535-537.

35] Joris, H., Le chasseur perdu dans la forêt, Elem. Math. **35** (1980) p.1–14.

[36] Kawohl, B., *Rearrangements and convexity of level sets in PDE.* Springer Lecture Notes in Math. **1150** (1985)

[37] Kawohl, B., H.A.Levine & W.Velte, Buckling eigenvalues for a clamped plate embedded in an elastic medium and related questions. SIAM J. Math. Anal. **224** (1993) p.327–340.

[38] Kawohl, B., Remarks on some old and current eigenvalue problems. in: *Partial differential equations of elliptic type.* Eds.: A.Alvino, E.Fabes and G.Talenti. Symposia Mathematica **XXXV**, Cambridge Univ. Press (1994) p.165–183.

[39] Kawohl, B., The opaque square and the opaque circle, in: *General Inequalities 7*, Int. Ser. Numer. Math. **123** (1997) p.339–346.

[40] Kawohl, B. & C.Schwab, Convergent finite elements for a class of nonconvex variational problems, IMA J. Numer. Anal. **18** (1998) p.133–149.

[41] Klötzler, R., Universale Rettungskurven I, Zeitschr. f. Analysis u. ihre Anwendungen, **5** (1986) p.27–38.

[42] Klötzler, R. & S.Pickenhain, Universale Rettungskurven II, Zeitschr. f. Analysis u. ihre Anwendungen, **6** (1987) p.363–369.

[43] Krahn. E., Über eine von Rayleigh formulierte Minimaleigenschaft des Kreises. Math. Ann. **94** (1924) p.97–100.

[44] Marcellini, P., Nonconvex Integrals of the Calculus of Variations. *Proceedings of "Methods of Nonconvex Analysis", Varenna 1989*, Ed.: A.Cellina, Springer Lecture Notes in Math. **1446** (1990) p.16–57.
[45] McShane, E.J., The calculus of variations from the beginning through optimal control theory. SIAM J. Control Optim. **27** (1989) p. 916–939.
[46] Mester, M., Das Newtonproblem und die Single Shock Bedingung für Kegel, Diploma Thesis, Cologne 1998
[47] Miele, A., *Theory of Optimum Aerodynamic Shapes*, Academic Press, New York (1965)
[48] Mohr, E., Über die Rayleighsche Vermutung: Unter allen Platten von gegebener Fläche und konstanter Dichte und Elastizität hat die kreisförmige den tiefsten Grundton. Ann. Mat. Pura Appl. Ser. IV **104** (1975) p.85–122 and **107** (1976) p.395.
[49] Nadirashvili, N.S., Rayleigh's conjecture on the principal frequency of the clamped plate. Arch. Ration. Mech. Anal. **129** (1995) p.1–10.
[50] Nicolai, F.: Un modello variazionale per l'ottimazione di profili aerodinamici, Tesi di Laurea, Pisa (1994)
[51] Oll, N., Thesis in preparation
[52] Payne, L.E., Inequalities for eigenvalues of membranes and plates. J. Ration. Mech. Anal **4** (1955) p.517–528.
[53] Payne, L.E., Isoperimetric inequalities and their applications. SIAM Review **9** (1967) p.453–488.
[54] Payne, L.E., G.Polya & H.F.Weinberger, On the ratio of two consecutive eigenvalues, J. Math. and Phys. **35** (1956) p.289–298.
[55] Polya, G. & G.Szegö, *Isoperimetric inequalities in Mathematical Physics,* Note F, Princeton Univ. Press, Princeton, N.J. 1951
[56] Serrin, J., A symmetry problem in potential theory, Arch. Ration. Mech. Anal. **43** (1971) p.304–318.
[57] Shermer, I., A counterexample to the algorithms for determining opaque minimal forests. Inform. Proc. Lett. **40** (1991) p.41-42.
[58] Shimakura, N., Le premiére valeur propre du Laplacien pour le probléme de Dirichlet. J. Math. Pures Appl. **62** (1983) p.129–152.
[59] Sholander, M., On certain minimum problems in the theory of convex curves. Trans. Amer. Math. Soc. **73** (1952) p.139–173.
[60] Smart, J.R. Searching for Mathematical Talent in Wisconsin II, Amer. Math. Monthly **73** (1996) p.401-406.
[61] Stewart, I., The great drain robbery, Scientific American, (Sept. 1995), p.166–167, (Dec. 1995), p.87 and (Feb 96), p.113.
[62] Stolz, A.D., Newtonscher Widerstand und "Single Shock" Bedingung, Diploma Thesis, Cologne 1999
[63] Szegö, G., On membranes and plates, Proc. Nat. Acad. Sci. USA **36** (1950) p.210–216 and **44** (1958) p.314–316.
[64] Talenti, G., On the first eigenvalue of the clamped plate. Ann. Mat. Pura Appl. Ser. IV **79** (1981) p.265–280.
[65] Tonelli, L., *Fondamenti di Calcolo delle Variazioni*. Zanichelli, Bologna 1923
[66] Whiteside, D.T., The Mathematical Papers of ISAAC NEWTON - Vol. VI, Cambridge Univ. Press, London 1974
[67] Wieners, C., A numerical existence proof of nodal lines for the first eigenfunction of the plate equation, Arch. Math. **66** (1996) p.420–427.
[68] Willms, N.B., An isoperimetric inequality for the buckling of a clamped plate, Lecture at the Oberwolfach meeting on "Qualitiative properties of PDE" (organized by H.Berestycki, B.Kawohl & G.Talenti), Feb. 1995

An Introduction to the Homogenization Method in Optimal Design

Luc Tartar

Department of Mathematical Sciences, CARNEGIE-MELLON University, Pittsburgh, PA 15213-3890, U.S.A.

1. Introduction

This course describes Optimization/Optimal Design problems which lead to Homogenization questions, together with the method to treat them that François MURAT and I had developed, mostly in the 70s. I adopt a chronological point of view, which I find best for describing how some new techniques (which are often misattributed nowadays) were introduced for overcoming some difficulties that we had encountered, or for generalizing a particular result that we had found useful in our search.

There are many ways to practice research in Mathematics, and I hope that the description of the way we thought in front of new questions could help some to experience for themselves the extraordinary feeling of discovery; it must be said, however, that one rarely finds oneself in exactly the same situations that others had experienced before. What one finds may also have been known to others, a fact that Ennio DE GIORGI had resumed in "Chi cerca trova, chi ricerca ritrova"[1].

In a few occasions it had probably been useful that I did not know the approach that others had followed before, as the one that I created appeared to be different and to offer new possibilities: the knowledge of an old method often makes it difficult to invent a new one, and this is an important obstacle in research.

Although the difference between exploration and exploitation is quite obvious for any oil engineer, many researchers in Mathematics spend their life

[1] From the sentence "Ask and it will be given to you; seek and you will find; knock and the door will be opened to you." in the gospels (Matthew 7:7, Luke 11:9), the middle part has given rise to the French saying "Qui cherche trouve", equivalent to the Italian "Chi cerca trova". The play on the prefix, "re" in French, "ri" in Italian, does not work as well in English (one could replace seek and find by search and discover in order to use research and rediscover).

exploiting methods invented by others, mimicking what they have read or heard. There is nothing really wrong about that behaviour for those who understand their role in the scientific community as that of soldiers of an army working for the benefit of the whole community. However, it is a potentially destructive behaviour that some pretend to have invented the ideas that they have read or that they have been taught directly by more creative people, even if they simply misattribute them to some of their friends instead, because too often they have not completely understood the whole potential of the ideas that they use, and they may transmit twisted informations with the main effect of misleading part of the army, and playing therefore the same role as traitors.

Having been raised in a religious background, I do not attribute my mathematical ability to my own efforts and therefore I consider it a duty to put the gift that I was given to the service of others. I have noticed that religious teachers from the past seem to have liked using parables, stories so simple that their students asked about their meaning, and oral tradition then transmitted us the initial story, the questions of the students, and one "example" given by the master; because it was so easy to remember, the teaching had therefore been transmitted intact by people who did not even understand it, unaware of the innumerable applications that the little story contained. Some mathematicians behave in this way too, writing general theorems but giving only one or two examples, or they may simply teach a general theory by explaining one example in detail, thinking that every trained mathematician will automatically see what the general idea is. I have done that "mistake" often, and after inventing a method applicable to all variational problems, I had been quite upset when I had found written that I had only solved the case of a diffusion equation, wondering how someone writing such a stupid statement could consider himself a mathematician, and why his coauthors had not jumped out of their seat at such an idiotic remark that could well be attributed to them too. Jealousy, the observation that mathematical ability is too sparsely distributed, or the saying "au pays des aveugles les borgnes sont rois",[2] come to my mind for explaining the behaviour of those who try to mislead students out of the right path, but it may well be the result of their complete lack of moral education.

For some mysterious reason, François MURAT and I were the first who put together the various pieces of the theory that I will describe, and I will explain what other people had done according to my information. I had taught similar lectures in 1983, before some people embarked in a systematic campaign of misattribution of ideas and results: everyone with a minimal mathematical ability may quickly identify the names of those who have indulged in intentional misleading, but this does not mean that they have not themselves had some genuine idea, in which case I will mention their name for that.

[2] In the land of the blind, a one-eyed man is king.

2. A counter-example of François MURAT

In 1970, François MURAT worked on an academic problem of optimization that had been proposed by Jacques-Louis LIONS [Li1], and he found that it had no solution [Mu1]. His result was unexpected,[3] and as we were sharing an office in Jussieu, we started a long and fruitful collaboration, first discussing some generalizations of his first idea [Mu2], and then embarking on the exploration that led us to (re)discover[4] the general theory of "Homogenization"[5] [Ta1], [Ta2], [Ta3], [Mu3]. The initial problem that François MURAT considered was to minimize the cost function

$$J(a) = \int_0^1 |y(x) - 1 - x^2|^2\, dx, \tag{1.1}$$

when the state y solves the equation of state

$$-\frac{d}{dx}\left(a\frac{dy}{dx}\right) + a\,y = 0 \text{ in } (0,1),\ y(0) = 1,\ y(1) = 2;\ y \in H^1(0,1), \tag{1.2}$$

and the control a lies in the following admissible control set

$$\mathcal{A} = \{a | a \in L^\infty(0,1), \alpha \le a \le \beta \text{ a.e. in } (0,1)\}, \tag{1.3}$$

and as he tried to apply the direct method of the Calculus of Variations, he noticed that for a sequence $a^n \in \mathcal{A}$ such that

$$a^n \rightharpoonup a_+ \text{ and } \frac{1}{a^n} \rightharpoonup \frac{1}{a_-} \text{ in } L^\infty(0,1) \text{ weak } \star, \tag{1.4}$$

---

[3] We had not heard about the work that Laurence C. YOUNG had done in the 40s [Yo]. I had first met him in 1971 in Madison, but I only learned many years after that he was the inventor of the objects which I was still using under the name of parametrized measures in my HERIOT-WATT lectures in 1978 [Ta8]; it was Ronald DIPERNA who then insisted that they should be called YOUNG measures.

[4] We were not aware of the earlier work on G-convergence of Sergio SPAGNOLO [Sp1], [Sp2], and his work with Antonio MARINO [Ma&Sp], or the work of Tullio ZOLEZZI [Zo]. It was only after having developed our own approach, which François MURAT named H-convergence in [Mu3], that we became aware of these works and that of Ennio DE GIORGI and Sergio SPAGNOLO [DG&Sp].

[5] As in the title of my PECCOT lectures in 1977, I have adopted the term Homogenization, first introduced by Ivo BABUŠKA [Ba], for describing our general approach of H-convergence; of course, no constraints of periodicity are considered. Some authors automatically associate the term with periodic structures, probably because they have not understood the general framework introduced by Sergio SPAGNOLO or by François MURAT and me!

then the corresponding sequence of solutions y_n of (1.2) converges in $H^1(0,1)$ weak to the solution y_∞ of

$$-\frac{d}{dx}\left(a_-\frac{dy_\infty}{dx}\right)+a_+\,y_\infty=0 \text{ in } (0,1),\ y_\infty(0)=1,\ y_\infty(1)=2;\ y_\infty\in H^1(0,1), \tag{1.5}$$

with

$$a^n\frac{dy_n}{dx}\to a_-\frac{dy_\infty}{dx} \text{ in } L^2(0,1) \text{ strong}, \tag{1.6}$$

and

$$J(a^n)\to\tilde{J}(a_-,a_+)=\int_0^1|y_\infty(x)-1-x^2|^2\,dx. \tag{1.7}$$

Indeed, y_n is bounded in $H^1(0,1)$ and $v_n=a^n\frac{dy_n}{dx}$ is bounded in $L^2(0,1)$, but as its derivative is $a^n\,y_n$ which is bounded in $L^2(0,1)$, v_n is actually bounded in $H^1(0,1)$. Therefore one can extract a subsequence such that y_m converges in $H^1(0,1)$ weak and in $L^2(0,1)$ strong to y_∞, and v_m converges in $L^2(0,1)$ strong to v_∞. Then $a^m\,y_m$ and $\frac{dy_m}{dx}=\frac{1}{a^m}v_m$ converge in $L^2(0,1)$ weak, respectively to a_+y_∞ and to $\frac{dy_\infty}{dx}=\frac{1}{a_-}v_\infty$, showing (1.6), and therefore (1.5); the fact that y_∞ is uniquely determined by (1.5) shows that the extraction of a subsequence is not necessary.

For $\alpha=\frac{\sqrt{2}-1}{\sqrt{2}},\beta=\frac{\sqrt{2}+1}{\sqrt{2}}$, François MURAT used the particular sequence

$$a^n(x)=\begin{cases}1-\sqrt{\frac{1}{2}-\frac{x^2}{6}} \text{ when } x\in\left(\frac{2k}{2n},\frac{2k+1}{2n}\right), k=0,\dots,n-1\\ 1+\sqrt{\frac{1}{2}-\frac{x^2}{6}} \text{ when } x\in\left(\frac{2k+1}{2n},\frac{2k+2}{2n}\right), k=0,\dots,n-1,\end{cases} \tag{1.8}$$

corresponding to

$$a_-=\frac{1}{2}+\frac{x^2}{6};\ a_+=1;\ y_\infty=1+x^2 \text{ in } (0,1), \tag{1.9}$$

showing that $\inf_{a\in\mathcal{A}}J(a)=0$. He checked then that it was not possible to have $y=1+x^2$ in (1.2) for some $a\in\mathcal{A}$, by considering (1.2) as a differential equation for a, and noticing that all nonzero solutions are unbounded, as they are $a=\frac{C}{\sqrt{x}}exp(\frac{x^2}{4})$.

We were naturally led to characterize all the possible pairs (a_-,a_+) which could appear in (1.4) and we found

$$\alpha\le a_-(x)\le a_+(x)\le\frac{a_-(x)(\alpha+\beta)-\alpha\,\beta}{a_-(x)}\le\beta \text{ a.e. } x\in(0,1), \tag{1.10}$$

or equivalently

$$\frac{1}{a_+(x)}\le\frac{1}{a_-(x)}\le\frac{\alpha+\beta-a_+(x)}{\alpha\,\beta} \text{ a.e. } x\in(0,1). \tag{1.11}$$

Our proof of (1.11) easily extended to the following more general situation,[6] and the characterization (1.11) corresponds to using the following Lemma, with a and $\frac{1}{a}$ being the components of U, K being the piece of hyperbola $U_1 U_2 = 1$ with $\alpha \leq U_1 \leq \beta$.

Lemma 1: Let $U^{(n)}$ be a sequence of measurable functions from an open set $\Omega \subset R^N$ into R^p satisfying $U^{(n)} \rightharpoonup U^{(\infty)}$ in $L^\infty(\Omega; R^p)$ weak $\star$ and $U^{(n)}(x) \in K$ a.e. $x \in \Omega$. For a bounded set K,[7] the characterization of all the possible limits $U^{(\infty)}$ is $U^{(\infty)}(x) \in \overline{conv(K)}$, the closed convex hull of K, a.e. $x \in \Omega$.

Proof: The closed convex hull of K is the intersection of all the closed half spaces which contain K, and a closed half space H_+ has an equation $\{\lambda | \lambda \in R^p, L(\lambda) \geq 0\}$ for some nonconstant affine function L, and if H_+ contains K one has $L(U^n) \geq 0$ a.e. $x \in \Omega$ and therefore $L(U^\infty) \geq 0$ a.e. $x \in \Omega$, i.e. $U^\infty(x) \in H_+$ a.e. $x \in \Omega$; the conclusion follows, if one is careful to write the closed convex hull as a countable intersection of closed half spaces containing K.

Let $V \in L^\infty(\Omega; R^p)$ be such that $V(x) \in clconv(K)$ a.e. $x \in \Omega$. For each m, one can cut R^p into small cubes of size $\frac{1}{m}$ and choose a point of $conv(K)$, the convex hull of K, in each cube intersecting $clconv(K)$ and that helps creating a function $W^{(m)} \in L^\infty(\Omega; R^p)$ such that $|V - W^{(m)}| \leq \frac{1}{m}$ a.e. $x \in \Omega$ and $W^{(m)}$ takes only a finite number of values in $conv(K)$. On a measurable subset ω of Ω where $W^{(m)}$ is constant, we want to construct a sequence of functions converging in $L^\infty(\omega; R^p)$ weak $\star$ to $W^{(m)}$ and taking their values in K, and putting these functions together will create a sequence converging in $L^\infty(\Omega; R^p)$ weak $\star$ to $W^{(m)}$, and then, as the weak $\star$ topology of $L^\infty(\Omega; R^p)$ is metrizable on bounded sets, this will ensure that one can approach V in that topology.

---

[6] As we had learned that weak convergence is not adapted to nonlinear problems, we were surprised to have found such a simple characterization, and I called our common thesis advisor, Jacques-Louis LIONS, to ask him if this was not known already. He suggested that I ask Ivar EKELAND, who told me that it had been implicitely used in some work of CASTAING and was related to a classical result of LYAPUNOV, valid for a set endowed with a nonnegative measure without atoms; indeed our proof extended easily to such a general case. I only learned in 1975 from Zvi ARTSTEIN about his simple proof of LYAPUNOV's result [Ar].

[7] If K is unbounded, one denotes K_M the elements of K of norm $\leq M$, and the characterization is that there exists M with $U^\infty(x) \in \overline{conv(K_M)}$ a.e. $x \in \Omega$. The functions $U^{(\infty)}$ such that $U^{(\infty)}(x) \in \overline{conv(K)}$ correspond to the weak $\star$ closure in $L^\infty(\Omega; R^p)$ of the functions taking (a.e.) their values in K, and one can easily find cases where these two sets of functions are different, showing that the weak $\star$ topology is not metrizable on these unbounded sets.

Let $W^{(m)} = \lambda \in conv(K)$ on ω, so that $\lambda = \sum_i \theta^i k^i$, with $k^i \in K$ and the sum is finite (with all $\theta^i \geq 0$, $\sum_i \theta^i = 1$). We cut now ω into measurable pieces of diameter at most $\frac{1}{n}$, then partition each of such pieces E into measurable subsets E^i with $meas(E^i) = \theta^i meas(E)$ and define the function Z^n to be equal to k^i on each such E^i. The claim is then that as n tends to ∞, the sequence Z^n converges in $L^\infty(\omega; R^p)$ weak $\star$ to λ; as Z^n only takes a finite number of values in K it is bounded, and it is enough to check that for every continuous function φ with compact support $\int_\omega \varphi\, Z^n\, dx \to \int_\omega \varphi\, \lambda\, dx$. As φ is uniformly continuous $|\varphi(x) - \varphi(y)| \leq \varepsilon$ when $|x - y| \leq \frac{1}{n}$, so if $e \in E$ one has $|\varphi(e) \int_E Z^n\, dx - \int_E \varphi\, Z^n\, dx| \leq \varepsilon\, M\, meas(E)$ and $|\varphi(e) \int_E \lambda\, dx - \int_E \varphi\, \lambda\, dx| \leq \varepsilon\, M\, meas(E)$ where M is a bound of the norm of R^p on K, but as $\int_E Z^n\, dx = \sum_i \int_{E_i} k^i\, dx = \sum_i \theta^i\, meas(E) k^i = meas(E)\lambda = \int_E \lambda\, dx$, one deduces $|\int_\omega \varphi\, Z^n\, dx - \int_\omega \varphi\, \lambda\, dx| \leq 2\varepsilon\, M\, meas(\omega)$.

We left aside the construction of the E^i from E, which is easy for sets in R^N. $\widetilde{L}$ being a nonzero linear function on R^N, the measure of $E \bigcap \{x | x \in R^N, \widetilde{L}(x) \geq t\}$ is a continuous function of t which grows from 0 to $meas(E)$ and one obtains the desired partition of E by cutting E by suitable hyperplanes $\widetilde{L}^{-1}(t_i)$. The construction can be generalized when Ω is any set equipped with a measure without atoms, as stated by a classical result of LYAPUNOV (in 1975, I learned from Zvi ARTSTEIN a method of his giving a simple proof of that result as well as bang-bang results in control theory [Ar]).■

3. The independent discoveries of others

Jean-Louis ARMAND had studied at Ecole Polytechnique in Paris a year ahead of me, but we only met almost fifteen years after graduating, not so much because we were then part time lecturers at Ecole Polytechnique in Palaiseau (he in Mechanics, I in Mathematics), but because he had learned about my work by going to visit Konstantin LUR'IE, in Leningrad. Jean-Louis ARMAND had been computing some Optimal Design problems and he had been puzzled by the fact that in the meetings that he attended, various engineers were showing results which were quite different, although they were supposed to solve the same problem; however, he seemed to have been alone in thinking that this was the sign of a serious theoretical difficulty. He had tried to find explanations in the litterature, and he had discovered an article by K. LUR'IE which seemed relevant; he had then traveled to visit him in Leningrad and, having been told about my work there, he had contacted me after his return to Paris and he had mentioned to me what K. LUR'IE had done and what he had told him.

K. LUR'IE had extended some ideas of PONTRYAGUIN to partial differential equations, and he had been able to obtain better necessary conditions of optimality than those given by a classical method.[8] However, he had been quite puzzled when he had discovered a situation with no function satisfying his necessary conditions [Lu]. I do not know if K. LUR'IE had already obtained the right intuition about what was going on before finding my article [Ta2], but he had mentioned to Jean-Louis ARMAND that it was in [Ta2] that he had found the missing ideas that he needed. I had not given the detail of my work with François MURAT in [Ta2], but I had mentioned the work on G-convergence of Sergio SPAGNOLO [Sp1], [Sp2], as well as the work of Henri SANCHEZ-PALENCIA [S-P1], [S-P2] , which had helped us understand that our work had something to do with the question of effective properties of mixtures; K. LUR'IE had then coined the term G-closure for describing the set of all possible effective tensors of admissible mixtures. I will describe later the intuitive ideas behind the necessary conditions of optimality obtained by K. LUR'IE, but it is important to understand that the reason why we were able to go further was that we were rediscovering and extending the ideas of Laurence C. YOUNG, while K. LUR'IE was following the ideas of PONTRYAGUIN and he could hardly have realized that he had taken the wrong track. As pointed out by Laurence C. YOUNG [Yo], if one has obtained some necessary conditions of optimality for an optimization problem, and one finds that only one function satisfies them, one cannot even assert that it is the solution of the problem, unless of course one has already proved that there exists at least one solution of the problem. PERRON's paradox seems too naive an example,[9] but the point of view of PONTRYAGUIN becomes indeed useless if the problem at hand has no solution. On the contrary, the point of view of Laurence C. YOUNG is adapted to that kind of situation, and it creates a relaxed problem which answers two questions: first it explains how the minimizing sequences may have their limit outside the initial space in the case of nonexistence of solutions, then it does give the necessary conditions obtained by following the point of view of PONTRYAGUIN, as they are just part of the necessary conditions of optimality for the relaxed problem. I will describe these questions on an elementary model.

Once we had become aware of the work of Sergio SPAGNOLO, we did find there some ideas which we had not thought about, and we checked that we

[8] It is an idea going back to HADAMARD to push an interface along its normal for computing derivatives of functionals, in order to obtain necessary conditions of optimality for example. Although many applications of that idea may be formal, François MURAT and Jacques SIMON have spent some time giving the method a rigourous framework [Mu&Si]. However, the classical approach only gives conditions that must be satisfied along the interface, while K. LUR'IE's approach as well as ours give conditions which must be satisfied everywhere.

[9] The paradox quoted in [Yo] is as follows: let n be the largest integer; a necessary condition of optimality is $n \geq n^2$, which leaves only two candidates, 0 or 1; therefore the largest integer must be 1!

could handle them with our methods, but for the question of Optimal Design which was our initial motivation we needed more precise results. For example, Antonio MARINO and Sergio SPAGNOLO had shown that in order to obtain all the materials with an effective tensor equal to a general symmetric tensor having its eigenvalues between α and β, it was sufficient to mix isotropic materials with tensor γI with $\gamma \in [\alpha', \beta']$ for some $\alpha' > 0$ and $\beta' < \infty$ [Ma&Sp], but in order to compute necessary conditions of optimality for questions of Optimal Design we needed a more precise characterization, and we were able to obtain one in dimension 2. As we will see the question of optimal bounds has two sides, one where one must prove inequalities that must be satisfied by every mixture using given proportions, and we had found a method for doing that, and one where one must build particular mixtures and compute their effective tensors, and we used repeated layering for that, as Antonio MARINO and Sergio SPAGNOLO had done, but they had not addressed the first question.

I am not sure if we had found reference to the work of Sergio SPAGNOLO before reading some work of Tullio ZOLEZZI, and one of his articles had puzzled us for a while, as we thought that one of his theorems contradicted some of ours [Zo]. The puzzling theorem stated that if a sequence a^n converges weakly to a_+ in $L^\infty(\Omega)$ then the corresponding sequence of solutions u_n converges weakly to the solution associated to a_+. François MURAT thought that some nuance in Italian might have tricked us in mistranslating what was meant, but as we were pondering if "debolmente" could mean anything else than weakly, it suddenly appeared that our mistake had been to read correctly weakly and to interpret it incorrectly as weakly $\star$, as indeed it was the first time that we had seen any use of the weak topology of $L^\infty(\Omega)$ in a concrete situation; we understood then why there was a reference to an article of Alexandre GROTHENDIECK, who had shown that convergence in $L^\infty(\Omega)$ weak implies strong convergence in $L^p_{loc}(\Omega)$ for every finite p.

In the early 70s, Jean CÉA and his team in Nice had been performing some numerical computations for similar problems of Optimal Design than the ones which I had been studying with François MURAT. We had been aware of some work by Denise CHENAIS [Ch], which means that if one imposes some kind of regularity condition on an interface between two materials then the set of corresponding characteristic functions belongs to a compact subset of $L^p(\Omega)$ for $p < \infty$, and therefore a classical optimal solution exists; our work had suggested that if one does not impose such a condition there may not exist any classical solution, in which case one has to use the generalized solutions that we had been studying, corresponding to mixtures.

In 1974, after my talk containing the necessary conditions of optimality described in [Ta2], I had not been able to convince Jean CÉA that the apparition of mixtures was a real possibility that one had to consider, and he might have been mistaken because of a result which he had obtained a few years before with K. MALANOWSKI [Cé&Ma], corresponding to (4.1)/(4.3) with $g(x, u, a) = f(x)u$, for which a special simplification occurs and a classical solution exists. For a given triangulation, one of the numerical methods that Jean CÉA had tested consists in choosing each triangle to be entirely made of only one material, and I could not convince him that if one refines triangulations enough one may start to see oscillations and that our analysis is important for that reason. Had the computers been more powerful in those days, he might indeed have discovered numerical oscillations in refining his triangulations, but the cost would have been prohibitive at the time and only coarse triangulations were used. The classical way for obtaining necessary conditions of optimality by pushing the interface in the direction of its normal, an idea of J. HADAMARD that is often only used at a formal level but which François MURAT and Jacques SIMON have put into a rigourous framework [Mu&Si], gives conditions that must be satisfied on the interface, while I was obtaining necessary conditions which are valid everywhere; in his method J. CÉA could switch any triangle from one material to another, and therefore there was no real interface in his approach and this led him to some kind of discretized necessary condition valid everywhere, and he might have been mistaken by what he may have seen as a similarity between our results.

After some discussions with Guy CHAVENT, who was studying the related problem of identifying the local permeability of an oil field from measurements at various points, I had proposed a numerical approach for solving numerically the type of optimization problem that we had been studying, but the numerical method that I had proposed, and that one of his students had implemented, did not work well at all. As we knew that the optimal mixture that we were looking for could be obtained locally as a layered medium, I had chosen to parametrize the various possible mixtures with a proportion $\theta \in [0,1]$ and with an angle describing the orientation of layers (as I was considering a 2-dimensional problem), but that method appeared to be quite unstable because when θ is 0 or 1, i.e. the material is isotropic, the orientation of the layers is not well defined. I did not try another numerical method, but I had learned that even when the solution of an optimization problem is on the boundary of a set, it might not be a good idea to move only along the boundary of this set in order to find the solution, and a better approach could be to cut through the set in order to arrive more quickly at the interesting points on the boundary.

In June 1980 in New York, Robert V. KOHN had told me that he had been to a meeting organized by Jean CÉA and E.J. HAUG, and as I knew that the approach of my work with François MURAT had probably not been mentioned at this meeting, I taught him our ideas about how Homogenization problems appear in Optimal Design problems.

4. An elementary model problem

This model problem was mentioned to me by Ivar EKELAND, and I believe that his answer involved huge abstract compactified spaces, while my solution is entirely based on our Lemma 1 (which corresponds to a small compactification).

We want to minimize the cost function

$$J(u) = \int_0^T (|y|^2 - |u|^2)\,dt, \tag{3.1}$$

where the control u belongs to

$$U_{ad} = \{u | u \in L^\infty(0,T),\ -1 \le u(t) \le 1,\ \text{a.e. } t \in (0,T)\}, \tag{3.2}$$

and the state y is defined by the equation of state

$$\frac{dy}{dt} = u \text{ a.e. on } (0,T);\ y(0) = 0. \tag{3.3}$$

We can consider here classical necessary conditions of optimality because U_{ad} is convex; the map $u \mapsto y$ is affine continuous and therefore the map $u \mapsto J$ is quadratic continuous, and therefore FRÉCHET differentiable, but the following arguments are valid in cases where only GÂTEAUX differentiability holds. Let $u_* \in U_{ad}$, corresponding to a state y_*, and let $\delta u \in L^\infty(0,T)$ be an admissible direction at u, i.e. $u = u_* + \varepsilon\,\delta u \in U_{ad}$ for $\varepsilon > 0$ small. Then $y = y_* + \varepsilon\,\delta y$ and $J(u) = J(u*) + \varepsilon\,\delta J + o(\varepsilon)$, where

$$\frac{d(\delta y)}{dt} = \delta u;\ \delta y(0) = 0, \tag{3.4}$$

and

$$\delta J = 2\int_0^T (y_*\,\delta y - u_*\,\delta u)\,dt \tag{3.5}$$

and the classical necessary conditions of optimality consist in writing that $\delta J \ge 0$ for all admissible δu. In order to eliminate δy so that δJ is expressed only in terms of δu, one introduces the adjoint state[10] p_* by

$$-\frac{dp_*}{dt} = y_*;\ p_*(T) = 0, \tag{3.6}$$

and a simple integration by parts gives

$$\int_0^T y_*\,\delta y\,dt = -\int_0^T \frac{dp_*}{dt}\,\delta y\,dt = \int_0^T p_*\,\frac{d\delta y}{dt}\,dt = \int_0^T p_*\,\delta u\,dt, \tag{3.7}$$

[10] I do not know who introduced that notion. It does play an important role in PONTRYAGUIN's approach, and he may have introduced it.

and therefore

$$\delta J = 2\int_0^T (p_* - u_*)\delta u\, dt. \tag{3.8}$$

The admissibility of δu means that $\delta u \geq 0$ where $u_* = -1$, $\delta u \leq 0$ where $u_* = +1$, and δu arbitrary where $-1 < u_* < 1$ (one first works on the set of points where $-1+\eta \leq u \leq 1-\eta$ for $\eta > 0$ and then one takes the union of these sets for all $\eta > 0$), and therefore one immediately deduces the classical necessary conditions of optimality

$$\begin{cases} u_* = -1 \text{ implies } p_* - u_* \geq 0, \text{ i.e. } p_* \geq -1 \\ -1 < u_* < +1 \text{ implies } p_* - u_* = 0, \text{ i.e. } p_* = u_* \\ u_* = +1 \text{ implies } p_* - u_* \leq 0, \text{ i.e. } p_* \leq +1, \end{cases} \tag{3.9}$$

which can be read as giving u_* as the following multivalued function of p_*

$$\begin{cases} p_* < -1 \text{ implies } u_* = +1 \\ -1 \leq p_* \leq +1 \text{ implies } u_* \in \{-1, p_*, +1\} \\ p_* > +1 \text{ implies } u_* = -1. \end{cases} \tag{3.10}$$

One can notice that the system of these classical necessary conditions of optimality, i.e. (3.3), (3.6) and (3.9)/(3.10), has at least the solution $u_* = 0$ on $(0,T)$, corresponding to $y_* = p_* = 0$ on $(0,T)$.

The point of view of PONTRYAGUIN for obtaining necessary conditions of optimality consists in comparing u_* to another control $w \in U_{ad}$ by noticing that any control which jumps from u_* to w is admissible. In the language of Functional Analysis it means

$$u = (1-\chi)u_* + \chi\, w \in U_{ad} \quad \text{for every characteristic function } \chi \text{ of a measurable subset of } (0,T), \tag{3.11}$$

and it is then natural to consider a sequence χ_n of characteristic functions such that

$$\chi_n \rightharpoonup \theta \text{ in } L^\infty(0,T) \text{ weak } \star, \tag{3.12}$$

with $0 \leq \theta \leq 1$ a.e. in $(0,T)$; Lemma 1 also tells us that any such θ can be obtained in this way, as can be checked easily directly. One notices then that the corresponding functions y_n, which satisfy a uniform LIPSCHITZ condition, converge uniformly to y_∞ solution of

$$\frac{dy_\infty}{dt} = (1-\theta)u_* + \theta\, w \text{ in } (0,T);\; y_\infty(0) = 0, \tag{3.13}$$

and, using the fact that $F\big((1-\chi)u_* + \chi w\big) = (1-\chi)F(u_*) + \chi F(w)$ for every function F and every characteristic function χ, one deduces that $J(u_n)$ converges to $\widetilde{J}(\theta)$ given by

$$\widetilde{J}(\theta) = \int_0^T \big(|y_\infty|^2 - (1-\theta)|u_*|^2 - \theta|w|^2\big)\, dt. \tag{3.14}$$

If J attains its minimum on U_{ad} at u_*, one deduces that $\widetilde{J}(0) = J(u_*) \leq \lim_n J(u_n) = \widetilde{J}(\theta)$, and therefore $\widetilde{J}$ attains its minimum at 0. One writes then the classical necessary conditions of optimality for $\widetilde{J}$, noticing that admissibility for $\delta\theta$ means $\delta\theta \geq 0$, that δy solves

$$\frac{d\delta y}{dt} = (w - u_*)\delta\theta; \; \delta y(0) = 0, \tag{3.15}$$

and, as $\theta = 0$ corresponds to $y_\infty = y_*$, that

$$\delta\widetilde{J} = \int_0^T (2y_*\,\delta y + (|u_*|^2 - |w|^2)\delta\theta)\,dt. \tag{3.16}$$

Using the same p_* as defined in (3.6), the integration by parts gives a different result because the equation for δy is different

$$\int_0^T 2y_*\,\delta y\,dt = -2\int_0^T \frac{dp_*}{dt}\,\delta y\,dt = \int_0^T 2p_*\,\frac{d\delta y}{dt}\,dt = \int_0^T 2p_*(w - u_*)\delta\theta\,dt, \tag{3.17}$$

and therefore

$$\delta\widetilde{J} = \int_0^T \Big(2p_*(w - u_*) + (|u_*|^2 - |w|^2)\Big)\delta\theta\,dt, \tag{3.18}$$

and the necessary conditions of optimality for $\widetilde{J}$ become

$$2p_*(w - u_*) + (|u_*|^2 - |w|^2) \geq 0 \text{ a.e. on } (0,T). \tag{3.19}$$

It is only now that one lets w vary in U_{ad} and, taking advantage of the fact that p_* is independent of the choice of w, (3.19) means

$$\begin{aligned} 2p_*\,u_* - |u_*|^2 &= \inf_{-1\leq w\leq 1}(2p_*\,w - |w|^2) \text{ a.e. on } (0,T) \\ &= -2|p_*| - 1 \text{ a.e. on } (0,T), \end{aligned} \tag{3.20}$$

and therefore

$$\begin{cases} u_* = -1 \text{ implies } p_* \geq 0 \\ -1 < u_* < +1 \text{ does not occur} \\ u_* = +1 \text{ implies } p_* \leq 0, \end{cases} \tag{3.21}$$

or

$$\begin{cases} p_* < 0 \text{ implies } u_* = +1 \\ p_* = 0 \text{ implies } u_* = \pm 1 \\ p_* > 0 \text{ implies } u_* = -1, \end{cases} \tag{3.22}$$

which are obviously more restrictive than (3.9)/(3.10).

The choice of the model problem comes from the simple direct observation that it has no solution, as I will show later. This fact by itself does not tell much about the existence of solutions for the system of necessary conditions of optimality, but the analysis of the relaxed problem that I will also introduce later will have as a consequence that no such solution exists.

However, one can see by a direct computation that there is no function $u_* \in U_{ad}$ for which the necessary conditions of optimality (3.21)/(3.22) hold, with y_* defined by (3.3) and p_* defined by (3.6). Indeed

$$0 = \int_0^T \frac{d(y_* p_*)}{dt}\, dt = \int_0^T (u_* p_* - |y_*|^2)\, dt = -\int_0^T (|p_*| + |y_*|^2)\, dt, \quad (3.23)$$

shows that one must have $y_* = p_* = 0$ a.e. in $(0,T)$, but $y_* = 0$ is incompatible with the condition $u_* = \pm 1$ a.e. in $(0,T)$.

The point of view of PONTRYAGUIN usually gives stronger necessary conditions of optimality than the classical ones.[11] If the necessary conditions of optimality (either the classical ones or those of PONTRYAGUIN) have no solution, then the minimization problem cannot have any solution, but there is then no obvious explanation of what minimizing sequences are doing, for example. Of course, the proof of the necessary conditions contains a hint about oscillating sequences,[12] and it is Laurence C. YOUNG's point of view to study directly such oscillating sequences, in order to create a relaxed problem.[13]

In order to show directly that our minimization problem has no solution, one first notices that

$$J(u) > -T \text{ for all } u \in U_{ad}, \quad (3.24)$$

because $|y|^2 - |u|^2 \geq -1$ a.e. on $(0,T)$ for every $u \in U_{ad}$ implies $J(u) \geq -T$, and because one cannot have $J(u) = -T$, which would require both $y = 0$ and $|u| = 1$ a.e. on $(0,T)$, in contradiction with the fact that $y = 0$ a.e. on $(0,T)$ implies $u = 0$ a.e. on $(0,T)$. Then one notices that

---

[11] If the problem is convex, it gives the same conditions as the classical ones. If the controls are imposed to take values in a discrete set, there is no natural differentiable path from one control to another and therefore one cannot obtain any classical necessary conditions of optimality. However, even for a convex set of admissible functions, if the equation of state has the form $y' = A(y,u)$ and the cost function has the form $J(u) = \int_0^T B(y,u)\, dt$, one requires differentiability of A and B in both y and u in order to obtain the classical necessary conditions of optimality, while one only requires differentiability of A and B in y for obtaining the necessary conditions of optimality of PONTRYAGUIN.

[12] The original proof of PONTRYAGUIN's principle, which Jacques-Louis LIONS had asked me to read in the late 60s, contains no Functional Analysis at all, but the idea of switching quickly from u_* to w is explicit there. I found the proof shown above much later, probably in the late 70s or early 80s, and it might have been used in this way before.

[13] Again, I am not really sure about who introduced the term relaxation, but I have probably heard it first in the seminar PALLU DE LA BARRIÈRE at IRIA in the late 60s, when I also heard about parametrized measures, now named YOUNG measures.

$$u_n \rightharpoonup 0 \text{ in } L^\infty(0,T) \text{ weak } \star \text{ and } |u_n| = 1 \text{ a.e. in } (0,T) \text{ imply } J(u_n) \to -T, \tag{3.25}$$

as $u_n \rightharpoonup 0$ in $L^\infty(0,T)$ weak $\star$ implies that y_n converges uniformly to 0; an example of such a sequence u_n belonging to U_{ad} is defined by $u_n(t) = sign(\cos n\,t)$ on $(0,T)$.

One sees also that any minimizing sequence, i.e. any sequence $u_n \in U_{ad}$ such that $J(u_n) \to -T = \inf_{u\in U_{ad}} J(u)$, must be such that $y_n \to 0$ in $L^2(0,T)$ strong and $u_n^2 \to 1$ in $L^1(0,T)$ strong . Because U_{ad} is bounded in $L^\infty(0,T)$, $y_n \to 0$ in $L^2(0,T)$ strong is equivalent to $u_n \rightharpoonup 0$ in $L^\infty(0,T)$ weak $\star$, and because $|u_n| \leq 1$ a.e. in $(0,T)$, $u_n^2 \to 1$ in $L^1(0,T)$ strong is equivalent to $u_n^2 \rightharpoonup 1$ in $L^\infty(0,T)$ weak $\star$.

The same analysis shows that if a sequence $u_n \in U_{ad}$ converges in $L^\infty(0,T)$ weak $\star$ to u, one can deduce that y_n converges uniformly to y given by (3.3), but one cannot deduce what the limit of $J(u_n)$ is. However, if one knows that

$$\begin{aligned} &u_n \rightharpoonup u \text{ in } L^\infty(0,T) \text{ weak } \star \\ &u_n^2 \rightharpoonup v \text{ in } L^\infty(0,T) \text{ weak } \star, \end{aligned} \tag{3.26}$$

then,

$$J(u_n) \to \widetilde{J}(u,v) = \int_0^T (|y|^2 - v)\,dt. \tag{3.27}$$

Lemma 1 characterizes the pairs (u,v) that one can obtain by (3.26) for a sequence $u_n \in U_{ad}$, choosing for K the piece of parabola $U_2 = U_1^2$ with $-1 \leq U_1 \leq 1$, and one can therefore introduce a relaxed problem defined on

$$\widetilde{U}_{ad} = \{(u,v) | -1 \leq u \leq 1;\ u^2 \leq v \leq 1 \text{ a.e. in } (0,T)\}, \tag{3.28}$$

the state y still being given by (3.3), and the cost function $\widetilde{J}$ being given by the formula in (3.27). The original problem is a subset of the new one as

$$\begin{aligned} &u \in U_{ad} \text{ if and only if } (u,u^2) \in \widetilde{U}_{ad} \\ &J(u) = \widetilde{J}(u,u^2) \text{ for all } u \in U_{ad}. \end{aligned} \tag{3.29}$$

By Lemma 1, for every $(u,v) \in \widetilde{U}_{ad}$ there exists a sequence u_n with $u_n \rightharpoonup u$ and $u_n^2 \rightharpoonup v$ in $L^\infty(0,T)$ weak $\star$, which imply $J(u_n) \to \widetilde{J}(u,v)$, and using (3.29) one deduces that

$$\inf_{u\in U_{ad}} J(u) = \inf_{(u,v)\in \widetilde{U}_{ad}} \widetilde{J}(u,v) \tag{3.30}$$

and

$$u_* \text{ minimizes } J \text{ on } U_{ad} \text{ if and only if } (u_*, u_*^2) \text{ minimizes } \widetilde{J} \text{ on } \widetilde{U}_{ad}. \tag{3.31}$$

Let us look for the classical necessary conditions of optimality for $(u_*, v_*) \in \widetilde{U}_{ad}$. Actually these conditions will be sufficient conditions of optimality as

$\widetilde{U}_{ad}$ is convex and $\widetilde{\mathcal{J}}$ is a convex function; as $\widetilde{\mathcal{J}}$ is strictly convex in u, u_* is known in advance to be unique, but although $\widetilde{\mathcal{J}}$ is affine in v, one deduces that $v_* = 1$ a.e. on $(0,T)$ from the observation that

$$\widetilde{\mathcal{J}}(u,v) \leq \widetilde{\mathcal{J}}(u,1) = K(u) - T \text{ with } K(u) = \int_0^T |y|^2\,dt, \tag{3.32}$$

with equality if and only if $v = 1$ a.e. on $(0,T)$. Obviously K attains its minimum at $u_* = 0$, but let us forget for a moment that $\widetilde{\mathcal{J}}$ attains its minimum only at $(0,1)$, and let us check what the necessary conditions of optimality for $\widetilde{\mathcal{J}}$ are at an arbitrary point $(u_*,v_*) \in \widetilde{U}_{ad}$. For an admissible direction $(\delta u, \delta v)$, δy is still given by (3.4), but

$$\delta\widetilde{\mathcal{J}} = \int_0^T (2y_*\,\delta y - \delta v)\,dt, \tag{3.33}$$

which, using p_* defined by (3.6) and the integration by parts (3.7) gives

$$\delta\widetilde{\mathcal{J}} = \int_0^T (2p_*\,\delta u - \delta v)\,dt. \tag{3.34}$$

As $\widetilde{U}_{ad}$ is convex, it is equivalent to restrict attention to the admissible directions of the form $(\delta u, \delta v) = (w - u_*, w^2 - v_*)\delta\eta$ with $w \in U_{ad}$ and $\delta\eta \in L^\infty(0,T)$ with $\delta\eta \geq 0$ a.e. in $(0,T)$, and therefore the necessary conditions of optimality can be read as

$$2p_*(w - u_*) - (w^2 - v_*) \geq 0 \text{ a.e. on } (0,T), \tag{3.35}$$

which in the case $v_* = u*^2$ coincide with the PONTRYAGUIN necessary conditions of optimality (3.19). Instead of (3.20), (3.35) implies

$$2p_*\,u_* - v_* = -2|p_*| - 1 \text{ a.e. on } (0,T), \tag{3.36}$$

and therefore

$$u_* \in -sign(p_*);\ v_* = 1 \text{ a.e. on } (0,T), \tag{3.37}$$

and the system of necessary conditions (3.3), (3.6), (3.37) gives then $u_* = 0$.

Instead of the above relaxed problem, I could have used a set much bigger than $\widetilde{U}_{ad}$ by introducing the set of YOUNG measures, which describe the possible weak $\star$ limits of sequences $F(u_n)$ for all continuous functions F. It would have appeared then that only the limits in $L^\infty(0,T)$ weak $\star$ of u_n and u_n^2 were important, i.e. the only useful functions F are the identity id, together with id^2. Therefore starting with a relaxed problem which is too big for the problem at hand, one has not lost information but one carries some unnecessary information; one can reduce the size of the relaxed problem by getting rid of a part of that unnecessary information.

The preceding analysis has identified a topology which is adapted to the initial problem, namely that defined by (3.26). As the weak $\star$ topology of

$L^\infty(0,T)$ is metrizable on bounded sets, one can define it using a distance d_0 for the set U_{ad}, and (3.26) corresponds to using the new topology associated to the distance d_1 defined by $d_1(f,g) = d_0(f,g) + d_0(f^2,g^2)$. The set U_{ad} is not complete for the metric d_1, but Lemma 1 describes the completion of U_{ad} which is $\widetilde{U}_{ad}$, and it also shows that $\widetilde{U}_{ad}$ is compact for the metric d_1. The function J is uniformly continuous for d_1, and therefore it extends in a unique way to the completion, and this extension is the function $\widetilde{J}$ defined in (3.27).

The preceding construction also fulfills the following requirements for a relaxed problem.

An initial problem is to minimize a function F on a set X,[14] a relaxed problem of it is a topological space $\widetilde{X}$, a mapping j from X into $\widetilde{X}$ and a lower semicontinuous function $\widetilde{F}$ defined on $\widetilde{X}$ satisfying the following properties

i) $\widetilde{F}(j(x)) = F(x)$ for all x in X,

ii) $j(X)$ is dense in $\widetilde{X}$,

iii) For every $y \in \widetilde{X}$ there exists a sequence x_n in X such that $j(x_n)$ converges in $\widetilde{X}$ to y, and $\widetilde{F}(y) = \lim_n F(x_n)$ in the case where the topology of $\widetilde{X}$ is metrizable. If the topology of $\widetilde{X}$ is not metrizable, for every $\varepsilon > 0$ and every neighbourhood V of y in $\widetilde{X}$ there exists $x \in X$ with $j(x) \in V$ and $|\widetilde{F}(y) - F(x)| \leq \varepsilon$.

By i) the relaxed problem contains a copy of the initial one; if ii) was not satisfied one would restrict oneself to the closure of $j(X)$; if iii) was not satisfied, one would replace $\widetilde{F}$ by a larger function by taking the lower semi-continuous envelope of the function equal to $F \circ j^{-1}$ on $j(X)$ and $+\infty$ on $\widetilde{X} \setminus j(X)$, eventually removing the point where this envelope is $+\infty$ if one wants to work with functions which are finite everywhere.

I could have used a smaller set than $\widetilde{U}_{ad}$ by working on U_{ad} but minimizing the function $K(u) - T$, and this would be consistent with the observation (3.32), and although one can still compute $\inf_{u \in U_{ad}} J(u)$ from this new problem, it cannot help discover what is the adequate topology for the initial problem, as defined by (3.26). This new problem is not a relaxed problem of the initial one, as i) is not satisfied, but the function $u \mapsto K(u) - T$ is the Γ-limit of J, i.e. the lower semi-continuous envelope of J, for a particular topology, the $L^\infty(0,T)$ weak $\star$ topology. Indeed, $u_n \rightharpoonup u$ in $L^\infty(0,T)$ weak $\star$ implies $\liminf_n J(u_n) \geq K(u) - T$ and for every $u \in U_{ad}$ there exists a sequence $u_n \in U_{ad}$ with $u_n \rightharpoonup u$ and $u_n^2 \rightharpoonup 1$ in $L^\infty(0,T)$ weak $\star$, and therefore $J(u_n) \to \widetilde{J}(u,1) = K(u) - T$.

This last point suggests to be careful with the use of Γ-limits, which are not always the right objects to characterize if one has not choosen the right

[14] If X has already a topology, one may forget about it, as j is usually not continuous from X into $\widetilde{X}$.

topology, which might not belong to the list of usual topologies that one is accustomed to use. Another reason to be careful lies in the difference between G-convergence and H-convergence,[15] the latter being more general and adapted to most of the situations occuring in Continuum Mechanics/Physics. As many problems of Optimal Design come from engineering, and often involve Elasticity, it is worth mentioning not only the inadequacy of linearized Elasticity, but the inadequacy of the Γ-convergence approach, which is not the same as Homogenization, to questions of Elasticity. Although an intensive propaganda has made many mathematicians believe that Nature minimizes Energy, it is obviously not so, and one must remember that "conservation of Energy" is the First Principle of Thermodynamics, which no one doubts (of course, one has to include all forms of Energy, including Heat, but in cases where some energy can be stored and released later, one might have to be careful in writing the balance of Energy). Unfortunately Thermodynamics should be called Thermostatics as it only deals with questions at equilibrium, and its Second Principle does not explain what are the possible exchanges of Energy under its different forms and only postulates the result of these exchanges, but it would be quite naive to believe that in an elastic material equilibrium is obtained instantaneously.[16]

---

[15] When I first heard Ennio DE GIORGI talk about Γ-convergence at the seminar of Jacques-Louis LIONS at Collège de France around 1977, I understood it as a natural generalization of his earlier work on G-convergence with Sergio SPAGNOLO [DG&Sp], but I had been impressed by the application that he had mentioned that energy localized on a surface could appear as the Γ-limit of a three-dimensional problem. The natural association which immediately came to my mind was that surface tension in liquids should be determined from three-dimensional laws, and that one should extend the idea of Γ-convergence to evolution problems in order to study that question.

[16] I heard a talk of Joseph KELLER at a meeting of the Institute for Mathematics and its Applications in Minneapolis in 1985, in which he explained damping in real elastic materials by the presence of inhomogeneities together with the effect of geometric nonlinearity. Elastic waves are scattered by the various inclusions in the material, or by the grain boundaries in the case of a polycrystal, but without the nonlinearity of geometric origin in the strain-stress relation there would be no coupling between different modes, and no possible explanation about why Energy gets trapped in higher and higher frequencies, which is the reason why one thinks that one has attained a macroscopic equilibrium.

5. H-convergence

After his one-dimensional counter-example, François MURAT had looked at the more general situation where u is the solution of

$$\mathcal{A}u = -div\Big(a\,grad(u)\Big) = f \text{ in } \Omega,\ u \in H_0^1(\Omega), \tag{4.1}$$

Ω being a bounded open set of R^N, with $f \in L^2(\Omega)$,[17] and

$$a \in A_{ad} = \{a | a \in L^\infty(\Omega), \alpha \le a \le \beta \text{ a.e. in } (\Omega)\}, \tag{4.2}$$

with $\alpha > 0$ (so that by LAX-MILGRAM lemma, the operator $\mathcal{A}$ is an isomorphism from $H_0^1(\Omega)$ onto its dual $H^{-1}(\Omega)$), with the intention of minimizing a functional J of the form

$$J(a) = \int_\Omega g\Big(x, u(x), a(x)\Big)\,dx, \tag{4.3}$$

and it is important in the sequel that $grad(u)$ does not enter explicitly in the functional, although some special dependence in $grad(u)$ can be allowed, like $g(u)grad(u)$, $g(u)a\,grad(u)$ or $g(u)\big(a\,grad(u).grad(u)\big)$, with some smoothness and growth properties imposed on g. He had noticed that he could solve explicitly the special case of "layered" media, involving sequences $a^n \in A_{ad}$ depending on x_1 alone for example, and as for (1.4) he assumed that for an interval I such that $\Omega \subset I \times R^{N-1}$

$$a^n \rightharpoonup a_+;\ \frac{1}{a^n} \rightharpoonup \frac{1}{a_-} \text{ in } L^\infty(I) \text{ weak } \star, \tag{4.4}$$

and he had deduced that

$$\begin{aligned} &u_n \rightharpoonup u_\infty \text{ in } H_0^1(\Omega) \text{ weak and } L^2(\Omega) \text{ strong} \\ &a^n \frac{\partial u_n}{\partial x_1} \rightharpoonup a_- \frac{\partial u_\infty}{\partial x_1} \text{ in } L^2(\Omega) \text{ weak} \\ &a^n \frac{\partial u_n}{\partial x_j} \rightharpoonup a_+ \frac{\partial u_\infty}{\partial x_j} \text{ in } L^2(\Omega) \text{ weak for } j = 2, \ldots, N, \end{aligned} \tag{4.5}$$

---

[17] More generally $f \in H^{-1}(\Omega)$, the dual of $H_0^1(\Omega)$. $H_0^1(\Omega)$ is the closure of smooth functions with compact support in Ω in the SOBOLEV space $H^1(\Omega)$, consisting of functions in $L^2(\Omega)$ having each of their partial derivatives in $L^2(\Omega)$. As Ω is bounded, POINCARÉ inequality holds on $H_0^1(\Omega)$, i.e. $\int_\Omega |u|^2\,dx \le C \int_\Omega |grad(u)|^2|\,dx$. $H_0^1(\Omega)$ is compactly imbedded in $L^2(\Omega)$ and $L^2(\Omega)$ is compactly imbedded in $H^{-1}(\Omega)$. $H^{-1}(\Omega)$ consists of distributions in Ω which are sums of derivatives of functions in $L^2(\Omega)$. If the boundary $\partial\Omega$ of Ω is smooth, there is a notion of trace on the boundary for functions in $H^1(\Omega)$, and $H_0^1(\Omega)$ consists then of those functions in $H^1(\Omega)$ having trace 0 on the boundary.

and therefore that u_∞ is the solution of

$$A^{eff} u_\infty = - \sum_{i,j=1,\dots,N} \frac{\partial}{\partial x_i}\left(A^{eff}_{ij} \frac{\partial u_\infty}{\partial x_j}\right) = f \text{ in } \Omega;\ u_\infty \in H^1_0(\Omega), \quad (4.6)$$

with

$$\begin{aligned} A^{eff}_{11} &= a_- \\ A^{eff}_{jj} &= a_+, j = 2, \dots, N \\ A^{eff}_{ij} &= 0, i \neq j. \end{aligned} \quad (4.7)$$

Of course, one first extracts a subsequence for which $u_n \rightharpoonup u_\infty$ in $H^1_0(\Omega)$ weak and $L^2(\Omega)$ strong, and as (4.5) has a unique solution, all the sequence converges. For $j \neq 1$ one has $a^n \frac{\partial u_n}{\partial x_j} = \frac{\partial (a^n u_n)}{\partial x_j}$ and $a^n u_n \rightharpoonup a_+ u_\infty$ in $L^2(\Omega)$ weak because $u_n \to u_\infty$ in $L^2(\Omega)$ strong. For computing the limit of $a^n \frac{\partial u_n}{\partial x_1}$,[18] François MURAT used a compactness argument that we had learned from Jacques-Louis LIONS [Li2]: writing $D^n_i = a^n \frac{\partial u_n}{\partial x_i}$, he assumed that $f \in L^2(\Omega)$, and for an interval I in x_1 and a cube ω in $(x_2, \dots, x_N)$ such that $I \times \omega \subset \Omega$ he observed that $div\, D^n = f$ implies that $\frac{\partial D^n_1}{\partial x_1}$ is bounded in $L^2(I; H^{-1}(\omega))$, and as D^n_1 is bounded in $L^2(I; L^2(\omega))$ and the injection of $L^2(\omega)$ into $H^{-1}(\omega)$ is compact, the compactness argument implies that D^n_1 stays in a compact set of $L^2(I; H^{-1}(\omega))$. A subsequence D^m_1 converges in $L^2(I; H^{-1}(\omega))$ strong to D^∞_1 and therefore for $\varphi \in H^1_0(\omega)$ the function v^m defined in I by $v^m(x_1) = \int_\omega D^m_1(x_1, x_2, \dots, x_N)\varphi(x_2, \dots, x_N)\, dx_2 \dots dx_N$ converges in $L^2(I)$ strong to v^∞ defined in a similar way from D^∞_1, and therefore $\frac{1}{a^m} v^m$ converges in $L^2(\omega)$ weak to $\frac{1}{a_-} v^\infty$, and this implies that $D^\infty_1 = a_- \frac{\partial u_\infty}{\partial x_1}$ in $I \times \omega$; varying the cube $I \times \omega$ in Ω gives the same result in Ω.

I noticed later that the analysis of effective properties of layered media can be greatly simplified by using the Div-Curl lemma,[19] which we only proved in

[18] He had considered the more general case where $a^n(x) = \prod_i f^i_n(x_i)$ with $0 < \alpha_i \leq f^i_n \leq \beta_i$, with $f^i_n \rightharpoonup f^i_+$ and $\frac{1}{f^i_n} \rightharpoonup \frac{1}{f^i_-}$ in L^∞ weak $\star$, and he had found that $a^n \frac{\partial u_n}{\partial x_j} \rightharpoonup A^{eff}_{jj} \frac{\partial u_\infty}{\partial x_j}$ in $L^2(\Omega)$ weak, with $A^{eff}_{ii} = f^i_- \prod_{j \neq i} f^j_+$ for $i = 1, \dots, N$, and $A^{eff}_{ij} = 0$ for $i \neq j$.

[19] Some, who want to avoid mentioning the Div-Curl lemma or the more general theory of Compensated Compactness which I also developed with François MURAT in 1976, do not hesitate to lengthen their proofs in order to use only older methods. The correct behaviour in Mathematics is to mention the shortest proof even if one does not follow it, usually because the writer finds it too difficult for himself/herself, and assumes that it would be the same for the reader. Failing to mention such generalizations is a good way to slow down the progress of Science. In this course, the general Compensated

1974 after having completed our basic approach described here, in an attempt to unify the cases for which we were able to compute explicitly the effective coefficient A^{eff}.[20]

Lemma 2: (Div-Curl lemma) Let Ω be an open set of R^N. Let

$$\begin{aligned} &E^n \rightharpoonup E^\infty \text{ in } L^2_{loc}(\Omega; R^N) \text{ weak} \\ &D^n \rightharpoonup D^\infty \text{ in } L^2_{loc}(\Omega; R^N) \text{ weak} \\ &div\, D^n \text{ stays in a compact set of } H^{-1}_{loc}(\Omega) \text{ strong} \\ &curl\, E^n \text{ stays in a compact set of } H^{-1}_{loc}(\Omega; R^{N(N-1)/2}) \text{ strong,} \end{aligned} \tag{4.8}$$

then

$$\int_\Omega \Big(\sum_{i=1}^N E_i^n\, D_i^n\Big)\varphi\, dx \to \int_\Omega \Big(\sum_{i=1}^N E_i^\infty\, D_i^\infty\Big)\varphi\, dx \text{ for every } \varphi \in C_c(\Omega), \tag{4.9}$$

where $curl\, E$ denotes the lists of all $\frac{\partial E_i}{\partial x_j} - \frac{\partial E_i}{\partial x_j}$ and $C_c(\Omega)$ is the space of continuous functions with compact support in Ω.[21]■

In 1974,[22] our first proof involved localization in x, FOURIER transform, LAGRANGE formula and PLANCHEREL formula, but in the case where $E^n = grad(u_n)$ with u_n converging weakly to u_∞ in $H^1_{loc}(\Omega)$, there is an easier proof by integration by parts, which we had already used: for $\varphi \in C^1_c(\Omega)$ one has $\int_\Omega(\sum_i E_i^n\, D_i^n)\varphi\, dx = -\langle div(D^n), u_n\, \varphi\rangle - \int_\Omega u_n(\sum_i D_i^n\, \partial_i\, \varphi)\, dx$ and one passes easily to the limit as $u_n\, \varphi$ converges in $H^1_0(\Omega)$ weak to $u_\infty\, \varphi$ and u_n converges in $L^2_{loc}(\Omega)$ strong to u_∞. Most applications of the Div-Curl lemma correspond to E^n being a gradient, but I will describe later

---

Compactness theory (and the theory of H-measures which I developed in the late 80s) will only be used in describing methods for obtaining bounds on effective coefficients.

[20] After learning the term Homogenization, introduced by Ivo BABUŠKA, we called these limiting coefficients "homogenized" coefficients, but after learning the term effective, from George PAPANICOLAOU, I decided to adopt it; it is often used by physicists, even if they have almost never defined it correctly.

[21] One cannot use for φ the characteristic function of a smooth set, for example, but I have noticed that one can develop the theory of H-measures with test functions in $L^\infty \bigcap VMO$, by using a commutation lemma of COIFMAN, ROCHBERG and WEISS, and therefore one can use $\varphi \in L^\infty \bigcap VMO$ in the Div-Curl lemma.

[22] Joel ROBBIN taught me afterwards how to interpret the Div-Curl lemma in terms of differential forms, and he showed me another proof, based on HODGE decomposition. In 1976, François MURAT and I developped the Compensated Compactness theory, following our original proof using FOURIER transform, and PLANCHEREL formula (Proposition 30, Theorem 31).

the Compensated Compactness theorem (Theorem 31), which I also used for questions of bounds for effective coefficients.

As a corrolary of the Div-Curl lemma, if a sequence of functions ψ_n which only depends upon x_1 converges in $L^2_{loc}(R)$ weak to ψ_∞, then $\psi_n\, D_1^n$ converges in $L^2_{loc}(\Omega)$ weak to $\psi_\infty\, D_1^\infty$ if ψ_n is bounded in L^∞; as a simplification, I say that if D^n satisfies the hypothesis in Lemma 2, then D_1^n does not oscillate in x_1.[23] Before we had solved the general problem, extending the notion of G-convergence introduced by Sergio SPAGNOLO but unaware of his work yet, François MURAT had obtained an explicit formula for the effective coefficients of a layered media in the general anisotropic case; later, in the Spring 1975, Louis NIRENBERG had shown me a preprint of W. MCCONNELL (maybe a preliminary version of [MC]), who had derived the general formula for layered media in linearized Elasticity, and it was the analogue of what François MURAT had done, but with more technical computations of Linear Algebra; a few years after I noticed a general approach for obtaining the effective behaviour of layered media,[24] based on the preceding corollary. The result of François MURAT can be stated as

$$\begin{aligned}
&E^n \rightharpoonup E^\infty \text{ in } L^2_{loc}(\Omega; R^N) \text{ weak;}\\
&D^n = A^n(x_1)E^n \rightharpoonup D^\infty \text{ in } L^2_{loc}(\Omega; R^N) \text{ weak}\\
&\mathit{div}\, D^n \text{ and the components of } \mathit{curl}\, E^n \text{ stay in a compact set of}\\
&\qquad H^{-1}_{loc}(\Omega) \text{ strong}\\
&\text{imply } D^\infty = A^{eff}(x_1)E^\infty \text{ a.e. in } \Omega,
\end{aligned} \tag{4.10}$$

where

[23] More precisely, a sequence v_n converging weakly in $L^2_{loc}(\Omega)$ does not oscillate in a direction ξ_0 if the H-measures associated with subsequences do not charge the direction ξ_0; a consequence is that $v_n\, f_n\big((\xi_0.\cdot)\big)$ converges weakly to $v_\infty\, f_\infty\big((\xi_0.\cdot)\big)$ if f_n converges weakly to f_∞ in $L^2_{loc}(R)$.

[24] In 1979, working with Georges DUVAUT as consultants for INRIA, we had been asked about an industrial application using layers of steel and rubber. I already knew the method shown here in the linear case, and I explained how to use it for nonlinear Elasticity, but I pointed out that there was no general theory of Homogenization for nonlinear Elasticity (this is still true, as the results based on Γ-convergence do not answer the right questions).

$$\frac{1}{A_{11}^n} \rightharpoonup \frac{1}{A_{11}^{eff}} \text{ in } L^\infty(\Omega) \text{ weak } \star$$

$$\frac{A_{1i}^n}{A_{11}^n} \rightharpoonup \frac{A_{1i}^{eff}}{A_{11}^{eff}} \text{ in } L^\infty(\Omega) \text{ weak } \star \text{ for } i = 2, \ldots, N$$

$$\frac{A_{i1}^n}{A_{11}^n} \rightharpoonup \frac{A_{i1}^{eff}}{A_{11}^{eff}} \text{ in } L^\infty(\Omega) \text{ weak } \star \text{ for } i = 2, \ldots, N$$

$$A_{ij}^n - \frac{A_{i1}^n A_{1j}^n}{A_{11}^n} \rightharpoonup A_{ij}^{eff} - \frac{A_{i1}^{eff} A_{1j}^{eff}}{A_{11}^{eff}} \text{ in } L^\infty(\Omega) \text{ weak } \star \text{ for } i, j = 2, \ldots, N. \tag{4.11}$$

The A^n are not assumed to be symmetric, and the proof actually shows that uniform ellipticity of the A^n is not necessary: in the case of layered media in x_1, the result holds if there exists $\alpha > 0$ such that $A_{11}^n \geq \alpha$ a.e. in Ω for all n. The basic idea is that D_1^n does not oscillate in x_1, but also $E_2^n, \ldots, E_N^n$, because of the information on $\partial_1 E_j^n - \partial_j E_1^n$ for $j \geq 2$; one forms the vector G^n with the "good" components $D_1^n, E_2^n, \ldots, E_N^n$, and one expresses the vector O^n of the "oscillating" components $E_1^n, D_2^n, \ldots, D_N^n$, and one has $O^n = B^n(x_1)G^n$ with $B^n = \Phi(A^n)$, and Φ is a well defined (involutive) nonlinear transformation; the corollary of the Div-Curl lemma shows that the weak limit of $B^n G^n$ is $B^\infty G^\infty$, and therefore the explanation of (4.11) is that whenever A^n only depends upon x_1 and $A_{11}^n \geq \alpha > 0$,

$$D^n = A^n E^n \text{ is the same as } \begin{pmatrix} E_1^n \\ D_2^n \\ \vdots \\ D_N^n \end{pmatrix} = \Phi(A^n) \begin{pmatrix} D_1^n \\ E_2^n \\ \vdots \\ E_N^n \end{pmatrix} \tag{4.12}$$

$$\Phi(A^n) \rightharpoonup \Phi(A^{eff}) \text{ in } L^\infty\Big(\Omega; \mathcal{L}(R^N, R^N)\Big) \text{ weak } \star.$$

In the case of linearized Elasticity, the relation is $\sigma_{ij}^n = \sum_{kl} C_{ijkl}^n(x_1)\varepsilon_{kl}^n$, with $\varepsilon_{kl}^n = (\partial_k u_l^n + \partial_l u_k^n)/2$, σ^n is the symmetric CAUCHY stress tensor, and one may always assume that $C_{ijkl}^n = C_{ijlk}^n = C_{jikl}^n$ for all i, j, k, l, but one does not need to use $C_{ijkl}^n = C_{klij}^n$ for all i, j, k, l (which correspond to hyperelastic materials); the equilibrium equations are $\sum_j \partial_j \sigma_{ij}^n = f_i$ for all i; for a direction ξ one defines the acoustic tensor $A^n(\xi)$ by $A_{ik}^n(\xi) = \sum_{jl} C_{ijkl}^n \xi_j \xi_l$ for all i, k, and in the case of layered media in x_1, the formula for layers holds if there exists $\alpha > 0$ such that $(A^n(e^1)\lambda.\lambda) \geq \alpha|\lambda|^2$ a.e. in Ω for all $\lambda \in R^N$ and all n; in that case the components of the good vector G^n are the σ_{i1}^n and σ_{1i}^n for all i and the ε_{kl}^n for $k, l \geq 2$, and the components of the oscillating vector O^n are the ε_{i1}^n and ε_{1i}^n for all i and the σ_{kl}^n for $k, l \geq 2$; one rewrites the constitutive relation $O_{ij}^n = \sum_{kl} \Gamma_{ijkl}^n G_{kl}^n$ and the formulas for layered material in linearized Elasticity that McCONNELL had derived consist in writing $\Gamma^n \rightharpoonup \Gamma^{eff}$.[25]

Sergio SPAGNOLO had introduced the notion of G-convergence in the late 60s, and unaware of it François MURAT and I had introduced a slightly different concept in the early 70s,[26] for which the name H-convergence was coined much later.[27] We were interested in sequences satisfying

$$\begin{aligned} &u_n \rightharpoonup u_\infty \text{ in } H^1_{loc}(\Omega) \text{ weak} \\ &-div\Big(A^n\, grad(u_n)\Big) = f_n \to f \text{ in } H^{-1}_{loc} \text{ strong,} \end{aligned} \tag{4.13}$$

where A^n satisfies

$$\begin{aligned} &A^n \text{ bounded in } L^\infty\Big(\Omega;\ \mathcal{L}(R^N,R^N)\Big);\ (A^n\,\lambda.\lambda) \geq \alpha|\lambda|^2 \text{ a.e. in } \Omega, \\ &\qquad \text{for all } \lambda \in R^N \text{ and all } n, \end{aligned} \tag{4.14}$$

with $\alpha > 0$; the bounds on u_n were deduced from an application of LAX-MILGRAM lemma, using specific boundary conditions (we had started with DIRICHLET conditions, but after having read that Sergio SPAGNOLO had noticed that A^{eff} is the same for different boundary conditions, we checked

[25] In the case of nonlinear Elasticity, the stress tensor used is the PIOLA-KIRCHHOFF stress tensor, which is not symmetric, and the strain F is defined by $F_{ij} = \delta_{ij} + \partial_j u_i$ where $u(x)$ is the displacement from the initial position x of a material point to its final position $x + u(x)$; for a material like steel which breaks if one stretches it more than 10%, F lies near the set of rotations $SO(3)$, but in linearized Elasticity one postulates (often wrongly) that it lies near I; in nonlinear Elasticity, the components of the good vector G^n are the σ^n_{i1} for all i and the F^n_{ij} for all i but only for $j \geq 2$, and the components of the oscillating vector O^n are the F^n_{i1} for all i and the σ^n_{ij} for all i but only for $j \geq 2$; in order to be able to compute the constitutive relation as $O^n = \Psi^n(G^n)$ in the case of hyperelastic materials, a natural condition to impose for the stored energy is the uniform convexity in all directions of the form $a \otimes e^1$.

[26] I had met Sergio SPAGNOLO at a CIME course in Varenna in 1970, and he had asked me if my interpolation results had something to do with his own results, but as soon as he had mentioned that he did not assume any regularity for the coefficients in his work I could tell him that what I had done could not help; however, I did not get a clear idea of what his results were.

[27] The name was choosen by François MURAT in the lectures that he gave in Algiers [Mu3], shortly after I had taught my PECCOT lectures in the Spring 1977, where under the title "Homogénéisation dans les équations aux dérivées partielles" I had described my method of oscillating test functions in Homogenization and the Compensated Compactness theorem, but the notion of H-convergence was indeed clear from our early work.

that this was also clear in our framework). I used notations from Electrostatics, denoting $E^n = grad(u_n)$ and $D^n = A^n\, grad(u_n)$, and after having extracted a subsequence so that $D^n \rightharpoonup D^\infty$ in $L^2_{loc}(\Omega; R^N)$ weak, the question was to identify what D^∞ was. If one showed that there exists A^{eff} such that $D^\infty = A^{eff} E^\infty$, then with usual boundary conditions u_∞ would be the solution of a similar boundary value problem, and this was what Sergio SPAGNOLO had done in the symmetric case; in the nonsymmetric case the knowledge of the inverse mapping $f \mapsto u_\infty$ does not characterize what A^{eff} is, but as nonsymmetric problems do not occur so often in applications, the main advantage of our approach is that after I had introduced my method of oscillating test functions,[28] it generalizes easily to all sort of linear partial differential equations or systems.[29]

In the early 70s, we had started by an abstract elliptic framework, where V is a real separable HILBERT space (corresponding to $H^1_0(\Omega)$ in the concrete example that we had in mind), using $||\cdot||$ for the norm on V, $||\cdot||_*$ for the norm on V', and $\langle\cdot,\cdot\rangle$ for the duality product between V' and V, and we had considered a bounded sequence $\mathcal{A}_n \in \mathcal{L}(V, V')$ (corresponding to $\mathcal{A}_n u = -div(A^n\, grad(u))$ in our example), satisfying a uniform V-ellipticity condition (corresponding to $(A^n(x)\xi.\xi) \geq \alpha|\xi|^2$, $|A^n(x)\xi| \leq M|\xi|$ for all $\xi \in R^N$, a.e. $x \in \Omega$ in our example), i.e. one assumes that there exist $0 < \alpha \leq M < \infty$ such that

$$\langle \mathcal{A}_n u, u\rangle \geq \alpha||u||^2 \text{ and } ||\mathcal{A}_n u||_* \leq M||u|| \text{ for all } u \in V. \tag{4.15}$$

By LAX-MILGRAM lemma, each $\mathcal{A}_n$ is an isomorphism from V onto V', and the first basic result in this abstract framework is the following lemma.

[28] The method was discovered independently by Leon SIMON [Si]; his student MCCONNELL had only done the layered case in linearized Elasticity, and he had looked himself at the problem in a general way; it was the referee, probably from the Italian school, who had mentioned to him my work. I first wrote about my method in [Ta3], but I had explained it to Jacques-Louis LIONS in the Fall 1975 in Marseille [Li3]; he had been convinced by Ivó BABUŠKA of the importance in engineering of problems with a periodic structure, but he had not thought of asking about the proof of our results which I had mentioned in a meeting that he had organized in June 1974 [Ta2], either to François MURAT who had stayed in Paris, or to me who had spent the year in Madison, where he actually came in the Spring.

[29] As I taught it in my PECCOT lectures, my method extends easily to some monotone situations, but as I mentioned at a meeting in Rio de Janeiro a few months after [Ta7], I could not find a good setting for Homogenization of nonlinear Elasticity. This is still the case, and the spreading errorr of mistaking Γ-convergence for Homogenization seems to come from the fact that those who commit it had not paid much attention to the difference between G-convergence and H-convergence.

Lemma 3: There exists a subsequence $\mathcal{A}_m$ and a linear continuous operator $\mathcal{A}_{eff}$ from V into V' such that for every $f \in V'$, the sequence of solutions u_m of $\mathcal{A}_m u_m = f$ converges in V weak to the solution u_∞ of $\mathcal{A}_{eff} u_\infty = f$, and $\mathcal{A}_{eff}$ satisfies

$$\langle \mathcal{A}_{eff} u, u\rangle \geq \alpha ||u||^2 \text{ and } ||\mathcal{A}_{eff} u||_* \leq \frac{M^2}{\alpha}||u|| \text{ for all } u \in V, \tag{4.16}$$

but $\frac{M^2}{\alpha}$ can be replaced by M if all the operators $\mathcal{A}_n$ are symmetric.

Proof: One has $||(\mathcal{A}_n)^{-1}||_{\mathcal{L}(V',V)} \leq \frac{1}{\alpha}$ as $\mathcal{A}_n u_n = f$ implies $\alpha||u_n||^2 \leq \langle \mathcal{A}_n u_n, u_n\rangle = \langle f, u_n\rangle \leq ||f||_*||u_n||$, so that $||u_n|| \leq \frac{1}{\alpha}||f||_*$. One can extract a subsequence u_m converging in V weak to u_∞, and repeating this extraction for f belonging to a countable dense family F of V' and using a diagonal subsequence, one can extract a subsequence $\mathcal{A}_m$ such that for every $f \in F$, the sequence $u_m = (\mathcal{A}_m)^{-1} f$ converges in V weak to a limit $S(f)$; the sequence $(\mathcal{A}_m)^{-1}$ being uniformly bounded and F being dense, $(\mathcal{A}_m)^{-1} f$ converges in V weak to a limit $S(f)$ for every $f \in V'$. S is a linear continuous operator from V' into V, with $||S f|| \leq \frac{1}{\alpha}||f||_*$ for all $f \in V'$, and in order to show that S is invertible the mere fact that the operators $(\mathcal{A}_n)^{-1}$ are uniformly bounded is not sufficient, because in any infinite dimensional HILBERT space, one can construct a sequence of symmetric surjective isometries converging weakly to 0 (in $L^2(0,1)$ for example, one can take the multiplication by $sign(\sin(n\cdot))$). However, the ellipticity condition prevents this difficulty. One notices that $\langle S f, f\rangle = \lim_m \langle u_m, f\rangle = \lim_m \langle \mathcal{A}_m u_m, u_m\rangle \geq \alpha \liminf_m ||u_m||^2$ (and $\leq \frac{1}{M}||f||_*^2$ in the symmetric case), and as $M||u_m|| \geq ||\mathcal{A}_m u_m||_* = ||f||_*$, one finds $\langle S f, f\rangle \geq \frac{\alpha}{M^2}||f||_*^2$, and therefore S is invertible by LAX-MILGRAM lemma and its inverse $\mathcal{A}_{eff}$ has a norm bounded by $\frac{M^2}{\alpha}$. As u_m converges in V weak to $S f$, one has $\liminf_m ||u_m||^2 \geq ||S f||^2$, so that $\langle S f, f\rangle \geq \alpha||S f||^2$ for every $f \in V'$, or equivalently, as we know now that S is invertible, $\langle \mathcal{A}_{eff} v, v\rangle \geq \alpha||v||^2$ for every $v \in V$.■

Of course, as most results in Functional Analysis, this lemma only gives a general framework and does not help much for identifying $\mathcal{A}_{eff}$ in concrete cases,[30] but one often uses the information that $\mathcal{A}_{eff}$ is invertible so that one can choose $f \in V'$ for which the weak limit u_∞ is any element of V prescribed in advance.

In our concrete example, the problem of G-convergence consists in showing that u_∞ solves an equation $-div(A^{eff}\, grad(u_\infty)) = f$, while the problem of H-convergence consists in showing that $A^m\, grad(u_m)$ converges in $L^2(\Omega; R^N)$ weak to $A^{eff}\, grad(u_\infty)$, and this is a different question in non-symmetric situations, because if one adds to A^n a constant antisymmetric matrix B (small enough in norm in order to retain the ellipticity condition),

---

[30] Even when all the operators $\mathcal{A}_n$ are differential operators, it may happen that $\mathcal{A}_{eff}$ is not a differential operator and in some cases nonlocal integral corrections must be taken into account.

one will not change the operator $\mathcal{A}_n$ and therefore the preceding abstract result cannot help identify the precise matrix A^{eff}, as it only uses the operators $\mathcal{A}_n$.

Lemma 3 points to a technical difficulty in the nonsymmetric case, because we started with a bound M for $\mathcal{A}_n$ and ended with the greater bound $\frac{M^2}{\alpha}$ for $\mathcal{A}_{eff}$, and this is solved by defining differently the bounds on the coefficients A^n.

Definition 4: For $0 < \alpha \leq \beta < \infty$, $M(\alpha,\beta;\Omega)$ will denote the set of $A \in L^\infty(\Omega;\mathcal{L}(R^N,R^N))$ satisfying $(A(x)\xi.\xi) \geq \alpha|\xi|^2$ and $(A(x)\xi.\xi) \geq \frac{1}{\beta}|A(x)\xi|^2$ for all $\xi \in R^N$ (or equivalently $(A^{-1}(x)\xi.\xi) \geq \frac{1}{\beta}|\xi|^2$ for all $\xi \in R^N$), a.e. $x \in \Omega$.[31] If A is independent of $x \in \Omega$, one writes $A \in M(\alpha,\beta)$.■

The reason for using the sets $M(\alpha,\beta;\Omega)$ is that they are compact for the topology of H-convergence which I define now.

Definition 5: We will say that a sequence $A^n \in M(\alpha,\beta;\Omega)$ H-converges to $A^{eff} \in M(\alpha',\beta';\Omega)$ for some $0 < \alpha' \leq \beta' < \infty$,[32] if for every $f \in H^{-1}(\Omega)$, the sequence of solutions $u_n \in H^1_0(\Omega)$ of $-div\big(A^n\,grad(u_n)\big) = f$ converges in $H^1_0(\Omega)$ weak to u_∞, and the corresponding sequence $A^n\,grad(u_n)$ converges in $L^2(\Omega;R^N)$ weak to $A^{eff}\,grad(u_\infty)$; u_∞ is therefore the solution of $-div\big(A^{eff}\,grad(u_\infty)\big) = f$ in Ω.■

H-convergence comes from a topology on $X = \bigcup_{n\geq 1} M\big(\frac{1}{n},n;\Omega\big)$, union of all $M(\alpha,\beta;\Omega)$, which is the coarsest topology that makes a list of maps continuous. For $f \in H^{-1}(\Omega)$ one such map is $A \mapsto u$ from X into $H^1_0(\Omega)$ weak and another one is $A \mapsto A\,grad(u)$ from X into $L^2(\Omega;R^N)$ weak, where u is the solution of $-div\big(A\,grad(u)\big) = f$. When one restricts that topology to $M(\alpha,\beta;\Omega)$, it is equivalent to consider only f belonging to a countable bounded set whose combinations are dense in $H^{-1}(\Omega)$; then u and $A\,grad(u)$ belong to bounded sets respectively of $H^1_0(\Omega)$ and $L^2(\Omega;R^N)$ which are metrizable for their respectuve weak topology, and therefore the restriction of that topology to $M(\alpha,\beta;\Omega)$ is defined by a countable number of semi-distances and is therefore defined by a semi-distance. That it is actually a distance can be seen by showing uniqueness of the limit: if a sequence A^n H-converges to both A^{eff} and to B^{eff}, then one deduces that $A^{eff}\,grad(u_\infty) = B^{eff}\,grad(u_\infty)$ a.e. $x \in \Omega$ for every $f \in H^{-1}(\Omega)$, and therefore for every $u_\infty \in H^1_0(\Omega)$; choosing then u_∞ to coincide successively with x_j, $j = 1,\ldots,N$, on an open subset ω with compact closure in Ω, one

[31] If $A \in \mathcal{L}(R^N,R^N)$ satisfies $(A\,\xi.\xi) \geq \frac{1}{\beta}|A\,\xi|^2$ for all $\xi \in R^N$, then $|A\,\xi| \leq \beta|\xi|$ for all $\xi \in R^N$, but if $A \in \mathcal{L}(R^N,R^N)$ satisfies $(A\,\xi.\xi) \geq \alpha|\xi|^2$ and $|A\,\xi| \leq M|\xi|$ for all $\xi \in R^N$, then one can only deduce $(A\,\xi.\xi) \geq \frac{\alpha}{M^2}|A\,\xi|^2$, if A is not symmetric, while of course one has $(A\,\xi.\xi) \geq \frac{1}{M}|A\,\xi|^2$ if A is symmetric.

[32] Theorem 6 shows that one can take $\alpha' = \alpha$ and $\beta' = \beta$.

must have $A^{eff} = B^{eff}$ a.e. $x \in \omega$. Of course, one never needs much from this topology, but some arguments do make use of the fact that $M(\alpha, \beta; \Omega)$ is metrizable.

However, if one wants to let α tend to 0, like some people do for domains with holes, one must be very careful because one cannot use arguments based on metrizability in that case.

Theorem 6: For any sequence $A^n \in M(\alpha, \beta; \Omega)$ there exists a subsequence A^m and an element $A^{eff} \in M(\alpha, \beta; \Omega)$ such that A^m H-converges to A^{eff}.
Proof: Using the same argument than in Lemma 3, F being a countable dense set of $H^{-1}(\Omega)$, we can extract a subsequence A^m such that for every $f \in F$ the sequence $u_m \in H_0^1(\Omega)$ of solutions of $-div\big(A^m\, grad(u_m)\big) = f$ converges in $H_0^1(\Omega)$ weak to $u_\infty = S(f)$ and $A^m\, grad(u_m)$ converges in $L^2(\Omega; R^N)$ weak to $R(f)$; the same is true then for all $f \in H^{-1}(\Omega)$, the operator S is invertible, and $R(f) = C\, u_\infty$ where C is a linear continuous operator from $H_0^1(\Omega)$ into $L^2(\Omega; R^N)$. It remains to show that C is local, of the form $C\, v = A^{eff}\, grad(v)$ for all $v \in H_0^1(\Omega)$, and that $A^{eff} \in M(\alpha, \beta; \Omega)$. We first show that for all $v \in H_0^1(\Omega)$ one has $\big(C\, v.grad(v)\big) \ge \alpha |grad(v)|^2$ and $\big(C\, v.grad(v)\big) \ge \frac{1}{\beta} |C\, v|^2$ a.e. $x \in \Omega$.

For $v \in H_0^1(\Omega)$, let $f = -div(C\, v)$, so that $u_\infty = v$, and let φ be a smooth function so that we may use $\varphi\, u_m$ and $\varphi\, v$ as test functions. One gets $\langle f, \varphi\, u_m\rangle = \int_\Omega \big(A^m\, grad(u_m).\varphi\, grad(u_m) + u_m\, grad(\varphi)\big)\, dx$, and as u_m converges strongly to v in $L^2(\Omega)$ because $H_0^1(\Omega)$ is compactly imbedded into $L^2(\Omega)$, one deduces that $\langle f, \varphi\, v\rangle = \lim_m \int_\Omega \varphi\big(A^m\, grad(u_m).grad(u_m)\big)\, dx + \int_\Omega \big(C\, v.v\, grad(\varphi)\big)\, dx$, but $\langle f, \varphi\, v\rangle = \int_\Omega \big(C\, v.\varphi\, grad(v) + v\, grad(\varphi)\big)\, dx$, and therefore one deduces that for every smooth function φ one has[33]

$$\int_\Omega \varphi\Big(A^m\, grad(u_m).grad(u_m)\Big)\, dx \to \int_\Omega \varphi\Big(C\, v.grad(v)\Big)\, dx. \tag{4.17}$$

Choosing now φ to be nonnegative, and using the first part of the definition of $M(\alpha, \beta; \Omega)$, we deduce that

$$\int_\Omega \varphi\Big(C\, v.grad(v)\Big)\, dx \ge \alpha \liminf_m \int_\Omega \varphi |grad(u_m)|^2\, dx \ge \alpha \int_\Omega \varphi |grad(v)|^2\, dx, \tag{4.18}$$

where the second inequality follows from the fact that $grad(u_m)$ converges in $L^2(\Omega; R^N)$ weak to $grad(v)$, and as this inequality holds for all smooth nonnegative functions φ, one obtains

$$\Big(C\, v.grad(v)\Big) \ge \alpha |grad(v)|^2 \text{ a.e. } x \in \Omega, \text{ for every } v \in H_0^1(\Omega). \tag{4.19}$$

Using the second part of the definition of $M(\alpha, \beta; \Omega)$, we deduce that

[33] One could deduce (4.17) directly from the Div-Curl lemma, but I am showing how we first argued, and we had proved this result before discovering the Div-Curl lemma.

$$\int_\Omega \varphi\Big(C\,v.grad(v)\Big)\,dx \geq \frac{1}{\beta}\liminf_m \int_\Omega \varphi|A^m\,grad(u_m)|^2\,dx \geq \alpha \int_\Omega \varphi|C\,v|^2\,dx, \tag{4.20}$$

as $A^m\,grad(u_m)$ converges in $L^2(\Omega; R^N)$ weak to $C\,v$, so that

$$\Big(C\,v.grad(v)\Big) \geq \frac{1}{\beta}|C\,v|^2 \text{ a.e. } x \in \Omega, \text{ for every } v \in H_0^1(\Omega). \tag{4.21}$$

From (4.21) one deduces

$$|C\,v| \leq \beta|grad(v)| \text{ a.e. } x \in \Omega, \text{ for every } v \in H_0^1(\Omega), \tag{4.22}$$

and as C is linear, (4.22) implies that

$$\text{if } grad(v) = grad(w) \text{ a.e. in an open subset } \omega, \text{ then } C\,v = C\,w \text{ a.e. in } \omega. \tag{4.23}$$

Writing Ω as the union of an increasing sequence ω_k of open subsets with compact closure in Ω, we define A^{eff} in the following way: for $\xi \in R^N$, we choose $v_k \in H_0^1(\Omega)$ such that $grad(v_k) = \xi$ on ω_k, and we define $A^{eff}\,\xi$ on ω_k to be the restriction of $C(v_k)$ to ω_k; this defines $A^{eff}\,\xi$ as a measurable function in Ω because $C\,v_k$ and $C\,v_l$ coincide on $\omega_k \bigcap \omega_l$ by (4.23); (4.23) also implies that A^{eff} is linear in ξ. If $w \in H_0^1(\Omega)$ is piecewise affine so that $grad(w)$ is piecewise constant, then (4.23) implies that $C\,w = A^{eff}\,grad(w)$ a.e. $x \in \Omega$. As piecewise affine functions are dense in $H_0^1(\Omega)$, for each $v \in H_0^1(\Omega)$ there is a sequence w_j of piecewise affine functions such that $grad(w_j)$ converges strongly in $L^2(\Omega; R^N)$ to $grad(v)$, and as $|C\,v - A^{eff}\,gradw_k| = |C\,v - C\,w_j| \leq \beta|grad(v - w_j)|$ a.e. $x \in \Omega$, one deduces $C\,v = A^{eff}\,grad(v)$ a.e. $x \in \Omega$. Having shown that $C\,v = A^{eff}\,grad(v)$ for a measurable A^{eff}, (4.19) and (4.21) imply that $A^{eff} \in M(\alpha, \beta; \Omega)$ as one can take v to be any affine function in an open subset ω with compact closure in Ω.■

At a meeting in Roma in the Spring of 1974,[34] I had apparently upset Ennio DE GIORGI by my claim that my method (which was actually the joint work with François MURAT) was more general than the method developed by the Italian school. I thought that MEYERS's regularity theorem which Sergio SPAGNOLO was using in his proof was based on the maximum principle [Me], but Ennio DE GIORGI had told me that it was not restricted to second order equations. Actually, it would have been difficult for me at that time to explain how to perform all the computations for higher order nonnecessarily symmetric equations or to systems like linearized Elasticity, and even more difficult to explain what to do for nonlinear elliptic equations, but less than a year after I had noticed that most of the important known properties of

[34] Umberto MOSCO had insisted that I should write something before I left, and instead of visiting Roma, I stayed in the hotel, so nicely located above Piazza di Spagna, writing [Ta1] and another short description concerning quasi-variational inequalities.

Homogenization of second order variational elliptic equations in divergence form could be obtained through repeated applications of the Div-Curl lemma, and the extension to linear elliptic systems in a variational framework (and some simple nonlinear systems of monotone type) really became straightforward. As my method, which has been wrongly called the "energy method" and which I prefer to call the "method of oscillating test functions", uses only a variational structure, it can be extended with minor changes[35] to most of the linear partial differential equations of Continuum Mechanics (not much is understood for nonlinear equations).

One starts with the same abstract analysis, Lemma 3 and the beginning of the proof of Theorem 6; one extracts a subsequence A^m for which there is a linear continuous operator C from $H_0^1(\Omega)$ into $L^2(\Omega; R^N)$ such that for every $f \in H^{-1}(\Omega)$ the sequence of solutions $u_m \in H_0^1(\Omega)$ of $-div(A^m\, grad(u_m)) = f$ converges in $H_0^1(\Omega)$ weak to u_∞ and $A^m\, grad(u_m)$ converges in $L^2(\Omega; R^N)$ weak to $R(f) = C(u_\infty)$. One uses then the Div-Curl lemma to give a new proof that C is a local operator of the form $C(v) = A^{eff}\, grad(v)$ with $A^{eff} \in L^\infty(\Omega; \mathcal{L}(R^N, R^N))$.

One constructs a sequence of oscillating test functions v_m satisfying

$$-div\Big((A^m)^T grad(v_m)\Big) \text{ converges in } H^{-1}_{loc}(\Omega) \text{ strong,} \tag{4.24}$$

where $(A^m)^T$ is the transposed operator of A^m and

$$v_m \rightharpoonup v_\infty \text{ in } H^1(\Omega) \text{ weak; } (A^m)^T grad(v_m) \rightharpoonup w_\infty \text{ in } L^2(\Omega; R^N) \text{ weak,} \tag{4.25}$$

and one passes to the limit in the identity

$$\Big(A^m\, grad(u_m).grad(v_m)\Big) = \Big(grad(u_m).(A^m)^T grad(v_m)\Big). \tag{4.26}$$

The Div-Curl lemma applies to the left side of (4.26) which converges in the sense of measures[36] to $\big(C(u_\infty).grad(v_\infty)\big)$, because $div(A^m\, grad(u_m))$ is

---

[35] Becoming a mathematician requires some ability with abstract concepts, and it is part of the training to check that one can apply a general method to particular examples, and it depends upon one's taste and one's own scientific stature to decide if it was an exercise or something worth publishing.

[36] A sequence s_n converges to s_∞ in the sense of distributions in Ω if for every $\varphi \in C_c^\infty(\Omega)$, the space of indefinitely differentiable functions with compact support in Ω, one has $\langle s_n, \varphi\rangle \to \langle s_\infty, \varphi\rangle$; the s_n must be distributions and according to the theory of distributions of Laurent SCHWARTZ the limit s_∞ is automatically a distribution. The sequence s_n converges to s_∞ in the sense of measures if the preceding convergence holds for every $\varphi \in C_c(\Omega)$, the space of continuous functions with compact support in Ω; the s_n must be RADON measures and s_∞ is automatically a RADON measure. If $s_n \in L^1_{loc}(\Omega)$, then $\langle s_n, \varphi\rangle$ means $\int_\Omega s_n\, \varphi\, dx$.

a fixed element of $H^{-1}(\Omega)$ and $grad(v_m)$ converges in $L^2(\Omega; R^N)$ weak to $grad(v_\infty)$; the Div-Curl lemma also applies to the right side of (4.26) which converges in the sense of measures to $(grad(u_\infty).w_\infty)$, because of (4.24) and (4.25), and this shows

$$\big(C(u_\infty).grad(v_\infty)\big) = (grad(u_\infty).w_\infty) \text{ a.e. in } \Omega. \tag{4.27}$$

One constructs a sequence v_m satisfying (4.24) and (4.25) by first choosing an open set Ω' containing the closure of Ω, then extending A^m in $\Omega' \setminus \Omega$ for example by $A^m(x) = \alpha I$ for $x \in \Omega' \setminus \Omega$, and then choosing $v_m \in H_0^1(\Omega')$ solution of

$$-div\Big((A^m)^T grad(v_m)\Big) = g \text{ in } \Omega', \tag{4.28}$$

for some $g \in H^{-1}(\Omega')$. One obtains a sequence v_m bounded in $H_0^1(\Omega')$ and therefore its restriction to Ω is bounded in $H^1(\Omega)$, and a subsequence satisfies (4.24) and (4.25). By Lemma 3 one can choose $g \in H^{-1}(\Omega')$ such that v_∞ is any arbitrary element of $H_0^1(\Omega')$, and in particular for each $j = 1, \ldots, N$, there exists $g_j \in H^{-1}(\Omega')$ such that $v_\infty = x_j$ a.e. in Ω, and using these N choices of g_j, (4.27) means $C\,u_\infty = A^{eff}\,grad(u_\infty)$ for some $A^{eff} \in L^2(\Omega; \mathcal{L}(R^N, R^N))$.

This method quickly gives only an intermediate result and is not very good for questions of bounds, and it must emphasized that questions of bounds are not yet very well understood for general equations or systems. As pointed out by François MURAT, one can easily show that $A^{eff} \in L^\infty(\Omega; \mathcal{L}(R^N, R^N))$ by using the following lemma based on the continuity of C, but the information $A^{eff} \in M(\alpha, \beta; \Omega)$ is not so natural in this approach.

Lemma 7: If $M \in L^2(\Omega; \mathcal{L}(R^N, R^N))$ and the operator C defined by $C(v) = M\,grad(v)$ for all $v \in H_0^1(\Omega)$ is a linear continuous operator from $H_0^1(\Omega)$ into $L^2(\Omega; R^N)$ of norm $\le \gamma$, then one has $M \in L^\infty(\Omega; \mathcal{L}(R^N, R^N))$ and $||M(x)||_{\mathcal{L}(R^N,R^N)} \le \gamma$ a.e. in Ω.

Proof: Let $\xi \in R^N \setminus 0$ and $\varphi \in C_c^1(\Omega)$, the space of functions of class C^1 with compact support in Ω. Define φ_n by $\varphi_n(x) = \varphi(x)\frac{\sin n(\xi.x)}{n}$ for $x \in \Omega$, which gives a bounded sequence in $H_0^1(\Omega)$ with

$$\lim_n ||grad(\varphi_n)||_{L^2(\Omega;R^N)} = \frac{1}{\sqrt{2}}||\xi\,\varphi||_{L^2(\Omega;R^N)} = \frac{|\xi|}{\sqrt{2}}||\varphi||_{L^2(\Omega)}.$$

$$C(\varphi_n) = M\big(\frac{\sin n(\xi.\cdot)}{n} grad(\varphi) + \varphi \cos n(\xi.\cdot)\xi\big),$$

and therefore

$$\lim_n ||C(\varphi_n)||_{L^2(\Omega;R^N)} = \frac{1}{\sqrt{2}}||\varphi\,M\,\xi||_{L^2(\Omega)},$$

because φ and $grad(\varphi)$ are bounded, so that $\varphi\,M\,\xi$ and $M\,grad(\varphi)$ belongs to $L^2(\Omega; R^N)$. Therefore one deduces that $||\varphi\,M\,\xi||_{L^2(\Omega)} \le \gamma|\xi|\,||\varphi||_{L^2(\Omega)}$ for every $\varphi \in C_c^1(\Omega)$, and therefore for every $\varphi \in L^2(\Omega)$ by density, and this means $||M\,\xi||_{L^\infty(\Omega;R^N)} \leq \gamma|\xi|$, and as this is valid for all ξ, the lemma is proved.■

Using the same approach,[37] one can derive a few useful properties of H-convergence, the main tool remaining the Div-Curl lemma.

Proposition 8: If a sequence $A^n \in M(\alpha,\beta;\Omega)$ H-converges to A^{eff}, then the transposed sequence $(A^n)^T$ H-converges to $(A^{eff})^T$. In particular if a sequence A^n H-converges to A^{eff} and if $A^n(x)$ is symmetric a.e. $x \in \Omega$ for all n, then $A^{eff}(x)$ is symmetric a.e. $x \in \Omega$.
Proof: $A \in M(\alpha,\beta;\Omega)$ implies (and therefore is equivalent to) $A^T \in M(\alpha,\beta;\Omega)$ as $A \in M(\alpha,\beta;\Omega)$ means $(A(x)\xi.\xi) \geq \alpha|\xi|^2$ and $(A^{-1}(x)\xi.\xi) \geq \frac{1}{\beta}|\xi|^2$ for all $\xi \in R^N$, a.e. $x \in \Omega$, and as $(A^{-1})^T = (A^T)^{-1}$, this is the same as $(A^T(x)\xi.\xi) \geq \alpha|\xi|^2$ and $((A^T)^{-1}(x)\xi.\xi) \geq \frac{1}{\beta}|\xi|^2$ for all $\xi \in R^N$, a.e. $x \in \Omega$. By Theorem 6 a subsequence $(A^m)^T$ H-converges to B^{eff}. For $f,g \in H^{-1}(\Omega)$, let us define the sequences $u_m, v_m \in H_0^1(\Omega)$ by

$$-div\Big(A^m\,grad(u_m)\Big) = f,\ -div\Big((A^m)^T\,grad(v_m)\Big) = g \text{ in } \Omega, \tag{4.29}$$

so that u_m and v_m converge in $H_0^1(\Omega)$ weak respectively to u_∞ and v_∞, $A^m\,grad(u_m)$ and $(A^m)^T grad(v_m)$ converge in $L^2(\Omega;R^N)$ weak respectively to $A^{eff}\,grad(u_\infty)$ and $B^{eff} grad(v_\infty)$, and one can use the Div-Curl lemma to take the limit of the identity

$$\Big(A^m\,grad(u_m).grad(v_m)\Big) = \Big(grad(u_m).(A^m)^T grad(v_m)\Big), \tag{4.30}$$

and obtain

$$\Big(A^{eff}\,grad(u_\infty).grad(v_\infty)\Big) = \Big(grad(u_\infty).B^{eff}\,grad(v_\infty)\Big), \tag{4.31}$$

and as u_∞ and v_∞ can be arbitrary elements of $H_0^1(\Omega)$ by Lemma 3, this implies $(A^{eff})^T = B^{eff}$ a.e. in Ω. The second part of the Proposition results from uniqueness of H-limits.■

The next result shows that H-convergence inside Ω is not related to any particular boundary condition imposed on $\partial\Omega$.

Proposition 9: If a sequence $A^n \in M(\alpha,\beta;\Omega)$ H-converges to A^{eff} and a sequence u_n converges in $H^1_{loc}(\Omega)$ weak to u_∞, and $div\big(A^n\,grad(u_n)\big)$ belongs to a compact set of $H^{-1}_{loc}(\Omega)$ strong, then the sequence $A^n\,grad(u_n)$ converges to $A^{eff}\,grad(u_\infty)$ in $L^2_{loc}(\Omega;R^N)$ weak.

[37] The preceding method is not adapted to nonlinear problems, but I developped a variant where the oscillating test functions satisfy the initial equation instead of the transposed equation, and it extends to the case of monotone operators.

Proof: Let $\varphi \in C_c^1(\Omega)$ so that φu_n converges in $H_0^1(\Omega)$ weak to φu_∞, $\varphi\, grad(u_n)$ converges in $L^2(\Omega; R^N)$ weak to $\varphi\, grad(u_\infty)$; $curl(\varphi\, grad(u_n))$ has its components bounded in $L^2(\Omega)$, as they are of the form $\frac{\partial \varphi}{\partial x_j}\frac{\partial u_n}{\partial x_k} - \frac{\partial \varphi}{\partial x_k}\frac{\partial u_n}{\partial x_j}$. As $div(\varphi A^n\, grad(u_n)) = \varphi\, div(A^n\, grad(u_n)) + (A^n\, grad(u_n).grad(\varphi))$, it belongs to a compact set of $H^{-1}(\Omega)$ strong, as multiplication by φ maps $H_{loc}^{-1}(\Omega)$ into $H^{-1}(\Omega)$ and $(A^n\, grad(u_n).grad(\varphi))$ is bounded in $L^2(\Omega)$. One extracts a subsequence such that $\varphi A^m\, grad(u_m)$ converges in $L^2(\Omega; R^N)$ weak to w_∞. For $f \in H^{-1}(\Omega)$, one defines $v_n \in H_0^1(\Omega)$ by $-div((A^n)^T grad(v_n)) = f$, so that v_n converges in $H_0^1(\Omega)$ weak to v_∞ and $(A^n)^T grad(v_n)$ converges in $L^2(\Omega; R^N)$ weak to $(A^{eff})^T grad(v_\infty)$ by Proposition 8.

One then passes to the limit in both sides of $(\varphi A^m\, grad(u_m).grad(v_m)) = (\varphi\, grad(u_m).(A^m)^T grad(v_m))$ by using the Div-Curl lemma, and one obtains the relation $(w_\infty.grad(v_\infty)) = (\varphi\, grad(u_\infty).(A^{eff})^T grad(v_\infty))$ a.e. in Ω. As v_∞ is an arbitrary element of $H_0^1(\Omega)$ by Lemma 3, $w_\infty = \varphi A^{eff}\, grad(u_\infty)$ a.e. in Ω, and as φ is arbitrary in $C_c^1(\Omega)$ and the limit does not depend upon which subsequence has been chosen, one deduces that all the sequence $A^n\, grad(u_n)$ converges in $L_{loc}^2(\Omega; R^N)$ weak to $A^{eff}\, grad(u_\infty)$.■

In the preceding proof, $div(A^n\, grad(\varphi u_n))$ may not belong to a compact set of $H^{-1}(\Omega)$ strong as it is $div(\varphi A^n\, grad(u_n)) + div(u_n A^n\, grad(\varphi))$ and $div(\varphi A^n\, grad(u_n))$ does indeed belong to a compact set of $H^{-1}(\Omega)$ strong as was already used, but it is not clear if $div(u_n A^n\, grad(\varphi))$ does, because $u_n A^n\, grad(\varphi)$ may only converge in $L^2(\Omega; R^N)$ weak. We have used then the complete form of the Div-Curl lemma and not only the special case where one only considers gradients.

Proposition 9 expresses that the boundary conditions used for u_n are not so important, as long as the solutions stay bounded, as had been noticed by Sergio SPAGNOLO in the case of G-convergence. We did define H-convergence by using DIRICHLET conditions, but the result inside Ω would be the same for other boundary conditions, if one can apply LAX-MILGRAM lemma for existence as we need to start by using Lemma 3. Using DIRICHLET conditions has the advantage that no smoothness assumption is necessary for the boundary of Ω. What happens on the boundary $\partial\Omega$ may depend upon the particular boundary condition used; the particular cases of nonhomogeneous DIRICHLET conditions, NEUMANN conditions, and other variational conditions can all be considered at once in the framework of variational inequalities, allowing actually some nonlinearity in the boundary conditions (the nonlinearity inside Ω is a different matter).

The next result states that H-convergence has a local character, extending the corresponding result of Sergio SPAGNOLO for G-convergence.

Proposition 10: If a sequence $A^n \in M(\alpha, \beta; \Omega)$ H-converges to A^{eff}, and ω is an open subset of Ω, then the sequence $M^n = A^n|_\omega$ of the restrictions of A^n

to ω H-converges to $M^{eff} = A^{eff}|_{\omega}$. Therefore if a sequence $B^n \in M(\alpha, \beta; \Omega)$ H-converges to B^{eff} and $A^n = B^n$ for all n, a.e. $x \in \omega$, then $A^{eff} = B^{eff}$ a.e. $x \in \omega$.

Proof: If all A^n belong to $M(\alpha, \beta; \Omega)$, then all M^n belong to $M(\alpha, \beta; \omega)$ and by Theorem 6 a subsequence M^m H-converges to some $M^{eff} \in M(\alpha, \beta; \omega)$. For $f \in H^{-1}(\omega)$ and $g \in H^{-1}(\Omega)$, let us solve $-div\big(M^m\, grad(u_m)\big) = f$ in ω and $-div\big((A^m)^T\, grad(v_m)\big) = g$ in Ω so that u_m converges in $H_0^1(\omega)$ weak to u_∞, $M^m\, grad(u_m)$ converges in $L^2(\omega; R^N)$ weak to $M^{eff}\, grad(u_\infty)$, v_m converges in $H_0^1(\Omega)$ weak to v_∞, and $(A^m)^T\, grad(v_m)$ converges in $L^2(\Omega; R^N)$ weak to $(A^{eff})^T\, grad(v_\infty)$. Extending u_n and u_∞ by 0 in $\Omega \backslash \omega$, one can apply the Div-Curl lemma in ω for the left side and in Ω for the right side of the equality $\big(M^m\, grad(u_m).grad(v_m)\big) = \big(grad(u_m).(A^m)^T\, grad(v_m)\big)$, and one obtains $\big(M^{eff}\, grad(u_\infty).grad(v_\infty)\big) = \big(grad(u_\infty).(A^{eff})^T\, grad(v_\infty)\big)$ a.e. in ω. As by Lemma 3 u_∞ can be arbitrary in $H_0^1(\omega)$ and v_∞ arbitrary in $H_0^1(\Omega)$, one deduces that $M^{eff} = A^{eff}$ a.e. in ω; as the H-limit is independent of the subsequence used, all the sequence M^n H-converges to $A^{eff}|_{\omega}$.■

Actually if for a measurable subset ω of Ω, one has $A^n = B^n$ for all n, a.e. $x \in \omega$, and the sequences $A^n, B^n \in M(\alpha, \beta; \Omega)$ H-converge respectively in Ω to A^{eff}, B^{eff}, then one has $A^{eff} = B^{eff}$ a.e. $x \in \omega$. This can be proved by applying the same regularity theorem of MEYERS [Me] that Sergio SPAGNOLO used in the symmetric case. It is equivalent to prove that $A^{eff} = B^{eff}$ a.e. in $\omega(\varepsilon)$ for each $\varepsilon > 0$, where $\omega(\varepsilon)$ is the set of points of ω at a distance at least ε from $\partial\Omega$. Defining v_n as above but choosing $f \in H^{-1}(\Omega)$ and $u_n \in H_0^1(\Omega)$ instead, the problem is to use $\chi_{\omega(\varepsilon)}$, the characteristic function of $\omega(\varepsilon)$, as a test function in the Div-Curl lemma in Ω. For obtaining that result one first takes $f, g \in W^{-1,p}(\Omega)$ with $p > 2$, and MEYERS's regularity theorem tells that $grad(u_n)$ and $grad(v_n)$ stay bounded in $L^{q(\varepsilon)}\big(\omega(\varepsilon)\big)$ for some $q(\varepsilon) \in (2, p]$, and therefore $\big(A^n\, grad(u_n).grad(v_n)\big)$ and $\big(grad(u_n).(B^n)^T\, grad(v_n)\big)$ (which are equal on ω) are bounded in $L^{q(\varepsilon)/2}\big(\omega(\varepsilon)\big)$, and converge in $L^{q(\varepsilon)/2}\big(\omega(\varepsilon)\big)$ weak to $\big(A^{eff}\, grad(u_\infty).grad(v_\infty)\big)$ and $\big(grad(u_\infty).(B^{eff})^T\, grad(v_\infty)\big)$ which are then equal a.e. in $\omega(\varepsilon)$. As $W^{-1,p}(\Omega)$ is dense in $H^{-1}(\Omega)$, one can pass to the limit in this equality in $\omega(\varepsilon)$ and obtain it for arbitrary $f, g \in H^{-1}(\Omega)$, i.e. for arbitrary $u_\infty, v_\infty \in H_0^1(\Omega)$, and that gives $A^{eff} = B^{eff}$ a.e. in $\omega(\varepsilon)$, and therefore a.e. in ω.

The argument of Proposition 10 is variational and extends therefore to all variational situations, while in order to extend the preceding argument to a general situation one would have to prove a regularity theorem like MEYERS's one, and I do not know if this has been done; it has been checked by Jacques-Louis LIONS that the analogous statement is valid for some linearized Elasticity systems.

6. Bounds on effective coefficients: first method

In the case $A^n = a^n I$, which we had investigated first, we knew that the $L^\infty(\Omega)$ weak $\star$ limits of a^n and $\frac{1}{a^n}$, denoted respectively by a_+ and $\frac{1}{a_-}$, are needed for expressing A^{eff} in the case where a^n only depends upon one variable. We had obtained sequences $E^n = grad(u_n)$ and $D^n = a^n E^n = a^n\, grad(u_n)$ converging in $L^2(\Omega; R^N)$ weak, respectively to E^∞ and D^∞, and the analogue of (4.17) had told us that $a^n|E^n|^2$ was converging in the sense of measures to $(D^\infty.E^\infty)$. We had then decided to look at the convex hull in R^{2N+3} of the set

$$K = \left\{\left(E, a\,E, a|E|^2, a, \frac{1}{a}\right) | E \in R^N, a \in [\alpha, \beta]\right\}, \tag{5.1}$$

with the goal of investigating what could be deduced if D^∞ and E^∞ satisfy the property

$$\left(E^\infty, D^\infty, (D^\infty.E^\infty), a_+, \frac{1}{a_-}\right) \in clconv(K). \tag{5.2}$$

I will show the necessary computations, which we did not carry out exactly as (5.34)/(5.40) in the early 70s, as we had noticed that a simple argument of convexity showed that

$$a_- I \le A^{eff} \le a_+ I. \tag{5.3}$$

Lemma 11: If a sequence v_n converges in $L^2(\Omega; R^N)$ weak to v_∞, if $M^n \in M(\alpha, \beta; \Omega)$ is symmetric a.e. $x \in \Omega$ and $(M^n)^{-1}$ converges in $L^\infty(\Omega; \mathcal{L}(R^N, R^N))$ weak $\star$ to $(M_-)^{-1}$, then for every nonnegative continuous function φ with compact support in Ω, one has

$$\liminf_n \int_\Omega \varphi(M^n\, v_n.v_n)\, dx \ge \int_\Omega \varphi(M_-\, v_\infty.v_\infty)\, dx, \tag{5.4}$$

i.e. if $(M^n\, v_n.v_n)$ converges to a RADON measure ν in the sense of measures (i.e. weakly $\star$), then $\nu \ge (M_-\, v_\infty.v_\infty)$ in the sense of measures in Ω.

Proof: If $\mathcal{L}_{s+}(R^N, R^N)$ denotes the convex cone of symmetric positive operators from R^N into itself, Lemma 11 follows from the fact that $(P, v) \mapsto (P^{-1}\, v.v)$ is convex on $\mathcal{L}_{s+}(R^N, R^N) \times R^N$. Indeed, one has

$$(P^{-1}\, v.v) = (P_0^{-1}\, v_0.v_0) + 2(P_0^{-1}\, v_0.v - v_0) - (P_0^{-1}(P - P_0)P_0^{-1}\, v_0.v_0) + R \tag{5.5}$$

and the remainder R is nonnegative for every $P_0 \in \mathcal{L}_{s+}(R^N, R^N)$ and $v_0 \in R^N$, as an explicit computation shows that

$$R = \left(P(P^{-1}\, v - P_0^{-1}\, v_0).(P^{-1}\, v - P_0^{-1}\, v_0)\right). \tag{5.6}$$

■

As an application of Lemma 11, we can deduce upper bounds as well as lower bounds for A^{eff}, improving then the result of Theorem 6; the bounds are expressed in terms of the weak $\star$ limit of A^n and the weak $\star$ limit of $(A^n)^{-1}$.

Proposition 12: Assume that a sequence $A^n \in M(\alpha,\beta;\Omega)$ H-converges to A^{eff}, if $(A^n)^T(x) = A^n(x)$ a.e. $x \in \Omega$ for all n, and satisfies

$$A^n \rightharpoonup A_+;\ (A^n)^{-1} \rightharpoonup (A_-)^{-1} \text{ in } L^\infty\Big(\Omega; \mathcal{L}(R^N, R^N)\Big) \text{ weak } \star. \tag{5.7}$$

Then one has

$$A_- \leq A^{eff} \leq A_+ \text{ a.e. } x \in \Omega. \tag{5.8}$$

Proof: In the proof of Theorem 6 we have constructed a sequence $grad(u_n)$ converging in $L^2(\Omega; R^N)$ weak to $grad(u_\infty)$, and such that $A^n\, grad(u_n)$ converges in $L^2(\Omega; R^N)$ weak to $A^{eff}\, grad(u_\infty)$; moreover we had shown that $\big(A^n\, grad(u_n).grad(u_n)\big)$ converges in the sense of measures to $\big(A^{eff}\, grad(u_\infty).grad(u_\infty)\big)$ (which can be deduced from the Div-Curl lemma).

By using Lemma 11 with $M^n = A^n$ and $v_n = grad(u_n)$ one obtains $\big(A^{eff}\, grad(u_\infty).grad(u_\infty)\big) \geq \big(A_-\, grad(u_\infty).grad(u_\infty)\big)$ in the sense of measures. As both sides of the inequality belong to $L^1(\Omega)$ the inequality is valid a.e. $x \in \Omega$. From the fact that u_∞ can be any element of $H^1_0(\Omega)$, one can choose $grad(u_\infty)$ to be any constant vector λ on an open subset ω with compact closure in Ω, so that one has proved that $(A^{eff}\lambda.\lambda) \geq (A_-\lambda.\lambda)$ for every $\lambda \in R^N$, and therefore $A^{eff} \geq A_-$ a.e. in Ω.

Similarly, $\big(A^{eff}\, grad(u_\infty).grad(u_\infty)\big) \geq \big((A_+)^{-1}A^{eff}\, grad(u_\infty).A^{eff}\, grad(u_\infty)\big)$ in the sense of measures, by applying Lemma 11 with $M^n = (A^n)^{-1}$ and $v_n = A^n\, grad(u_n)$, and as both sides of the inequality belong to $L^1(\Omega)$ the inequality is valid a.e. $x \in \Omega$, and choosing $grad(u_\infty) = \lambda$ on an open subset ω with compact closure in Ω, one obtains $(A^{eff}\,\lambda.\lambda) \geq (A_+)^{-1}A^{eff}\,\lambda.A^{eff}\,\lambda)$ for every $\lambda \in R^N$, and therefore $A^{eff} \geq A^{eff}(A_+)^{-1}A^{eff}$, or equivalently $(A^{eff})^{-1} \geq (A_+)^{-1}$ or $A^{eff} \leq A_+$ a.e. in Ω.■

There is an important logical point to be emphasized here, as this kind of result may easily be attributed to a few different persons. It would be interesting to check if those who either claim to have proved it before François MURAT and I had proved it in the early 70s, or claim that it had been proved a long time ago by such or such a pioneer in Continuum Mechanics or Physics, could show that there was a clear definition of what one was looking for in any of these "proofs". When one says that something is well known, it only means that it is well known by those who know it well, and in Roma in 1974, I think that Ennio DE GIORGI did not know about such an inequality; I believe that he would have quickly found a proof if I had asked him the question instead of saying that I had already proved the result. It is difficult to explain how anyone could have proved the result before there was a definition of what effective coefficients were, i.e. before the work of Sergio SPAGNOLO in the late 60s or the work of François MURAT and me in the early 70s.

Many would probably argue that they knew about effective coefficients much before there was a definition, and indeed some had a good intuition about that question, but many just had a fuzzy idea of what it was about, and I could observe that at a meeting at the Institute of Mathematics and its Applications in Minneapolis in the Fall of 1995 when one of the speakers challenged the mathematicians by saying that he had proved a result that mathematicians had not proved. Of course that could well happen and I have always been ready to learn from engineers about anything that they may know on interesting scientific matters, but it did not start well because the speaker was working with linearized Elasticity, and if there are still a few mathematicians who do not know about the defects of linearized Elasticity it is not really my fault because I have been a strong advocate of mentioning the known defects of models, and those of linearized Elasticity are well known by now, and I expected a better understanding about this kind of questions from an engineer anyway. The speaker pretended then to have proved bounds for (linear) effective elastic coefficients using inclusions of (linear) elastic materials that were not elliptic, and as he was claiming that his bounds only involved proportions I had mentioned to my neighbour and then loud enough to be heard by all that it was false;[38] indeed my comment was heard by many but after the talk there was only one person in the audience interested in clarifying the question, as John WILLIS came to tell me that the speaker had not really meant to say nonelliptic, because the materials that he wanted to use were actually strongly elliptic (by opposition to very strongly elliptic), and certainly an engineer who does not know the definition of ellipticity should avoid challenging mathematicians in public,[39] but that was not the only problem in the statement. The speaker had only obtained

[38] For a diffusion equation, the question is very similar to using a function of one complex variable in the style of the work of David BERGMAN [Be], and extending it for small negative real values, and one cannot expect to use materials which are not elliptic without imposing something on the interface, as can be checked for the checkerboard pattern, according to a formula of Joseph KELLER [Ke], which George PAPANICOLAOU pointed out to me in 1980 after I had proved the same result. In the early 80s, Stefano MORTOLA and Sergio STEFFÉ had shown that using regularity of interfaces one can indeed use some "materials" with nonelliptic coefficients, but there is a relation between a measure of the regularity of the interface and the amount of nonellipticity allowed [Mo&St1].

[39] He may also have mistaken as mathematicians some people who do not hesitate to publish without correct attribution some results that they have learned from others, without realizing that it may soon become apparent that they do not understand very well the subject that they are talking about;

results for DIRICHLET conditions and he would have been in great trouble for showing that his results were local and could be obtained for all kinds of variational boundary conditions. Some people might argue that boundary conditions are not of such importance in Elasticity, as they may have in other questions,[40] but the problem is that the materials used in the mathematical approach do not exist in the real world, and if engineers misuse their knowledge and intuition about the real world by considering unrealistic situations and pretending that they know the mathematical answer to some questions, they may just be wrong: I do not see much reason why engineers should have a better intuition than mathematicians about problems which are completely unrealistic.[41]

According to the work of Sergio SPAGNOLO in the lates 60s or the work of François MURAT and me in the early 70s, homogenized/effective properties are "local" properties of a mixture of materials, and in this course about Op-

---

the lack of reaction of the audience might have been a sign that many did not care much about publishing wrong results.

[40] I remember a Chemistry teacher showing us a piece of phosphorus in a container full of oil, and he explained why it was kept in oil by taking a small piece out, and it quickly burst into flames; the boundary conditions on a piece of phosphorus are important because of chemical reactions taking place precisely at the boundary.

[41] After this incident in Minneapolis, I asked my student Sergio GUTIÉRREZ to look if one could extend the theory of Homogenization in linearized Elasticity to some materials which are strongly elliptic but not very strongly elliptic. As he showed as part of his PhD thesis [Gu], a local theory englobing the very strongly elliptic materials and for which the formula of layers is valid cannot englobe any (isotropic) material which is strongly elliptic but not very strongly elliptic, except perhaps for the limiting cases. I considered that result as a fact that it is unlikely that such materials may exist, and I conjectured that the (linearized) evolution equation with a single interface with one of these materials could be ill posed. Mort GURTIN has pointed out that one can obtain some of these materials by linearization around an unstable equilibrium in (nonlinear) Elasticity, and it suggests then that it is unlikely that one could avoid DIRICHLET conditions if one uses some of these materials, except perhaps by putting enormous forces at the boundary in order to avoid these materials to become unstable. It is interesting to notice that the publication of Sergio GUTIÉRREZ's result has created a strange reaction from a referee, who thought that it was contradicting a result on Γ-convergence; I have not been able to convince the editor that this was irrelevant and proved that the referee did not understand what Homogenization is about if he/she thinks that Γ-convergence is Homogenization, that the correctness of the computations of Sergio GUTIÉRREZ is quite easy to check and that there is no reason to impose on him the burden of explaining the errors that others may have committed elsewhere in unrelated subjects.

timal Design it is extremely important to use local properties and to avoid any restriction like periodic situations for example, because we are looking for the best design and we should not postulate what we would like it to be. It is a different question to consider "global" properties, like how much energy is located in a container, and some pioneers might have understood effective coefficients only in this restricted way. People who are interested in the question of how much energy a given domain contains without being interested about where this energy is located precisely and how this energy moves around,[42] often drift to quite unrealistic questions, and some still use names like Elasticity for these unrealistic questions, luring a few naive mathematicians out of the scientific path.

Let us go back to deriving bounds on effective coefficients, and look at the compatibility of H-convergence with the usual preorder relation on $\mathcal{L}(R^N, R^N)$; let us recall that for $A, B \in \mathcal{L}(R^N, R^N)$, $A \leq B$ means that for every $\xi \in R^N$ one has $(A\,\xi.\xi) \leq (B\xi.\xi)$; this preorder is not an order on $\mathcal{L}(R^N, R^N)$, but it is a partial order if one restricts attention to symmetric operators.

Proposition 13: [DG&Sp] If a sequence $A^n \in M(\alpha, \beta; \Omega)$ satisfies $(A^n)^T = A^n$ for all n, a.e. $x \in \Omega$ and H-converges to A^{eff}, if a sequence u_n converges to u_∞ in $H^1_0(\Omega)$ weak and if $\varphi \geq 0$ in Ω with $\varphi \in C_c(\Omega)$, then one has

$$\liminf_n \int_\Omega \varphi\Big(A^n\, grad(u_n).grad(u_n)\Big)\,dx \geq \int_\Omega \varphi\Big(A^{eff}\, grad(u_\infty).grad(u_\infty)\Big)\,dx. \tag{5.9}$$

For every $u_\infty \in H^1_0(\Omega)$ there exists a sequence v_n converging to u_∞ in $H^1_0(\Omega)$ weak and such that for every $\varphi \in C_c(\Omega)$ one has

$$\lim_n \int_\Omega \varphi\Big(A^n\, grad(v_n).grad(v_n)\Big)\,dx = \int_\Omega \varphi\Big(A^{eff}\, grad(u_\infty).grad(u_\infty)\Big)\,dx. \tag{5.10}$$

Proof. Let $f = -div\big(A^{eff}\, grad(u_\infty)\big)$ and let $v_n \in H^1_0(\Omega)$ be the solution of $-div\big(A^n\, grad(v_n)\big) = f$ in Ω, then v_n converges to some v_∞ in $H^1_0(\Omega)$ weak and as v_∞ is solution of $-div\big(A^{eff}\, grad(v_\infty)\big) = f$ in Ω, one must have $v_\infty = u_\infty$ a.e. in Ω; therefore v_n converges to u_∞ in $H^1_0(\Omega)$ weak

[42] Many mathematicians still seem to believe in a world described by stationary equations, where energy is minimized, as if they had never heard about the First Principle of "conservation of energy", and although it is a difficult question to explain how some energy is transformed into heat and how good the Second Principle is, it certainly seems utterly unrealistic to believe that after a short time every system finds its point of minimum potential energy.

and $A^n\, grad(v_n)$ converges to $A^{eff}\, grad(u_\infty)$ in $L^2(\Omega; R^N)$ weak. Then one computes

$$\liminf_n \int_\Omega \varphi\Big(A^n\big(grad(u_n) - grad(v_n)\big).grad(u_n) - grad(v_n)\Big)\, dx, \quad (5.11)$$

which is a nonnegative number. One term is $\int_\Omega \varphi(A^n\, grad(u_n).grad(u_n))\, dx$ whose $\liminf_n$ is what we are interested in; then, because of the symmetry of A^n, the other terms are $\int_\Omega \varphi(A^n\, grad(v_n).grad(-2u_n + v_n))\, dx$, and the Div-Curl lemma applies so the limit is $-\int_\Omega \varphi(A^{eff}\, grad(u_\infty).grad(u_\infty))\, dx$, and this gives (5.10). Of course, (5.11) is obtained by using the sequence v_n just constructed and applying the Div-Curl lemma.■

The preceding result is due to Ennio DE GIORGI and Sergio SPAGNOLO [DG&Sp], who were using characteristic functions of measurable sets for φ, because the use of MEYERS's regularity result permits to prove the preceding result for $\varphi \geq 0$, $\varphi \in L^\infty(\Omega)$. Notice that the convexity argument of Lemma 11 gives A_- on the right side of (5.11) instead of A^{eff}, so Proposition 13 is more precise then that Lemma 11 as $A_- \leq A^{eff}$ by Proposition 12.

Proposition 14: If a sequence $A^n \in M(\alpha, \beta; \Omega)$ satisfies $(A^n)^T = A^n$ for all n, a.e. $x \in \Omega$ and H-converges to A^{eff}, if a sequence $B^n \in M(\alpha, \beta; \Omega)$ H-converges to B^{eff} and if $A^n \leq B^n$ a.e. in Ω for all n, then one has $A^{eff} \leq B^{eff}$ a.e. $x \in \Omega$.

Proof: Let $g \in H^{-1}(\Omega)$ and let $v_n \in H^1_0(\Omega)$ be the sequence of solutions of $-div(B^n\, grad(v_n)) = g$ in Ω, which converges in $H^1_0(\Omega)$ weak to v_∞, and $B^n\, grad(v_n)$ converges in $L^2(\Omega; R^N)$ weak to $B^{eff}\, grad(v_\infty)$. Then for $\varphi \geq 0$, $\varphi \in C_c(\Omega)$, one passes to the limit in the inequality

$$\int_\Omega \varphi\Big(B^n\, grad(v_n).grad(v_n)\Big)\, dx \geq \int_\Omega \varphi\Big(A^n\, grad(v_n).grad(v_n)\Big)\, dx. \quad (5.12)$$

The left side converges to $\int_\Omega \varphi(B^{eff}\, grad(v_\infty).grad(v_\infty))\, dx$ by the Div-Curl lemma; the $\liminf_n$ of the right side is $\geq \int_\Omega \varphi(A^{eff}\, grad(v_\infty).grad(v_\infty))\, dx$ by Proposition 13, and one obtains

$$\int_\Omega \varphi\Big(B^{eff}\, grad(v_\infty).grad(v_\infty)\Big)\, dx \geq \int_\Omega \varphi\Big(A^{eff}\, grad(v_\infty).grad(v_\infty)\Big)\, dx. \quad (5.13)$$

As v_∞ is arbitrary in $H^1_0(\Omega)$, one obtains $\varphi\, B^{eff} \geq \varphi\, A^{eff}$ a.e. in Ω, and varying φ gives the result.■

The second part of Proposition 13 is valid without any symmetry requirement, but the first part is not always true without symmetry. Similarly,

Proposition 14 is not true for a general sequence A^n, even if all the B^n are symmetric instead. Furthermore, in Proposition 12 we compared A^{eff} with A_+, and the symmetry hypothesis on A^n is also important there. Let us construct a counter-example for $N \geq 2$ (as every operator is symmetric if $N = 1$); define A^n by

$$A^n = I + \psi_n(x_1)(e_1 \otimes e_2 - e_2 \otimes e_1), \tag{5.14}$$

where

$$\psi_n \rightharpoonup \Psi_1;\ (\psi_n)^2 \rightharpoonup \Psi_2 \text{ in } L^\infty(R) \text{ weak } \star, \tag{5.15}$$

and choose the sequence ψ_n so that $\Psi_2 > (\Psi_1)^2$. The formula for layers (4.11) shows that

$$A^n \text{ H-converges to } A^{eff} = I + \Psi_1(e_1 \otimes e_2 - e_2 \otimes e_1) + \Big(\Psi_2 - (\Psi_1)^2\Big) e_2 \otimes e_2. \tag{5.16}$$

As $A_+ = I + \Psi_1(e_1 \otimes e_2 - e_2 \otimes e_1)$, this gives an example where $A^{eff} \leq A_+$ is false. One has $A^n \leq I$ for all n, while one does not have $A^{eff} \leq I$, and one has instead $A^{eff} \geq I$, as it must be from Proposition 14 because one has $A^n \geq I$ for all n. Finally, taking $u_n = u_\infty$ for all n, one has $\int_\Omega \varphi(A^n\, grad(u_n).grad(u_n))\, dx = \int_\Omega \varphi |\, grad(u_\infty)|^2\, dx$ but

$$\int_\Omega \Big(A^{eff}\, grad(u_\infty).grad(u_\infty)\Big)\, dx = \int_\Omega \varphi\Big(|grad(u_\infty)|^2 + (\Psi_2 - \Psi_1^2)\Big|\frac{\partial u_\infty}{\partial x_2}\Big|^2\Big)\, dx \tag{5.17}$$

showing that in this case (5.9) is not true.

Although in this course on Optimal Design all the problems considered are symmetric, I find useful to describe general results valid without symmetry assumptions: it is a good training in Mathematics to learn how to deal with general problems first so that one can easily deduce what to do on simpler problems, and it is usually difficult for those who have been only trained on simple special cases to understand what to do when they encounter a new situation. For that reason, I describe estimates which are useful for questions of perturbations and continuous dependence of H-limits with respect to parameters.

Lemma 15: Let $A \in M(\alpha, \beta; \Omega)$ and $D \in L^\infty\Big(\Omega; \mathcal{L}(R^N, R^N)\Big)$ with

$$||D||_{L^\infty(\Omega;\mathcal{L}(R^N,R^N))} \leq \delta < \alpha, \tag{5.18}$$

then

$$A + D \in M\Big(\alpha - \delta, \frac{\alpha\beta - \delta^2}{\alpha - \delta}; \Omega\Big). \tag{5.19}$$

Proof: Of course $((A + D)\xi.\xi) \geq \alpha|\xi|^2 - |D\xi|.|\xi| \geq (\alpha - \delta)|\xi|^2$. If A and D are symmetric, one has immediately $A + D \in M(\alpha - \delta, \beta + \delta; \Omega)$, but in the general case the replacement for β requires more technical computations. One first notices that $(A\xi.\xi) \geq \frac{1}{\beta}|A\xi|^2$ means $\left|A\xi - \frac{\beta}{2}\xi\right| \leq \frac{\beta}{2}|\xi|$, and drawing a picture in a Euclidean plane containing ξ and $A\xi$ helps understand how to obtain the above bound and also see why it is optimal. Analytically, defining L by $2L = \frac{\alpha\beta-\delta^2}{\alpha-\delta}$ one wants to show that for all ξ one has $|(A+D)\xi - L\xi| \leq L|\xi|$, and this will be a consequence of $|A\xi - L\xi| \leq (L-\delta)|\xi|$, i.e. of $|A\xi|^2 - 2L(A\xi.\xi) \leq (-2\delta L + \delta^2)|\xi|^2$, and by the definition of L one has $-2\delta L + \delta^2 = (\beta - 2L)\alpha$; then one notices that $|A\xi|^2 - 2L(A\xi.\xi) \leq (\beta - 2L)(A\xi.\xi)$ which is $\leq (\beta - 2L)\alpha|\xi|^2$ as $\beta - 2L \leq 0$.■

Proposition 16: If sequences $A^n \in M(\alpha,\beta;\Omega)$ and $B^n \in M(\alpha',\beta';\Omega)$ H-converge respectively to A^{eff} and B^{eff}, and $|B^n - A^n|_{\mathcal{L}(R^N,R^N)} \leq \varepsilon$ a.e. $x \in \Omega$ for all n, then

$$||B^{eff} - A^{eff}||_{L^\infty(\Omega;\mathcal{L}(R^N,R^N))} \leq \varepsilon\sqrt{\frac{\beta\,\beta'}{\alpha\,\alpha'}}. \tag{5.20}$$

Proof: For $f, g \in H^{-1}(\Omega)$, one solves

$$-div\Big(A^n\,grad(u_n)\Big) = f \text{ in } \Omega;\ -div\Big((B^n)^T\,grad(v_n)\Big) = g \text{ in } \Omega, \tag{5.21}$$

so u_n, v_n converge in $H^1_0(\Omega)$ weak respectively to u_∞, v_∞, and $A^n\,grad(u_n)$ and $(B^n)^T\,grad(v_n)$ converge in $L^2(\Omega;R^N)$ weak to $A^{eff}\,grad(u_\infty)$ and $(B^{eff})^T\,grad(v_\infty)$ respectively. By the Div-Curl lemma one knows that $\big(A^n\,grad(u_n).grad(v_n)\big)$ and $\big(grad(u_n).(B^n)^T\,grad(v_n)\big)$ converge in the sense of measures respectively to

$\big(A^{eff}\,grad(u_\infty), grad(v_\infty)\big)$ and $\big(grad(u_\infty).(B^{eff})^T\,grad(v_\infty)\big)$,

and so for every $\varphi \in C_c(\Omega)$ one has

$$\lim_n \int_\Omega \varphi\Big((B^n - A^n)\,grad(u_n).grad(v_n)\Big)\,dx = \int_\Omega \varphi\Big((B^{eff} - A^{eff})\,grad(u_\infty).grad(v_\infty)\Big)\,dx. \tag{5.22}$$

Choosing moreover $\varphi \geq 0$, and defining

$$X = \left|\int_\Omega \varphi\Big((B^{eff} - A^{eff})\,grad(u_\infty).grad(v_\infty)\Big)\,dx\right| \tag{5.23}$$

one deduces

$$X \leq \varepsilon \limsup_n \int_\Omega \varphi|grad(u_n)|\,|grad(v_n)|\,dx. \tag{5.24}$$

Then $|grad(u_n)|\,|grad(v_n)| \leq a\,\alpha|grad(u_n)|^2 + b\,\alpha'|grad(v_n)|^2$ when $4a\,b\,\alpha\,\alpha' \geq 1$, and as $A^n \in M(\alpha,\beta;\Omega)$ and $B^n \in M(\alpha',\beta';\Omega)$ one has

$$X \leq \varepsilon \limsup_n \int_\Omega \varphi \Big[a\Big(A^n\, grad(u_n).grad(u_n)\Big) + b\Big(B^n\, grad(v_n).grad(v_n)\Big)\Big]\, dx, \tag{5.25}$$

which gives

$$X \leq \varepsilon \int_\Omega \varphi \Big[a\Big(A^{eff}\, grad(u_\infty).grad(u_\infty)\Big) + b\Big(B^{eff}\, grad(v_\infty).grad(v_\infty)\Big)\Big]\, dx, \tag{5.26}$$

and therefore

$$X \leq \varepsilon\, a\, \beta \int_\Omega \varphi |grad(u_\infty)|^2\, dx + \varepsilon\, b\, \beta' \int_\Omega \varphi |grad(v_\infty)|^2\, dx. \tag{5.27}$$

As this inequality is true for every $\varphi \geq 0$, $\varphi \in C_c(\Omega)$, one deduces

$$\Big|\Big((B^{eff} - A^{eff})\, grad(u_\infty).grad(v_\infty)\Big)\Big| \leq \varepsilon\Big(a\,\beta |grad(u_\infty)|^2 + b\,\beta' |grad(v_\infty)|^2\Big) \tag{5.28}$$

a.e. in Ω, and optimizing on rationals a and b satisfying $4a\, b\, \alpha\, \alpha' \geq 1$, one obtains

$$\Big|\Big((B^{eff} - A^{eff})\, grad(u_\infty).grad(v_\infty)\Big)\Big| \leq \varepsilon \sqrt{\frac{\beta\,\beta'}{\alpha\,\alpha'}} |grad(u_\infty)|\, |grad(v_\infty)|, \text{ in } \Omega, \tag{5.29}$$

and as u_∞ and v_∞ are arbitrary, one obtains (5.20).■

Proposition 17: Let P be an open set of R^p. Let A^n be a sequence defined on $\Omega \times P$, such that $A^n(\cdot,p) \in M(\alpha,\beta;\Omega)$ for each $p \in P$, and such that the mappings $p \to A^n(\cdot,p)$ are of class C^k (or real analytic) from P into $L^\infty(\Omega; \mathcal{L}(R^N, R^N))$, with bounds of derivatives up to order k independent of n. Then there exists a subsequence A^m such that for every $p \in P$ the sequence $A^m(\cdot,p)$ H-converges to $A^{eff}(\cdot,p)$ and $p \to A^{eff}(\cdot,p)$ is of class C^k (or real analytic) from P into $L^\infty(\Omega; \mathcal{L}(R^N, R^N))$.
Proof: One considers a countable dense set Π of P and, using a diagonal subsequence, one extracts a subsequence A^m such that for every $p \in \Pi$ the sequence $A^m(\cdot,p)$ H-converges to a limit $A^{eff}(\cdot,p)$. Using the fact that A is uniformly continuous on compact subsets of P and Proposition 16, one deduces then that $p \to A^{eff}(\cdot,p)$ is continuous from P into $L^\infty(\Omega; \mathcal{L}(R^N, R^N))$ and that for every $p \in P$ the sequence $A^m(\cdot,p)$ H-converges to $A^{eff}(\cdot,p)$.

Defining the operators $\mathcal{A}_m(p)$ from $V = H^1_0(\Omega)$ into $V' = H^{-1}(\Omega)$ by $\mathcal{A}_m(p)v = -div\big(A^m(\cdot,p)\, grad(v)\big)$, one finds that the mappings $p \to \mathcal{A}_m(p)$ are of class C^k (or real analytic) from P to $\mathcal{L}(V, V')$ and similarly

$p \to \left(\mathcal{A}_m(p)\right)^{-1}$ are of class C^k (or real analytic) from P to $\mathcal{L}(V',V)$, and finally the operators R_m defined by $R_m v = A^m(\cdot,p)\, grad(v_m)$ with v_m defined by $\mathcal{A}_m(v_m) = \mathcal{A}_{eff}(v)$ are of class C^k (or real analytic) from P into $\mathcal{L}(V; L^2(\Omega; R^N))$; all the bounds of derivatives up to order k being independent of m so that the limit inherits of the same bounds and as $R^{eff}v = A^{eff}(\cdot,p)\, grad(v)$, one deduces therefore that $p \to A^{eff}(\cdot,p)$ is of class C^k (or real analytic) from P into $L^\infty(\Omega; \mathcal{L}(R^N, R^N))$.■

As I mentioned, some of the preceding results will not be used in this course but they serve as a more general training on questions of Homogenization, but let us now come back to the method that François MURAT and I were following in the early 70s in order to obtain some information on effective coefficients: we had computed the convex hull of K defined in (5.1), we had used the relation (5.2) (where a first version of the Div-Curl lemma had been used), and we concluded that A^{eff} must satisfy (5.3); then we had noticed a quicker way to prove the result by the convexity argument of Lemma 11, but I will describe in (5.34)/(5.40) how to carry out the computations of the convex hull in a way that will be useful for further generalizations. We were considering the special case where $a^n = \chi_n\alpha + (1-\chi_n)\beta$, with $0 < \alpha < \beta < \infty$, and χ_n being a sequence of characteristic functions converging in $L^\infty(\Omega)$ weak $\star$ to θ, so that $\theta(x)$ represent the proportion of the material of conductivity α near the point x; in that case we denote

$$\lambda_+(\theta) = \theta\,\alpha + (1-\theta)\beta; \quad \frac{1}{\lambda_-(\theta)} = \frac{\theta}{\alpha} + \frac{1-\theta}{\beta}, \tag{5.30}$$

and the analogue of Proposition 12 told us that the eigenvalues of A^{eff} were in the interval $[\lambda_-(\theta), \lambda_+(\theta)]$. In dimension $N=2$, the eigenvalues λ_1, λ_2 of A^{eff} must satisfy

$$\lambda_-(\theta) = \frac{\alpha\,\beta}{\theta\,\beta + (1-\theta)\alpha} \le \lambda_1, \lambda_2 \le \theta\,\alpha + (1-\theta)\beta = \lambda_+(\theta), \tag{5.31}$$

defining a square in the (λ_1,λ_2) plane, and varying θ between 0 and 1, the union of these squares gave us the constraint that any mixture of two isotropic materials with conductivity α and β must satisfy the inequalities

$$\begin{aligned} \alpha \le \frac{\alpha\,\beta}{\alpha+\beta-\max\{\lambda_1,\lambda_2\}} &\le \min\{\lambda_1,\lambda_2\} \le \\ &\le \max\{\lambda_1,\lambda_2\} \le \alpha+\beta - \frac{\alpha\,\beta}{\min\{\lambda_1,\lambda_2\}} \le \beta, \end{aligned} \tag{5.32}$$

and we showed that the characterization (5.32) is optimal (while the characterization (5.31) is not optimal). For this we used the formula for layers, which in dimension N creates a tensor A^{eff} with one eigenvalue equal to $\lambda_-(\theta)$ with the eigenvector orthogonal to the layers, and $(N-1)$ eigenvalues equal to $\lambda_+(\theta)$ with the eigenvectors parallel to the layers, so that if $N=2$ all the points on the boundary of the set (5.32) are obtained by layered materials. Then we took one anisotropic material with eigenvalues (γ,δ) and changing the orientation of the material gave two diagonal tensors $A_1 = (\gamma,\delta)$ and $A_2 = (\delta,\gamma)$, and using layers orthogonal to the x_1 axis, using proportion η of material A_1 and proportion $(1-\eta)$ of material A_2 gave materials with diagonal tensors $\left(\frac{\gamma\,\delta}{\eta\,\delta+(1-\eta)\gamma}, \eta\,\delta + (1-\eta)\gamma\right)$, having therefore determinant equal to $\gamma\,\delta$, so that by taking $\gamma = \lambda_-(\theta)$ and $\delta = \lambda_+(\theta)$ showed that the piece of equilateral hyperbola joining the two (non isotropic) diagonal corners of the square defined by (5.31) were attainable by mixtures and the union of these pieces of hyperbolas covered the set (5.32) (independently, Alain BAMBERGER had noticed that in dimension $N = 2$ if one mixes materials with $det(A) = c$ then one has $det(A^{eff}) = c$.

In doing so, we had followed the intuition that if one mixes a few materials which themselves have been obtained as mixtures of some initial materials, then the result can be obtained by mixing directly the initial materials in an adapted way. Mathematically, it is here that the metrizability property mentioned before is important: one tries to define the closure for a metrizable topology of a set containing the tensors of the form $(\chi\,\alpha + (1-\chi)\beta)I$ with χ being the characteristic function of an arbitrary measurable set (or of a smoother set like an open set, for example), and one identifies then some first generation sets contained in the sequential closure of the initial set, then one identifies some second generation sets contained in the sequential closure of some first generation sets, and one repeats the process finitely many times, and because the topology is metrizable every set constructed is included in the sequential closure of the initial set. However, the local character of H-convergence proved in Proposition 10 has also been used: if ω_j, $j \in J$, is a countable collection of disjoint open[43] sets such that Ω is the union of all the ω_j plus a set of measure 0, and if for each j one has a sequence $A_{jn} \in M(\alpha,\beta;\omega_j)$ which H-converges to A_j^{eff}, then one can glue these pieces together, defining $A_n = \sum_j \chi_{\omega_j} A_{jn}$, which belongs to $M(\alpha,\beta;\Omega)$ and by Theorem 6 a subsequence H-converges to some A^{eff}, and Proposition 10 asserts that the restriction of A^{eff} to ω_j is A_j^{eff}, and therefore all the sequence H-converges to $A^{eff} = \sum_j \chi_{\omega_j} A_j^{eff}$. The preceding arguments enabled us to create in the desired H-closure (or G-closure, as we were working with symmetric tensors) any measurable tensor taking a constant value belonging to the set (5.32) for each of the disjoint open sets ω_j, and the conclusion came from the remark that if a sequence $A^n \in M(\alpha,\beta;\Omega)$ converges almost everywhere to A^∞ then it H-converges to A^∞.

This is the result which I used in 1974 in order to compute necessary conditions of optimality for classical solutions [Ta2]. Although François MURAT and I had followed the same type of construction that Antonio MARINO and

[43] In his work on G-convergence, Sergio SPAGNOLO uses MEYERS's regularity theorem [Me], and he can use disjoint measurable sets.

Sergio SPAGNOLO had used in [Ma&Sp], the reason why we had been able to go further was that we had obtained the necessary condition of Proposition 12, which had given us in dimension $N = 2$ the conditions (5.31) and (5.32). We were not able at the time to obtain the characterization in dimension $N \geq 3$, or even in the case $N = 2$ we could not find the optimal characterization improving (5.31), when one imposes to use given proportions. The first step towards the solution of these more general questions was my introduction at the end of 1977 of a new method for obtaining bounds for effective coefficients [Ta6]; this method makes use of the notion of correctors in Homogenization and it requires the choice of adapted functionals for which one checks the hypotheses by applying the Compensated Compactness theory. Before describing these new ingredients, I want to show what a more precise analysis of (5.2) with the definition (5.1) gives.[44]

In the case $A^n = a^n I$ which we had investigated first, we knew that the $L^\infty(\Omega)$ weak $\star$ limits of a^n and $\frac{1}{a^n}$, denoted respectively by a_+ and $\frac{1}{a_-}$, were needed in the case where a^n only depends upon one linear combination of coordinates. We had constructed sequences $E^n = grad(u_n)$ and $D^n = a^n E^n = a^n grad(u_n)$ converging in $L^2(\Omega; R^N)$ weak, respectively to E^∞ and D^∞, and we had shown by an integration by parts (instead of the Div-Curl lemma which we discovered later) that $a^n|E^n|^2$ converges in the sense of measures to $(D^\infty.E^\infty)$. Therefore we decided to look at the convex hull in R^{2N+3} of the set K defined by (5.1), and one may wonder if it changes much to add other functions of a^n to the list and look at the convex hull of the subset $\widetilde{K}$ of R^{2N+k+1} defined by

$$\widetilde{K} = \Big\{\Big(E, a\,E, a|E|^2, f_1(a), \ldots, f_k(a)\Big) | E \in R^N, a \in [\alpha, \beta]\Big\}, \tag{5.33}$$

and $f_1, \ldots, f_k$ are k given continuous functions on $[\alpha, \beta]$. The computations will show that the $L^\infty(\Omega)$ weak $\star$ limits of a^n and $\frac{1}{a^n}$ appear naturally. One characterization of the closed convex hull of $\widetilde{K}$ requires considering quantities of the form $(E.u_1) + (a\,E.u_2) + C_0\,a\,|E|^2 + \sum_{i=1}^k C_i f_i(a)$, where $u_1, u_2 \in R^N$ and $C_0, \ldots, C_k \in R$, in order to compute their infimum for $E \in R^N$ and $a \in [\alpha, \beta]$. One has then to consider the infimum of $(E.u_1) + (a\,E.u_2) + C_0\,a\,|E|^2$ for $E \in R^N$, and this infimum is $-\infty$ if $C_0 < 0$, or if $C_0 = 0$ and either u_1 or u_2 is not 0; therefore the typical formula is

[44] Although we could have done the following computations in the early 70s, I only noticed Lemma 18 and Lemma 19 while I was working on a set of lecture notes for my CBMS-NSF course in Santa Cruz in the Summer 1993 (I have abandoned this project since), and for preparing my lecture for a meeting in Nice in 1995, where I applied our method for the case of mixing arbitrary anisotropic materials [Ta14]; I will describe later this extension (Lemma 42), which is based on Lemma 18.

$$\min_{E\in R^N}\Big(a|E|^2-2(E.v)-2(a\,E.w)\Big)=-\frac{|v+a\,w|^2}{a}=$$
$$=-\frac{|v|^2}{a}-2(v.w)-a\,|w|^2 \text{ for } v,w\in R^N. \tag{5.34}$$

Taking the limit of

$$a^n|E^n|^2-2(E^n.v)-2(a^n\,E^n.w)\geq-\frac{|v|^2}{a^n}-2(v.w)-a^n\,|w|^2, \tag{5.35}$$

one obtains

$$(D^\infty.E^\infty)-2(E^\infty.v)-2(D^\infty.w)\geq-\frac{|v|^2}{a_-}-2(v.w)-a_+|w|^2, \tag{5.36}$$

and that inequality is true for every $v,w\in R^N$. The best choice for v and w in (5.36) is obtained by solving the system

$$\frac{v}{a_-}+w=E^\infty$$
$$v+a_+w=D^\infty, \tag{5.37}$$

and there is a difficulty if $a_-=a_+$, as one needs to have $D^\infty=a_+E^\infty$; this is not surprising as one always has $a_-\leq a_+$, with equality if and only if a^n converges in $L^1_{loc}(\Omega)$ strong to a_+ (i.e. in $L^p_{loc}(\Omega)$ strong for every $p<\infty$, because a^n is bounded in $L^\infty(\Omega)$). If $a_-<a_+$, the solution of (5.37) is given by

$$v=\frac{a_-(a_+E^\infty-D^\infty)}{a_+-a_-}$$
$$w=\frac{D^\infty-a_-E^\infty}{a_+-a_-}, \tag{5.38}$$

and (5.34) leads to

$$(a_+-a_-)(D^\infty.E^\infty)-$$
$$-a_-\Big(E^\infty.(a_+E^\infty-D^\infty)\Big)-\Big(D^\infty.(D^\infty-a_-E^\infty)\Big)\geq 0, \tag{5.39}$$

i.e.

$$(D^\infty-a_-E^\infty.D^\infty-a_+E^\infty)\leq 0, \tag{5.40}$$

which means that D^∞ belongs to the closed ball with diameter $[a_-E^\infty,a_+E_\infty]$, and the formula is still valid if $a_-=a_+$. Having shown already that $D^\infty=A^{eff}E^\infty$ for a symmetric matrix A^{eff}, the validity of (5.40) for every $E^\infty\in R^N$ is equivalent to A^{eff} having all its eigenvalues between a_- and a_+ (defining $M=A^{eff}-\frac{\alpha+\beta}{2}I$, the condition becomes $(M\,z.z)\leq\frac{\beta-\alpha}{2}|z|^2$ for all $z\in R^N$, equivalent to M having all its eigenvalues with modulus $\leq\frac{\beta-\alpha}{2}$).

A slightly different point of view for dealing with (5.40) is that if $0 < b \le a < \infty$ and two vectors $E, D \in R^N$ satisfy $(D - a\,E.D - b\,E) \le 0$, then there exists a symmetric B having its eigenvalues between b and a such that $D = B\,E$. Indeed, assuming $b < a$, let $\gamma = \frac{a+b}{2}, \delta = \frac{a-b}{2}$ and $F = \frac{1}{\delta}(D - \gamma\,E)$, one has $|F| \le |E|$ and one wants to find a symmetric C of norm ≤ 1, equal to $\frac{1}{\delta}(B - \gamma\,I)$, such that $C\,E = F$, and there is actually such a C satisfying $||C|| \le \frac{|F|}{|E|}$ if $E \neq 0$; in the case of two unit vectors e_1, e_2 which are not parallel, the basic construction for finding a symmetric contraction mapping e_1 onto e_2 is to consider the symmetry which has $e_1 \pm e_2$ as eigenvectors with eigenvalues ± 1 and if $N > 2$ the subspace orthogonal to e_1 and e_2 as eigenspace with an eigenvalue between -1 and $+1$. If $N \ge 2$, the preceding construction shows that if $E \neq 0$ and $(D - a\,E.D - b\,E) = 0$ then $D = B\,E$ for a symmetric B having one eigenvalue b and the $N-1$ other eigenvalues equal to a, and such a case appears when one uses layerings.

This helps proving the following result.

Lemma 18: If $0 < \alpha' \le b^n \le a^n \le \beta' < \infty$ a.e. in Ω, $E^n, D^n \in L^2(\Omega; R^N)$ for all n, and

$$(D^n - a^n\,E^n.D^n - b^n\,E^n) \le 0 \text{ a.e. in } \Omega, \tag{5.41}$$

with

$$\begin{aligned} &a^n \rightharpoonup a^\infty \text{ in } L^\infty(\Omega) \text{ weak } \star,\\ &\frac{1}{b^n} \rightharpoonup \frac{1}{b^\infty} \text{ in } L^\infty(\Omega) \text{ weak } \star,\\ &E^n \rightharpoonup E^\infty \text{ in } L^2(\Omega; R^N) \text{ weak},\\ &D^n \rightharpoonup D^\infty \text{ in } L^2(\Omega; R^N) \text{ weak},\\ &(E^n.D^n) \rightharpoonup (E^\infty.D^\infty) \text{ in the sense of measures,} \end{aligned} \tag{5.42}$$

then one has

$$(D^\infty - a^\infty\,E^\infty.D^\infty - b^\infty\,E^\infty) \le 0 \text{ a.e. in } \Omega. \tag{5.43}$$

Proof: By the preceding analysis, one has $D^n = B^n\,E^n$ a.e. in Ω, with B^n symmetric and having its eigenvalues in $[b^n, a^n]$. Therefore for $v, w \in R^N$ one has

$$\begin{aligned} &(D^n.E^n) - 2(E^n.v) - 2(D^n.w) = (B^n\,E^n.E^n) - 2(E^n.v + B^n\,w) \ge\\ &\ge -\Big(((B^n)^{-1}(v + B^n\,w).(v + B^n\,w)\Big) =\\ &= -\Big(((B^n)^{-1}v.v\Big) - 2(v.w) - (B^n\,w.w) \ge -\frac{1}{b^n}|v|^2 - 2(v.w) - a^n|w|^2, \end{aligned} \tag{5.44}$$

a.e. in Ω; after using test functions $\varphi \in C_c(\Omega)$ with $\varphi \ge 0$ in Ω, one obtains

$$\begin{aligned} &(D^\infty.E^\infty) - 2(E^\infty.v) - 2(D^\infty.w) \ge\\ &\qquad \ge -\frac{1}{b^\infty}|v|^2 - 2(v.w) - a^\infty|w|^2 \text{ a.e. in } \Omega, \end{aligned} \tag{5.45}$$

and as this is the same inequality than (5.36) with a_- replaced by b^∞ and a_+ replaced by a^∞, one deduces the analogue of (5.40), which is (5.43).■

Lemma 18 can also be derived as a consequence of the following result.

Lemma 19: Define the real function Φ on $R^N \times R^N \times (0,\infty) \times (0,\infty)$ by

$$\Phi(E,D,a,b) = \begin{cases} \frac{1}{a-b}(D - a\,E.D - b\,E) \text{ if } b < a, \\ 0 \text{ if } b = a \text{ and } D = a\,E, \\ +\infty \text{ otherwise,} \end{cases} \tag{5.46}$$

then Φ is a convex function in $\big(E, D, (E.D), a, \frac{1}{b}\big)$, and more precisely

$$\Phi(E,D,a,b) = \sup_{v,w\in R^N} \Big(-(D.E) + 2(E.v) + 2(D.w) - \frac{1}{b}|v|^2 - 2(v.w) - a|w|^2 \Big). \tag{5.47}$$

Proof: Indeed in the case $0 < b < a$ the quadratic form $-\frac{1}{b}|v|^2 - 2(v.w) - a|w|^2$ is negative definite, and the supremum is attained when (v,w) solves the analogue of (5.37), $\frac{v}{b} + w = E$ and $v + a\,w = D$, which gives the analogue of (5.38), $v = \frac{b(a\,E-D)}{a-b}$ and $w = \frac{D-b\,E}{a-b}$, and the value of the supremum is the quantity defined in (5.46). If $b > a > 0$ the quadratic form is not definite and the supremum is $+\infty$. If $b = a > 0$, the quantity to maximize is $-(D.E) + 2(D - a\,E.w) + 2(E.v + a\,w) - \frac{1}{a}|v + a\,w|^2$, and the supremum is $+\infty$ if $D - a\,E \neq 0$, and if $D = a\,E$ it is $-(D.E) + a|E|^2$, which is 0 in that case.■

7. Correctors in Homogenization

In 1975, I heard Ivo BABUŠKA mention the importance of amplification factors: in real elastic materials there is a threshold above which nonelastic effects (plastic deformation or fracture) usually appear. In a mixture it is not the average stress which is relevant but the local stress, and therefore one must know an amplification factor for computing the local stresses from the average stress; I did keep this comment in mind when I defined my correctors.[45] I first consider the case of layered media, for which one can prove a stronger result than for the general case of H-convergence.[46]

[45] I had then heard Jacques-Louis LIONS describe his computations for the periodically modulated case, which he had studied with Alain BENSOUSSAN and George PAPANICOLAOU [Be&Li&Pa], and I found that his notation with χ_{ij} created an unnecessary chaos with indices, which I decided to avoid.

[46] I did the general framework in 1975 or 1976, but I only noticed later the stronger result for the case of layered media, because of a lecture at a meeting in Luminy in the Summer 1993 [Ta13], where I considered functionals depending upon the gradient, and for the reasons mentioned in footnote 44.

Proposition 20: Let a sequence $A^n \in M(\alpha,\beta;\Omega)$ be such that it only depends upon x_1 and that it H-converges to A^{eff}. If a sequence u_n converges in $H^1_{loc}(\Omega)$ weak to u_∞ and $div(A^n\, grad(u_n))$ stays in a compact of $H^{-1}_{loc}(\Omega)$ strong, then

$$\begin{aligned}&\Big(A^n\, grad(u_n)\Big)_1 \to \Big(A^{eff}\, grad(u_\infty)\Big)_1 \text{ in } L^2_{loc}(\Omega) \text{ strong},\\ &\frac{\partial u_n}{\partial x_i} \to \frac{\partial u_\infty}{\partial x_i} \text{ in } L^2_{loc}(\Omega) \text{ strong, for } i=2,\ldots,N.\end{aligned} \tag{6.1}$$

If one defines the sequence $P_n \in L^\infty(\Omega;\mathcal{L}(R^N,R^N))$ by

$$\begin{aligned}(P_n)_{11} &= \frac{(A^{eff})_{11}}{(A^n)_{11}},\\ (P_n)_{1j} &= \frac{(A^{eff})_{1j}}{(A^{eff})_{11}} - \frac{(A^n)_{1j}}{(A^n)_{11}} \text{ for } j=2,\ldots,N,\\ (P_n)_{ij} &= \delta_{ij} \text{ for } i=2,\ldots,N, \text{ and } j=1,\ldots,N,\end{aligned} \tag{6.2}$$

then one has

$$grad(u_n) - P_n\, grad(u_\infty) \to 0 \text{ in } L^2_{loc}(\Omega;R^N) \text{ strong.} \tag{6.3}$$

Proof: If one denotes $E^n = grad(u_n)$ and $D^n = A^n\, grad(u_n)$ and if one uses the vector G^n and the tensor $B^n = \Phi(A^n)$ introduced after (4.11), the statement (6.1) means that G^n converges in $L^2_{loc}(\Omega;R^N)$ strong to G^∞. In order to prove this statement, one first notices that

$$\Big(B^n(G^n - G^\infty).G^n - G^\infty\Big) \text{ converges to 0 in the sense of measures.} \tag{6.4}$$

Indeed, $(B^n\, G^n.G^n) = (D^n.E^n)$ which converges in the sense of measures to $(D^\infty.E^\infty) = (B^\infty\, G^\infty.G^\infty)$ by the Div-Curl lemma, and as G^n does not oscillate in x_1 and B^n only depends upon x_1 and converges in $L^\infty(\Omega;\mathcal{L}(R^N,R^N))$ weak $\star$ to $B^\infty = \Phi(A^{eff})$, both $(B^n\, G^n.G^\infty)$ and $(B^n\, G^\infty.G^n)$ converge in $L^1_{loc}(\Omega)$ weak to $(B^\infty\, G^\infty.G^\infty)$. Then one notices that there exists $\gamma > 0$ such that

$$(B^n\, \lambda.\lambda) \geq \gamma|\lambda|^2 \text{ for all } \lambda \in R^N, \tag{6.5}$$

as one may take $\gamma = \frac{\alpha}{\beta^2+1}$ for example, as $(B^n\, G^n.G^n) = (A^n\, E^n.E^n) \geq \alpha|E^n|^2$ and $|G^n|^2 \leq |D^n|^2 + |E^n|^2 \leq (\beta^2+1)|E^n|^2$. Then (6.1) follows from writing

$$\frac{\partial u_n}{\partial x_1} = \frac{1}{(A^n)_{11}}\left[\Big(A^n\, grad(u_n)\Big)_1 - \sum_{j=2}^N (A^n)_{1j}\frac{\partial u_n}{\partial x_j}\right], \tag{6.6}$$

and using the fact that $\frac{1}{(A^n)_{11}}$ is uniformly bounded by $\frac{1}{\alpha}$.■

I have mentioned after (4.11) that the formula for computing the effective properties of a layered material is valid under a much weaker hypothesis than the one used for the general theory of H-convergence, namely A^n bounded and $A^n_{11} \geq \alpha > 0$ a.e. in Ω for layers in the direction x_1.[47] The main reason for using LAX-MILGRAM lemma in the general framework is that it enables to construct sequences $E^n = grad(u_n)$ converging in $L^2(\omega, R^N)$ weak to a constant vector with $D^n = (A^n)^T E^n$ such that $div(D^n)$ stays in a compact of $H^{-1}_{loc}(\Omega)$; such an abstract construction is not needed in the case of layers because one can immediately write down explicitly a similar sequence; indeed taking G to be a constant vector and defining $O^n = B^n(x_1)G$, gives a vector E^n which is a gradient as $(E^n)_1$ only depends upon x_1 and E_i is constant for $i = 2, \ldots, N$, and the vector $D^n = A^n(x_1)E^n$ has divergence 0 as $(D^n)_1$ is constant and $(D^n)_i$ only depends upon x_1 for $i = 2, \ldots, N$; the sequence E^n indeed converges in $L^2(\Omega; R^N)$ weak to a limit E^∞, and E^∞ is not constant as $(E^\infty)_1$ may indeed be a nonconstant function of x_1, but it is not so important that the limit be constant as what is necessary for the argument to work is that one can construct N such sequences with limits which are linearly independent, and this is true here.

However, the strong convergence result of Proposition 20 cannot be true without assuming an ellipticity condition: let A be a constant tensor with $A_{11} > 0$ and $(A\,\xi.\xi) = 0$ for some $\xi \neq 0$ (and $\xi \neq e_1$, as $A_{11} > 0$), then the sequence u_n defined by $u_n(x) = \frac{1}{n}\sin\big(n(\xi.x)\big)$ satisfies $div\big(A\,grad(u_n)\big) = 0$ and G^n converges in $L^2(\Omega; R^N)$ weak to 0 but it does not converge in $L^2_{loc}(\Omega; R^N)$ strong. It is therefore natural to assume that $(A\,\xi.\xi)$ does not vanish and the hypothesis $A^n \in M(\alpha, \beta; \Omega)$ appears then as a natural restriction when one wants the argument to apply for every direction of layers.

In the general framework of H-convergence, one cannot prove a result as strong as Proposition 20, and the basic result is the following.

Theorem 21: Let a sequence $A^n \in M(\alpha, \beta; \Omega)$ H-converge to A^{eff}. Then there is a subsequence A^m and an associated sequence P^m of correctors such that

$$
\begin{aligned}
&P^m \rightharpoonup I \text{ in } L^2\Big(\Omega; \mathcal{L}(R^N, R^N)\Big) \text{ weak},\\
&A^m\, P^m \rightharpoonup A^{eff} \text{ in } L^2\Big(\Omega; \mathcal{L}(R^N, R^N)\Big) \text{ weak},\\
&\text{curl}(P^m\,\lambda) = 0 \text{ in } \Omega \text{ for all } \lambda \in R^N,\\
&div(A^m\, P^m\,\lambda) \text{ stays in a compact of } H^{-1}_{loc}(\Omega) \text{ strong}, \forall\lambda \in R^N.
\end{aligned}
\tag{6.7}
$$

For any sequence u_m converging in $H^1_{loc}(\Omega)$ weak to u_∞ with $div\big(A^m\,grad(u_m)\big)$ staying in a compact of $H^{-1}_{loc}(\Omega)$ strong, one has

[47] It applies to the hyperbolic equation $\frac{\partial}{\partial t}\Big(\rho_n(x)\frac{\partial u_n}{\partial t}\Big) - \frac{\partial}{\partial x}\Big(a^n(x)\frac{\partial u_n}{\partial x}\Big) = f$ if $a^n \geq \alpha$ and $\rho_n \geq \alpha$ a.e. for all n: if ρ_n converges to ρ_∞ and $\frac{1}{a^n}$ converges in $L^\infty(\Omega)$ weak $\star$ to $\frac{1}{a^{eff}}$, then the effective equation has the same form, with ρ_n and a^n replaced by ρ_∞ and a^{eff}.

$$grad(u_m) - P^m\, grad(u_\infty) \to 0 \text{ in } L^1_{loc}(\Omega; R^N) \text{ strong.} \tag{6.8}$$

Proof. For an open set Ω' of R^N containing $\overline{\Omega}$, one extends A^n by $\alpha\, I$ in $\Omega'\setminus\overline{\Omega}$ and one extracts a subsequence A^m which H-converges to a limit on Ω'; one also denotes this limit A^{eff}, and by Proposition 10 it must be an extension to Ω' of the H-limit already defined on Ω. Then for $i = 1, \ldots, N$, one chooses a function $\varphi_i \in H^1_0(\Omega')$ such that $grad(\varphi_i) = e_i$ on Ω, and one defines $P^m\, e_i = grad(v_m)$ in Ω, where $v_m \in H^1_0(\Omega')$ is the solution of $div\big(A^m\, grad(v_m) - A^{eff}\, grad(\varphi_i)\big) = 0$ in Ω'. By this construction v_m converges in $H^1_0(\Omega')$ weak to v_∞, solution of $div\big(A^{eff}\, grad(v_\infty) - A^{eff}\, grad(\varphi_i)\big) = 0$ in Ω', i.e. $v_\infty = \varphi_i$, and therefore $grad(v_m)$ and $A^m\, grad(v_m)$ converge in $L^2(\Omega'; R^N)$ weak, respectively to $grad(\varphi_i)$ and $A^{eff}\, grad(\varphi_i)$, i.e. $P^m\, e_i$ and $A^m\, P^m\, e_i$ converge in $L^2(\Omega; R^N)$ weak, respectively to e_i and $A^{eff}\, e_i$. By repeating this construction for $i = 1, \ldots, N$, one obtains a sequence P^m satisfying (6.7). Actually the construction gives P^m satisfying a more precise condition than (6.7), as one has $div(A^m\, P^m\, \lambda - A^{eff}\, \lambda) = 0$ in Ω for all $\lambda \in R^N$, but it is useful to impose only (6.7) as there may be slightly different definitions for P^n that may not satisfy this supplementary requirement.[48] The sequence $P^m\, grad(u_\infty)$ is bounded in $L^1(\Omega; R^N)$ because the sequence of correctors P^m is bounded in $L^2\big(\Omega; \mathcal{L}(R^N, R^N)\big)$. In order to prove (6.8), one chooses $g \in C(\Omega; R^N)$, $\varphi \in C_c(\Omega)$, and one computes the limit of

$$X_m = \int_\Omega \varphi\Big(A^m(grad(u_m) - P^m\, g).grad(u_m) - P^m\, g\Big)\, dx. \tag{6.9}$$

Writing $g = \sum_k g_k e_k$, one expands the integrand in X_m and by (6.7) the Div-Curl lemma applies to each term: $\big(A^m\, grad(u_m).grad(u_m)\big)$ converges in the sense of measures to $\big(A^{eff}\, grad(u_\infty).grad(u_\infty)\big)$, and similarly $(A^m\, grad(u_m).P^m\, e_l)$ converges to $(A^{eff}\, grad(u_\infty).e_l)$, $\big(A^m\, P^m\, e_k.grad(u_m)\big)$ converges to $\big(A^{eff}\, e_k.grad(u_\infty)\big)$ and $(A^m\, P^m\, e_k.P^m\, e_l)$ converges to $(A^{eff}\, e_k.e_l)$; as each g_k is continuous and φ has compact support, one deduces that

$$X_m \to X_\infty = \int_\Omega \varphi\Big(A^{eff}(grad(u_\infty) - g).grad(u_\infty) - g\Big)\, dx. \tag{6.10}$$

If $u_\infty \in C^1(\Omega)$ one can take $g = grad(u_\infty)$, and one deduces that $X_m \to 0$; by taking $0 \le \varphi \le 1$ and $\varphi = 1$ on a compact K of Ω, one deduces that $grad(u_m) - P^m\, grad(u_\infty) \to 0$ in $L^2(K; R^N)$ strong for every compact of Ω. If $u_\infty \in H^1(\Omega)$, one cannot use $g = grad(u_\infty)$ in general, and therefore one approaches $grad(u_\infty)$ by $g \in C(\Omega; R^N)$ in order to have

[48] In the case of layers for example, it is more natural to look for P^m depending only upon x_1, and the correctors defined in (6.2) satisfy $div(A^m\, P^m\, \lambda) = 0$ in Ω, even if $div(A^{eff}\, \lambda)$ does depend upon x_1. In the periodic case, it is more natural to ask for P^m to be periodic.

$$||grad(u_\infty) - g||_{L^2(\Omega;R^N)} \le \varepsilon, \tag{6.11}$$

implying

$$X_\infty \le \beta \int_\Omega |grad(u_\infty) - g|^2\, dx \le \beta\, \varepsilon^2. \tag{6.12}$$

By (6.10) and (6.12) one has

$$\limsup_m \int_K \alpha |grad(u_m) - P^m\, g|^2\, dx \le \beta\, \varepsilon^2, \tag{6.13}$$

from which one deduces

$$\limsup_m \int_K |grad(u_m) - P^m\, g|\, dx \le \varepsilon \sqrt{\frac{\beta\, meas(K)}{\alpha}}. \tag{6.14}$$

Using (6.11) one deduces that

$$\limsup_m \int_K |grad(u_m) - P^m\, grad(u_\infty)|\, dx \le \varepsilon \sqrt{\frac{\beta\, meas(K)}{\alpha}} + C\, \varepsilon, \tag{6.15}$$

where C is an upper bound for the norm of P^m in $L^2(\Omega; \mathcal{L}(R^N, R^N))$; therefore $grad(u_m) - P^m\, grad(u_\infty) \to 0$ in $L^1(K; R^N)$ strong and as K is an arbitrary compact of Ω, one obtains the desired result (6.8).■

Using better integrability property for $grad(u_\infty)$, and MEYERS's regularity theorem [Me], one can prove that $grad(u_m) - P^m\, grad(u_\infty)$ converges in $L^p_{loc}(\Omega; R^N)$ strong to 0 for some $p > 1$. If P^m is bounded in $L^q_{loc}(\Omega; \mathcal{L}(R^N, R^N))$ for some $q > 2$, as can be shown using MEYERS's regularity theorem [Me], or directly as in the case of layers where one can take $q = \infty$, and $grad(u_\infty) \in L^r(\Omega; R^N)$ for some $r \ge 2$, then one can take $g \in L^s(\Omega; R^N)$ in (6.9) and (6.10) with $s = \frac{2q}{q-2}$, and if g is near $grad(u_\infty)$ in $L^r(\Omega; R^N)$ or equal to $grad(u_\infty)$ if $r \ge s$, then $P^m(grad(u_\infty) - g)$ is small in $L^t_{loc}(\Omega; R^N)$ with $t = \frac{qr}{q+r}$; one can then choose $p = \min\{s, t\}$.

Although the following results will not be used in this course, it is important to realize that correctors are important even for finding the effective equation for similar equations obtained by adding lower order terms; for simplicity, I will ignore the advantage of using MEYERS's regularity theorem [Me].

Proposition 22: Let a sequence $A^n \in M(\alpha, \beta; \Omega)$ H-converge to A^{eff}, let c^n be a sequence bounded in $L^p(\Omega; R^N)$ with $p > N$ if $N \ge 2$, $p = 2$ if $N = 1$, and assume that a sequence u_n converges in $H^1_{loc}(\Omega)$ weak to u_∞ and satisfies

$$-div\Big(A^n\, grad(u_n)\Big) + \Big(c^n.grad(u_n)\Big) \to f \text{ in } H^{-1}_{loc}(\Omega) \text{ strong}. \tag{6.16}$$

Then u_∞ satisfies the equation

$$-div\Big(A^{eff}\, grad(u_\infty)\Big) + \Big(c^{eff}.grad(u_\infty)\Big) = f \text{ in } \Omega, \tag{6.17}$$

with an effective coefficient c^{eff} such that for a subsequence

$$(P^m)^T c^m \rightharpoonup c^{eff} \text{ in } L^{2p/(p+2)} \text{ weak} \\ \text{if } N \geq 2, \text{ in the sense of measures if } N = 1, \tag{6.18}$$

for a sequence of correctors P^m.

Proof: One extracts a subsequence such that $(P^m)^T c^m$ converges in $L^{2p/(p+2)}(\Omega)$ weak to c^{eff} if $N \geq 2$ and in the sense of measures if N =1.[49] Using SOBOLEV's imbedding theorem, $\big(c^m.grad(u_m)\big)$ stays in a compact of $H^{-1}_{loc}(\Omega)$ and therefore Theorem 21 implies that $grad(u_m) - P^m\, grad(u_\infty)$ converges in $L^1_{loc}(\Omega; R^N)$ strong to 0, and one wants to prove that

$$\Big(c^m.grad(u_m)\Big) \rightharpoonup \Big(c^{eff}.grad(u_\infty)\Big) \text{ in } L^{2p/(p+2)} \text{ weak} \\ \text{if } N \geq 2, \text{ in the sense of measures if } N = 1. \tag{6.19}$$

Indeed $(c^m.P^m g)$ converges to $(c^{eff}.g)$ if $g \in C(\Omega; R^N)$, and if g satisfies (6.11), then (6.13) implies that both $(c^m.grad(u_m) - P^m g)$ and $(c^{eff}.grad(u_\infty) - g)$ have a small norm in L^1, so that one deduces (6.19).∎

One could add a term $d^n u_n$ in the equation, with d^n being a bounded sequence in $L^q(\Omega)$ with $q > \frac{N}{2}$ for $N \geq 2$, $q = 1$ for $N = 1$. The term $d^n u_n$ stays in a compact of $H^{-1}_{loc}(\Omega)$ strong and a subsequence converges to $d^\infty u_\infty$ if for that subsequence d^m converges in $L^q(\Omega)$ weak to d^∞ for $N \geq 2$ or in the sense of measures if $N = 1$. Without loss of generality, this term can be put into the right hand side converging in $H^{-1}_{loc}(\Omega)$ strong to a known limit.

I conclude by some computations of François MURAT giving other properties of the correctors and showing how to treat cases with terms converging only in $H^{-1}_{loc}(\Omega)$ weak.

Proposition 23: Let a sequence $A^n \in M(\alpha, \beta; \Omega)$ H-converge to A^{eff}, let b^n be a sequence bounded in $L^2(\Omega; R^N)$ and let u_n be a sequence converging in $H^1_{loc}(\Omega)$ weak to u_∞ and satisfying

$$-div\Big(A^n\, grad(u_n) + b^n\Big) \rightarrow f \text{ in } H^{-1}_{loc}(\Omega) \text{ strong}. \tag{6.20}$$

Then u_∞ satisfies

$$-div\Big(A^{eff}\, grad(u_\infty) + b^{eff}\Big) = f \text{ in } \Omega, \tag{6.21}$$

with an effective term $b^{eff} \in L^2(\Omega; R^N)$ such that

[49] There could be different subsequences of $(P^m)^T c^m$ converging to different limits, but Proposition 22 shows that all these limits give the same value for $\big(c^{eff}.grad(u_\infty)\big)$.

$$(\Pi^m)^T b^m \rightharpoonup b^{eff} \text{ in the sense of measures,} \tag{6.22}$$

for a subsequence for which Π^m denotes the corresponding correctors associated to $(A^m)^T$.[50]

Proof: One extracts a subsequence such that (6.22) holds and

$$A^m\, grad(u_m) + b^m \text{ converges in } L^2(\Omega; R^N) \text{ weak to } \xi, \tag{6.23}$$

and one shows that

$$\xi = A^{eff}\, grad(u_\infty) + b^{eff} \text{ in } \Omega. \tag{6.24}$$

For $v_\infty \in C_c^1(\Omega)$, let $v_n \in H_0^1(\Omega)$ be the sequence of solutions of

$$div\Big((A^m)^T grad(v_m) - (A^{eff})^T grad(v_\infty)\Big) = 0 \text{ in } \Omega, \tag{6.25}$$

so that

$$\begin{aligned} &v_m \rightharpoonup v_\infty \text{ in } H_0^1(\Omega) \text{ weak},\\ &(A^m)^T grad(v_m) \rightharpoonup (A^{eff})^T grad(v_\infty) \text{ in } L^2(\Omega; R^N) \text{ weak},\\ &grad(v_m) - \Pi^m\, grad(v_\infty) \to 0 \text{ in } L^2_{loc}(\Omega; R^N) \text{ strong,}\end{aligned} \tag{6.26}$$

and one computes the limit of $\big(A^m\, grad(u_m) + b^m.grad(v_m)\big)$, which is $\big(\xi.grad(v_\infty)\big)$, by using the Div-Curl lemma. As $\big(grad(u_m).(A^m)^T grad(v_m)\big)$ has limit $\big(grad(u_\infty).(A^{eff})^T grad(v_\infty)\big)$ and as $\big(b^m.grad(v_m)\big)$ has the same limit as $\big(b^m.\Pi^m\, grad(v_\infty)\big)$, i.e. $\big(b^{eff}.grad(v_\infty)\big)$, one has proved that

$$\Big(\xi - A^{eff}\, grad(u_\infty) - b^{eff}.grad(v_\infty)\Big) = 0 \text{ in } \Omega, \tag{6.27}$$

and by density of $C_c^1(\Omega)$ in $H_0^1(\Omega)$, one obtains (6.24).■

François MURAT also noticed that although P^m may only be bounded in $L^p(\Omega; \mathcal{L}(R^N, R^N))$ for some $p < \infty$ (with $p \in (2, \infty)$ if one uses MEYERS's regularity theorem [Me], or $p = 2$ if one does not use it), it is nevertheless true that for $q \in [2, \infty]$ and any sequence β_n bounded in $L^q(\Omega)$ and any $i, j = 1, \ldots, N$, all the limits of subsequences of $(P^m)_{ij}\, \beta_m$ in the sense of measures actually belong to $L^q(\Omega)$. For $q = 2$ this is what (6.22) asserts for Π^m instead of P^m by taking $b^m = \beta_m e_j$. For $q > 2$, assume that $(P^m)_{ij}\, \beta_m$ converges in $L^{2q/(q+2)}(\Omega)$ weak to ξ and β_m^2 converges in $L^{q/2}(\Omega)$ weak to β_∞^2 with $\beta_\infty \in L^q(\Omega)$; then one uses the fact that $(A^m\, P^m\, \lambda.P^m\, \lambda)$ converges in the sense of measures to $(A^{eff}\, \lambda.\lambda)$ by the Div-Curl lemma, and therefore the limit of any $(P^m)_{ij}^2$ in the sense of measures belongs to $L^\infty(\Omega)$. For $\varphi \in C_c(\Omega)$ with $-1 \le \varphi \le 1$ in Ω, one has

[50] From the explicit computation (6.2) of P^n in the case of layers, one sees that Π^m is not in general equal to $(P^m)^T$; of course, if $(A^n)^T = A^n$ for all n one can choose $\Pi^m = P^m$.

$$\int_{\Omega} \varphi \xi \, dx = \lim_{m} \int_{\Omega} \varphi (P^m)_{ij} \beta_m \, dx \le \lim_{m} \int_{\Omega} |\varphi| \Big(\frac{\varepsilon(x)}{2} (P^m)^2_{ij} + \frac{1}{2\varepsilon(x)} \beta^2_m \Big) \, dx \tag{6.28}$$

for every function $\varepsilon > 0$; therefore $\int_{\Omega} \varphi \xi \, dx \le \int_{\Omega} |\varphi| \big(\frac{\varepsilon}{2} C^2 + \frac{1}{2\varepsilon} \beta^2_\infty \big) \, dx$ and then taking the infimum in ε gives $\int_{\Omega} \varphi \xi \, dx \le \int_{\Omega} |\varphi| C \beta_\infty \, dx$, i.e. $|\xi| \le C\beta_\infty$ a.e. $\in \Omega$ by varying φ.

Proposition 24: Under the hypotheses of Proposition 23, one has

$$grad(u_m) - P^m \, grad(u_\infty) - r_m \to 0 \text{ in } L^1_{loc}(\Omega) \text{ strong}, \tag{6.29}$$

for some r_m (constructed explicitly) which satisfies

$$\begin{aligned} & r_m \rightharpoonup 0 \text{ in } L^2(\Omega; R^N) \text{ weak}, \\ & A^m \, r_m + b^m \rightharpoonup b^{eff} \text{ in } L^2(\Omega; R^N) \text{ weak}. \end{aligned} \tag{6.30}$$

Proof. Let $\rho_n \in H^1_0(\Omega)$ be the solution of

$$div\Big(A^n \, grad(\rho_n) + b^n\Big) = 0 \text{ in } \Omega, \tag{6.31}$$

so that ρ_n is bounded in $H^1_0(\Omega)$ and by Proposition 23 a subsequence ρ_m converges in $H^1_0(\Omega)$ weak to ρ_∞, solution of

$$div\Big(A^{eff} \, grad(\rho_\infty) + b^{eff}\Big) = 0 \text{ in } \Omega. \tag{6.32}$$

Then one notices that $div(A^m \, grad(u_m) - A^m \, grad(\rho_m)) \to f$ in $H^{-1}_{loc}(\Omega)$ strong so that $grad(u_m) - grad(\rho_m) - P^m (grad(u_\infty) - grad(\rho_\infty)) \to 0$ in $L^1_{loc}(\Omega; R^N)$ strong, and therefore one has (6.29) and (6.30) by taking

$$r_m = grad(\rho_m) - P^m \, grad(\rho_\infty). \tag{6.33}$$

■

As a corollary, if a sequence $A^n \in M(\alpha, \beta; \Omega)$ H-converges to A^{eff}, if b^n is bounded in $L^2(\Omega; R^N)$, if c^n is bounded in $L^p(\Omega; R^N)$ with $p > N$ if $N \ge 2$, $p = 2$ if $N = 1$, and if u_n is a sequence converging in $H^1_{loc}(\Omega)$ weak to u_∞ and satisfies

$$-div\Big(A^n \, grad(u_n) + b^n\Big) + \Big(c^n . grad(u_n)\Big) \to f \text{ in } H^{-1}_{loc}(\Omega) \text{ strong}, \tag{6.34}$$

then, using the definitions of c^{eff} and b^{eff} given by (6.18) and (6.22), u_∞ satisfies

$$-div\Big(A^{eff} \, grad(u_\infty) + b^{eff}\Big) + \Big(c^{eff} . grad(u_\infty)\Big) + e_\infty = f \text{ in } \Omega, \tag{6.35}$$

where

$$(c^n . r_n) \rightharpoonup e_\infty \text{ in } L^{2p/(p+2)}(\Omega) \text{ weak}. \tag{6.36}$$

As many seem to believe that Homogenization means periodicity, it is important to notice that in the framework of G-convergence that Sergio SPAGNOLO had developped in the late 60s or in the framework of H-convergence that François MURAT and I had developed in the early 70s, there were no conditions of periodicity. As François MURAT and I were looking at questions of Optimal Design, there was no reason for thinking that periodicity had anything to do with our problem, and when we discovered that Henri SANCHEZ-PALENCIA had been working on asymptotic methods for periodic structures [S-P1], [S-P2], it helped us understand that what we had been doing was related to effective properties of mixtures, but it was not more useful for our purpose. In the Fall of 1974, after I had described my work in Madison, Carl DE BOOR had mentioned some work by Ivo BABUŠKA; this work was restricted to some engineering applications where periodicity is natural, and when I first met Ivo BABUŠKA in the Spring 1975 [Ba], I did learn from him about some practical questions, quite unrelated to those that we were interested in our work.[51] In the Fall of 1975, at a IUTAM meeting in Marseille, I learned that Jacques-Louis LIONS had been convinced by Ivo BABUŠKA of the importance of Homogenization for periodic structures and had worked with Alain BENSOUSSAN and George PAPANICOLAOU, and I showed him my method of oscillating test functions associated with the Div-Curl lemma, and the first mention of it appears then in the article which he wrote for the proceedings [Li3]. It was only on the occasion of my lectures on our method at Bréau-sans-Nappe in the Summer 1983 [Mu&Ta1] that George PAPANICOLAOU told me that he finally understood why I had insisted so much about working without periodicity assumptions. Although I taught about general questions of Homogenization in my PECCOT lectures in the Spring 1977, many who attended these lectures but specialized in questions with periodic structures seem to have forgotten to either quote that they were using my method or that my method was not restricted to periodic situations: it might be for that reason that Olga OLEINIK rediscovered my method by considering first quasi-periodic situations and then general situations.

Some people, who seem to try to avoid mentioning either the name of Sergio SPAGNOLO for the introduction of G-convergence in the late 60s or the names of François MURAT and me for the introduction of H-convergence in the early 70s, often state that it is enough to consider periodic media; they may be unaware that such a statement is perfectly meaningless for someone

[51] I could imagine some real situations where our work could be useful, at least after we would have made some progress on the question of characterization of effective coefficients. I think that it was on this occasion that I learned from Ivo BABUŠKA about the importance of amplification factors for stress, but I do not recall ever hearing him mention that the defects of linearized Elasticity were quite worse for mixtures than for homogeneous materials, and I only realized that many years after.

who does not know that there exists a general theory; they may not realize either that for those who know about the general theory it clearly shows that they have been unable to understand the general framework. It seems that many who started by studying the special case of periodic structures have had some trouble learning about the general framework, while for all those who have started by learning the general framework, the case of periodic structures appears as the following simple exercise.

In the periodic setting, one starts with a period cell Y, generated by N linearly independent vectors $y_1, \ldots, y_N$, of R^N i.e.

$$Y = \Big\{ y | y \in R^N, y = \sum_{i=1}^{N} \xi_i \, y_i, 0 \leq \xi_i \leq 1 \text{ for } i = 1, \ldots, N \Big\}, \tag{6.37}$$

and one says that a function g defined on R^N is Y-periodic if

$$g(y + y_i) = g(y) \text{ a.e. } y \in R^N, \text{ for } i = 1, \ldots, N. \tag{6.38}$$

For $A \in M(\alpha, \beta; R^N)$ and Y-periodic, one defines A^n by

$$A^n(x) = A\Big(\frac{x}{\varepsilon_n}\Big) \text{ a.e. } x \in \Omega, \tag{6.39}$$

where ε_n tends to 0.

Proposition 25: The whole sequence A^n defined by (6.39) H-converges to a constant A^{eff}, independent of the particular sequence ε_n used, and A^{eff} can be computed in the following way. For $\lambda \in R^N$, let $w_\lambda \in H^1_{loc}(R^N)$ be the Y-periodic solution (defined up to addition of a constant) of

$$div\Big(A(grad(w_\lambda) + \lambda)\Big) = 0 \text{ in } R^N, \tag{6.40}$$

and let $P \in H^1_{loc}(R^N; \mathcal{L}(R^N, R^N))$ be the Y-periodic function defined by

$$P\lambda = grad(w_\lambda) + \lambda \text{ a.e. in } R^N. \tag{6.41}$$

Then

$$A^{eff}\,\lambda = \frac{1}{meas(Y)} \int_Y A(grad(w_\lambda) + \lambda)\, dy \text{ for every } \lambda \in R^N, \tag{6.42}$$

and a sequence of corrector is defined by

$$P^n(x) = P\Big(\frac{x}{\varepsilon_n}\Big) \text{ a.e. } x \in R^N. \tag{6.43}$$

Proof: The sequence u_n defined by

$$u_n(x) = (\lambda . x) + \varepsilon_n w_\lambda\Big(\frac{x}{\varepsilon_n}\Big), \text{ a.e. } x \in R^N, \tag{6.44}$$

satisfies $u_n \in H^1_{loc}(R^N)$ and $div(A^n\, grad(u_n)) = 0$ in R^N. The sequence u_n converges in $H^1_{loc}(R^N)$ weak to u_∞, defined by $u_\infty(x) = (\lambda.x)$, $grad(u_n)$ is the rescaled version of $\lambda + grad(w_\lambda)$ which is Y-periodic and therefore it converges in $L^2_{loc}(R^N; R^N)$ weak to its average on the period cell Y, i.e. to $grad(u_\infty) = \lambda$, and $A^n\, grad(u_n) \in L^2_{loc}(R^N; R^N)$ is the rescaled version of $A(\lambda + grad(w_\lambda))$ which is Y-periodic and therefore it converges in $L^2_{loc}(R^N; R^N)$ weak to its average on Y, i.e. to the value $A^{eff}\,\lambda$ as defined by (6.42); using N linearly independent $\lambda \in R^N$ characterizes the H-limit of A^n as A^{eff}.

As $grad(u_n) = P^n\, grad(u_\infty)$ a.e., and P_n satisfies the conditions (6.7), the sequence P_n gives acceptable correctors.■

Once correctors had become natural objects for studying Homogenization, it was very natural to use them for obtaining bounds on effective coefficients.

8. Bounds on effective coefficients: second method

A first difference between this new method that I introduced at the end of 1977 in [Ta7] and the preceding one that I had used with François MURAT in the early 70s based on (5.1) and (5.2) (after having used an earlier version of the Div-Curl lemma), is that instead of considering one sequence of solutions one considers N linearly independent sequences of solutions which are the columns of the sequence of correctors P^n. A second difference is that the Div-Curl lemma had to be replaced by the more general theory of Compensated Compactness that I had just developed with François MURAT in the meantime [Mu4], [Ta4], [Ta5], [Ta6], [Ta8].[52] If $A^n \in M(\alpha, \beta; \Omega)$ H-converges to A^{eff}, then any sequence of correctors P^m has the property that

[52] Jacques-Louis LIONS had asked François MURAT to generalize our Div-Curl lemma, and he had given him an article by SCHULENBERGER and WILCOX which he thought related. François MURAT first proved a bilinear theorem: a sequence U^n converged weakly to U^∞ and satisfied a list of differential constraints, another sequence V^n converged weakly to V^∞ and satisfied another list of differential constraints, and he characterized which bilinear forms B had the property that $B(U^n, V^n)$ automatically converged in the sense of measures to $B(U^\infty, V^\infty)$. I told him that the bilinear setting looked artificial and that a quadratic setting was more natural: for a sequence U^n converging weakly to U^∞ and satisfying a list of differential constraints, he then characterized which quadratic forms Q are such that $Q(U^n)$ automatically converges in the sense of measures to $Q(U^\infty)$. While he was giving a talk about his results at the seminar that Jacques-Louis LIONS was organizing at Institut Henri POINCARÉ, it suddenly occurred to me that the right question was to look at quadratic forms Q such that if $Q(U^n)$ converges in the sense of measures to ν then one automatically has $\nu \geq Q(U^\infty)$ and before

$$\begin{aligned} &P^m \rightharpoonup P^\infty = I \text{ in } L^2\big(\Omega; \mathcal{L}(R^N, R^N)\big) \text{ weak},\\ &curl(P^m\,\lambda) \text{ stays in a compact of } H^{-1}_{loc}\big(\Omega; \mathcal{L}_a(R^N, R^N)\big) \text{ strong} \\ &\qquad\qquad\qquad\qquad\text{for all } \lambda \in R^N, \end{aligned} \tag{7.1}$$

where $\mathcal{L}_a(R^N, R^N)$ is the space of antisymmetric matrices, and if one defines the sequence Q^m by

$$Q^m = A^m\, P^m, \tag{7.2}$$

then Q^m has the property that

$$\begin{aligned} &Q^m \rightharpoonup Q^\infty = A^{eff} \text{ in } L^2\big(\Omega; \mathcal{L}(R^N, R^N)\big) \text{ weak},\\ &div(Q^m\,\lambda) \text{ stays in a compact of } H^{-1}_{loc}(\Omega) \text{ strong for all } \lambda \in R^N. \end{aligned} \tag{7.3}$$

Of course each column of P^m plays the role of a vector E^m and each column of Q^m plays the role of a vector D^m for which the Div-Curl lemma applies, and this means that $(Q^m)^T P^m$ converges in the sense of measures to $(Q^\infty)^T P^\infty = (A^{eff})^T$, but the Compensated Compactness theorem creates a few other interesting inequalities. While I was visiting the Mathematics Research Center in Madison in the Fall 1977, I had found a crucial additive to the Compensated Compactness theorem, as I had discovered a way to use a formal computation based on "entropies" for passing to the limit in general systems.[53] As I was wondering what I should talk about at a meeting in Versailles in December 1977, and I had thought of improving bounds on effective coefficients, it was then natural that I tried to use more general functionals, not necessarily quadratic.

---

the end of the talk I had checked that the same method that we had used for the Div-Curl lemma gave me the right characterization, and I did not even need the hypothesis of constant rank that François MURAT had to impose, because of a slightly different method of proof.

[53] This is the improvement which I call the Compensated Compactness Method, on which I based my lectures at HERIOT-WATT University in the Summer 1978 [Ta8]. Of course, my framework was never restricted to hyperbolic systems, and I had already explained in [Ta5] how to use it for minimization problems, and I had described again the same example in [Ta8] in order to show that my approach based on characterizing YOUNG measures associated to a given list of differential constraints was better than the programme that others preferred of looking only at sequentially weakly lower semi-continuous functionals. Of course, "entropies" were never specific to hyperbolic situations, and before discussing the case of hyperbolic systems, I had explained how "entropies" explain the sequential weak continuity of Jacobian determinants of size larger than 2 as examples of the Compensated Compactness theorem. In [Ta6] I had advocated a different fact, that "entropy conditions" were also necessary for stationary solutions of Elasticity.

Theorem 26: Assume that F is a continuous function on $\mathcal{L}(R^N, R^N) \times \mathcal{L}(R^N, R^N)$ which has the property that

$$\begin{aligned}
&\widetilde{P}^m \rightharpoonup \widetilde{P}^\infty \text{ in } L^2\Big(\Omega; \mathcal{L}(R^N, R^N)\Big) \text{ weak,}\\
&\widetilde{Q}^m \rightharpoonup \widetilde{Q}^\infty \text{ in } L^2\Big(\Omega; \mathcal{L}(R^N, R^N)\Big) \text{ weak,}\\
&curl(\widetilde{P}^m \lambda) \text{ stays in a compact of } H^{-1}_{loc}\Big(\Omega; \mathcal{L}_a(R^N, R^N)\Big) \text{ strong } \forall \lambda \in R^N,\\
&div(\widetilde{Q}^m \lambda) \text{ stays in a compact of } H^{-1}_{loc}(\Omega) \text{ strong } \forall \lambda \in R^N,
\end{aligned} \tag{7.4}$$

imply

$$\begin{aligned}
&\liminf_{m\to\infty} \int_\Omega F(\widetilde{P}^m, \widetilde{Q}^m)\varphi\, dx \geq\\
&\qquad \geq \int_\Omega F(\widetilde{P}^\infty, \widetilde{Q}^\infty)\varphi\, dx \text{ for all } \varphi \in C_c(\Omega), \varphi \geq 0.
\end{aligned} \tag{7.5}$$

One defines the function g on $\mathcal{L}(R^N, R^N)$, possibly taking the value $+\infty$, by

$$g(A) = \sup_{P \in \mathcal{L}(R^N, R^N)} F(P, A\, P). \tag{7.6}$$

Then if $A^n \in M(\alpha, \beta; \Omega)$ H-converges to A^{eff}, then one has

$$\liminf_{n\to\infty} \int_\Omega g(A^n)\varphi\, dx \geq \int_\Omega g(A^{eff})\varphi\, dx \text{ for all } \varphi \in C_c(\Omega), \varphi \geq 0. \tag{7.7}$$

Proof: Of course, one assumes that the left side of (7.7) is $< +\infty$, one extracts a subsequence A^m for which $\liminf_m$ is a limit and a sequence of correctors P^m exists. For $X \in C^1(\Omega; \mathcal{L}(R^N, R^N))$, the sequences $\widetilde{P}^m = P^m X$ and $\widetilde{Q}^m = Q^m X$ satisfy (7.4) with $\widetilde{P}^\infty = X$ and $\widetilde{Q}^\infty = A^{eff} X$, and therefore by (7.5) one has

$$\begin{aligned}
&\liminf_{m\to\infty} \int_\Omega F(P^m X, A^m P^m X)\varphi\, dx \geq\\
&\qquad \geq \int_\Omega F(X, A^{eff} X)\varphi\, dx \text{ for all } \varphi \in C_c(\Omega), \varphi \geq 0,
\end{aligned} \tag{7.8}$$

and as $F(P^m X, A^m P^m X) \leq g(A^m)$ by (7.6), one deduces that

$$\begin{aligned}
&\liminf_{m\to\infty} \int_\Omega g(A^m)\varphi\, dx \geq\\
&\qquad \geq \int_\Omega F(X, A^{eff} X)\varphi\, dx \text{ for all } X \in C^1\Big(\Omega; \mathcal{L}(R^N, R^N)\Big),
\end{aligned} \tag{7.9}$$

and for all $\varphi \in C_c(\Omega)$, $\varphi \geq 0$. For $X \in L^\infty(\Omega; \mathcal{L}(R^N, R^N))$, there exists a sequence $X_n \in C^1(\Omega; \mathcal{L}(R^N, R^N))$ such that X_n stays bounded in $L^\infty(\Omega;\mathcal{L}(R^N,R^N))$ and converges a.e. to X, and by LEBESGUE dominated convergence theorem $F(X_n, A^{eff} X_n)$ converges in $L^1(\Omega)$ strong to $F(X, A^{eff} X)$ and therefore (7.9) is true for all $X \in L^\infty(\Omega;\mathcal{L}(R^N,R^N))$.

For $r < \infty$ let

$$g_r(A) = \sup_{||P|| \le r} F(P, A\,P), \tag{7.10}$$

which is continuous, as F is uniformly continuous on bounded sets. For $\varepsilon > 0$ let M_ε be a measurable function taking only a finite number of distinct values in $\mathcal{L}(R^N,R^N)$ and such that $||M_\varepsilon - A^{eff}|| \le \varepsilon$ a.e. in Ω. Then one can choose a measurable X_ε taking only a finite number of distinct values in $\mathcal{L}(R^N,R^N)$, such that $||X_\varepsilon|| \le r$ and $F(X_\varepsilon, M_\varepsilon X_\varepsilon) = g_r(M_\varepsilon)$ a.e. in Ω, and as $g_r(M_\varepsilon)$ converges uniformly to $g_r(A^{eff})$ as ε tends to 0, one deduces from the inequality (7.9) for X_ε that one has

$$\begin{aligned}&\liminf_{m\to\infty} \int_\Omega g(A^m)\varphi\,dx \ge \int_\Omega g_r(A^{eff})\varphi\,dx\\ &\qquad\text{for all } r < +\infty \text{ and all } \varphi \in C_c(\Omega), \varphi \ge 0.\end{aligned} \tag{7.11}$$

Then $g_r(A^{eff})$ increases and converges to $g(A^{eff})$ as r increases to $+\infty$, and one deduces (7.7) by BEPPO-LEVI's theorem.■

Of course, the (quadratic) theorem of Compensated Compactness, which I will state and prove a little later, provides an analytic characterization of all the homogeneous quadratic functions F which are such that (7.4) implies (7.5), namely it is true if and only if

$$\begin{aligned}&F(\eta\otimes\xi, Q_\xi) \ge 0\\ &\qquad\forall \eta,\xi \in R^N \text{ and all } Q_\xi \in \mathcal{L}(R^N,R^N) \text{ satisfying } Q_\xi\,\xi = 0.\end{aligned} \tag{7.12}$$

It was only in June 1980, while I was visiting the COURANT Institute at New York University, that I tried to find which F would be suitable for the case of mixing isotropic materials, restricting myself to the case where A^{eff} would also be isotropic, i.e. equal to $a^{eff}\,I$, and I decided then to look for functions F satisfying (7.12) which would also be invariant under a change of orthonormal basis. Of course as a consequence of the Div-Curl lemma the functions $F^\pm_{ij}(P,Q) = \pm(Q\,P^T)_{ij} = \pm\sum_k Q_{ik}\,P_{jk}$ do satisfy (7.12), and therefore $F^\pm(P,Q) = \pm trace(Q\,P^T)$ give two such invariant functions F satisfying (7.12). As $trace(P^T\,P)$, $\big(trace(P)\big)^2$, $trace(Q^T\,Q)$ and $\big(trace(Q)\big)^2$ are invariant under a change of orthonormal basis, I checked which linear combinations of these particular functions would satisfy (7.12). It is obvious that $F_1(P) = trace(P^T\,P) - \big(trace(P)\big)^2$ does satisfy (7.12), because if $P = \xi\otimes\eta$ then $trace(P^T\,P) = |\xi|^2\,|\eta|^2$ and $trace(P) = (\xi.\eta)$, and therefore $trace(P^T\,P) \ge \big(trace(P)\big)^2$ by CAUCHY-SCHWARZ's inequality. Then I found that $F_2(Q) = (N-1)trace(Q^T\,Q) - \big(trace(Q)\big)^2$ also satisfies (7.12), by applying the following lemma to Q_ξ whose rank is at most $N-1$.

Lemma 27: If $M \in \mathcal{L}(R^N, R^N)$ then

$$rank(M)\, trace(M^T M) - \big(trace(M)\big)^2 \geq 0. \tag{7.13}$$

Proof: If $rank(M) = k$, one chooses an orthogonal basis such that the range of M is spanned by the first k vectors of the basis, and then $trace(M) = \sum_i M_{ii}$ and $trace(M^T M) = \sum_{i,j} M_{ij}^2 \geq \sum_i M_{ii}^2$, which is $\geq \frac{1}{k}(\sum_i M_{ii})^2$ by CAUCHY-SCHWARZ's inequality.■

Among the combinations of these particular functions, I quickly selected two simple ones, corresponding to the following two lemmas. In June 1980, I only computed $g(A)$ for $A = \lambda I$, but as François MURAT suggested in the Fall that the same functionals would also give an optimal result for anisotropic A, we did together the computations for general symmetric A, and I show this general computation below.

Lemma 28: If

$$F_1(P,Q) = \alpha\Big[trace(P^T P) - \big(trace(P)\big)^2\Big] - trace(Q\,P^T) + 2trace(P), \tag{7.14}$$

then for $A \in \mathcal{L}(R^N, R^N)$ with $A^T = A$ and $A \geq \alpha I$, and denoting $\lambda_1, \ldots, \lambda_N$, the eigenvalues of A, one has

$$g_1(A) = \frac{\tau}{1+\alpha\tau}, \text{ with } \tau = \sum_{j=1}^N \frac{1}{\lambda_j - \alpha}. \tag{7.15}$$

Proof: Of course, if α is an eigenvalue of A then $\tau = \infty$ and $g_1(A) = \frac{1}{\alpha}$. One chooses an orthonormal basis where A is diagonal, and the form of $F_1(P, A\,P)$ is unchanged, and one must compute

$$\sup_P\Big(\alpha\sum_{i,j=1}^N P_{ij}^2 - \alpha(\sum_{i=1}^N P_{ii})^2 - \sum_{i,j=1}^N \lambda_i P_{ij}^2 + 2\sum_{i=1}^N P_{ii}\Big), \tag{7.16}$$

and for $i \neq j$ a good choice for P_{ij} is 0 (it does not really matter what P_{ij} is if $\lambda_i = \alpha$), and one must then compute

$$\sup_P\Big(\sum_{i=1}^N(\alpha - \lambda_i)P_{ii}^2 - \alpha(\sum_{i=1}^N P_{ii})^2 + 2\sum_{i=1}^N P_{ii}\Big). \tag{7.17}$$

If $\sum_i P_{ii}$ is a given value t, then in the case where $\lambda_i > \alpha$ for all i, maximizing $\sum_i(\alpha - \lambda_i)P_{ii}^2$ is obtained by taking $P_{ii} = \frac{C}{\lambda_i - \alpha}$ for all i, so that $t = C\,\tau$, and one finds t by maximizing $-C^2\,\tau - \alpha\,t^2 + 2t$, i.e. by maximizing $-\frac{t^2}{\tau} - \alpha\,t^2 + 2t$, which gives the value of t and the maximum equal to $\frac{\tau}{1+\alpha\,\tau}$. If $\lambda_i = \alpha$ for some i, the best is to take $P_{ii} = t$ and $P_{jj} = 0$ for $j \neq i$, and then the best value of t and the maximum are equal to $\frac{1}{\alpha}$.■

Lemma 29: If

$$\begin{aligned} F_2(P,Q) =&(N-1)trace(Q^T Q)- \\ &-\Big(trace(Q)\Big)^2 - \beta(N-1)trace(Q\,P^T) + 2trace(Q), \end{aligned} \tag{7.18}$$

then for $A \in \mathcal{L}(R^N, R^N)$ with $A^T = A$ and $A \le \beta\, I$, and denoting $\lambda_1, \ldots, \lambda_N$ the eigenvalues of A, one has

$$g_2(A) = \frac{\sigma}{\sigma + N - 1}, \text{ with } \sigma = \sum_{j=1}^N \frac{\lambda_j}{\beta - \lambda_j}. \tag{7.19}$$

Proof: Of course, if β is an eigenvalue of A then $\sigma = \infty$ and $g_2(A) = 1$. One chooses an orthonormal basis where A is diagonal, and the form of $F_2(P, A\,P)$ is unchanged, and one must compute

$$\sup_P\Big((N-1)\sum_{i,j=1}^N \lambda_i^2 P_{ij}^2 - (\sum_{i=1}^N \lambda_i\, P_{ii})^2 - \beta(N-1)\sum_{i,j=1}^N \lambda_i\, P_{ij}^2 + 2\sum_{i=1}^N \lambda_i\, P_{ii}\Big), \tag{7.20}$$

and for $i \ne j$ a good choice for P_{ij} is 0 (it does not really matter what P_{ij} is if $\lambda_i = \beta$), and one must then compute

$$\sup_P\Big((N-1)\sum_{i=1}^N(\lambda_i - \beta)\lambda_i\, P_{ii}^2 - (\sum_{i=1}^N \lambda_i\, P_{ii})^2 + 2\sum_{i=1}^N \lambda_i\, P_{ii}\Big). \tag{7.21}$$

If $\sum_i \lambda_i\, P_{ii}$ is a given value s, then in the case where $\lambda_i < \beta$ for all i, maximizing $\sum_i(\lambda_i - \beta)\lambda_i\, P_{ii}^2$ is obtained by taking $P_{ii} = \frac{C}{\beta - \lambda_i}$ for all i, so that $s = C\,\sigma$, and one finds s by maximizing $-(N-1)C^2\,\sigma - s^2 + 2s$, i.e. by maximizing $-\frac{N-1}{\sigma}s^2 - s^2 + 2s$, which gives the value of s and the maximum equal to $\frac{\sigma}{\sigma+N-1}$. If $\lambda_i = \beta$ for some i, the best is to take $P_{ii} = \frac{s}{\lambda_i}$ and $P_{jj} = 0$ for $j \ne i$, and then the best value of s and the maximum are equal to 1.■

Of course, I had also considered more general combinations like

$$\begin{aligned} F_3(P,Q) = &-trace(Q\,P^T) + a\Big[trace(P^T\,P) - \Big(trace(P)\Big)^2\Big] + \\ &+b\Big[(N-1)trace(Q^T\,Q) - \Big(trace(Q)\Big)^2\Big] + \\ &+ 2c\,trace(P) + 2d\,trace(Q) \text{ with} a, b \ge 0, \end{aligned} \tag{7.22}$$

for which the computation of $g_3(\gamma\, I)$ requires to compute

$$\begin{aligned} \sup_P\Big[&(-\gamma + a + b(N-1)\gamma^2)trace(P^T\,P) - \\ &- (a + b\gamma^2)\Big(trace(P)\Big)^2 + 2(c + \gamma\, d)trace(P)\Big]. \end{aligned} \tag{7.23}$$

In order to have $g_3(\gamma I) < +\infty$, one needs to have $-\gamma + a + b(N-1)\gamma^2 \leq 0$, and one can then choose all non diagonal coefficients of P equal to 0; for $trace(P)$ given one wants to minimize $trace(P^T P)$, and therefore one only considers $P = p I$, and one wants then to maximize $\big(-\gamma + a + b(N-1)\gamma^2 - N(a + b\gamma^2)\big)p^2 + 2(c + \gamma d)p\big)$, and one obtains

$$g_3(\gamma I) = \frac{(c+\gamma d)^2}{(N-1)a + \gamma + b\,\gamma^2} \text{ if } a, b \geq 0 \text{ and } -\gamma + a + b(N-1)\gamma^2 \leq 0. \tag{7.24}$$

As it was not so easy to handle, I had choosen the simplification of considering either $b = d = 0$, which corresponds to Lemma 28, or $a = c = 0$, which corresponds to Lemma 29. I was interested in characterizing the possible effective tensors A^{eff} of mixtures obtained by using proportion θ of an isotropic material with tensor αI and proportion $1-\theta$ of an isotropic material with tensor βI, i.e. I considered $A^n = (\chi_n \alpha + (1-\chi_n)\beta)I$ with a sequence of characteristic functions χ_n converging in $L^\infty(\Omega)$ weak $\star$ to θ, and A^n H-converging to A^{eff}. I already knew (5.3); in order to show explicitly the dependence in θ, (5.3) means that the eigenvalues $\lambda_1, \ldots, \lambda_N$ of A^{eff} satisfy

$$\lambda_-(\theta) \leq \lambda_j \leq \lambda_+(\theta), j = 1, \ldots, N \text{ a.e. in } \Omega, \tag{7.25}$$

where, as in (5.3)

$$\lambda_+(\theta) = \theta\,\alpha + (1-\theta)\beta,\ \frac{1}{\lambda_-(\theta)} = \frac{\theta}{\alpha} + \frac{1-\theta}{\beta}, \tag{7.26}$$

Theorem 26 asserts that

$$g(A^{eff}) \leq \theta\, g(\alpha I) + (1-\theta) g(\beta I) \text{ a.e. in } \Omega, \tag{7.27}$$

whenever g is associated to a function F for which (7.4) implies (7.5). For the particular function g_1 given by Lemma 28, one has $g_1(\alpha I) = \frac{1}{\alpha}$ and $g_1(\beta I) = \frac{N/(\beta-\alpha)}{1+\alpha N/(\beta-\alpha)} = \frac{N}{(N-1)\alpha+\beta}$, and therefore (7.27) means that $\frac{\tau^{eff}}{1+\alpha\tau^{eff}} \leq \frac{\theta}{\alpha} + \frac{(1-\theta)N}{(N-1)\alpha+\beta} = \frac{(N-\theta)\alpha+\theta\,\beta}{\alpha\big((N-1)\alpha+\beta\big)}$, which gives for τ^{eff} the upper bound

$$\tau^{eff} = \sum_{j=1}^{N} \frac{1}{\lambda_j - \alpha} \leq \frac{(N-\theta)\alpha + \beta}{(1-\theta)\alpha(\beta-\alpha)}. \tag{7.28}$$

Equality occurs for the case of layers, which according to (4.11) corresponds to A^{eff} having one eigenvalue equal to $\lambda_-(\theta)$ and the $N-1$ others equal to $\lambda_+(\theta)$, i.e

$$\frac{1}{\lambda_-(\theta)-\alpha} + \frac{N-1}{\lambda_+(\theta)-\alpha} = \frac{(N-\theta)\alpha+\beta}{(1-\theta)\alpha(\beta-\alpha)}. \tag{7.29}$$

For the particular function g_2 given by Lemma 29, one has $g_2(\alpha I) = \frac{N\alpha/(\beta-\alpha)}{N\alpha/(\beta-\alpha)+N-1} = \frac{N\alpha}{\alpha+(N-1)\beta}$ and $g_2(\beta I) = 1$, and therefore (7.27) means that $\frac{\sigma^{eff}}{\sigma^{eff}+N-1} \leq \frac{\theta\, N\, \alpha}{\alpha+(N-1)\beta} + (1-\theta) = \frac{(\theta\, N+1-\theta)\alpha+(1-\theta)(N-1)\beta}{\alpha+(N-1)\beta}$, which gives for σ^{eff} the upper bound

$$\sigma^{eff} = \sum_{j=1}^{N} \frac{\lambda_j}{\beta-\lambda_j} \leq \frac{(\theta\, N+1-\theta)\alpha+(1-\theta)(N-1)\beta}{\theta(\beta-\alpha)}, \tag{7.30}$$

and equality occurs for the case of layers, i.e

$$\frac{1}{\beta-\lambda_-(\theta)} + \frac{N-1}{\beta-\lambda_+(\theta)} = \frac{(\theta\, N+1-\theta)\alpha+(1-\theta)(N-1)\beta}{\theta(\beta-\alpha)}. \tag{7.31}$$

I discuss now the basic result of Compensated Compactness theory, which has been used for Lemma 29 through the condition (7.12); Lemma 28 is more easy, and actually follows from the Div-Curl lemma, stated in (4.8)/(4.9) in Lemma 2, and mostly used in the case of gradients for which a simple proof by integration by parts has been shown, except for an application in Proposition 9. The necessity of a condition like (7.12) is easy and the general result is not even restricted to quadratic functionals [Ta8].

Proposition 30: Assume that Ω is an open subset of R^N, $A_{ijk}, i = 1,\dots,q, j = 1,\dots,p, k = 1,\dots,N$, are real constants and F is a continuous real function on R^p such that, whenever

$$\begin{aligned} &U^n \rightharpoonup U^\infty \text{ in } L^\infty(\Omega; R^p) \text{ weak } \star \\ &F(U^n) \rightharpoonup V^\infty \text{ in } L^\infty(\Omega) \text{ weak } \star \\ &\sum_{j=1}^{p}\sum_{k=1}^{N} A_{ijk} \frac{\partial U_j^n}{\partial x_k} = 0 \text{ for } i = 1,\dots,q, \end{aligned} \tag{7.32}$$

one can deduce that

$$V^\infty \geq F(U^\infty) \text{ a.e. in } \Omega. \tag{7.33}$$

Then F is Λ-convex, i.e.

$$t \mapsto F(a+t\lambda) \text{ is convex for all } a \in R^p \text{ and all } \lambda \in \Lambda, \tag{7.34}$$

where

$$\begin{aligned} \Lambda = \Big\{\lambda | \lambda \in R^p : &\text{ there exists } \xi \in R^N \setminus 0, \\ &\sum_{j=1}^{p}\sum_{k=1}^{N} A_{ijk}\lambda_j \xi_k = 0 \text{ for } i = 1,\dots,q\Big\}. \end{aligned} \tag{7.35}$$

Proof: Let $\lambda \in \Lambda$ and $\xi \in R^N \setminus 0$ satisfy the condition in (7.35), then if one takes

$$U^n(x) = a + \lambda f^n\Big((\xi.x)\Big), \tag{7.36}$$

with f^n smooth, one has

$$\sum_{j=1}^{p}\sum_{k=1}^{N} A_{ijk}\frac{\partial U_j^n}{\partial x_k} = \Big(\sum_{j=1}^{p}\sum_{k=1}^{N} A_{ijk}\lambda_j\xi_k\Big)(f^n)'\Big((\xi.x)\Big) = 0. \tag{7.37}$$

One chooses a sequence χ_n of characteristic functions converging in $L^\infty(R)$ weak $\star$ to θ and a regularization $f^n = \rho_n \star \chi_n$ such that $f^n - \chi_n$ converges almost everywhere to 0, one has

$$\begin{aligned} &U^n \approx \chi_n(a+\lambda) + (1-\chi_n)a,\ F(U^n) \approx \chi_n F(a+\lambda) + (1-\chi_n)F(a)\\ &U^\infty = \theta(a+\lambda) + (1-\theta)a,\ V^\infty = \theta\, F(a+\lambda) + (1-\theta)F(a). \end{aligned} \tag{7.38}$$

By hypothesis, one has $V^\infty \geq F(U^\infty)$ and (7.34) follows by varying $\theta \in (0,1)$, $a \in R^N$, and $\lambda \in \Lambda$.■

The sufficiency of a condition like (7.12) comes from applying a general result valid for quadratic functionals, which I often call the quadratic theorem of Compensated Compactness [Ta8].

Theorem 31: Let Q be a real homogeneous quadratic form on R^p which is Λ-convex, with Λ defined in (7.35), or equivalently

$$Q(\lambda) \geq 0 \text{ for all } \lambda \in \Lambda. \tag{7.39}$$

If

$$\begin{aligned} &U^n \rightharpoonup U^\infty \text{ in } L^2_{loc}(\Omega; R^p) \text{ weak}\\ &Q(U^n) \rightharpoonup \nu \text{ in the sense of measures}\\ &\sum_{j=1}^{p}\sum_{k=1}^{N} A_{ijk}\frac{\partial U_j^n}{\partial x_k} \text{ stays in a compact of } H^{-1}_{loc}(\Omega) \text{ strong for } i=1,\ldots,q, \end{aligned} \tag{7.40}$$

then one has

$$\nu \geq Q(U^\infty) \text{ in the sense of measures.} \tag{7.41}$$

Proof. $U^n - U^\infty$ satisfies (7.40) with U^∞ replaced by 0, and ν replaced by $\nu - Q(U^\infty)$, and one may then assume that $U^\infty = 0$ with the goal of proving that $\nu \geq 0$. For $\varphi \in C^1_c(\Omega)$, let $W^n = \varphi U^n$, which is extended by 0 outside Ω and let us prove that

$$\liminf_{n\to\infty} \int_{R^N} Q(W^n)\, dx \geq 0. \tag{7.42}$$

This shows that $\langle \nu, \varphi^2\rangle \geq 0$ for all $\varphi \in C^1_c(\Omega)$, and by density for all $\varphi \in C_c(\Omega)$, and as every nonnegative function in $C_c(\Omega)$ is a square, one deduces that $\nu \geq 0$ in the sense of measures.

If $Q(U) = \sum_{ij} q_{ij}\, U_i U_j$ with $q_{ij} = q_{ji}$ for all $i,j = 1,\ldots,p$, I still denote Q the Hermitian extension to C^p, i.e. $Q(U) = \sum_{ij} q_{ij}\, U_i \overline{U_j}$, and by PLANCHEREL formula, (7.42) is equivalent to

$$\liminf_{n\to\infty} \int_{R^N} Q(\mathcal{F}W^n)\, d\xi \geq 0, \tag{7.43}$$

where $\mathcal{F}$ denotes FOURIER transform, for which I use Laurent SCHWARTZ's notations

$$\mathcal{F}W^n(\xi) = \int_{R^N} W^n(x)\, e^{-2i\pi(x.\xi)}\, dx. \tag{7.44}$$

One can replace Q by $\Re Q$ in (7.43), because the integral in (7.42) and therefore in (7.43) is real, and one notices that (7.39) is equivalent to

$$\Re Q(\lambda) \geq 0 \text{ for all } \lambda \in \Lambda + i\,\Lambda \subset C^p. \tag{7.45}$$

As W^n converges in $L^2(R^N; R^p)$ weak to 0 and keeps its support in a fixed compact set K of R^N, its FOURIER transform converges pointwise to 0 and is uniformly bounded and therefore by LEBESGUE dominated convergence theorem it converges in $L^2_{loc}(R^N)$ strong to 0, and the problem for proving (7.43) lies in the behaviour of $\mathcal{F}W^n$ at infinity. Information at infinity is given by the partial differential equations satisfied by W^n, and because $\sum_j \sum_k A_{ijk} \frac{\partial W_j^n}{\partial x_k}$ must converge in $H^{-1}(R^N)$ strong to 0 for $i = 1, \ldots, q$, one deduces that

$$\sum_{i=1}^{q} \int_{R^N} \frac{1}{1+|\xi|^2} \Big| \sum_{j=1}^{p} \sum_{k=1}^{N} A_{ijk}\, \mathcal{F}W_j^n(\xi)\xi_k \Big|^2 d\xi \to 0. \tag{7.46}$$

For $|\xi|$ large $\frac{\xi_k}{\sqrt{1+|\xi|^2}} \approx \frac{\xi_k}{|\xi|}$, and (7.46) tells that near infinity $\mathcal{F}W^n$ is near $\Lambda + i\,\Lambda$, where $\Re Q \geq 0$, and a proof of (7.43) follows from the fact that for every $\varepsilon > 0$ there exists C_ε such that

$$\Re Q(Z) \geq -\varepsilon |Z|^2 - C_\varepsilon \sum_{i=1}^{q} \Big| \sum_{j=1}^{p} \sum_{k=1}^{N} A_{ijk}\, Z_j \frac{\xi_k}{|\xi|} \Big|^2 \text{ for all } Z \in C^p, \xi \in R^N \setminus 0. \tag{7.47}$$

Applying (7.47) to $Z = \mathcal{F}W^n(\xi)$ and integrating in ξ for $|\xi| \geq 1$, gives a lower bound for $\int_{|\xi|\geq 1} \Re Q(\mathcal{F}W^n)\, d\xi$ where the coefficient of $-\varepsilon$ is bounded as W^n is bounded in $L^2(R^N; R^p)$ and the coefficient of C_ε tends to 0 by (7.46), and therefore one deduces that $\liminf_n \int_{R^N} \Re Q(\mathcal{F}W^n)\, d\xi \geq -M\,\varepsilon$, and letting ε tend to 0 proves (7.43). The inequality (7.47) is proved by contradiction: if there exists $\varepsilon_0 > 0$ and a sequence $Z^n \in C^N, \xi^n \in R^N \setminus 0$ such that $\Re Q(Z^n) < -\varepsilon_0 |Z^n|^2 - n \sum_i \big| \sum_{jk} A_{ijk}\, Z_j^n \frac{\xi_k^n}{|\xi^n|} \big|^2$, then after normalizing Z^n to $|Z^n| = 1$, and extracting a subsequence such that Z^n converges to Z^∞ and $\eta^n = \frac{\xi^n}{|\xi^n|}$ converges to η_∞, one finds that $\Re Q(Z^\infty) \leq -\varepsilon_0$, which contradicts the fact that (Z^∞, η^∞) satisfies the conditions of the definition of Λ in (7.35), showing that $Z^\infty \in \Lambda + i\,\Lambda$ and therefore implying $\Re Q(Z^\infty) \geq 0$ by (7.45).■

9. Computation of effective coefficients

I computed the bounds (7.28) and (7.30) in the Fall 1980 with François MURAT, but in June 1980 in New York I had only done the corresponding computations for the case where $A^{eff} = a^{eff}\, I$, and having shown my new bounds to George PAPANICOLAOU he had suggested that I compare them with the HASHIN-SHTRIKMAN bounds [Ha&Sh], which I was hearing about for the first time. As for the first method for obtaining bounds for effective coefficients that I had used before with François MURAT, that second method which I had developed in [Ta7] was not restricted to symmetric tensors, and I could not understand what would replace in more general problems the particular minimization formulation that Zvi HASHIN and S. SHTRIKMAN had used, but there was an obvious gap in their "proof", and at the time I could not find a mathematical argument which could explain their computation.[54] However, the bounds which I had just found were indeed the same as the formal bounds that they had derived, and I had therefore given the first mathematical proof that the HASHIN-SHTRIKMAN bounds are indeed valid for mixtures of two isotropic materials in the case where the effective tensor is isotropic. I had not yet thought of showing that my bounds were optimal and could be attained (which I would have tried with the method of successive layerings, which was the only simple explicit construction that I knew), and as the construction of coated spheres that Zvi HASHIN and S. SHTRIKMAN had used was clear enough to me, I easily transformed it into a correct mathematical argument, but I did not try to compare with what the repeated formula for layerings would have given.

When I went back to Paris after spending the Summer at the Mathematics Research Center in Madison, I showed my computations to François MURAT an he suggested that the same functionals might also give optimal bounds for anisotropic effective tensors, and therefore we computed (7.28) and (7.30) and we tried to show that the bounds were attained by a construction of coated confocal ellipsoids. Edward FRAENKEL was visiting Paris in the Fall of 1980, and as I had mentioned to him our plan, he had given us some advice about the way to compute with ellipsoids, but we could not follow precisely what he had told us. We tried then families of general surfaces, and in order to simplify a very technical computation, we made a simplifying assumption, and that gave us exactly the case of confocal ellipsoids, but using different

[54] In their argument, Zvi HASHIN and S. SHTRIKMAN used something that did not make any (mathematical) sense at the time, and it seems now related to H-measures, which I only introduced in the late 80s for a different purpose [Ta12]; the new method for deriving bounds which I wrote in [Ta12], generalizing my earlier approach of [Ta7], has actually some analogy with the argument of Zvi HASHIN and S. SHTRIKMAN, but it is not restricted to minimization problems.

formulas than the ones that Edward FRAENKEL had advocated. Indeed the set of bounds (7.25), (7.28), (7.30) gave the characterization of the effective tensors of mixtures using exactly proportion θ of an isotropic material with tensor αI and proportion $1-\theta$ of an isotropic material with tensor βI. I will not describe the computations for confocal ellipsoids, for which I refer to [Ta9], as I will describe a simpler approach later.

I described our results at a meeting at New York University in June 1981, and they gave the missing link in the method that I had partially described in 1974 [Ta2]. As I suggested that the case of mixing more than two isotropic materials would probably be very similar with a construction like the HASHIN-SHTRIKMAN coated spheres in the isotropic case, with materials of increasing conductivity from inside out or from outside in depending upon which bound was considered, I was surprised to hear a comment by a young participant that even for three materials it was not so, and that hiding the best conductor in the middle was sometimes giving a better effective conductivity than if the best conductor was put outside. The comment was coming from a young Australian physicist, still a graduate student at the time, who has since imposed himself as the best specialist for questions of bounds of effective coefficients, Graeme MILTON.[55]

In the Spring 1982, I gave an introductory course to Homogenization at Ecole Polytechnique, and as I thought that our construction with confocal ellipsoids could not be avoided, I had asked two students, Philippe BRAIDY and Didier POUILLOUX, to make a numerical study comparing the materials that we had constructed by using confocal ellipsoids and those that could be constructed by successive layerings, which was the method that we had used for the results quoted in [Ta2]. Contrary to my mistaken expectations, they reported that the numerical computations showed that the two sets were the same, and a few days after they had a proof of it, using N layerings in orthogonal directions, where in each layering the direction orthogonal to the layers is a common eigenvector for the two materials being mixed [Br&Po]. I immediately checked that the repeated layering construction has the same restricted "generalized BERGMAN function" than our construction with confocal ellipsoids. For example, in my interpretation of the construction with coated spheres of Zvi HASHIN and S. SHTRIKMAN in June 1980, for a parameter $\theta \in [0,1]$ I used a sequence of VITALI coverings with smaller and smaller coated spheres, all showing the same proportion of volume between the inside spheres and the outside spherical coats; the geometry being given I imagined

[55] In the early 90s, while I visited Graeme MILTON in New York, he gave me a physical explanation of why it is sometime good to hide the best conductor available; he argued that if one has a spherical core of a very poor conductor, the electric current tries to avoid it and that creates a high concentration of field lines near the surface of the poor conductor and therefore it is there that the best conductor is more useful.

all the interior spheres filled with an isotropic material with tensor αI and all the outside spherical coats filled with an isotropic material with tensor βI, and then I showed that the sequence A^n defined in this way H-converges to $\Phi_0(\alpha,\beta)\, I$, where $\Phi_0(\alpha,\beta)$ is one of the two HASHIN-SHTRIKMAN bounds, corresponding to equality in (7.28).[56] This is exactly the type of functions used by David BERGMAN when one mixes two isotropic conductors and one expects the resulting effective material to be isotropic whatever the ratio of the two conductivities are, and therefore one has $\Phi_0(\alpha,\beta) = \alpha\, F\big(\frac{\beta}{\alpha}\big)$. David BERGMAN made the important observation that F extends to the complex upper half plane into a holomorphic function satisfying $\Im(F(z)) \geq 0$ [Be].[57] What I call a "generalized BERGMAN function" is a similar situation where a sequence of geometries is given for mixing r materials with proportions $\theta_1,\ldots,\theta_r$, and if the r materials used have tensors $M_1,\ldots,M_r$, then the resulting effective tensor is $\Phi(M_1,\ldots,M_r)$. One assumes that for $j=1,\ldots,r$, $M_j \in M(\alpha_j,\beta_j;\Omega)$, and that one has r sequences of characteristic functions of measurable sets from a partition χ_j^n, $j=1,\ldots,r$, satisfying $\chi_j^n\chi_k^n = 0$ for $j \neq k$, $\sum_j \chi_j^n = 1$ a.e. in Ω, and one assumes that χ_j^n converges in $L^\infty(\Omega)$ weak $\star$ to θ_j for $j=1,\ldots,r$; then one uses Proposition 17 in order to show that there is a subsequence for which for all such $M_1,\ldots,M_r$, $\sum_j \chi_j^m M_j$ H-converges to an element $\Phi(M_1,\ldots,M_r)$ of $M(\alpha,\beta;\Omega)$ with $0<\alpha=\min\{\alpha_1,\ldots,\alpha_r\} \leq \beta = \max\{\beta_1,\ldots,\beta_r\} < \infty$. I use the qualificative restricted for expressing the fact that one restricts attention to a special class

[56] My proof relied on the fact that, despite the huge arbitrariness in the choice of the VITALI coverings, for any $\lambda \in R^N$ I could write explicit solutions of $div(A^n\, grad(u_n)) = 0$ for which $grad(u_n)$ converges weakly to λ and I could compute the limit of $A^n\, grad(u_n)$. My construction was local, and I followed the computation that I had just read in the article of Zvi HASHIN and S. SHTRIKMAN [Ha&Sh], i.e. the explicit solution of $div(A\, grad(v)) = 0$ in a coated sphere domain with affine boundary conditions on the outside coat; this computation is a by-product of the formula for the change in electric field created by an isotropic spherical conducting inclusion in an infinite isotropic medium, a classical formula for physicists who associate it with various names, some as famous as MAXWELL, but it must have been known to GAUSS and to DIRICHLET, who seems to be credited for a similar formula for an ellipsoid (and he may therefore have known the formulas that François MURAT and I (re)discovered for our construction with coated ellipsoids).

[57] The idea may have been used before, and I think that I had heard such an idea attributed to PRAGER. In dimension $N=2$, the function F also satisfies the relation $F(z)F\big(\frac{1}{z}\big) = 1$ by an argument of Joseph KELLER [Ke], and in the early 80s Graeme MILTON showed me that all such functions can be obtained.

of $M_1, \ldots, M_r$, for example isotropic tensors $m_1 I, \ldots, m_r I$,[58] and I do not know how to compute the generalized BERGMAN function for a geometry of coated spheres or confocal ellipsoids, and it may be dependent of other properties of the VITALI covering used.[59] In their computation, Philippe BRAIDY and Didier POUILLOUX had used the same method that Antonio MARINO and Sergio SPAGNOLO or François MURAT and I had used in the early 70s, and the reiteration of the layering formula was simple enough because at each step the direction orthogonal to the layers was a common eigenvector of both (symmetric) tensors which were mixed, and actually the two tensors had a common basis of eigenvectors. During the Spring 1983, while I was visiting the Mathematical Sciences Research Institute in Berkeley, I tried to compute the formula for mixing arbitrary materials in arbitrary directions, having in mind to reiterate the procedure. I wanted to rewrite formula (4.11) in a more intrinsic way, and I could easily deduce what the formula (4.11) would become if I used layers orthogonal to a vector e, i.e. A^n depending only upon $(x.e)$, but that did not change much, and it was a different idea that simplified the computation. Using layers orthogonal to e for mixing two materials with tensors A and B, with respective proportions θ and $1-\theta$, the simplification came by considering θ small, and because the formula appeared to have the form $B + \theta\, F(A, B, e) + o(\theta)$, it suggested to write a differential equation $B' = F(A, B, e)$ and integrate it. In other terms, for e fixed, increasing the proportion of A from 0 to 1 creates a curve going from B to A in the space of matrices, and I first computed that curve by considering it as the trajectory of a differential equation, which was easy to write down. One can first rewrite formula (4.11) for layers orthogonal to e.

---

58 Ken GOLDEN and George PAPANICOLAOU have studied functions of r complex variables F appearing when one imposes the restriction $\Phi(m_1 I, \ldots, m_r I) = F(m_1, \ldots, m_r)I$ for all $m_1, \ldots, m_r > 0$.

59 The computations of Zvi HASHIN and S. SHTRIKMAN for coated spheres and diffusion equation consists in looking for solutions of the form $x_j f(r)$ and one finds that f must satisfy a differential equation; they also used the same construction of coated spheres for linearized Elasticity with isotropic materials, and they could compute the effective bulk modulus because it corresponds to applying a uniform pressure and the displacement has the form $x\, g(r)$, and one finds that g must satisfy a differential equation. I do not know how to compute the effective shear modulus for the geometry of coated spheres, and it may depend upon which VITALI covering is used (if I understood correctly what Graeme MILTON told me a few years ago, he knew that it does depend upon the covering). Gilles FRANCFORT and François MURAT have computed in [Fr&Mu] the complete effective elasticity tensors of mixtures, but following the method of multiple layerings, adapting the extension that I had given in [Ta9] of the computation of Philippe BRAIDY and Didier POUILLOUX.

$$
\begin{aligned}
&\frac{1}{(A^n e.e)} \rightharpoonup \frac{1}{(A^{eff} e.e)} \text{ in } L^\infty(\Omega) \text{ weak } \star\\
&\frac{(A^n f.e)}{(A^n e.e)} \rightharpoonup \frac{(A^{eff} f.e)}{(A^{eff} e.e)} \text{ in } L^\infty(\Omega) \text{ weak } \star \text{ for every } f \perp e\\
&\frac{(A^n e.g)}{(A^n e.e)} \rightharpoonup \frac{(A^{eff} e.g)}{(A^{eff} e.e)} \text{ in } L^\infty(\Omega) \text{ weak } \star \text{ for every } g \perp e\\
&(A^n f.g) - \frac{(A^n f.e)(A^n e.g)}{(A^n e.e)} \rightharpoonup (A^{eff} f.g) - \frac{(A^{eff} f.e)(A^{eff} e.g)}{(A^{eff} e.e)}\\
&\qquad\qquad \text{ in } L^\infty(\Omega) \text{ weak } \star \text{ for every } f \perp e, g \perp e,
\end{aligned}
\tag{8.1}
$$

where I have used the Euclidean structure of R^N.[60] Mixing A with a small proportion θ and B with proportion $1-\theta$ in layers orthogonal to e gives then

$$
\begin{aligned}
&\frac{1}{(A^{eff} e.e)} = \frac{1-\theta}{(B\, e.e)} + \frac{\theta}{(A\, e.e)}\\
&(A^{eff} e.e) = (B\, e.e) + \theta\Big[(B\, e.e) - \frac{(B\, e.e)^2}{(A\, e.e)}\Big] + o(\theta),
\end{aligned}
\tag{8.2}
$$

and then for f and g orthogonal to e

$$
\begin{aligned}
&\frac{(A^{eff} f.e)}{(A^{eff} e.e)} = \frac{(1-\theta)(B\, f.e)}{(B\, e.e)} + \frac{\theta(A\, f.e)}{(A\, e.e)}\\
&(A^{eff} f.e) = (B\, f.e) + \theta\Big[\frac{(A\, f.e)(B\, e.e)}{(A\, e.e)} - \frac{(B\, f.e)(B\, e.e)}{(A\, e.e)}\Big] + o(\theta)
\end{aligned}
\tag{8.3}
$$

$$
\begin{aligned}
&\frac{(A^{eff} e.g)}{(A^{eff} e.e)} = \frac{(1-\theta)(B\, e.g)}{(B\, e.e)} + \frac{\theta(A\, e.g)}{(A\, e.e)}\\
&(A^{eff} e.g) = (B\, e.g) + \theta\Big[\frac{(A\, e.g)(B\, e.e)}{(A\, e.e)} - \frac{(B\, e.g)(B\, e.e)}{(A\, e.e)}\Big] + o(\theta)
\end{aligned}
\tag{8.4}
$$

$$
\begin{aligned}
&(A^{eff} f.g) - \frac{(A^{eff} f.e)(A^{eff} e.g)}{(A^{eff} e.e)} =\\
&\quad = (1-\theta)(B\, f.g) - \frac{(1-\theta)(B\, f.e)(B\, e.g)}{(B\, e.e)} + \theta(A\, f.g) - \frac{\theta(A\, f.e)(A\, e.g)}{(A\, e.e)}\\
&(A^{eff} f.g) = (B\, f.g) +\\
&\quad + \theta\Big[(A\, f.g) - (B\, f.g) - \frac{\Big((B\, f.e) - (A\, f.e)\Big)\Big((B\, e.g) - (A\, e.g)\Big)}{(A\, e.e)}\Big] + o(\theta).
\end{aligned}
\tag{8.5}
$$

The form of (8.5) suggests that one has

---

[60] It can be avoided by denoting $\mathcal{E}$ the ambient vector space, taking e as an element of the dual $\mathcal{E}'$, and considering the tensors A, B, as elements of $\mathcal{L}(\mathcal{E}', \mathcal{E})$, as well as $e \otimes e$ which appears in some formulas.

$$A^{eff} = B + \theta\Big[A - B - (B - A)\frac{e \otimes e}{(A\,e.e)}(B - A)\Big] + o(\theta), \tag{8.6}$$

and indeed this is compatible with (8.2)/(8.4). When e and A are given, formula (8.6) corresponds to a differential equation

$$M' = A - M - (M - A)\frac{e \otimes e}{(A\,e.e)}(M - A). \tag{8.7}$$

The integral curve corresponds to the formula for layering with a material with tensor A, with layers orthogonal to e. Formula (8.1), which corresponds to the first lines of (8.2)/(8.5) when one mixes two materials with tensors A and B says that the integral curves become straight lines if one performs the change of variable $A \mapsto \left(\frac{1}{(A\,e.e)}, \frac{(A\,f.e)}{(A\,e.e)}, \frac{(A\,e.g)}{(A\,e.e)}, (A\,f.g) - \frac{(A\,f.e)(A\,e.g)}{(A\,e.e)}\right)$ when f, g span the subspace orthogonal to e. In the early 70s, we knew that if one does not pay attention to the proportions used of various materials, formula (4.11) means that the set of effective tensors has the property that all its images by maps like the one mentioned above are automatically convex. This condition gives a geometric characterization of the sets that one cannot enlarge by layering, at least for the case where one is not allowed to rotate the materials used, in the case where one starts with some anisotropic materials; in realistic problems, one must also allow for rotations of the materials used, i.e. the set must be stable by mappings $A \mapsto P^T A\,P$ for $P \in SO(N)$, with $N = 2$ or $N = 3$ usually.

Assuming that $M - A$ is invertible, (8.7) can be written as

$$[(M - A)^{-1}]' = -(M - A)^{-1}A'(M - A)^{-1} = (M - A)^{-1} + \frac{e \otimes e}{(A\,e.e)}, \tag{8.8}$$

whch is a linear equation in $(M - A)^{-1}$. Using τ as variable, and assuming that $\tau = 0$ corresponds to B, the solution of (8.8) is

$$(M - A)^{-1} = -\frac{e \otimes e}{(A\,e.e)} + e^{\tau}\Big((B - A)^{-1} + \frac{e \otimes e}{(A\,e.e)}\Big), \tag{8.9}$$

and if M corresponds to having used proportion $\eta(\tau)$ of A and $1 - \eta(\tau)$ of B, then for θ small $\eta(\tau + \theta) = \theta + (1 - \theta)\eta(\tau) + o(\theta)$ gives $\eta' = 1 - \eta$ and therefore $\eta = 1 - e^{-\tau}$ or equivalently $e^{\tau} = \frac{1}{1-\eta}$ for proportion η of A, giving

$$(M - A)^{-1} = \frac{(B - A)^{-1}}{1 - \eta} + \frac{\eta}{1 - \eta}\frac{e \otimes e}{(A\,e.e)} \text{ for proportion } \eta \text{ of } A. \tag{8.10}$$

If $(B - A)z = 0$ for a nonzero vector z, then (8.7) shows that $(M - A)z = 0$, and in this case one must reinterpret (8.10). Of course, exchanging the role of A and B and changing η into $1 - \eta$, (8.10) is replaced by

$$(M - B)^{-1} = \frac{(A - B)^{-1}}{\eta} + \frac{1 - \eta}{\eta}\frac{e \otimes e}{(B\,e.e)} \text{ for proportion } \eta \text{ of } A. \tag{8.11}$$

With formula (8.10) at hand I could easily reiterate the layering process with various directions of layers, with the condition that each layering uses the material with tensor A, and it gave the following generalization of the formula which had been obtained by Philippe BRAIDY and Didier POUILLOUX in the special case where A and B have a common basis of eigenvectors and each e is one of these common eigenvectors.

Proposition 32: For $\eta \in (0,1)$, let $\xi_1, \ldots, \xi_p$ be p positive numbers with $\sum_j \xi_j = 1-\eta$, let $e_1, \ldots, e_p$ be p nonzero vectors of R^N, then using proportion η of material with tensor A and proportion $1-\eta$ of material with tensor B, one can construct by multiple layerings the material with tensor M such that

$$(M-B)^{-1} = \frac{(A-B)^{-1}}{\eta} + \frac{1}{\eta}\Big(\sum_{j=1}^{p} \xi_j \frac{e_j \otimes e_j}{(B\,e_j.e_j)}\Big). \tag{8.12}$$

Proof: Of course, one assumes that $B-A$ is invertible, as the formula must be reinterpreted if $B-A$ is not invertible. One starts from $M_0 = A$ and by induction one constructs M_j by layering M_{j-1} and B in proportions η_j and $1-\eta_j$, with layers orthogonal to e_j. Formula (8.11) gives

$$(M_j - B)^{-1} = \frac{(M_{j-1}-B)^{-1}}{\eta_j} + \frac{1-\eta_j}{\eta_j}\frac{e \otimes e}{(B\,e.e)} \text{ for } j = 1, \ldots, p, \tag{8.13}$$

which is adapted to reiteration and provides (8.12) with

$$\begin{aligned} &\eta = \eta_1 \cdots \eta_p \\ &\xi_1 = 1-\eta_1, \xi_j = \eta_1 \cdots \eta_{j-1}(1-\eta_j) \text{ for } j = 1, \ldots, p, \end{aligned} \tag{8.14}$$

which gives $\xi_1 + \ldots + \xi_j = 1 - \eta_1 \cdots \eta_j$ for $j = 1, \ldots, p$, and this defines in a unique way η_j for $j = 1, \ldots, p$.■

The preceding computations did not require any symmetry assumption for A or B. The characterization of the sum $\sum_j \xi_j \frac{e_j \otimes e_j}{(B\,e_j.e_j)}$ for all $\xi_j > 0$ with sum $1-\eta$ and all nonzero vectors e_j depends only on the symmetric part of B (and of η).

Lemma 33: If B is symmetric positive definite then for $\xi_1, \ldots, \xi_p > 0$ and nonzero vectors $e_1, \ldots, e_p$, one has

$$\begin{aligned} &\sum_{j=1}^{p} \xi_j \frac{e_j \otimes e_j}{(B\,e_j.e_j)} = B^{-1/2}\,K\,B^{-1/2}, \\ &\qquad \text{with } K \text{ symmetric nonnegative and } trace(K) = \sum_{j=1}^{p} \xi_j, \end{aligned} \tag{8.15}$$

and conversely any such K can be obtained in this way.

Proof: Putting $e_j = B^{-1/2} f_j$ for $j = 1, \ldots, p$, one has $K = \sum_j \xi_j \frac{f_j \otimes f_j}{|f_j|^2}$, and each $\frac{f_j \otimes f_j}{|f_j|^2}$ is a nonnegative symmetric tensor with trace 1, and (8.15) follows. Conversely if K is a symmetric nonnegative tensor with trace equal to S, then there is an orthonormal basis of eigenvectors $f_1, \ldots, f_N$, with $K f_j = \kappa_j f_j$ and $\kappa_j \geq 0$ for $j = 1, \ldots, N$, and $\sum_j \kappa_j = S$, so that $K = \sum_j \kappa_j f_j \otimes f_j$.■

Using Proposition 32 and Lemma 33, with $A = \alpha I$ and $B = \beta I$, one can construct materials with a symmetric tensor M with eigenvalues $\lambda_1, \ldots, \lambda_N$, and (8.12) and Lemma 33 mean that

$$\begin{aligned} &\frac{1}{\lambda_j - \beta} \geq \frac{1}{\eta(\alpha - \beta)} \text{ for } j = 1, \ldots, N \\ &\sum_{j=1}^{N} \frac{1}{\lambda_j - \beta} = \frac{N}{\eta(\alpha - \beta)} + \frac{1-\eta}{\eta\,\beta}, \end{aligned} \tag{8.16}$$

i.e. $\lambda_j \leq \lambda_+(\eta)$ for $j = 1, \ldots, N$, and equality in (7.30), which implies $\lambda_j \geq \lambda_-(\eta)$ for $j = 1, \ldots, N$, because of (7.31). Exchanging the roles of A and B one can obtain another part of the boundary of possible effective tensors with equality in (7.28), and filling the interior of the set is then easy.

After I had mentioned these new results to Robert KOHN, who was also visiting MSRI at the time, he wondered if one could find a more direct proof, and I therefore proved again the formulas (8.10)/(8.11) directly.

Lemma 34: Mixing materials with tensor A and B with respective proportions η and $1-\eta$ in layers orthogonal to e gives an effective tensor A^{eff} given by

$$A^{eff} = \eta\, A + (1-\eta) B - \eta(1-\eta)(B-A)\frac{e \otimes e}{(1-\eta)(A\, e.e) + \eta(B\, e.e)}(B-A). \tag{8.17}$$

Proof: One considers a sequence of characteristic functions χ_n converging in $L^\infty(R)$ weak $\star$ to η and depending only upon $(x.e)$, and one chooses $A^n = \chi_n A + (1-\chi_n) B$. For an arbitrary vector $E^\infty \in R^N$, one constructs a sequence $E^n = grad(u_n)$ converging in $L^2_{loc}(R^N; R^N)$ weak to E^∞, depending only upon $(x.e)$ and satisfying $div(A^n\, grad(u_n)) = 0$, and one computes the limit in $L^2_{loc}(R^N; R^N)$ weak of $D^n = A^n\, grad(u_n)$, which will be $D^\infty = A^{eff} E^\infty$, with A^{eff} given by (8.17).

One looks for $E_A, E_B \in R^N$ such that one can take

$$\begin{aligned} E^n &= \chi_n\, E_A + (1-\chi_n) E_B \\ D^n &= \chi_n\, A\, E_A + (1-\chi_n) B\, E_B \\ \eta\, E_A &+ (1-\eta) E_B = E^\infty, \end{aligned} \tag{8.18}$$

and the constraints $curl(E^n) = div(D^n) = 0$ become

$$\begin{aligned} &E_B - E_A = c\,e \\ &(B\,E_B - A\,E_A.e) = 0, \end{aligned} \tag{8.19}$$

and then one should have

$$\eta\,A\,E_A + (1-\eta)B\,E_B = A^{eff}\,E^\infty. \tag{8.20}$$

One chooses then

$$E_A = E^\infty + c_A\,e;\ E_B = E^\infty + c_B\,e;\ \eta\,c_A + (1-\eta)c_B = 0, \tag{8.21}$$

and (8.19) requires that

$$\Big((B-A)E^\infty.e\Big) + c_B(B\,e.e) - c_A(A\,e.e) = 0, \tag{8.22}$$

and (8.21)/(8.22) give

$$\begin{aligned} &\Big((1-\eta)(A\,e.e) + \eta(B\,e.e)\Big)c_A = (1-\eta)\Big((B-A)E^\infty.e\Big) \\ &\Big((1-\eta)(A\,e.e) + \eta(B\,e.e)\Big)c_B = -\eta\Big((B-A)E^\infty.e\Big), \end{aligned} \tag{8.23}$$

and therefore (8.20) becomes

$$\begin{aligned} A^{eff}\,E^\infty &= (\eta\,A + (1-\eta)B)E^\infty + \\ &+ \frac{\Big((B-A)E^\infty.e\Big)}{(1-\eta)(A\,e.e) + \eta(B\,e.e)}\Big(\eta(1-\eta)A\,e - \eta(1-\eta)B\,e\Big), \end{aligned} \tag{8.24}$$

and as (8.24) is true for every $E^\infty \in R^N$, one deduces formula (8.17) for A^{eff}.■

One deduces then (8.10)/(8.11) from (8.17) by applying a result of Linear Algebra.

Lemma 35: If $M \in \mathcal{L}(\mathcal{E},\mathcal{F})$ is invertible, and if $a \in \mathcal{F}, b \in \mathcal{E}'$, then $M + a\otimes b$ is invertible if $(M^{-1}a.b) \neq -1$ and

$$(M + a\otimes b)^{-1} = M^{-1} - \frac{1}{1 + (M^{-1}a.b)}M^{-1}(a\otimes b)M^{-1}. \tag{8.25}$$

Proof: One wants to solve $(M + a\otimes b)x = y$, i.e. $M\,x + a(b.x) = y$, and therefore $x = M^{-1}\,y - t\,M^{-1}\,a$ with $t = (b.x)$, but one needs then to have $t = (b.M^{-1}\,y) - t(M^{-1}\,a.b)$, which is possible because $(M^{-1}\,a.b) \neq -1$, and gives $x = M^{-1}\,y - M^{-1}\,a\frac{(b.M^{-1}\,y)}{1+(M^{-1}\,a.b)}$, and as y is arbitrary it gives (8.25).■

A few years ago, working with François MURAT on the relation between YOUNG measures and H-measures,[61] we computed the analog of formula (8.17) when one mixes r different materials.

Lemma 36: Mixing r materials with tensors $M_1, \ldots, M_r$, with respective proportions $\eta_1, \ldots, \eta_r$, in layers orthogonal to e, gives an effective tensor M^{eff} given by

$$
\begin{aligned}
M^{eff} &= \sum_{i=1}^{r} \eta_i M_i - \sum_{1 \le i < j \le r} \eta_i \eta_j (M_i - M_j) R_{ij} (M_i - M_j) \\
R_{ij} &= \frac{1}{(M_i\, e.e)} \frac{e \otimes e}{H} \frac{1}{(M_j\, e.e)} \text{ for } i,j = 1, \ldots, r \qquad (8.26) \\
H &= \sum_{k=1}^{r} \frac{\eta_k}{(M_k\, e.e)}.
\end{aligned}
$$

Proof: As for the proof of Lemma 34, one uses

$$
E_i = E^\infty + c_i\, e \text{ in the layers of material } \#i, \quad \text{for } i = 1, \ldots, r, \text{ and } \sum_{i=1}^{r} \eta_i\, c_i = 0, \qquad (8.27)
$$

and one must have $(M_i\, E_i.e) = (M_j\, E_j.e)$ if there is an interface between material $\#i$ and material $\#j$, and therefore there exists a constant C such that

$$
(M_i\, E_i.e) = C \text{ for } i = 1, \ldots, r. \qquad (8.28)
$$

With the definition of $E_i, i = 1, \ldots, r$, (8.28) implies

$$
c_i = \frac{C - (M_i\, E^\infty.e)}{(M_i\, e.e)} \text{ for } i = 1, \ldots, r, \qquad (8.29)
$$

and the condition $\sum_i \eta_i c_i = 0$ gives

$$
H\, C = \sum_{i=1}^{r} \eta_i \frac{(M_i\, E^\infty.e)}{(M_i\, e.e)}, \qquad (8.30)
$$

with H given in (8.26). Using (8.28) one obtains

[61] I have used in various publications not related to the purpose of this course our construction of admissible pairs of a YOUNG measure and a H-measure associated with a sequence, the first time at a meeting for the 600th anniversary of the University of Ferrara in 1991. Our redaction of these results is still a draft which we have not looked at for years, but I have nevertheless given it to a few persons, and I wonder how many will have claimed our results as theirs.

$$H\,(M_i\,e.e)c_i = \Big(\sum_{j=1}^{r}\eta_j\frac{(M_j\,E^\infty.e)}{(M_j\,e.e)}\Big) - \frac{(M_i\,E^\infty.e)}{(M_i\,e.e)} = \\ = \sum_{j=1}^{r}\eta_j\frac{\Big(\big(M_j - M_i\big)\,E^\infty.e\Big)}{(M_j\,e.e)} \text{ for } i = 1,\ldots,r. \tag{8.31}$$

This gives

$$M^{eff}\,E^\infty = \\ = \Big(\sum_{i=1}^{r}\eta_i\,M_i\Big)E^\infty + \frac{1}{H}\sum_{i=1}^{r}\frac{\eta_i}{(M_i\,e.e)}\Big(\sum_{j=1}^{r}\eta_j\frac{\Big(\big(M_j - M_i\big)\,E^\infty.e\Big)}{(M_j\,e.e)}\Big)M_i\,e = \\ = \Big(\sum_{i=1}^{r}\eta_i\,M_i\Big)E^\infty - \frac{1}{2H}\sum_{i,j=1}^{r}\eta_i\eta_j\frac{\Big(\big(M_i - M_j\big)E^\infty.e\Big)}{(M_i\,e.e)(M_j\,e.e)}(M_i - M_j)e = \\ = \Big(\sum_{i=1}^{r}\eta_i\,M_i\Big)E^\infty - \frac{1}{H}\sum_{i<j}\eta_i\eta_j(M_i - M_j)\frac{e\otimes e}{(M_i\,e.e)(M_j\,e.e)}(M_i - M_j)E^\infty, \tag{8.32}$$

proving (8.26).■

I have shown the derivation of the differential equation (8.7) because it has some intrinsic interest. I noticed later that by using relaxation techniques related to Lemma 1, one can replace $\frac{e\otimes e}{(A\,e.e)}$ in (8.7) by any convex combination $\sum_j \theta_j \frac{e_j\otimes e_j}{(A\,e_j.e_j)}$, or more generally $\int_{S^{N-1}} \frac{e\otimes e}{(A\,e.e)}\,d\pi(e)$ for a probability measure π on the sphere S^{N-1}, and this gives a differential analogue of formula (8.12).[62]

As I will explain in the next chapter, I also discovered in the Spring 1983 that the characterization obtained with François MURAT was not absolutely necessary for solving the problems that we had in mind. There was an obvious generalization to mixing an arbitrary number of isotropic materials, but there were some technical details for mixing anisotropic materials, and I only noticed the following results much later (see footnote 44), and Lemma 18 played a crucial role. Although the two methods for obtaining bounds on effective coefficients that I have described in chapters 5 and 7 are valid for nonnecessarily symmetric operators (as is the third one that I have mentioned in footnote 54, based on H-measures), the only applications that I know use

[62] One can also use a convex combination in $(A.e)$, and I had hoped that this trick would give more characterization of effective coefficients. I did talk about this method at a meeting in Minneapolis in 1985, but I only mentioned it in writing for a meeting in Los Alamos in 1987.

symmetric operators,[63] and because one allows for arbitrary rotations the set of effective operators corresponding to mixtures using precise proportions of each constituent is a set of matrices defined in terms of their eigenvalues, and that point is worth discussing in detail.

One starts from a finite number of materials with symmetric anisotropic tensors $M_i, i = 1, \ldots, r$, and mixing them with local proportions $\eta_i, i = 1, \ldots, r$, with $\sum_i \eta_i = 1$ a.e. in Ω, means that one considers sequences χ_i^n of characteristic functions of disjoint measurable sets for $i = 1, \ldots, r$, i.e. $\chi_i^n \chi_j^n = 0$ whenever $i \neq j$, such that

$$\begin{aligned} &\chi_i^n \rightharpoonup \eta_i \text{ in } L^\infty(\Omega) \text{ weak} \star \\ &A^n = \sum_{i=1}^r \chi_i^n (R^n)^T M_i R^n, \text{ with } R^n \in SO(N) \text{ a.e. in } \Omega \\ &A^n \text{ H-converges to } A^{eff}, \end{aligned} \tag{8.33}$$

and one writes

$$A^{eff} \in \mathcal{K}(\eta_1, \ldots, \eta_r; M_1, \ldots, M_r) \text{ a.e. in } \Omega, \tag{8.34}$$

and the claim is that the set $\mathcal{K}(\eta_1, \ldots, \eta_r; M_1, \ldots, M_r)$ of effective materials obtained by mixing the materials $M_1, \ldots, M_r$ with "exact proportions" $\eta_1, \ldots, \eta_r$ only depends upon A^{eff} through its eigenvalues $\lambda_1^{eff}, \ldots, \lambda_N^{eff}$, which satisfy

$$(\lambda_1^{eff}, \ldots, \lambda_N^{eff}) \in \Lambda(\eta_1, \ldots, \eta_r; M_1, \ldots, M_r) \text{ a.e. in } \Omega, \tag{8.35}$$

where $\Lambda(\eta_1, \ldots, \eta_r; M_1, \ldots, M_r)$ is a subset of R^N which is invariant by permutation of the coordinates, because one has not imposed any rule for ordering the eigenvalues. Like for the discussion following Proposition 17, the statement above is clear from the intuitive understanding of what mixing is about, but I have not given any precise mathematical definition of what the

[63] Around 1984, I generalized a formula of Joseph KELLER valid for dimension $N = 2$ [Ke], and I later learned from Graeme MILTON that he had also discovered the same result (and he had kindly proposed that I cosign an article where he was using these formulas; I do not know the precise reference of his article). He had been led to these formulas by studying HALL effect, and if my memory is correct it is a nonzero value of a magnetic field which is responsible for the appearance of a nonsymmetric tensor, and this is the only instance of a non symmetric situation which I have heard about in real problems. Later I tried to use these formulas with Michel ARTOLA in order to prove a conjecture of Stefano MORTOLA and Sergio STEFFÉ [Mo&St2] (Sergei KOZLOV told me in 1993 in Trieste that the conjecture is false, but he did not provide enough information, so that I do not know if he had proved it or if he just thought that it was wrong).

set $\mathcal{K}(\eta_1,\ldots,\eta_r;M_1,\ldots,M_r)$ is yet. In 1983, Robert KOHN had asked me a question showing that he was concerned about a problem of this kind: in my work with François MURAT, obtained for mixing two isotropic materials, we had found a necessary condition of the type $A^{eff}\in S(\theta)$ for a set of matrices $S(\theta)$ which was our candidate for $\mathcal{K}(\theta,1-\theta;\alpha\,I,\beta\,I)$, and because this set is convex it is easy to approach a function (θ,A) such that $A(x)\in S\big(\theta(x)\big)$ a.e. $x\in\Omega$ by a piecewise constant function satisfying the same constraint, as after decomposing Ω into small open cubes (plus a set of measure 0), one can replace θ on any such small cube ω by its average $\overline{\theta}$ on ω and replace A by its projection on $S(\overline{\theta})$, giving a new function $(\widetilde{\theta},\widetilde{A})$ satisfying the same constraint with $\widetilde{\theta}$ piecewise constant; then on each small cube one approaches $\widetilde{A}$ by a piecewise constant function (on much smaller cubes) taking its values in $S(\overline{\theta})$. One also uses the fact that the HAUSDORFF distance from $S(\theta_1)$ to $S(\theta_2)$ is $O(|\theta_1-\theta_1|$. However, the convexity of each $S(\theta)$ is not really important, and one can give a precise meaning of (8.34), for the case (8.33) or for other still more general situations, in the following way.

Definition 37: For nonnegative real numbers $\theta_1,\ldots,\theta_r$, with $\sum_i\theta_i=1$, and $P\in\mathcal{L}(R^N,R^N)$, one says that P belongs to $\mathcal{K}(\theta_1,\ldots,\theta_r;M_1,\ldots,M_r)$ if and only if there exist sequences χ_i^n of characteristic functions of disjoint measurable sets for $i=1,\ldots,r$, and a sequence of rotations $R^n\in SO(N)$ such that (8.33) holds, and moreover such that there exists $x_*\in\Omega$, LEBESGUE point of A^{eff} and of $\eta_1,\ldots,\eta_r$, with $\eta_i(x_*)=\theta_i$ for $i=1,\ldots,r$, and $A^{eff}(x_*)=P$.■

Of course, because the set of LEBESGUE points of any vector valued function is dense, the fact that (8.34) is valid is now merely the statement of Definition 37, but one must show that this definition is consistent with the intuitive idea of mixing materials. The proof makes use of an obvious fact, that H-convergence commutes with translations and dilations. Commuting with translations means that if A^n H-converges to A^{eff} in Ω and if for $a\in R^N$ one has $B^n(x)=A^n(x+a)$ a.e. $x\in\Omega-a$, then B^n H-converges to B^{eff} in $\Omega-a$ and $B^{eff}(x)=A^{eff}(x+a)$ a.e. in $\Omega-a$; commuting with dilations, or rescaling, means that if for $s\neq 0$ one has $C^n(x)=A^n(s\,x)$ a.e. $x\in s^{-1}\,\Omega$, then C^n H-converges to C^{eff} in $s^{-1}\,\Omega$ and $C^{eff}(x)=A^{eff}(s\,x)$ a.e. in $s^{-1}\,\Omega$. More generally, one has the following result about changing variables in H-convergence, identical to the formula first proved by Sergio SPAGNOLO in the case of G-convergence.

Lemma 38: If φ is a diffeomorphism from Ω onto $\varphi(\Omega)$ and A^n H-converges to A^{eff} in Ω, and B^n is defined on $\varphi(\Omega)$ by

$$B^n\big(\varphi(x)\big)=\frac{1}{det\big(\nabla\varphi(x)\big)}\nabla\varphi(x)A^n(x)\nabla\varphi^T(x)\text{ a.e. }x\in\Omega,\tag{8.36}$$

then B^n H-converges in $\varphi(\Omega)$ to B^{eff}, and

$$B^{eff}\big(\varphi(x)\big)=\frac{1}{det\big(\nabla\varphi(x)\big)}\nabla\varphi(x)A^{eff}(x)\nabla\varphi^T(x)\text{ a.e. }x\in\Omega.\tag{8.37}$$

Proof: If $-div\big(A^n\, grad(u_n)\big) = f$ in Ω, one defines v_n in $\varphi(\Omega)$ by $v_n(y) = u_n\big(\varphi^{-1}(y)\big)$ a.e. $y \in \varphi(\Omega)$, or equivalently $u_n(x) = v_n\big(\varphi(x)\big)$ a.e. $x \in \Omega$, so that $grad(u_n)(x) = \nabla\varphi^T(x)\, grad(v_n)\big(\varphi(x)\big)$ a.e. $x \in \Omega$. Writing then the equation in variational form $\int_\Omega \big(A^n\, grad(u_n).grad(w)\big)\, dx = \int_\Omega f\, w\, dx$ for all $w \in C_c^1(\Omega)$, one finds that $-div\big(B^n\, grad(v_n)\big) = g$ in $\varphi(\Omega)$, with B^n given by (8.36) and $g\big(\varphi(x)\big) = \frac{1}{det\big(\nabla\varphi(x)\big)} f(x)$ a.e. $x \in \Omega$, in the case $f \in L^2(\Omega)$, with straightforward generalization in the case $f \in H^{-1}(\Omega)$. Then one relates the weak limits of $grad(v_n)$ and of $B^n\, grad(v_n)$ in $\varphi(\Omega)$ to the weak limits of $grad(u_n)$ and of $A^n\, grad(v_n)$ in Ω, and one finds that B^{eff} is defined by (8.37).■

Lemma 39: For nonnegative real numbers $\theta_1, \ldots, \theta_r$, with $\sum_i \theta_i = 1$, and $P \in \mathcal{K}(\theta_1, \ldots, \theta_r; M_1, \ldots, M_r)$, there exist sequences χ_i^n of characteristic functions of disjoint measurable sets for $i = 1, \ldots, r$, and a sequence $R^n \in L^\infty\big(\Omega; SO(N)\big)$ such that (8.33) holds, with $\eta_i = \theta_i$ a.e. in Ω for $i = 1, \ldots, r$, and $A^{eff} = P$ a.e. in Ω.

Proof: One proves the Lemma with Ω replaced by a cube Q centered at 0 and large enough to contain Ω; then one restricts the result to Ω, using Proposition 10. By Definition 37 there exists a point x_* and sequences χ_i^n, $i = 1, \ldots, r$, R^n, which are not yet the ones needed in the Lemma, and one obtains the desired ones by translation and rescaling. For an integer k large enough so that $x_* + \frac{1}{k}Q \subset \Omega$, one defines $\chi_i^{n,k}$, $i = 1, \ldots, r$, $R^{n,k}$, $A^{n,k}$ in Q by $\chi_i^{n,k}(x) = \chi_i^n\big(x_* + \frac{x}{k}\big)$ a.e. $x \in Q$, $i = 1, \ldots, r$, $R^{n,k}(x) = R^n\big(x_* + \frac{x}{k}\big)$, $A^{n,k}(x) = A^n\big(x_* + \frac{x}{k}\big)$ a.e. $x \in Q$. By Lemma 38, for k fixed $A^{n,k}$ H-converges in Q to $A^{eff,k}$, defined by $A^{eff,k}(x) = A^{eff}\big(x_* + \frac{x}{k}\big)$ a.e. $x \in Q$, and because x_* is a LEBESGUE point of A^{eff}, $A^{eff,k}$ converges in $L^\infty(Q)$ weak $\star$ and $L^1(Q)$ strong to the constant tensor $P = A^{eff}(x_*)$, and therefore $A^{eff,k}$ H-converges to P in Q. Similarly, for k fixed $\chi_i^{n,k}$ converges in $L^\infty(Q)$ weak $\star$ to $\chi^{\infty,k}$ defined by $\chi^{\infty,k}(x) = \eta_i\big(x_* + \frac{x}{k}\big)$ a.e. $x \in Q$ for $i = 1, \ldots, r$, and because x_* is a LEBESGUE point of each η_i, $\chi^{\infty,k}$ converges in $L^\infty(Q)$ weak $\star$ and $L^1(Q)$ strong to the constant function $\theta_i = \eta_i(x_*)$, for $i = 1, \ldots, r$. Using the metrizability of H-convergence restricted to $M(\alpha, \beta; Q)$, when $0 < \alpha \le \beta < \infty$ have been choosen so that $M_i \in M(\alpha, \beta)$ for $i = 1, \ldots, r$, and the metrizability of $L^\infty(Q)$ weak $\star$ convergence on bounded sets, there exists a diagonal subsequence indexed by n', k' such that $A^{n',k'}$ H-converges in Q to the constant tensor P, and $\chi_i^{n',k'}$ converges in $L^\infty(Q)$ weak $\star$ to the constant function θ_i, for $i = 1, \ldots, r$.■

Lemma 40: If for $i = 1, \ldots, r$, $\eta_i \in L^\infty(\Omega)$, $\eta_i \ge 0$ almost everywhere in Ω, with $\sum_i \eta_i = 1$ a.e. in Ω, and $P \in L^\infty\big(\Omega; \mathcal{L}(R^N, R^N)\big)$ with $P(x) \in \mathcal{K}(\eta_1(x), \ldots, \eta_r(x); M_1, \ldots, M_r)$ a.e. $x \in \Omega$, then there exists sequences χ_i^n of characteristic functions of disjoint measurable sets for $i = 1, \ldots, r$, and a sequence $R^n \in L^\infty\big(\Omega; SO(N)\big)$ such that (8.33) holds and $A^{eff} = P$ in Ω.

Proof. Let $g \in L^1(\Omega; R^p)$ and for $\varepsilon > 0$ let ρ_ε be defined by $\rho_\varepsilon(x) = \frac{1}{\varepsilon^N}\rho_1\left(\frac{x}{\varepsilon}\right)$ with $\rho_1 \in L^1(R^N)$, nonnegative and with compact support; then $\int_{\Omega\times R^N} |g(x) - g(x-y)|\rho_\varepsilon(y)\,dx\,dy$ tends to 0 as ε tends to 0 (as it is $\leq 2||g||_{L^1}||\rho_1||_{L^1}$, it is enough to prove the result for a dense subspace, and for $g \in C_c(R^N)$ it is immediate as g is uniformly continuous). For ρ_1 the characteristic function of the cube $(0,1)^N$, and $\delta = \frac{1}{k}$ one can then choose ε small enough to have $\int_{|x-z|\leq\varepsilon} |g(x) - g(z)|\,dx\,dz \leq \delta\,\varepsilon^N$. Then decomposing R^N into disjoint cubes ω_j of size $\frac{\varepsilon}{\sqrt{N}}$ (plus a set of measure 0), one chooses for each cube ω_j a point $z^j \in \omega_j$ such that $\int_{\omega_j} |g(x) - g(z_j)|\,dx \leq \frac{2}{\varepsilon^N}\int_{\omega_j\times\omega_j} |g(x) - g(z)|\,dx\,dz \leq \frac{2}{\varepsilon^N}\int_{x\in\omega_j,|x-z|\leq\varepsilon} |g(x) - g(z)|\,dx\,dz$, and if g^k denotes the function equal to $g(z_j)$ on ω_j, then g is piecewise constant and one has $||g - g^k||_{L^1(R^N)} \leq \frac{1}{k}$.

One applies the preceding analysis to $g = (\eta_1,\ldots,\eta_r,P)$, and $g^k = (\eta_1^k,\ldots,\eta_r^k,P^k)$ and one has $P^k \in \mathcal{K}(\eta_1^k,\ldots,\eta_r^k; M_1,\ldots,M_r)$ on each cube ω_j. Using Lemma 39, as well as Proposition 10 in order to glue the pieces together, there exist sequences $\chi_i^{n,k}$ of characteristic functions of disjoint measurable sets for $i = 1,\ldots,r$, and a sequence $R^{n,k} \in L^\infty(\Omega; SO(N))$ such that (8.33) holds, with η_i replaced by η_i^k for $i = 1,\ldots,r$, and A^{eff} replaced by P^k. As k tends to ∞, η_i^k converges in $L^\infty(\Omega)$ weak $\star$ and $L^1(\Omega)$ strong to η^k for $i = 1,\ldots,r$, and P^k converges in $L^\infty(\Omega)$ weak $\star$ and $L^1(\Omega)$ strong to P, and therefore P^k H-converges to P in Ω. Using the metrizability of H-convergence restricted to $M(\alpha,\beta;Q)$, and the metrizability of $L^\infty(Q)$ weak $\star$ convergence on bounded sets, there exists a diagonal subsequence indexed by n',k' such that $\chi_i^{n',k'}$ converges in $L^\infty(\Omega)$ weak $\star$ to η_i, for $i = 1,\ldots,r$, and $A^{n',k'}$ H-converges in Ω to P.■

Because of the local character of H-convergence expressed by Proposition 10, there is no mention of a particular open set in Definition 37, which is actually valid without any hypothesis of symmetry and without the use of rotations R^n (depending measurably in x) in (8.33). However, it is precisely the symmetry of M_i, $i = 1,\ldots,r$, and the use of these rotations which implies that (8.34) has the form (8.35), and this is seen by using Lemma 38 with $\varphi(x) = R\,x$ with $R \in SO(N)$: it shows that if $P \in \mathcal{K}(\eta_1,\ldots,\eta_r; M_1,\ldots,M_r)$ then $R\,P\,R^T \in \mathcal{K}(\eta_1,\ldots,\eta_r; M_1,\ldots,M_r)$ for any $R \in SO(N)$, and as P is symmetric, all tensors with the same eigenvalues than P belong to the set $\mathcal{K}(\eta_1,\ldots,\eta_r; M_1,\ldots,M_r)$, and one deduces (8.35).[64]

[64] The first method for obtaining bounds on effective coefficients, which I developed with François MURAT in the early 70s, and which I have described in chapter 5, is not restricted to symmetric operators, but I do not know what kind of transformations to impose in the nonsymmetric case. I have not read carefully the article of Graeme MILTON mentioned in footnote 63 as the only instance of a nonsymmetric situation that I have heard of in a realistic situation, but knowing that the HALL effect is concerned with electrical cur-

I must say that prior to writing these notes I had not been so interested in writing down the detail of the analysis which I just showed, considering that these are uninteresting details which everyone with a good knowledge of measure theory can check, and that the important questions of Homogenization lie elsewhere. I have heard some mention of a work of Gianni DAL MASO and Robert KOHN which might have addressed this question in general, or it may have been the answer to the question that Robert KOHN was asking me in 1983, which was related to linearized Elasticity, which is not a frame indifferent theory, as he already knew; his concern was to identify which are the invariant quantities replacing the eigenvalues in the case of fourth order tensors C_{ijkl}. I do not see any special difficulty in adapting Definition 37 to this case, but as I consider that questions in linearized Elasticity are too often spoiled by unrealistic effects which deprive the mathematical results of much of their value, I have preferred not to be involved in such questions.

The set $\mathcal{K}(\eta_1,\ldots,\eta_r;M_1,\ldots,M_r)$ is not necessarily convex: in dimension $N=2$ with only one material M_1 symmetric with distinct eigenvalues $\alpha<\beta$, $\mathcal{K}(1;M_1)$ is the set of symmetric matrices with eigenvalues $\lambda_1,\lambda_2\in[\alpha,\beta]$ with $\lambda_1\,\lambda_2=\alpha\,\beta$. However, some projections of $\mathcal{K}(\eta_1,\ldots,\eta_r;M_1,\ldots,M_r)$ are convex [Ta14], and this is not dependent on symmetry requirements or use of rotations.

Lemma 41: Consider $\eta_1,\ldots,\eta_r$, nonnegative numbers adding up to 1, and $M_1,\ldots,M_r$, tensors from $M(\alpha,\beta)$; then for every set of $N-1$ vectors $a^1,\ldots,a^{N-1}$ of R^N,

$$\{(P\,a^1,\ldots,P\,a^{N-1})|P\in\mathcal{K}(\eta_1,\ldots,\eta_r;M_1,\ldots,M_r)\}\text{ is convex.}\tag{8.38}$$

Moreover

$$\begin{aligned}&P_1,P_2\in\mathcal{K}(\eta_1,\ldots,\eta_r;M_1,\ldots,M_r)\text{ and }rank(P_2-P_1)\le N-1\\&\text{imply }\theta\,P_1+(1-\theta)P_2\in\mathcal{K}(\eta_1,\ldots,\eta_r;M_1,\ldots,M_r)\text{ for all }\theta\in(0,1).\end{aligned}\tag{8.39}$$

Proof: For $P_1,P_2\in\mathcal{K}(\eta_1,\ldots,\eta_r;M_1,\ldots,M_r)$ and $\theta\in(0,1)$ one exhibits $P_3\in\mathcal{K}(\eta_1,\ldots,\eta_r;M_1,\ldots,M_r)$ such that

$$P_3\,a^i=\theta\,P_1\,a^i+(1-\theta)P_2\,a^i\text{ for }i=1,\ldots,N-1,\tag{8.40}$$

and this is done by layering materials with tensors P_1 and P_2, respectively with proportions η and $1-\eta$, and this obviously creates a tensor belonging to $\mathcal{K}(\eta_1,\ldots,\eta_r;M_1,\ldots,M_r)$, and then one uses formula (8.37) of Lemma 34. The nonzero vector e orthogonal to the layers must satisfy

$$\Big((P_2-P_1)a^i.e\Big)=0\text{ for }i=1,\ldots,N-1,\tag{8.41}$$

rents in thin ribbons, the macroscopic direction of the current is obviously imposed and therefore the situation is not subject to frame indifference.

and this is possible because there are only $N-1$ vectors a_i, but if $P_2 - P_1$ does not have full rank, one can take e orthogonal to the range of $P_2 - P_1$, and (8.39) follows.■

It means that if one considers the first $N-1$ columns of elements from $\mathcal{K}(\eta_1,\ldots,\eta_r;M_1,\ldots,M_r)$ in a basis $a^1,\ldots,a^N$, then one has a convex set. If the basis is orthogonal and all M_i are symmetric, then the choice (8.41) for e in the proof of Lemma 41 means that $(P_2-P_1)e$ is proportional to a^N and the last term in the formula (8.37) is then proportional to $a^N\otimes a^N$ and therefore if one considers the N^2-1 entries of elements P of $\mathcal{K}(\eta_1,\ldots,\eta_r;M_1,\ldots,M_r)$ obtained by deleting the entry P_{NN}, then one has a convex set. Even though $\mathcal{K}(\eta,1-\eta;\alpha I,\beta I)$ is characterized by explicit inequalities (7,25), (7.28), (7.30), I do not know a simple description of what all these convex sets are in this special case, but if one just wants to identify one column of elements of $\mathcal{K}(\eta_1,\ldots,\eta_r;M_1,\ldots,M_r)$, then one has the following result in the symmetric case using rotations [Ta14].

Lemma 42: Consider $\eta_1,\ldots,\eta_r$, nonnegative numbers adding up to 1, and $M_1,\ldots,M_r$, symmetric tensors from $M(\alpha,\beta)$, with eigenvalues $\lambda_j(M_i)$, $i=1,\ldots,r$, $j=1,\ldots,N$ and $N\geq 2$; then for every $E\in R^N$,

$$\{(P\,E|P\in\mathcal{K}(\eta_1,\ldots,\eta_r;M_1,\ldots,M_r)\}\text{ is the closed ball with diameter }[b\,E,a\,E]$$

$$a=\sum_{i=1}^{r}\eta_i\max_j\{\lambda_j(M_i)\}\qquad(8.42)$$

$$\frac{1}{b}=\sum_{i=1}^{r}\frac{\eta_i}{\min_j\{\lambda_j(M_i)\}}.$$

Proof: Let $\lambda_-(M_i)=\min_j\{\lambda_j(M_i)\}$ and $\lambda_+(M_i)=\max_j\{\lambda_j(M_i)\}$. Let $e_1,\ldots,e_N$ be an orthonormal basis, and rotate the materials in such a way that it becomes a basis of eigenvector for each $R_i^T\,M_i\,R_i$, with $R_i^T\,M_i\,R_i\,e_1=\lambda_-(M_i)e_1$ and $R_i^T\,M_i\,R_i\,e_N=\lambda_+(M_i)e_N$; layering in direction orthogonal to e_1 these materials $R_i^T\,M_i\,R_i$ with proportion η_i gives by formula (4.11) a tensor $P\in\mathcal{K}(\eta_1,\ldots,\eta_r;M_1,\ldots,M_r)$ such that $P\,e_1=b\,e_1$ and $P\,e_N=a\,e_N$. If E belongs to the plane spanned by e_1 and e_N, then $P\,E=b(E.e_1)e_1+a(E.e_N)e_N$, so that $P\,E-b\,E=(a-b)(E.e_N)e_N$ and $P\,E-a\,E=(b-a)(E.e_1)e_1$, and therefore $(P\,E-b\,E.P\,E-a\,E)=0$, showing that $P\,E$ belongs to the sphere with diameter $[b\,E,a\,E]$. By choosing all possible orientations, the set of $P\,E$ contains at least the sphere with diameter $[b\,E,a\,E]$, and as by Lemma 41, the set that we are looking for is convex, it must contain the closed ball with diameter $[b\,E,a\,E]$.

That the desired set is included in the closed ball with diameter $[b\,E,a\,E]$ follows from Proposition 12 and Proposition 14 (or from Lemma 18). Assume that $A^n=\sum_i\chi_i^n(R^n)^T\,M_i\,R^n$ H-converges to A^{eff} in Ω, where the

χ_i^n are characteristic functions of disjoint measurable sets with χ_i^n converging in $L^\infty(\Omega)$ weak $\star$ to η_i for $i = 1, \ldots, r$. As $(R^n)^T M_i R^n$ has eigenvalues between $\lambda_-(M_i)$ and $\lambda_+(M_i)$ then one has $B^n = \left(\sum_i \chi_i^n \lambda_-(M_i)\right) I \leq A^n \leq \left(\sum_i \chi_i^n \lambda_+(M_i)\right) = C^n$, and if one extracts a subsequence such that B^m H-converges to B^{eff} and C^m H-converges to C^{eff}, then by Proposition 14 one has $B^{eff} \leq A^{eff} \leq C^{eff}$. By Proposition 12, an upper bound for C^{eff} is C_+, the limit in $L^\infty(\Omega; \mathcal{L}(R^N, R^N))$ weak $\star$ of C^n, which by (8.42) is $a\,I$, and a lower bound for B^{eff} is B_-, where $(B_-)^{-1}$ is the limit in $L^\infty(\Omega; \mathcal{L}(R^N, R^N))$ weak $\star$ of $(B^n)^{-1}$, which by (8.42) is $\frac{1}{b}\,I$. One has then $b\,I \leq A^{eff} \leq a\,I$, and therefore (as was shown before Lemma 18), $A^{eff}\,E$ belongs to the closed ball with diameter $[b\,E, a\,E]$, a.e. in Ω.■

10. Necessary conditions of optimality: first step

At a meeting in New York in 1981, I had described the characterization found with François MURAT of all effective materials obtained by mixing two isotropic materials, i.e. (7.25) (Proposition 10, described in [Ta1]) and (7.28), (7.30), obtained by following my second method for obtaining bounds satisfied by effective coefficients (Theorem 26, described in [Ta7], and Lemmas 28, 29). It was clear that our program for questions of Optimal Design initialized in the early 70s was completed, but I did not try immediately to extend the computation of necessary conditions that I had done in [Ta2].

During the Spring 1983, I was invited at a Midwest PDE Seminar in Madison, and I heard Michael RENARDY talk about his work with Daniel JOSEPH on POISEUILLE flows of two immiscible fluids.[65] For a cylinder with arbitrary cross section Ω, there are infinitely many possible POISEUILLE flows, but when Ω is a disc one only observes the flow where the less viscous fluid occupies an annular region near the boundary, and they had imagined that it was related to a maximization of dissipation. This situation is described by (4.1) with $f = 1$ (a is the viscosity of the fluid, independent of the direction along the pipe, and the gradient of the pressure is constant, pointing in the direction of the pipe), and one wants to maximize $\int_\Omega a|grad(u)|^2\,dx = \int_\Omega u\,dx$, and therefore it corresponds to $g(x, u, a) = -u$ in (4.3). There is indeed a classical solution in the case of a circular cross section, which is as described, but as they were conjecturing the existence of a classical solution in general, I told Michael RENARDY that on the contrary I expected that no classical solution would exist in general.[66] In order to check about this question, I looked at the necessary conditions of optimality, as I had done in [Ta2], but using now the

[65] In applications, it either meant melted polymers or crude oil and water, the water being added in small quantity in a pipeline for lubricating it.

[66] The idea that turbulent flows are trying to optimize something has been suggested before, as I have read in a book by Daniel JOSEPH [Jo], but it is not

characterization that François MURAT and I had obtained two years before. Surprisingly, I discovered that our precise characterization could be ignored completely, and that the same necessary conditions could be deduced from the crude bounds that we had obtained almost ten years earlier.[67] Assuming that a classical solution exists, and that the interface between the two materials is smooth enough, the necessary conditions of optimality imply that some DIRICHLET conditions and some NEUMANN conditions must be satisfied on the interface, which is quite unlikely in general. On my way back to France I stopped in New York, and discussing about this matter in the lounge at the COURANT Institute, the precise argument for rejecting that possibility was mentioned to me by Joel SPRUCK, who reminded me of a result of James SERRIN that I had heard at a conference in Jerusalem in 1972 (and he told me about a quicker proof by Hans WEINBERGER); this result assumes the domain to be simply connected, and shows that among simply connected domains a classical solution only exists for circular domains.[68]

In July 1983, I described completely our method for a Summer Course CEA - EDF - INRIA on Homogenization at Bréau sans Nappe, which gave François MURAT and I the occasion to write [Mu&Ta1]. As this first text had been written in French, we wrote a summary in English for a conference at Isola d'Elba in the Fall [Mu&Ta2]. In November, I took the occasion of a meeting in Paris dedicated to Ennio DE GIORGI for writing down the details of the characterization obtained with François MURAT in 1980 [Ta9], characterization which I had first described at a meeting in New York in 1981 but only quoted in [Mu&Ta1], and in [Ta9] I gave our initial construction with coated ellipsoids generalizing the idea of using coated spheres that I had learned in the original work of Zvi HASHIN and S. SHTRIKMAN [Ha&Sh], just after it had been pointed out to me by George PAPANICOLAOU, and for the computations with multiple layerings I gave the generalization that I had obtained in the Spring 1983 in Berkeley of the work initially done the year before by Philippe BRAIDY and Didier POUILLOUX [Br&Po].

In the early 70s, François MURAT and I had developed the Homogenization approach in order to describe a relaxed problem, and this is a way to

---

clear if fluids are trying to optimize something, and what it could be; however, this idea is consistent with the Homogenization approach that in order to optimize some criteria one might have to create adequate microstructures.

[67] As I learned later, this point had already been noticed in the late 70s by RAITUM [Ra].

[68] Of course, when Ω is not a disc this negative result does not tell what configurations will be chosen by a mixture of fluids in a pipe, as the question of what mixtures of fluids really try to optimize must be studied further. Daniel JOSEPH, Michael RENARDY and Yuriko RENARDY have also studied the stability of the solution for circular domains, and for such questions of stability one must go back to the full NAVIER-STOKES equation, of course.

prove existence of generalized solutions. For the sake of generality, I describe here the more general framework that I developed later in [Ta14], instead of the restricted problem of mixing only two isotropic materials, as we had done in [Mu&Ta1]. For a bounded open set Ω of R^N, one considers a mixture of r symmetric possibly anisotropic materials $M_1, \ldots, M_r \in M(\alpha, \beta)$, in finite quantity $\kappa_1, \ldots, \kappa_r$, assuming that $\sum_i \kappa_i \geq meas(\Omega)$, of course. Then one defines

$$A = \sum_{i=1}^{r} \chi_i \, R^T \, M_i \, R \text{ in } \Omega, \tag{9.1}$$

where χ_i, $i = 1, \ldots, r$, are characteristic functions of disjoint measurable sets with union equal to Ω (up to a set of measure 0), and R is measurable and takes values in $SO(N)$, a.e. in Ω, and one assumes that

$$\int_\Omega \chi_i \, dx \leq \kappa_i \text{ for } i = 1, \ldots, r. \tag{9.2}$$

Then one solves the state equation

$$-div\Big(A \, grad(u)\Big) = f \text{ in } \Omega, \text{ with } u \in H_0^1(\Omega), \tag{9.3}$$

with $f \in H^{-1}(\Omega)$, and I have choosen homogeneous DIRICHLET conditions for u as an example. Finally one wants to minimize the cost function J defined by

$$J(\chi_1, \ldots, \chi_r, R) = \int_\Omega \sum_{i=1}^{r} \chi_i \, F_i\Big(x, u(x)\Big) \, dx, \tag{9.4}$$

so that there is no cost for rotating materials, and one adds regularity and growth hypotheses on $F_1, \ldots, F_r$, in order to have

$$u \mapsto F_i(\cdot, u) \text{ is sequentially continuous from } H_0^1(\Omega) \text{ weak into } L^1(\Omega) \text{strong, for } i = 1, \ldots, r, \tag{9.5}$$

which is usually obtained by assuming that each F_i is a CARATHÉODORY function, and that $\sum_i |F_i(x, v)| \leq \mu(x) + C|v|^p$ for all $v \in R$ a.e. $x \in \Omega$, with $\mu \in L^1(\Omega)$ and $p < \frac{2N}{N-2}$ for $N \geq 2$ (but for $N = 2$ one can assume instead that $\sum_i |F_i(x, v)| \leq \mu(x) + G(|v|)$ and allow some exponential growth for G), and for $N = 1$ one assumes that $\sum_i \sup_{|v| \leq r} |F_i(x, v)| \leq \mu_r(x)$ with $\mu_r \in L^1(\Omega)$ for all r. In the case where $\sum_i \kappa_i > meas(\Omega)$, one can also add in J a term of the form $G(\int_\Omega \chi_1 \, dx, \ldots, \int_\Omega \chi_r \, dx)$.

The set of admissible A is not empty, because $\sum_i \kappa_i \geq meas(\Omega)$, and it is included in some $M(\alpha, \beta; \Omega)$, so that all the u obtained belong to a bounded set of $H_0^1(\Omega)$. For a sequence A^n, for example a minimizing sequence, one can extract a subsequence which H-converges to A^{eff}, and also such that for $i = 1, \ldots, r$, the sequence χ_i^n converges in $L^\infty(\Omega)$ weak $\star$ to a nonnegative function η_i, which satisfies and

$$\int_\Omega \eta_i\,dx \le \kappa_i \text{ for } i=1,\ldots,r, \tag{9.6}$$

with $\sum_i \eta_i = 1$ in Ω, and therefore by Definition 37

$$A^{eff} \in \mathcal{K}(\eta_1,\ldots,\eta_r;M_1,\ldots,M_r) \text{ a.e. in } \Omega, \tag{9.7}$$

and also such that u_n converges in $H^1_0(\Omega)$ weak to u_∞, and therefore (9.5) implies

$$J(\chi_1^n,\ldots,\chi_r^n,R^n) \to \int_\Omega \Big(\sum_{i=1}^r \eta_i F_i(x,u_\infty)\Big)\,dx. \tag{9.8}$$

By Lemma 40, if r nonnegative functions η_i, $i=1,\ldots,r$, satisfy $\sum_i \eta_i = 1$ in Ω and (9.6), and $P \in \mathcal{K}(\eta_1,\ldots,\eta_r;M_1,\ldots,M_r)$ a.e. in Ω, there exist sequences χ_i^n of characteristic functions of disjoint measurable sets converging in $L^\infty(\Omega)$ weak $\star$ to η_i for $i=1,\ldots,r$, and A^n H-converges to P, but there is a small technical point to resolve, because the sequence created in the proof of Lemma 40 might not satisfy (9.2). In that case one has $\int_\Omega \chi_i^n\,dx \le \kappa_i + \varepsilon_n$ for $i=1,\ldots,r$, and ε_n tends to 0 because $\int_\Omega \eta_i\,dx \le \kappa_i$ for $i=1,\ldots,r$. For each i one can change χ_i^n on a set of measure at most ε_n in order to create sequences $\widetilde{\chi}_i^n$ of characteristic functions of disjoint measurable sets converging in $L^\infty(\Omega)$ weak $\star$ to η_i for $i=1,\ldots,r$ and satisfying (9.2), and $\widetilde{A}^n = \sum_i \widetilde{\chi}_i^n (R^n)^T M_i R^n$ H-converges to $\widetilde{P}$, and one must show that $\widetilde{P} = P$ a.e. in Ω. As the perturbations are not small in L^∞ norm, one cannot apply Proposition 16, but one can control the effect of small perturbations of the coefficients in L^s norm with $s<\infty$ (with the coefficients staying in $M(\alpha,\beta;\Omega)$) by using MEYERS's regularity theorem [Me]; however, one can also construct a more direct proof based on Proposition 10 as follows. If for $i=1,\ldots,r$, one chooses a small cube ω_i such that $\int_{\omega_i} \eta_i\,dx > 0$, and one chooses n large enough so that $\int_{\omega_i} \chi_i^n\,dx \ge \varepsilon_n$ for $i=1,\ldots,r$, then one can manage to make the changes only inside $\bigcup_i \omega_i$, and as $\widetilde{A}^n = A^n$ in $\Omega \setminus (\bigcup_i \overline{\omega_i})$, Proposition 10 implies that $\widetilde{P} = P$ in $\Omega \setminus (\bigcup_i \overline{\omega_i})$; the ω_i can be taken as small as one wants (around a density point of η_i), and therefore $\widetilde{P} = P$ a.e. in Ω.

One has therefore identified a relaxed problem (as defined in the end of chapter 3): the new set of controls is the set of $(\eta_1,\ldots,\eta_r,A)$ satisfying

$$\begin{aligned}
&0 \le \eta_i \text{ a.e. in } \Omega \text{ for } i=1,\ldots,r,\ \sum_{i=1}^r \eta_i = 1 \text{ a.e. in } \Omega\\
&\int_\Omega \eta_i\,dx \le \kappa_i \text{ for } i=1,\ldots,r\\
&A \in \mathcal{K}(\eta_1,\ldots,\eta_r;M_1,\ldots,M_r) \text{ a.e. in } \Omega,
\end{aligned} \tag{9.9}$$

the state u is still given by (9.3), and the cost function $\widetilde{J}$ to be minimized is given by

$$\widetilde{J}(\eta_1,\ldots,\eta_r,A) = \int_\Omega \Big[\sum_{i=1}^r \eta_i\, F_i\big(x,u(x)\big)\Big]\,dx. \tag{9.10}$$

The initial problem corresponds to the case where each η_i is a characteristic function.

Lemma 43: The function $\widetilde{J}$ given by (9.10) with u defined by (9.3) attains its minimum on the set described by (9.9).

Proof: The topology on the new control set described by (9.9) is the $L^\infty(\Omega)$ weak $\star$ topology for each η_i and the H-convergence topology for A, and this topology is metrizable. The new control set is the completion of the initial control set, and it is compact for this topology (this uses Theorem 6, Lemma 40, and the simple trick shown above for enforcing (9.2), together with classical results of Functional Analysis for dealing with the $L^\infty(\Omega)$ weak $\star$ topology). Moreover, $\widetilde{J}$ is continuous for that topology. A consequence of all these facts is that the minimum of $\widetilde{J}$ is attained.■

The questions are rather different for more general functionals depending upon $grad(u)$, as we already knew in the early 70s; I have described some very partial results on this question in [Ta13]. However, the property (9.4)/(9.5) is still true for some functionals depending upon $grad(u)$ in a "fake" way. For example, (9.3) implies that $\int_\Omega (A\,grad(u).grad(u))\,dx = \int_\Omega f\,u\,dx$ if $f \in L^2(\Omega)$, so that one should not really say that the first integral depends upon $grad(u)$! There are other situations of this kind like the following ones, where one cannot make $grad(u)$ disappear completely; for simplicity, I do not try to use MEYERS's regularity theorem [Me]. Assume that for some j, one has $J(\chi_1,\ldots,\chi_r,R) = \int_\Omega g(u)\frac{\partial u}{\partial x_j}v\,dx$, with g continuous and having limited growth $|g(u)| \le C(1+|u|^p)$ and $p < \frac{N}{N-2}$ if $N \ge 2$, with $v \in L^q(\Omega)$ and $p(\frac{1}{2}-\frac{1}{N})+\frac{1}{2}+\frac{1}{q} \le 1$, so that $q < \infty$ and the integral makes sense by an application of SOBOLEV imbedding theorem. If v is not regular, one cannot make $grad(u)$ disappear entirely by integration by parts, but one can decompose v as $v_1 + v_2$ with v_1 small in $L^q(\Omega)$ and $v_2 \in C^1_c(\Omega)$, and one notices that $\int_\Omega g(u)\frac{\partial u}{\partial x_j}v_2\,dx = -\int_\Omega G(u)\frac{\partial v_2}{\partial x_j}\,dx$, where $G(s) = \int_0^s g(\sigma)\,d\sigma$; because the last integral is sequentially continuous on $H^1_0(\Omega)$ weak, one finds that the same is true for J as a uniform limit of sequentially weakly continuous functionals. One can prove the same result differently, and it also applies to the case where one has $J(\chi_1,\ldots,\chi_r,R) = \int_\Omega g(u)\big(A\,grad(u)\big)_j v\,dx$, with the same hypotheses on g and v. Because the sequences considered are such that $A^n\,grad(u_n)$ converges in $L^2(\Omega;R^N)$ weak to $A^{eff}\,grad(u_\infty)$, it is enough to show that $g(u_n)v$ converges in $L^2(\Omega)$ strong to $g(u_\infty)v$, and using another decomposition of v as $v_3 + v_4$ with v_3 small in $L^q(\Omega)$ and $v_4 \in L^\infty(\Omega)$, it is enough to show that $g(u_n)$ converges in $L^2(\Omega)$ strong to $g(u_\infty)$, and this follows from SOBOLEV imbedding theorem and the fact that the injection of $H^1_0(\Omega)$ into $L^2(\Omega)$ is compact.

The next step is to write necessary conditions of optimality, and therefore one wants to use differentiable paths between the candidate to optimality and any other control. One assumes then that each F_i is continuously differentiable in u, that $\frac{\partial F_i(\cdot,u)}{\partial u}$ is a CARATHÉODORY function, and that some growth condition is satisfied so that

$$u \mapsto \frac{\partial F_i(\cdot,u)}{\partial u} \text{ is continuous from } H_0^1(\Omega) \text{ into } L^s(\Omega) \subset H^{-1}(\Omega), \quad (9.11)$$

for example $|\frac{\partial F_i}{\partial u}| \leq g(x) + C|u|^q$ with $g \in L^r(\Omega)$ and $r \geq \frac{2N}{N+2}$ if $N \geq 3$ or $r > 1$ if $N = 2$ (and $r = 1$ for $N = 1$), and $q \leq \frac{N+2}{N-2}$ if $N \geq 3$ or $q < \infty$ if $N \leq 2$. The condition of optimality will be simplified by using an adjoint state, defined by

$$-div\Big(A^T grad(p)\Big) = \sum_{i=1}^r \eta_i \frac{\partial F_i(\cdot,u)}{\partial u} \text{ in } \Omega,\ p \in H_0^1(\Omega), \quad (9.12)$$

and although I am dealing with symmetric A here, I have used A^T in (9.12) in order to show what is needed in a nonsymmetric situation. In 1983, I was checking the conditions of optimality for the special case of mixing two anisotropic materials, using the characterization obtained with François MURAT, i.e. (7.25), (7.28) and (7.30), and as these conditions define a convex set of matrices, I could use straight segments for mixtures corresponding to the same local proportions. I discovered that the precise form of (7.28) and (7.30) is not really important, and generalizing to mixing more than two isotropic materials became an easy exercise, but I only understood the general case of mixing anisotropic materials in given proportions when I observed the property of Lemma 42. For the generalization, which I show directly without starting by the particular case that I had done first, I will use the notation

$$\begin{aligned} &\mathcal{B}(\eta_1,\ldots,\eta_r;M_1,\ldots,M_r) = \{A | b\,I \leq A \leq a\,I\} \\ &\frac{1}{b} = \sum_{i=1}^r \frac{\eta_i}{\min_j\{\lambda_j(M_i)\}} \\ &a = \sum_{i=1}^r \max_j\{\lambda_j(M_i)\}, \end{aligned} \quad (9.13)$$

which defines a convex set $\mathcal{B}(\eta_1,\ldots,\eta_r;M_1,\ldots,M_r)$ containing $\mathcal{K}(\eta_1,\ldots,\eta_r;M_1,\ldots,M_r)$, but although they are different sets, Lemma 42 tells that some of their projections are identical.

Lemma 44: For $N \geq 2$, let $\eta_1^*,\ldots,\eta_r^*,A^*$ satisfy (9.9), and

$$\widetilde{J}(\eta_1^*,\ldots,\eta_r^*,A^*) \leq \widetilde{J}(\eta_1^*,\ldots,\eta_r^*,A) \text{ for all } A \in \mathcal{K}(\eta_1^*,\ldots,\eta_r^*,M_1,\ldots,M_r), \quad (9.14)$$

with $\widetilde{J}$ defined by (9.10). Then, if u^* and p^* are the corresponding solutions of respectively (9.3) and (9.12), one has

$$\int_\Omega \Big(A^* \, grad(u^*).grad(p^*)\Big)\, dx \geq \int_\Omega \Big(B \, grad(u^*).grad(p^*)\Big)\, dx \qquad (9.15)$$
$$\text{for all } B \in \mathcal{B}(\eta_1^*, \ldots, \eta_r^*, M_1, \ldots, M_r).$$

Proof. For $B \in \mathcal{B}(\eta_1^*, \ldots, \eta_r^*; M_1, \ldots, M_r)$, and $\varepsilon \in (0,1)$, one defines

$$A(\varepsilon) = (1-\varepsilon)A^* + \varepsilon\, B = A^* + \varepsilon\, \delta A, \qquad (9.16)$$

and $A(\varepsilon) \in \mathcal{B}(\eta_1^*, \ldots, \eta_r^*; M_1, \ldots, M_r)$ because it is convex and contains $\mathcal{K}(\eta_1^*, \ldots, \eta_r^*, M_1, \ldots, M_r)$. As $A(\varepsilon)$ is analytic in ε with values in $M(\alpha, \beta; \Omega)$, the operator $\mathcal{A}_\varepsilon = -div\big(A(\varepsilon) grad(\cdot)\big)$ is analytic with values in $\mathcal{L}\big(H_0^1(\Omega), H^{-1}(\Omega)\big)$ and invertible by LAX-MILGRAM lemma, and therefore its inverse $\mathcal{A}_\varepsilon^{-1}$ is analytic with values in $\mathcal{L}\big(H^{-1}(\Omega), H_0^1(\Omega)\big)$, so that the corresponding solution $u(\varepsilon)$ is analytic in ε with values in $H_0^1(\Omega)$, and its derivative δu at $\varepsilon = 0$ satisfies

$$\begin{aligned} &u(\varepsilon) = u^* + \varepsilon\, \delta u + o(\varepsilon) \text{ in } H_0^1(\Omega) \\ &div\Big(A^* \, grad(\delta u) + \delta A \, grad(u^*)\Big) = 0, \end{aligned} \qquad (9.17)$$

where $\delta A = B - A^*$. The function $\widetilde{J}\big(\eta_1^*, \ldots, \eta_r^*, A(\varepsilon)\big)$ is therefore differentiable in ε, and its derivative $\delta\widetilde{J}$ at $\varepsilon = 0$ satisfies

$$\begin{aligned} &\widetilde{J}\Big(\eta_1^*, \ldots, \eta_r^*, A(\varepsilon)\Big) = \widetilde{J}(\eta_1^*, \ldots, \eta_r^*, A^*) + \varepsilon\, \delta\widetilde{J} + o(\varepsilon) \\ &\delta\widetilde{J} = \int_\Omega \Big(\sum_{i=1}^r \eta_i^* \frac{\partial F_i(x, u^*)}{\partial u}\Big) \delta u \, dx, \end{aligned} \qquad (9.18)$$

and using the definition (9.12) of the adjoint state p^* and (9,17), one finds

$$\begin{aligned} \delta\widetilde{J} &= \int_\Omega \Big((A^*)^T grad(p^*).grad(\delta u)\Big)\, dx = \\ &= \int_\Omega \Big(A^* \, grad(\delta u).grad(p^*)\Big)\, dx = -\int_\Omega \Big(\delta A \, grad(u^*).grad(p^*)\Big)\, dx = \\ &= \int_\Omega \Big((A^* - B)\, grad(u^*).grad(p^*)\Big)\, dx. \end{aligned} \qquad (9.19)$$

If one had $A(\varepsilon) \in \mathcal{K}(\eta_1^*, \ldots, \eta_r^*, M_1, \ldots, M_r)$ for all $\varepsilon \in (0,1)$, one would have $\widetilde{J}\Big(\eta_1^*, \ldots, \eta_r^*, A(\varepsilon)\Big) \geq \widetilde{J}(\eta_1^*, \ldots, \eta_r^*, A^*)$ by (9.14), and therefore $\delta\widetilde{J} \geq 0$ by (9.18), and that would give (9.15) for the particular choice of B. However, almost everywhere in Ω, $A(\varepsilon) grad\big(u(\varepsilon)\big)$ belongs to the closed ball with diameter $[b\, grad\big(u(\varepsilon)\big), a\, grad\big(u(\varepsilon)\big)]$, and therefore by Lemma 42

$$\begin{aligned} &\text{there exists } M(\varepsilon) \in \mathcal{K}(\eta_1^*, \ldots, \eta_r^*, M_1, \ldots, M_r): \\ &\qquad A(\varepsilon) grad\Big(u(\varepsilon)\Big) = M(\varepsilon) grad\Big(u(\varepsilon)\Big) \text{ a.e. in } \Omega. \end{aligned} \qquad (9.20)$$

Because $A(\varepsilon)$ and $M(\varepsilon)$ create then the same state $u(\varepsilon)$, one has

$$\tilde{J}\Big(\eta_1^*,\ldots,\eta_r^*,A(\varepsilon)\Big) = \tilde{J}\Big(\eta_1^*,\ldots,\eta_r^*,M(\varepsilon)\Big) \geq \tilde{J}(\eta_1^*,\ldots,\eta_r^*,A^*), \tag{9.21}$$

and therefore the conclusion $\delta\tilde{J} \geq 0$ is valid.■

In the derivation of (9.20), there is a small technical difficulty, because Lemma 42 was given without any dependence upon $x \in \Omega$, and one must then check that there is a measurable $M(\varepsilon)$ satisfying (9.20) ($\varepsilon > 0$ being fixed). Of course, one can make Lemma 42 more precise by constructing an explicit lifting, which maps $\eta_1,\ldots,\eta_r, B \in \mathcal{B}(\eta_1,\ldots,\eta_r; M_1,\ldots,M_r), e \in R^N$ to $A \in \mathcal{K}(\eta_1,\ldots,\eta_r; M_1,\ldots,M_r)$ with $A\,e = B\,e$, but a natural construction relies on different cases, e being 0 or not, $B\,e$ being parallel to e or not, $B\,e$ being on the sphere with diameter $[b(\eta)e, a(\eta)e]$ or not, and so on, and in each case one only checks continuity of the lifting, so that the constructed A is measurable. One may also avoid this question of measurability altogether by restricting one's attention to specific mixtures obtained by layering in arbitrary directions, and in the end it gives the same condition (because in Lemma 42 the elements on the boundary of the ball are created by layerings). One chooses an orthonormal basis $e_1,\ldots,e_N$, and one orients material $\#i$ so that e_j is an eigenvector for the eigenvalue $\lambda_j(M_i)$, with eigenvalues increasing with j (so that $\lambda_1(M_j) = \min_j \lambda_j(M_i)$ and $\lambda_N(M_i) = \max_j \lambda_j(M_i)$), and then one uses layers orthogonal to e_1, with proportions $\eta_1^*,\ldots,\eta_r^*$, and one obtains a material with tensor

$$\begin{gathered}P = \sum_{j=1}^N \pi_j\, e_j \otimes e_j \in \mathcal{K}(\eta_1^*,\ldots,\eta_r^*; M_1,\ldots,M_r)\\ b = \pi_1 \leq \ldots \leq \pi_N = a \text{ a.e. in } \Omega.\end{gathered} \tag{9.22}$$

For a measurable subset ω of Ω one keeps $A(\varepsilon) = A^*$ in $\Omega \setminus \omega$, and in ω one defines $A(\varepsilon)$ by mixing the material with tensor A^* and the material with tensor P, in layers orthogonal to e and with respective proportions $1-\varepsilon$ and ε. Of course, one uses Lemma 40 for showing that such an $A(\varepsilon)$ is allowed in (9.14), and formula (8.17) gives

$$A(\varepsilon) = A^* + \varepsilon\,\delta A + o(\varepsilon) \text{ in } \omega \text{ with } \delta A = P - A^* - (P - A^*)\frac{e \otimes e}{(P\,e.e)}(P - A^*), \tag{9.23}$$

and this different definition of δA corresponds to a different δu to use in (9.17) and (9.18), and the last equality in (9.19) changes because of the different definition of δA, and by varying ω one deduces that (9.9) and (9.14) imply that for all $e \neq 0$ one has

$$\begin{aligned}\Big(A^*\,grad(u^*).grad(p^*)\Big) &\geq \Big(P\,grad(u^*).grad(p^*)\Big) -\\ &- \frac{\Big((P - A^*)grad(u^*).e\Big)\Big((P - A^*)grad(p^*).e\Big)}{(P\,e.e)},\end{aligned} \tag{9.24}$$

a.e. in Ω, and by choosing $e \neq 0$ orthogonal to $(P - A^*)grad(u^*)$, one finds

$$\Big(A^*\, grad(u^*).grad(p^*)\Big) \geq \Big(P\, grad(u^*).grad(p^*)\Big) \quad \text{for all } P \text{ satisfying (9.22), a.e. in } \Omega. \tag{9.25}$$

From (9.15) or (9.25) one deduces the following necessary condition of optimality.

Proposition 45: For $N \geq 2$, let $\eta_1^*, \ldots, \eta_r^*, A^*$ satisfy (9.9) and (9.14), then at almost every point where $|grad(u^*)|\,|grad(p^*)| \neq 0$ one has

$$\text{if } grad(p^*) = c\, grad(u^*) \text{ with } c > 0, \text{ then} \quad A^*\, grad(u^*) = a\, grad(u^*), A^*\, grad(p^*) = a\, grad(p^*), \tag{9.26}$$

$$\text{if } grad(p^*) = c\, grad(u^*) \text{ with } c < 0, \text{ then} \quad A^*\, grad(u^*) = b\, grad(u^*), A^*\, grad(p^*) = b\, grad(p^*), \tag{9.27}$$

if $grad(p^*)$ is not parallel to $grad(u^*)$ then

$$\begin{aligned} A^*\, grad(u^*) &= \frac{a+b}{2} grad(u^*) + \frac{a-b}{2} \frac{|grad(u^*)|}{|grad(p^*)|} grad(p^*) \\ A^*\, grad(p^*) &= \frac{a+b}{2} grad(p^*) + \frac{a-b}{2} \frac{|grad(p^*)|}{|grad(u^*)|} grad(u^*). \end{aligned} \tag{9.28}$$

Proof: As $|grad(u^*)| \neq 0$, when one changes the basis $e_1, \ldots, e_N$, the vector $P\, grad(u^*)$ spans the sphere with diameter $[b\, grad(u^*), a\, grad(u^*)]$, and the quantity $\big(P\, grad(u^*).grad(p^*)\big)$ attains its maximum at a unique point of the sphere; as $A^*\, grad(u^*)$ belongs to the closed ball with diameter $[b\, grad(u^*), a\, grad(u^*)]$, $A^*\, grad(u^*)$ must be equal to the value of $P\, grad(u^*)$ for the P for which the maximum is attained, and this gives the formulas for $A^*\, grad(u^*)$ in (9.26)/(9.28). The corresponding formulas for $A^*\, grad(p^*)$ follow by a simple remark from Linear Algebra, applied to $A^* - \frac{a+b}{2} I$, namely that if M is symmetric with norm $\leq \gamma$ and if for a vector $E \neq 0$ one has $M\,E = F$ and $|F| = \gamma |E|$, then $M\,F = \gamma^2\, E$ (because $|M\,F - \gamma^2\, E|^2 = |M\,F|^2 - 2\gamma^2 (F.M\,E) + \gamma^4 |E|^2 \leq \gamma^2(\gamma^2|E|^2 - |F|^2) = 0$).■

On the set Ω_+ where neither $grad(u^*)$ nor $grad(p^*)$ vanish, one may therefore replace A^* by P which is obtained by layering the materials with the same proportions $\eta_1^*, \ldots, \eta_r^*$, and one has the same cost because one can keep the same u^*, as $A^*\, grad(u^*) = P\, grad(u^*)$ a.e. in Ω_+. On the set Ω_u where $grad(u^*) = 0$, one can also replace A^* by any material without changing $A^*\, grad(u^*)$ (which remains 0), and in particular one can replace A^* by a material obtained by layering the materials with the same proportions. However, the situation is different for the set Ω_p where $grad(p^*) = 0$ and $grad(u^*) \neq 0$, but one can adapt an argument of RAITUM [Ra] for replacing A^* by materials obtained by layering but with different proportions (if $A^*\, grad(u^*)$ is not on the sphere with diameter $[b\, grad(u^*), a\, grad(u^*)]$, it may only belong to the boundary of another sphere obtained by changing b or a, and therefore one needs to change some of the proportions). Let

$$\Omega_c = \Big\{x | x \in \Omega, \Big(A^*grad(u^*) - b\, grad(u^*).A^*grad(u^*) - a\, grad(u^*)\Big) < 0\Big\}, \tag{9.29}$$

i.e. the set where $A^*grad(u^*)$ cannot be obtained by a layered material so that $grad(u^*) \neq 0$, and the preceding analysis shows that on Ω_c one must have $grad(p^*) = 0$, i.e. $\Omega_c \subset \Omega_p$. One defines $b(\eta)$ and $a(\eta)$ for

$$\eta = (\eta_1, \dots, \eta_r) \in L^\infty(\Omega_c; R^r) \text{ such that } \eta_i \geq 0 \text{ a.e. in } \Omega_c \\ \text{for } i = 1, \dots, r, \sum_{i=1}^{r} \eta_i = 1 \text{ a.e. in } \Omega_c, \tag{9.30}$$

by the formulas

$$\frac{1}{b(\eta)} = \sum_{i=1}^{r} \frac{\eta_i}{\min_j \lambda_j(M_i)}, \; a(\eta) = \sum_{i=1}^{r} \eta_i \max_j \lambda_j(M_i), \tag{9.31}$$

and one imposes the two constraints

$$\Big(A^*grad(u^*) - a(\eta)grad(u^*).A^*grad(u^*) - b(\eta)grad(u^*)\Big) \leq 0 \text{ a.e. in } \Omega_c, \tag{9.32}$$

$$\int_{\Omega_c} \eta_i \, dx = \int_{\Omega_c} \eta_i^* \, dx, i = 1, \dots, r, \tag{9.33}$$

and one wants to minimize

$$J_c(\eta) = \int_{\Omega_c} \Big(\sum_{i=1}^{r} \eta_i \, F_i(x, u^*)\Big) \, dx. \tag{9.34}$$

The set of η satisfying (9.30), (9.32)/(9.33) contains η^* (as (9.32) is true by Lemma 42). The condition (9.30) defines a convex set, which is $L^\infty(\Omega_c; R^r)$ weak $\star$ compact; the constraint (9.32) defines a $L^\infty(\Omega_c; R^2)$ weak $\star$ closed convex set in $\big(a, \frac{1}{b}\big)$ by Lemma 19, and by (9.31) the set of η satisfying (9.32) is therefore convex, and $L^\infty(\Omega_c; R^r)$ weak $\star$ closed, and similarly for the constraint (9.33). As J_c is linear and $L^\infty(\Omega_c; R^r)$ weak $\star$ continuous, it attains its minimum on at least one extreme point of the $L^\infty(\Omega_c; R^r)$ weak $\star$ compact convex set defined by (9.30) and (9.32)/(9.33). One shows then that for such an extreme point one must have equality in (9.32) a.e. in Ω_c, by using an argument of Zvi ARTSTEIN [Ar]. Indeed, assume that $\big(A^*grad(u^*) - a(\eta)grad(u^*).A^*grad(u^*) - b(\eta)grad(u^*)\big) \leq -\varepsilon < 0$ on a subset $K \subset \Omega_c$ with positive measure; one may assume that one can find $r + 1$ disjoint subsets $K_k \subset K$ with positive measure and on each such set K_k one can find two distinct indices $i(k), j(k) \in \{1, \dots, r\}$ such that $\eta_{i(k)}, \eta_{j(k)} \geq \delta_k > 0$ a.e. in K_k (if it was not true, a.e. in K_k there would only be one η_i different from 0 and thus equal to 1, in which case $a(\eta) = b(\eta)$

and one could not be in Ω_c); for $|c_k| \leq \delta_k$, one can add c_k to $\eta_{i(k)}$ in K_k and subtract c_k to $\eta_{j(k)}$ in K_k and still satisfy (9.30), and by restricting a little more $|c_k|$ one can still satisfy (9.32); then one has $r+1$ small arbitrary constants that one can play with and only r linear constraints (9.33) that they must satisfy, and it leaves the possibility to move both ways in at least one direction while satisfying all the constraints, contradicting the assumption that one has choosen an extreme point. Therefore one has found an optimal solution of the initial problem which can be realized by layerings inside Ω_c.

At least one solution of the optimization problem corresponds then to a material obtained by layerings (with varying proportions and directions), and therefore at the end the Homogenization questions seem to disappear from the analysis, and one may wonder if it was really necessary to study Homogenization in the first place.

There is an analogous question about Functional Analysis. One wants to show the existence of a characteristic function $\chi \in L^\infty(\Omega)$ which minimizes $\int_\Omega \chi\, f\, dx$, where f is given in $L^1(\Omega)$ and where the minimization is only considered for characteristic functions of measurable subsets of Ω satisfying a finite number of constraints $\int_\Omega \chi\, g_i\, dx = \alpha_i$ for $i = 1, \ldots, m$, where $g_1, \ldots, g_m$ are given in $L^1(\Omega)$; of course, one assumes that at least one such χ exists. The proof of existence that I know was taught to me by Zvi ARTSTEIN in 1975 [Ar],[69] and it consists in minimizing $L(\theta) = \int_\Omega \theta\, f\, dx$ among functions $\theta \in C = \{\theta | \theta \in L^\infty(\Omega), 0 \leq \theta \leq 1 \text{ a.e. in } \Omega, \int_\Omega \theta\, g_i\, dx = \alpha_i \text{ for } i = 1, \ldots, m\}$; as C is a nonempty compact set for the $L^\infty(\Omega)$ weak $\star$ topology (which is metrizable when restricted to C), and L is continuous for this topology, the minimum of L is attained, and because C is convex and L is affine, the minimum is actually attained on a (nonempty) convex compact subset C_0 of C, and C_0 has at least one extreme point θ_0 by KREIN-MILMAN theorem. Finally, Zvi ARTSTEIN argues that θ_0 belongs to a finite dimensional face of the bigger convex set $C_1 = \{\theta | \theta \in L^\infty(\Omega), 0 \leq \theta \leq 1 \text{ a.e. in } \Omega\}$, and using the fact that the LEBESGUE measure has no atoms, he shows that θ_0 must be a characteristic function, which is therefore optimal among characteristic functions from C.

In both situations, one wants to minimize on a set X some function F_i, belonging to a family of such functions indexed by $i \in I$, and one has constructed a compact space $\widehat{X}$ containing X as a dense subspace, and each F_i becomes the restriction to X of a function $\widehat{F_i}$ which is continuous on $\widehat{X}$. Of course, each function $\widehat{F_i}$ attains its minimum on $\widehat{X}$, but one exhibits a subset Y such that $X \subset Y \subset \widehat{X}$ (and equal to X in the second case), which has the property that for each $i \in I$ the set of minima of $\widehat{F_i}$ intersects Y. Of course, as X is dense in $\widehat{X}$, Y is also dense in $\widehat{X}$, and therefore not compact if $Y \neq \widehat{X}$. It is not clear if there exists another topology for which Y is compact with

[69] The following argument is precisely Zvi ARTSTEIN's proof of LYAPUNOV's theorem.

X dense in Y and each F_i is the restriction to X of a function G_i which is lower semi-continuous on Y.

I do not know any obvious abstract framework which explains how to avoid mentioning Functional Analysis or Homogenization in these examples, and although it could be useful to develop special results for the growing number of students who hope to use only elementary tools in Mathematics, it seems impossible to maintain such a goal if one is interested in practical problems. It is indeed often emphasized by those who have dealt with practical problems of Optimization, that one is rarely given a very precise function to minimize, and that one must reassert the goal in view of preliminary results. For what concerns Homogenization appearing in the problems that I have been considering, an obvious remark is that in the realization of such an optimal solution including mixtures in some areas, one needs to reconsider the function to be minimized by adding the cost of creating these mixtures.[70]

Actually, as was pointed out later by Robert KOHN, there is another natural class of functionals where the optimal solution seems to require more precise information on optimal bounds, and more general designs than simple layerings. In identification problems, one does a few experiments with the same arrangement of materials, and one tries to estimate some coefficient by using a finite number of measurements. As a simple model of this kind, one considers q equations of the form

$$-div\Big(A\, grad(u_i)\Big) = f_i \text{ in } \Omega,\ u_i \in H_0^1(\Omega), \text{ for } i = 1,\dots,q, \tag{9.35}$$

and all these equations use the same $A \in M(\alpha,\beta;\Omega)$ but different functions $f_i \in H^{-1}(\Omega), i = 1,\dots,q$, and q functions $v_i \in H_0^1(\Omega)$ (or simply $v_i \in L^2(\Omega)$) are given for $i = 1,\dots,q$, corresponding to measurements in a material using an unknown value A_0 that one wants to identify.[71] One idea would be to

[70] It is unfortunately often the case that some people dealing with questions in Elasticity not only forget to mention the defects of Linearized Elasticity but only consider extremely particular functionals. On the contrary, on the engineering side, as was emphasized by Martin BENDSØE at a meeting in Trieste in 1993, it is important to have an interactive point of view when dealing with Optimal Design problems; one rarely needs to compute with high precision which mixtures will appear in connection with minimizing a particular functional, because one may well abandon this particular functional during the interactive part of the procedure, and one may end up minimizing something different, in principle more adapted to the engineering application.

[71] Practical problems may include elecrical or heat conduction questions or permeability of oil reservoirs, and they do not correspond to homogeneous DIRICHLET data. Instead of using different f_i one may use different nonhomogeneous DIRICHLET data on some parts of the boundary and NEUMANN data on other parts of the boundary, and the different measurements may give the value of u_i or the flux $(A\, grad(u_i).n)$ on various other parts of the

minimize a cost function J of the form

$$J(A) = \int_{\Omega} \Big(\sum_{i=1}^{q} |u_i - v_i|^2 \Big)\, dx, \tag{9.36}$$

and if one knows that A is one of the materials $M_1, \ldots, M_r$ rotated in an arbitrary way (a.e. in Ω), one gets a problem where the Homogenization approach will appear, but if one considers functionals of the form

$$J_1(A) = \int_{\Omega} \Big(\sum_{i=1}^{q} |grad(u_i - v_i)|^2 \Big)\, dx, \tag{9.37}$$

the situation is quite different [Ta13]. In the case of (9.36), minimizing sequences will make some $A^{eff} \in \mathcal{K}(\eta_1, \ldots, \eta_r; M_1, \ldots, M_r)$ appear, and there exists then an optimal homogenized A, and one may wonder what its relation with the actual A_0 is, for example if in the case where $M_i = \mu_i I$ for $i = 1, \ldots, r$ and therefore A_0 is locally an isotropic material, but one finds that all the optimal A are anisotropic in some regions. Of course, if the measured values were exactly those associated with A_0 when one uses (9.35), then one would have $J(A_0) = 0$, but measurements are not always entirely accurate and this explains why, even when A_0 only takes isotropic values, it may do so on relatively small pieces, and A_0 may well be near an anisotropic A in a distance corresponding to H-convergence (mentioned after Definition 5). If one decides then to minimize (9.36) and one wants to write necessary conditions of optimality, an increase δA creates increases δu_i, $i = 1, \ldots, q$, solutions of

$$-div\Big(A\, grad(\delta u_i) + \delta A\, grad(u_i)\Big) = 0 \text{ in } \Omega,\ \delta u_i \in H_0^1(\Omega), i = 1, \ldots, q, \tag{9.38}$$

and an increase δJ given by

$$\delta J = \int_{\Omega} 2\Big(\sum_{i=1}^{q} (u_i - v_i)\delta u_i \Big)\, d\dot{x}, \tag{9.39}$$

and in order to express δJ in terms of δA, one introduces q adjoint states $p_1, \ldots, p_q$, solutions of

$$-div\Big(A\, grad(p_i)\Big) = u_i - v_i \text{ in } \Omega,\ p_i \in H_0^1(\Omega), i = 1, \ldots, q, \tag{9.40}$$

---

boundary. Another important class of problems arising in applications deals with eigenvalues, in which case one may measure a few of the lowest eigenvalues corresponding to some unknown A_0, which one tries to identify from these measurements. It is not difficult however to adapt the methods described in this course (for some academic models) in order to deal with these more realistic variants.

and the necessary condition of optimality becomes

$$\delta J = -2\int_\Omega \Big[\sum_{i=1}^{q}\Big(\delta A\, grad(u_i).grad(p_i)\Big)\Big]\,dx \geq 0 \text{ for admissible } \delta A. \quad (9.41)$$

In the first step of the obtention of necessary conditions that I have described, i.e. keeping the proportions $\eta_1^*,\ldots,\eta_r^*$ fixed, one sees that it would be useful to characterize the set of $\big(A\,grad(u_1),\ldots,A\,grad(u_q)\big)$ for $A \in \mathcal{K}(\eta_1,\ldots,\eta_r;M_1,\ldots,M_r)$, but for $q \geq 2$ the analogue of Lemma 42 is not known. Even for the case of mixing two isotropic materials, i.e. $M_1 = \alpha I, M_2 = \beta I$, for which I had obtained the characterization (7.25), (7.28), (7.30) of $\mathcal{K}(\theta,1-\theta;\alpha I,\beta I)$ with François MURAT in 1980, I do not know a simple characterization of this set, although Lemma 41 asserts that it is convex if $q \leq N-1$. It was for this case that Robert KOHN had pointed out that the full characterization seems necessary, but I do not know if the necessary condition (9.41) has given rise to a precise analysis of what optimal solutions look like.

11. Necessary conditions of optimality: second step

We can now look at the second step of the derivation of necessary conditions of optimality, which consists in varying $\eta_1,\ldots,\eta_r$, and is therefore more classical. I begin by treating two very special examples, corresponding to $p = u$ or to $p = -u$, where one can avoid Homogenization almost completely. The special examples correspond to minimizing either the cost function J_1 or the cost function J_2 defined by

$$\begin{aligned} J_1(\chi_1,\ldots,\chi_r,R) &= \int_\Omega f\,u\,dx \\ J_2(\chi_1,\ldots,\chi_r,R) &= -\int_\Omega f\,u\,dx, \end{aligned} \quad (10.1)$$

and it is important that in (10.1) one uses the same function f which appears in (9.3). The reason why these two functionals are special is that the definition (9.12) of the adjoint state in the case of minimizing J_1 gives $p = u$, and therefore $A^*\,grad(u^*) = a(\eta^*)grad(u^*)$ by (9.26) for $N \geq 2$, while in the case of minimizing J_2 it gives $p = -u$, and therefore $A^*\,grad(u^*) = b(\eta^*)grad(u^*)$ by (9.27) (which is valid for $N \geq 1$, and also for minimizing J_1 in the case $N = 1$), where the functions a and b are defined by (9.13), repeated as (9.31). Another important fact for the argument is that, because A is symmetric a.e. in Ω, solving (9.3) amounts to a minimization problem, namely

$$Min_{v\in H_0^1(\Omega)} \int_\Omega \Big[\Big(A\, grad(v).grad(v)\Big) - 2f\, v\Big]\, dx \text{ is attained for } v = u$$

defined by (9.3), and the value of the minimum is $-\int_\Omega f\,u\,dx.$

(10.2)

Lemma 46: The minimization of J_1 is equivalent to the following min-max problem

$$Max_\eta Min_{v\in H_0^1(\Omega)} \int_\Omega \Big(a(\eta)|grad(v)|^2 - 2f\,v\Big)\,dx \text{ if } N \geq 2, \tag{10.3}$$

$$Max_\eta Min_{v\in H_0^1(\Omega)} \int_\Omega \Big(b(\eta)|grad(v)|^2 - 2f\,v\Big)\,dx \text{ if } N = 1, \tag{10.4}$$

the set of η on which one maximizes is given by (9.6), repeated in (9.9), and in both cases u^* is defined in a unique way.
Proof: As already mentioned, for any optimal solution u^* one has $p^* = u^*$, and therefore one does not need the argument of RAITUM described after Proposition 45. For $N \geq 2$ one has (9.26) at points where $grad(u^*) \neq 0$, and as it holds automatically at points where $grad(u^*) = 0$, the formula (10.2) gives (10.3). For $N = 1$ one automatically has $A^{eff} = b(\eta)I$ and therefore (9.27) holds, and the formula (10.2) gives (10.4).

That the problems in (10.3) or (10.4) have at least one solution follows from classical min-max theorems, because by using the information $A \in M(\alpha,\beta;\Omega)$ one may restrict the minimization in v to a large enough closed ball of $H_0^1(\Omega)$, which is a compact convex set for the weak topology of $H_0^1(\Omega)$, the set of η is a compact convex set for the $L^\infty(\Omega;R^r)$ weak $\star$ topology, and the functional is convex lower semi-continuous in v and concave upper semi-continuous in η (it is actually linear continuous in the case $N \geq 2$). Moreover, because the functional is strictly convex in v, the solution u^* is unique. ■

Of course, if without knowing anything about Homogenization one has guessed that the problem (10.3) is a way to solve the initial problem of minimizing J_1, one still has to use Homogenization in order to interpret what a solution η^* means, and that it can be created by layerings. Jean CÉA and K. MALANOWSKI had unknowingly taken advantage of a similar miracle in [Cé&Ma], but they had started from accepting all possible γI with $\alpha \leq \gamma \leq \beta$, and they did not even have to explain by Homogenization the (classical) solution that they had obtained.

Lemma 47: The minimization of J_2 is equivalent to the following minimization problem

$$Min_\eta Min_{v\in H_0^1(\Omega)} \int_\Omega \Big(b(\eta)|grad(v)|^2 - 2f\,v\Big)\,dx, \tag{10.5}$$

and the set of η on which one minimizes is given by (9.6), repeated in (9.9), and $b(\eta^*)grad(u^*)$ is defined in a unique way.

Proof: As already mentioned, for any optimal solution u^* one has $p^* = -u^*$, and therefore one does not need the argument of RAITUM described after Proposition 45. For $N \geq 2$ one has (9.27) at points where $grad(u^*) \neq 0$, and as it holds automatically at points where $grad(u^*) = 0$, the formula (10.2) gives (10.5).

That the problem in (10.5) has a solution follows from classical theorems in Optimization, because by using the information $A \in M(\alpha, \beta; \Omega)$ one may restrict the minimization in v to a large enough closed ball of $H_0^1(\Omega)$, which is a compact convex set for the weak topology of $H_0^1(\Omega)$, the set of η is a compact convex set for the $L^\infty(\Omega; R^r)$ weak $\star$ topology, and the functional is convex in (v, η) and lower semi-continuous by Lemma 11 with $M = b(\eta)\, I$; moreover, as $P = M^{-1}$ in formula (5.6), one sees that the functional is strictly convex in $b(\eta) grad(v)$ and therefore $b(\eta^*) grad(u^*)$ is unique.■

Of course, if without knowing anything about Homogenization one has guessed that the problem (10.5) is a way to solve the initial problem of minimizing J_2, one still has to use Homogenization in order to interpret what a solution η^* means, and that it can be created by layerings.

These two examples are not always instances of an intermediate problem in the abstract setting mentioned after (9.34), and they do not always fit the framework of relaxed problems described in chapter 3 either, because the set X for which the minimization of J_1 or J_2 is considered (which is the set of characteristic functions $\chi_1, \ldots, \chi_r$ of disjoints sets satisfying (9.2), together with a measurable rotation R so that A is defined by (9.1)) is not always included in the set Y on which (10.3) or (10.4) are considered (which is the set of $\eta_1, \ldots, \eta_r$ satisfying (9.6), and A is $a(\eta) I$ for minimizing J_1 for $N \geq 2$, or $b(\eta) I$ for minimizing J_1 for $N = 1$ or for minimizing J_2). X can only be considered as a subset of Y if one can avoid the use of the rotation R, i.e. in the case where $M_i = \mu_i I$ for $i = 1, \ldots, r$; however, even in this case, the problem is not a completion, except for $N = 1$, where there is a formula for A^{eff}, as $A^{eff} = b(\eta)$.

Coming back to the general case, one wants now to write that the optimal mixture using the proportions $\eta_1^*, \ldots, \eta_r^*$ (with A^* corresponding to a layered material) is better than any other mixture using different proportions $\eta_1, \ldots, \eta_r$, and for this task one follows the arguments used in Lemma 44 (or the variant before Proposition 45), but one starts now from $A \in \mathcal{K}(\eta_1, \ldots, \eta_r; M_1, \ldots, M_r)$, with proportions $\eta_1, \ldots, \eta_r$ satisfying (9.9). By mixing then A with A^* in an adapted way one will create an admissible differentiable path $\eta(\varepsilon) = \eta^* + \varepsilon\, \delta\eta + o(\varepsilon)$, $A(\varepsilon) = A^* + \varepsilon\, \delta A + o(\varepsilon)$, $u(\varepsilon) = u^* + \varepsilon\, \delta u + o(\varepsilon)$, $\widetilde{J}(\eta(\varepsilon), A(\varepsilon)) = \widetilde{J}(\eta^*, A^*) + \varepsilon \delta \widetilde{J} + o(\varepsilon)$, with δu and δA still related by (9.17), but with $\delta\widetilde{J}$ given now by

$$\delta \widetilde{J} = \int_\Omega \Big[\sum_{i=1}^r \delta\eta_i\, F_i(x, u^*) + \Big(\sum_{i=1}^r \eta_i^* \frac{\partial F_i(x, u^*)}{\partial u}\Big) \delta u\Big]\, dx, \tag{10.6}$$

and using the definition (9.12) of the adjoint state p^* and (9.17), one finds

$$\delta\widetilde{J} = \int_\Omega \Big[\sum_{i=1}^r \delta\eta_i\, F_i(x,u^*) - \Big(\delta A\, grad(u^*).grad(p^*)\Big)\Big]\, dx. \tag{10.7}$$

Proposition 48: Let $\eta_1^*,\dots,\eta_r^*,A^*$ satisfy (9.9) and

$$\widetilde{J}(\eta_1^*,\dots,\eta_r^*,A^*) \le \widetilde{J}(\eta_1,\dots,\eta_r,A) \text{ for all } \eta_1,\dots,\eta_r,A \text{ satisfying (9.9)}, \tag{10.8}$$

with $\widetilde{J}$ defined by (9.10). For $N \ge 2$, (10.8) implies

$$\int_\Omega \Big[\sum_{i=1}^r \eta_i^*\, F_i(x,u^*) - \Big(A^*\, grad(u^*).grad(p^*)\Big)\Big]\, dx \le$$
$$\int_\Omega \Big[\sum_{i=1}^r \eta_i\, F_i(x,u^*) - \Big(B\, grad(u^*).grad(p^*)\Big)\Big]\, dx \tag{10.9}$$

for all $\eta_1,\dots,\eta_r$ satisfying (9.9), $B \in \mathcal{B}(\eta_1,\dots,\eta_r;M_1,\dots,M_r)$,

with $u^*,p^*,\mathcal{B}(\eta_1,\dots,\eta_r;M_1,\dots,M_r)$ defined by (9.3) (using A^*), (9.12) (using A^*), and (9.13). For $N=1$, one has $a^* = b(\eta^*)$ and (10.8) implies

$$\int_\Omega \Big[\sum_{i=1}^r \Big(F_i(x.u^*) + \frac{(a^*)^2}{\lambda(M_i)}\frac{du^*}{dx}\frac{dp^*}{dx}\Big)\eta_i^*\Big]\, dx \le$$
$$int_\Omega \Big[\sum_{i=1}^r \Big(F_i(x.u^*) + \frac{(a^*)^2}{\lambda(M_i)}\frac{du^*}{dx}\frac{dp^*}{dx}\Big)\eta_i\Big]\, dx \tag{10.10}$$

for all $\eta_1,\dots,\eta_r$ satisfying (9.9).

Proof: Let $\eta_1,\dots,\eta_r$ satisfy (9.9), and $B \in \mathcal{B}(\eta_1,\dots,\eta_r;M_1,\dots,M_r)$. For $\varepsilon \in (0,1)$, one defines $\eta(\varepsilon)$ by

$$\eta_i(\varepsilon) = (1-\varepsilon)\eta_i^* + \varepsilon\,\eta_i, \tag{10.11}$$

so that $\eta(\varepsilon)$ satisfies (9.9). Using Lemma 42 (and preceding remarks for constructing measurable liftings), there exists A such that

$$A\, grad(u^*) = B\, grad(u^*),\ A \in \mathcal{K}(\eta_1,\dots,\eta_r;M_1,\dots,M_r), \text{ a.e. } x \in \Omega. \tag{10.12}$$

For a nonvanishing $e \in L^\infty(\Omega;R^N)$ one defines $A(\varepsilon)$ by layering A^* and A with respective proportions $1-\varepsilon$ and ε, in layers orthogonal to e, and Lemma 34 gives

$$A(\varepsilon) = (1-\varepsilon)A^* + \varepsilon\, A - \varepsilon(1-\varepsilon)(A-A^*)\frac{e\otimes e}{\varepsilon(A^*\, e.e) + (1-\varepsilon)(A\, e.e)}(A-A^*), \tag{10.13}$$

and by construction one has $A(\varepsilon) \in \mathcal{K}(\eta_1(\varepsilon), \ldots, \eta_r(\varepsilon); M_1, \ldots, M_r)$ a.e. in Ω. Of course, (10.8) implies that $\delta\widetilde{J} \geq 0$, for which one uses (10.7); the value of $\delta\eta_i$ following from (10.11) is $\delta\eta_i = \eta_i - \eta_i^*$, and the value of δA following from (10.13) is $\delta A = A - A^* - (A - A^*)\frac{e\otimes e}{(A\,e.e)}(A - A^*)$, so that $\big(\delta A\,grad(u^*).grad(p^*)\big) = \big((A - A^*)grad(u^*).grad(p^*)\big) - \frac{1}{(A\,e.e)}\big((A - A^*)grad(u^*).e\big)\big((A - A^*)e.grad(p^*)\big)$, and by choosing e orthogonal to $(A - A^*)grad(u^*)$, one has $\big(\delta A\,grad(u^*).grad(p^*)\big) = \big((A - A^*)grad(u^*).grad(p^*)\big)$, which by (10.12) is $\big((B - A^*)grad(u^*).grad(p^*)\big)$, and $\delta\widetilde{J} \geq 0$ is then exactly (10.9).

In the case $N = 1$, $A \in \mathcal{K}(\eta_1, \ldots, \eta_r; M_1, \ldots, M_r)$ means $A = b(\eta)$, with b defined in (9.13), and therefore $\delta A = \sum_i \frac{\partial b}{\partial \eta_i}\delta\eta_i$; (9.13) implies that $\frac{\partial b}{\partial \eta_i} = -\frac{b^2}{\lambda(M_i)}$ (where $\lambda(M_i)$ is the value of the coefficient for the material M_i), and therefore one has

$$\delta\widetilde{J} = \int_\Omega \Big[\sum_{i=1}^r \Big(F_i(x.u^*) + \frac{(a^*)^2}{\lambda(M_i)}\frac{du^*}{dx}\frac{dp^*}{dx}\Big)\delta\eta_i\Big]\,dx, \tag{10.14}$$

from which one immediately deduces (10.10). ∎

Inequality (10.9) consists in minimizing a linear functional on a convex set, and it can be further simplified by noticing that $B \in \mathcal{B}(\eta_1, \ldots, \eta_r; M_1, \ldots, M_r)$ only enters (10.9) through $D = B\,grad(u^*)$, and when B runs through $\mathcal{B}(\eta_1, \ldots, \eta_r; M_1, \ldots, M_r)$, D spans the closed ball of diameter $[b(\eta)\,grad(u^*), a(\eta)\,grad(u^*)]$, and therefore (10.9) is equivalent to

$$\int_\Omega \Big[\sum_{i=1}^r \eta_i^*\,F_i(x,u^*) - \big(A^*\,grad(u^*).grad(p^*)\big)\Big]\,dx \leq$$
$$\int_\Omega \Big[\sum_{i=1}^r \eta_i\,F_i(x,u^*) - \big(D.grad(p^*)\big)\Big]\,dx \text{ for all } \eta_1, \ldots, \eta_r$$
$$\text{satisfying (9.9)}, \big(D - b(\eta)\,grad(u^*).D - a(\eta)\,grad(u^*)\big) \leq 0, \tag{10.15}$$

and the convexity of the set of $\eta_1, \ldots, \eta_r, D$ used in (10.15) follows from Lemma 19, and the convexity of the functions a and $\frac{1}{b}$, consequence of their definition in (9.13).

One can go further in the analysis of the necessary conditions obtained, (10.9) or (10.15) for the case $N \geq 2$, or (10.10) for the case $N = 1$, as they consist in minimizing linear functionals on convex sets defined by linear constraints, and LAGRANGE multipliers can then be used for making these conditions more precise, but I will not describe this question here as it is a more classical subject.

12. Conclusion

I have now completed the description of the particular subject of this course, which was to show how questions of Homogenization appear in Optimal Design problems, following the work which I had pioneered in the early 70s with François MURAT.

It is time now that I describe the intuitive ideas behind the necessary conditions of optimality obtained by Konstantin LUR'IE, as they were explained to me in the early 80s by Jean-Louis ARMAND, after he had visited LUR'IE in Leningrad, where he had been told about my work [Ta2]. As I do not read much, I do not know where these ideas have appeared in print,[72] and although the idea is quite natural, it seems hard to transform into sound mathematical estimates.[73] Assume that we consider a problem involving two (isotropic) materials, with imposed global proportions, and that we want to test the optimality of a given classical design, with a smooth interface between the two materials; the classical idea, which goes back to HADAMARD,[74] consists in pushing the interface along its normal of variable amounts (with one constraint related to the global proportion imposed), and computing the change in the cost function leads to a necessary condition of optimality valid along the interface; LUR'IE's first idea was to work away from the interface, taking a small sphere imbedded in one material and an identical one imbedded in the other material and exchanging their content, and computing the change in the cost function leads to a necessary condition of optimality valid everywhere;[75] LUR'IE's second idea was to consider ellipsoids of the same volume instead of spheres, and his new necessary conditions of optimality were

[72] Having learned about some ideas of LUR'IE, I do not need a published reference in order to attribute these ideas to him (and I could not remember of anyone else claiming them as his/hers, although I have not checked the work of Richard DUFFIN, as I mention in footnote 75). I may be alone in thinking that if a new idea is only mentioned orally by someone who does not put it in print immediately it should be attributed to this person, eventually with the names of those who would have found the same idea later but independently, but not with the names of those who had heard about the idea and had put it in print under their name, expecting to acquire fame for an idea which was not theirs.

[73] I wonder then if LUR'IE had been able to carry out these computations in a mathematician's way, or if he had just acquired convincing evidence that some formulas must hold, as nonmathematicians often do.

[74] I have already mentioned the precise analysis along this line of thought, carried out by François MURAT and Jacques SIMON in [Mu&Si].

[75] In the late 80s, my late colleague Richard DUFFIN had mentioned to me that he had worked on questions of Optimal Design, and he used to call such necessary conditions a "principle of democracy". Unfortunately, as I knew about such questions, I failed to enquire about his precise results, so that I

stronger; as he could now play with the orientation of the ellipsoids and the ratio of the axes, he realized that it was better to take them very slender, and in the limit he could understand that layered materials were important for his problem. As I have mentioned earlier, this was a quite good extension of the ideas of PONTRYAGUIN to a setting of partial differential equations, but not quite the right way to discover the analysis which I had performed with François MURAT, which was in some way a good extension of the better ideas of Laurence C. YOUNG.

The description of the method developed in this course, which corresponds in part to results which I had obtained with François MURAT in the 70s, is analogous to what I had already taught in 1983 at the CEA - EDF - INRIA Summer course at Bréau-sans-Nappe, written in [Mu&Ta1], and similar to what I had taught again in 1986 in Durham [Ta11]. I have included here much more of the basic results on Homogenization, which I had only alluded to before, and actually the description of the original method which I had followed with François MURAT in the early 70s had never appeared in print before, the reason being that it had been greatly simplified by my method of oscillating test functions which extends easily to all (linear) variational formulations; because I had decided to use a chronological point of view in this course, in order to show how new ideas had appeared, it was natural that I should describe first our original ideas, even though I had improved them later. I have added a few simplifications, which I had first written in 1995 in [Ta14], and which were therefore not included in my previous courses on the subject.

As I have mentioned earlier, some people have led a campaign of misattribution of my ideas which seems to have intensified around 1983. I may have inadvertently added to the confusion by forgetting to mention the reference [Ta7] of my second method for obtaining bounds on effective coefficients,[76] and I did not think that it had any importance because I could not imagine that there were people ready to steal an idea that they would have heard if they thought that it had not been written yet.[77] Unfortunately, the confusion may also have increased because Graeme MILTON called my method the "translation method",[78] and many have used the possibility of quoting my method by this new name, without attributing it correctly.

---

do not know if his conditions were of the HADAMARD type, or the LUR'IE type, and I do not know when he had first obtained such results.

[76] The reason for not giving the reference of that 1977 conference in Versailles was that the organizers had forgotten to send me a copy of the proceedings, and therefore I did not know the exact reference of my article.

[77] It had not been my intention to hide the existence of a written reference in order to confront later those who were ready to steal my ideas.

[78] I do not find that name so well adapted to what my method is about. On the other hand, when I was a student (in Paris in the late 60s), this precise

I have been told that Alexandre GROTHENDIECK has analyzed in an unpublished book a few ways in which misattribution of ideas is organized,[79] I have not read it and I cannot assert that my observations on this unfortunate aspect of academic behaviour coincide or not with his own. This "book" had first been mentioned to me in 1984 by Jean LERAY, who had pointed out that it was a good sign that my ideas were stolen, and that many who steal others ideas would probably like that some of their ideas be stolen too, as it would prove that they had had some of their own; Jean LERAY also had to face such an adverse behaviour, but if some of his ideas had been "borrowed" by a famous mathematician who had shown enough creativity of his own, for political reasons which were not too dissimilar to those which I had encountered myself more than thirty years later, I have not found myself as fortunate and many who use my ideas without saying it present such a distorted view that one does not have to be a very good student in Mathematics for performing the small detective work of identifying those who have stolen ideas that they do not even understand well enough.

This being said, I must say that I think that the worst sin of a teacher is to induce students in error, and I do consider it actually a minor sin to forget to name the inventor of an idea,[80] but a major sin to give a bad explanation of what an idea is, or to forget mentioning an important idea on a subject. In consequence, if someone would feel such a pressure for avoiding to mention the author of one of the ideas that I have described in this course, it would be better if he/she would start by learning well the content of this course, and then teach an improved version of it.

---

term was used for describing a method of Louis NIRENBERG for proving regularity of solutions of elliptic equations in smooth domains.

[79] Laurent SCHWARTZ told me recently that the publication of GROTHENDIECK's book, "Récoltes et Semailles", was not possible because of the numerous personal attacks that it contains. It is nevertheless available on the Internet, in Russian translation!

[80] My religious upbringing forbids me to steal, but my memory is not perfect, and I may have forgotten to quote some authors for their ideas. If I realized that I had made such a mistake, either by being told about it or by finding it myself, I would certainly try to give a corrected statement on the next occasion where I would write on the subject (and I hope that a second lapse of memory would not occur at that time).

13. Acknowledgements

I want to thank Arrigo CELLINA and António ORNELAS for their invitation to teach in the CIME / CIM Summer course on Optimal Design, held in Troia in June 1998. The first meeting that I ever attended was a CIME course in Varenna in 1970 (where my advisor, Jacques-Louis LIONS, was one of the main speakers), and I had enjoyed very much the working atmosphere of that course, which was also the occasion of my first visit to Italy. I had not been able to attend another CIME course since, and as I had not been able to accept a previous invitation to teach in such a course, it was a great pleasure to lecture in a CIME course, and a surprise that such a course would actually be held in Portugal.

My research is supported by CARNEGIE-MELLON University, and the National Science Foundation (grant DMS-97.04762, and through a grant to the Center for Nonlinear Analysis). It is a pleasure to acknowledge the support of the Max PLANCK Institute for Mathematics in the Sciences in Leipzig for my sabbatical year 1997-98.

References

[Ar] ARTSTEIN Z., "Look for the extreme points," SIAM Review **22** (1980), 172–185.

[Ba] BABUŠKA I., "Homogenization and its application. Mathematical and computational problems," *Numerical Solution of Partial Differential Equations-III, (SYNSPADE 1975, College Park MD, May* 1975), *Academic Press, New York*, 1976, 89–116.

[Be&Li&Pa] BENSOUSSAN A. & LIONS J.-L. & PAPANICOLAOU G., *Asymptotic Analysis for Periodic Structures*, Studies in Mathematics and its Applications 5. North-Holland, Amsterdam 1978.

[Be] BERGMAN D., "Bulk physical properties of composite media," *Homogenization methods: theory and applications in physics (Bréau-sans-Nappe, 1983*), 1–128, *Collect. Dir. Etudes Rech. Elec. France*, 57, *Eyrolles, Paris*, 1985.

[Br&Po] BRAIDY P. & POUILLOUX D., Mémoire d'Option, Ecole Polytechnique, 1982, unpublished.

[Cé&Ha] *Optimization of distributed parameter structures, Vol. I, II, J. CÉA & E. J. HAUG eds.* Proceedings of the NATO Advanced Study Institute on Optimization of Distributed Parameter Structural Systems, Iowa City, May 20-June 4, 1980. NATO advanced Study Institute Series E: Applied Sciences, 49, 50, *Martinus Nijhoff Publishers, The Hague*, 1981.

[Cé&Ma] CÉA J. & MALANOWSKI K., "An example of a max-min problem in partial differential equations," SIAM J. Control **8** 1970, 305–316.

[Ch] CHENAIS D., "Un ensemble de variétés à bord Lipschitziennes dans R^N: compacité, théorème de prolongements dans certains espaces de Sobolev," C. R. Acad. Sci. Paris Sér. A-B **280** (1975), Aiii, A1145–A1147.
[DG&Sp] DE GIORGI E. & SPAGNOLO S., "Sulla convergenza degli integrali dell'energia per operatori ellitici del 2 ordine," Boll. Un. Mat. Ital. (4) **8** (1973), 391–411.
[Fr&Mu] FRANCFORT G. & MURAT F., "Homogenization and Optimal Bounds in Linear Elasticity", Arch. Rational Mech. Anal., **94**, 1986, 307–334.
[Gu] GUTIÉRREZ S., "Laminations in Linearized Elasticity and a Lusin type Theorem for Sobolev Spaces," PhD thesis, CARNEGIE-MELLON University, May 1997.
[Ha&Sh] HASHIN Z. & SHTRIKMAN S., "A variational approach to the theory of effective magnetic permeability of multiphase materials", J. Applied Phys. **33**, (1962) 3125–3131.
[Jo] JOSEPH D., *Stability of fluid motions* I, II, Springer Tracts in Natural Philosophy, Vol. 27, 28, Springer-Verlag, Berlin-New York, 1976.
[Ke] KELLER J., "A theorem on the conductivity of a composite medium," J. Mathematical Phys. **5** (1964) 548–549.
[Li1] LIONS J.-L., *Contrôle optimal de systèmes gouvernés par des équations aux dérivées partielles*, Dunod - Gauthier-Villars, Paris, 1968.
[Li2] LIONS J.-L. , *Quelques méthodes de résolution des problèmes aux limites non linéaires*, Dunod; Gauthier-Villars, Paris, 1969.
[Li3] LIONS J.-L., "Asymptotic behaviour of solutions of variational inequalities with highly oscillating coefficients," *Applications of methods of functional analysis to problems in mechanics (Joint Sympos., IUTAM/IMU, Marseille, 1975)* , pp. 30–55. Lecture Notes in Math., 503. *Springer, Berlin*, 1976.
[Lu] LUR'IE K. A., "On the optimal distribution of the resistivity tensor of the working substance in a magnetohydrodynamic channel," J. Appl. Mech. (PMM) **34** (1970), 255–274.
[Ma&Sp] MARINO A. & SPAGNOLO S., "Un tipo di approssimazione dello operatore $\sum_{ij} D_i(a_{ij}D_j)$ con operatori $\sum_j D_j(b\,D_j)$," Ann. Scuola Norm. Sup. Pisa Cl. Sci. (3) **23** (1969), 657–673.
[MC] MCCONNELL W., "On the approximation of elliptic operators with discontinuous coefficients," Ann. Scuola Norm. Sup. Pisa Cl. Sci. (4) **3** (1976), no. 1, 123–137.
[Me] MEYERS N., "An L^p-estimate for the gradient of solutions of second order elliptic divergence equations," Ann. Scuola Norm. Sup. Pisa (3) **17** (1963), 189–206.
[Mo& St1] MORTOLA S. & STEFFÉ S., Unpublished report, Scuola Normale Superiore, Pisa.
[Mo& St2] MORTOLA S. & STEFFÉ S., "A two-dimensional homogenization problem," Atti Accad. Naz. Lincei Rend. Cl. Sci. Fis. Mat. Natur. (8) **78** (1985), no. 3, 77–82.

[Mu1] MURAT F., "Un contre-exemple pour le problème du contrôle dans les coefficients," C. R. Acad. Sci. Paris, Sér. A-B **273** (1971), A708–A711.

[Mu2] MURAT F., "Théorèmes de non-existence pour des problèmes de contrôle dans les coefficients," C. R. Acad. Sci. Paris, Sér. A-B **274** (1972), A395–A398.

[Mu3] MURAT F., "H-convergence," *Séminaire d'analyse fonctionnelle et numérique*, Université d'Alger, 1977-78. Translated into English as MURAT F. & TARTAR L., "H-convergence," *Topics in the mathematical modelling of composite materials*, 21–43, Progr. Nonlinear Differential Equations Appl., 31, *Birkhäuser Boston, Boston, MA*, 1997.

[Mu4] MURAT F., "Compacité par compensation," Ann. Scuola Norm. Sup. Pisa Cl. Sci. (4), **5** (1978), 489–507.

[Mu&Si] MURAT F. & SIMON J., "Sur le contrôle par un domaine géométrique" *Publication 76015, Laboratoire d'Analyse Numérique, Université Paris 6*, 1976.

[Mu&Ta1] MURAT F. & TARTAR L., "Calcul des variations et homogénéisation," *Homogenization methods: theory and applications in physics (Bréau-sans-Nappe, 1983)*, 319–369, Collect. Dir. Etudes Rech. Elec. France, 57, *Eyrolles, Paris*, 1985. Translated into English as "Calculus of variations and homogenization," *Topics in the mathematical modelling of composite materials*, 139–173, Prog. Nonlinear Differential Equations Appl., 31, *Birkhäuser Boston, M A*, 1997.

[Mu&Ta2] MURAT F. & TARTAR L., "Optimality conditions and homogenization," *Nonlinear variational problems (Isola d'Elba, 1983)*, 1–8, Res. Notes in Math., 127, *Pitman, Boston, Mass. -London*, 1985.

[Ra] RAITUM U.E., "The extension of extremal problems connected with a linear elliptic equation," Soviet Math. Dokl. **19**, (1978), 1342-1345.

[S-P1] SANCHEZ-PALENCIA E., "Solutions périodiques par rapport aux variables d'espaces et applications," C. R. Acad. Sci. Paris, Sér. A-B **271** (1970), A1129–A1132.

[S-P2] SANCHEZ-PALENCIA E., "Equations aux dérivées partielles dans un type de milieux hétérogènes," C. R. Acad. Sci. Paris, Sér. A-B **272** (1972), A395–A398.

[Si] SIMON L., "On G-convergence of elliptic operators," Indiana Univ. Math. J. **28** (1979), no. 4, 587–594.

[Sp1] SPAGNOLO S., "Sul limite delle soluzioni di problemi di Cauchy relativi all'equazione del calore," Ann. Scuola Norm. Sup. Pisa (3) **21** (1967), 657–699.

[Sp2] SPAGNOLO S., "Sulla convergenza di soluzioni di equazioni paraboliche ed ellitiche," Ann. Scuola Norm. Sup. Pisa (3) **22** (1968), 571–597.

[Ta1] TARTAR L., "Convergence d'opérateurs différentiels," *Atti Giorni Analisi Convessa e Applicazioni* (*Roma*, 1974), 101–104.

[Ta2] TARTAR L., "Problèmes de contrôle des coefficients dans des équations aux dérivées partielles," *Control theory, numerical methods and computer sys-*

tems modelling (Internat. Sympos., IRIA LABORIA, Rocquencourt, 1974), *pp.* 420–426. *Lecture Notes in Econom. and Math. Systems, Vol.* 107, *Springer, Berlin,* 1975. Translated into English as MURAT F. & TARTAR L., "On the control of coefficients in partial differential equations," *Topics in the mathematical modelling of composite materials,* 1–8, Prog. Nonlinear Differential Equations Appl., 31, *Birkhäuser Boston, MA,* 1997.

[Ta3] TARTAR L., "Quelques remarques sur l'homogénéisation," *Functional analysis and numerical analysis (Tokyo and Kyoto,* 1976), 469–481, *Japan Society for the Promotion of Science, Tokyo,* 1978.

[Ta4] TARTAR L., "Weak convergence in nonlinear partial differential equations," *Existence Theory in Nonlinear Elasticity,* 209–218. *The University of Texas at Austin,* 1977.

[Ta5] TARTAR L., "Une nouvelle méthode de résolution d'équations aux dérivées partielles non linéaires," *Journées d'Analyse Non Linéaire (Proc. Conf., Besançon, 1977),* pp. 228–241, Lecture Notes in Math., 665, *Springer, Berlin,* 1978.

[Ta6] TARTAR L., "Nonlinear constitutive relations and homogenization," *Contemporary developments in continuum mechanics and partial differential equations (Proc. Internat. Sympos., Inst. Mat., Univ. Fed. Rio de Janeiro, Rio de Janeiro)* , pp. 472–484, North-Holland Math. Stud., 30, *North-Holland, Amsterdam-New York,* 1978.

[Ta7] TARTAR L., "Estimations de coefficients homogénéisés," *Computing methods in applied sciences and engineering (Proc. Third Internat. Sympos., Versailles,* 1977), *I,* pp. 364–373, *Lecture Notes in Math.,* 704, *Springer, Berlin,* 1979. Translated into English as "Estimations of homogenized coefficients," *Topics in the mathematical modelling of composite materials,* 9–20, Prog. Nonlinear Differential Equations Appl., 31, *Birkhäuser Boston , MA,* 1997.

[Ta8] TARTAR L., "Compensated compactness and applications to partial differential equations," *Nonlinear analysis and mechanics: Heriot-Watt Symposium, Vol. IV,* pp. 136–212. *Res. Notes in Math.,* 39, *Pitman, San Francisco, Calif.,* 1979.

[Ta9] TARTAR L., "Estimations fines des coefficients homogénéisés," *Ennio De Giorgi colloquium (Paris, 1983),* 168–187, Res. Notes in Math., 125, *Pitman, Boston, MA-London,* 1985.

[Ta10] TARTAR L., "Remarks on homogenization," *Homogenization and effective moduli of materials and media (Minneapolis, Minn., 1984/1985),* 228–246, IMA Vol. Math. Appl., 1, *Springer, New York-Berlin,* 1986.

[Ta11] TARTAR L., "The appearance of oscillations in optimization problems," *Nonclassical continuum mechanics (Durham, 1986),* 129–150, London Math. Soc. Lecture Note Ser., 122, *Cambridge Univ. Press, Cambridge-New York,* 1987.

[Ta12] TARTAR L., "H-measures, a new approach for studying homogenisation, oscillations and concentration effects in partial differential equations," *Proc. Roy. Soc. Edinburgh Sect. A* **115**, (1990), no. 3-4, 193–230.
[Ta13] TARTAR L., "Remarks on Optimal Design Problems," *Calculus of variations, homogenization and continuum mechanics (Marseille, 1993)*, 279–296, Ser. Adv. Math. Appl. Sci., 18, *World Sci. Publishing, River Edge, NJ*, 1994.
[Ta14] TARTAR L., "Remarks on the homogenization method in optimal design problems," *Homogenization and applications to material sciences (Nice, 1995)*, 393–412, GAKUTO Internat. Ser. Math. Sci. Appl., 9, *Gakkokotosho, Tokyo*, 1995 .
[Yo] YOUNG, L. C., *Lectures on the Calculus of Variation and Optimal Control Theory*, (W. B. Saunders, Philadelphia: 1969).
[Zo] ZOLEZZI T., "Teoremi d'esistensa per problemi di controllo ottimo retti da equazioni ellittiche o paraboliche," Rend. Sem. Mat. Padova, **44** (1970), 155–173.

Shape Analysis and Weak Flow

Jean-Paul Zolésio

Directeur de Recherche au CNRS - Centre de Mathématiques Appliquées (CMA) Ecole des Mines de Paris, 2004 route des lucioles - 06904 Sophia Antipolis Cedex - France - e-mail: jean-paul.zolesio@cma.inria.fr

1. Introduction

In these lectures we consider domain variations in Partial Differential Equations and we give a flavor of several results and techniques developed at the Centre de Mathématiques Appliquées (CMA) de l'Ecole des Mines in Sophia Antipolis. We adopt the Eulerian view point: the moving domain is the image of a given domain by the flow of a non autonomous vector field. This topic was introduced in the years following 1976 and it was presented in the book "Introduction to Shape Optimization" SCM vol.16,Springer Verlag 1992. For the references, see for example the bibliography of the book "Shape and geometries: analysis, differential calculus and optimization" written with M.Delfour (to appear).

i) In the first part we recall the large evolution of domains (using the "large flow mapping" associated with a smooth vector field) and we briefly introduce this setting in the non classical situation of non smooth fields. These fields are divergence free and have only a L^2 regularity, so that the moving domain can change its topology. To illustrate the use of such large weak evolution of sets we present some results concerning free boundaries associated with large elastic displacements and the variational formulation of the Euler equation. Concerning the shape continuity and compactness families we will recall some results obtained respectively with D. Bucur and with M. Delfour before 95.

ii) In the second part, using Hausdorff complementary topology and capacitary constraints, we derive continuity results for the usual Dirichlet problem.

iii) In the third part, using more regularity on the domains (based on the oriented distance function and the intrinsic geometry approach) we derive continuity results for a much more general class of boundary value problems.

iv) In the fourth part we present topics on shape differentiability concerning fluid-shell systems (obtained with F. Desaint in 94) in a problem in connection with sloshing tanks in satellites devices.

v) In the fifth part we consider an academic introduction to shape control in outer Navier Sokes flow obtained with J.C. Aguilar in 96 . The objective is to show the possibility to develop a fluid-shell intrinsic model for the boundary layer, in view of developing a shape sensitivity analysis by the techniques developed in the previous part.

vi) In the last part we make various uses of the differentiation and shape derivative via Min Max analysis. We consider boundary functional governed by steady linearized versions of Navier Stokes flow (worked out with Y. Guido in 96) . The main points are the fact that the Lagrangian should have saddle points on "fixed spaces" (spaces which are fixed when the domains varies), say $E \times F$. This last point forces to perform changes of function spaces in the Min Max formulation. There are several ways to do this: we underline here three basic techniques. The interest of the last part is mainly academic: it shows, on a simple example, the use of Min Max in order to bypass any derivative of the "state equation" in deriving both the continuous and discrete shape gradient of a non trivial shape functional of fluid dynamics.

2. Large evolution of domains

2.1 Introduction

Shape Optimization in PDE is a field of analysis which arose twenty five years ago from numerical analysis. After having been able to compute approximations of solutions of boundary value problems the question was to optimize the geometrical boundaries in order to improve a given cost functional.

The problem was not to investigate the existence of a minimum, but to be able to develop algorithms in order to decrease the chosen functional. In the beginning we considered penalty approaches in which the moving domain is hidden in the coefficients trough its characteristic function. Tools for shape topological optimization were developed, comparable to the "fictitious domain" approach, as it is called today. In order to build "large deformations" of the domain, we introduced, past 1975, the global flow transformation of a non autonomous vector field $V \in C^0([0,\infty[, C^k(D,R^N))\cap C^0([0,\infty[, L^\infty(D,R^N))$ and, in order to build a "domain Lyapounov trajectory", we considered Shape differential equation in the form $V(t,.) + A.G(\Omega_t(V)) = 0,\ \ t > 0$, where A is a positive (duality) operator and G a continuous selection of the upper Euler derivative (when that cone is reduced to a single element we call it the shape gradient, a distribution supported by the boundary of the domain with zero transverse order). Global existence results for that shape differential equation where derived under boundedness assumptions on the distribution G.

In this approach the topology of the moving domain is given: Ω_t is homeomorphic to the initial domain Ω_0. An objective is to relax in this setting the spatial regularity of the non autonomous vector field V, namely not to assume that the field $V(t,.)$ is lipschitzian, so that the flow transformation is not classically defined. Nevertheless, the dynamical evolution of the characteristic function is preserved, as we show that the dynamical system $\chi_t + \nabla\chi.V(t,) = 0$ possesses a unique solution $\chi(t)$, which is a characteristic function when the field $V(t,.)$ is divergence free and has a Sobolev regularity in time. We exchange the space regularity of the field V for a time regularity, which is less restrictive, since, in control theory setting, it means that the acceleration term $\dot{V} = \frac{\partial}{\partial t}V$ will be the control in $L^2([0,\tau], L^2(D, R^N))$.

We recover a shape topological identification tool which in fact unifies the approaches. We present an example of a first eigenvalue for which the week shape evolution enables us to build the decaying domain evolution, and we present the associated shape differential equation.

2.2 Non cylindrical evolution problem

Let us first recall some geometrical basic considerations concerning the non cylindrical domains and the unitary normal field ν. For each time t a domain Ω_t of R^3 is given, n_t is the usual normal to its boundary Γ_t. Let us consider the non cylindrical evolution domain:

$$Q = \bigcup_{t_0<t<\tau} (\{t\} \times \Omega_t) \text{ and its lateral boundary : } \Sigma = \bigcup_{t_0<t<\tau} (\{t\} \times \Gamma_t).$$

We will say that Q is the tube filled by the physical fluid flow during the time-interval(t_0, τ); Ω_t (resp.Γ_t) are the domains (resp. the boundary of the domains) occupied by the fluid at time t. The domain Ω_t is assumed to be contained in a smooth and bounded three dimensional hold-all D for any $t \in [0, \tau]$ and its volume, $|\Omega_t|$, is constant in the time t. The lateral boundary Σ being smooth enough, let ν be its unitary normal field (out-going to Q).

The non cylindrical domain Q verifies: $Q \subset [t_0, \tau] \times D \subset R^4$, the normal field ν has the following form : $\nu = (\nu_t, \nu_x) \in R^4$, where the "horizontal" vector ν_x has the form $\nu_x = \beta\ \mathbf{n}_t(x) \in R^3$ for some positive real number β, $0 \leq \beta \leq 1$. If β is identically equal to 1 then the domain is cylindrical (and the time component ν_t is identically zero). As the field ν is unitary we have the equivalent writing: $\nu = (\sqrt{1+v_\nu^2})^{-1}(-v_\nu, n_t) \in \mathbb{R} \times \mathbb{R}^{\#}$ where $v_\nu \in R$, and $-(\sqrt{1+v_\nu^2})^{-1}v_\nu(t,x) = \nu_t$ is the time-component of ν.

When $v_\nu \geq 0$ (resp. ≤ 0), then the domain Ω_t is monotonic increasing (resp. decreasing) For any smooth tube Q there exists a vector field W that "builds " Q as follows: $\Omega_t = T_t(W)(\Omega_0)$, where $T_t(W)$ is the flow mapping of W. To check this property it is sufficient for W to verify, at each time t, the condition$< W(t,.), n_t(.) >= v_\nu(t)(.)$ onΓ_t, where $v - \nu$ is related to the time-component of the normal field ν. An example of such a field is given by $W(t,x) = (v_\nu(t,.)\ n_t(.)\ )op_t$, where p_t stands for the projection mapping onto Γ_t; this field is defined in a neighborhood of Σ and its extension to $[t_0, \tau] \times D$ is arbitrary. We shall refer to this specific field as the normal field that builds the tube.

Indeed in the sequel we assume that the given field V verifies the condition: $\forall x \in \Gamma_t$, $< V(t,x), n_t(x) >= v_\nu(t,x)$. The normal component $< V(t,x), n_t(x) >$ is then a geometrical data, while the tangent component is arbitrary, given with the field V. Moreover the vector field V satisfying div $V = 0$ in D and $V.n_D = 0$ on ∂D is given in $H^1((0,\tau);\mathrm{H}^m(D,\mathbb{R}^N))$, $(m > 5/2)$. The flow follows the non-cylindrical Navier-Stokes equations

$$\partial_t u - \eta \Delta u + Du.u + \nabla p = f \quad \text{in } Q \tag{2.1}$$

$$\operatorname{div} u = 0 \quad \text{in } Q \tag{2.2}$$

$$u = V \quad \text{on } \Sigma, \tag{2.3}$$

where η is the coefficient of cinematic viscosity of the fluid.
A domain Ω and a function u_0 being given, we assume that

$$\text{at} \quad t = t_0, \quad \Omega_0 = \Omega \quad \text{and} \quad u(t_0) = u_0 \text{in } \Omega. \tag{2.4}$$

2.3 Flow Transformation of the Non Autonomous Vector Field V

2.3.1 Continuous field. We first consider the situation in which the vector field V is smooth, precisely we shall consider a hold all D in $\mathbb{R}^N$ which is an open set with lipschitzian boundary ∂D. The domain D is not necessarily bounded here; later, for simplicity, we shall assume it to be bounded. We consider the Frechet linear space

$$C^{k,\infty}(D,R^N) = \{\phi \,|\, \phi \in C^k(\bar{D},R^N)\,,\ \phi \in L^\infty(D,\mathbb{R}^N)\,,\ < \phi, \ltimes\} \tag{2.5}$$

partial vector field V is now taken in the following linear space. Given an integer k, $k \geq 1$,

$$E^{k,\infty}(D) = C^0([0,\infty[, C^{k,\infty}(D,\mathbb{R}^N)) \tag{2.6}$$

A sequence V_n converges to an element V in$E^{k,\infty}(D)$ if and only if for each compact subset K of D, any $\tau > 0$ and any integer $\alpha, \alpha \leq k$, we have the uniform convergence of $\nabla^\alpha V_n(.,.)$ on the compact set$[0,\tau] \times K$ and also the uniform convergence over$[0,\tau] \times D$ of the sequence $V_n(.,.)$. We denote here $V(t,x)$ the value of the field $V(t)$ at the point X of D, that is we set

$$V(t,x) = [V(t)](x).$$

The first global existence result we quote here, from [50], is the following .

Theorem 2.1. *Let V belong to $E^{k,\infty}(D)$. Then the flow mapping $T(V)$: $t \longrightarrow T_t(V)$ is defined from $[0,\infty[$ in $C^0(\bar{D},\bar{D})$ and we have the following regularity:*

$$T(V) \in E^{1,k,\infty}(D) \tag{2.7}$$

$$E^{1,k,\infty}(D) = \left\{ T \mid (t \to T_t)\, and (t \to \frac{d}{dt}(T_t) \in C^0([0,\infty[, C^{k,\infty}(\bar{D},R^N)) \right\} \tag{2.8}$$

When D is bounded the previous Frechet linear spaces $C^{k,\infty}(D,R^N)$ are Banach spaces.

2.3.2 Field in L^p. We turn now to the situation when the vector field V has less regularity both in time variable t and in space variable x. Given a domain D bounded in $\mathbb{R}^N$ and $p > 1$, we consider the following linear space:

$$\mathcal{E}^p(D) = \Big\{ V \in L^0(0,\infty, Lip(D,\mathbb{R}^N)) \mid \text{ there exist } C_V, \rho_V, \tag{2.9}$$
$$\forall t \int_t^{t+\rho} ||V(s)||^p_{Lip(\bar{D},\bar{D})} ds \leq C\,, \quad < V(t,.), n_{\partial D} >= 0 \; a.e.t \Big\}$$

Theorem 2.2. *Let $p > 1$ and a vector field V in $\mathcal{E}^p(D)$ be given : then the flow mapping $T(V)$ is defined over $Q = [0,\infty[\times \bar{D}$ and we have for all τ, $\tau > 0$:*

$$T(V) \in L^\infty(0,\tau,\; Lip(D,\mathbb{R}^N)) \tag{2.10}$$

The mapping $T_t(V)$is bijective from D onto itself and the inverse mapping is a flow mapping associated to the following vector field:

$$\forall t \in [0,\tau]\,, \forall s \in [0,t]\,,\; V_t(s,y) = -V(t-s,y) \tag{2.11}$$

$$T_t(V)^{-1} = T_t(V_t) \in Lip(D,D) \tag{2.12}$$

Proof. In a first step we chose τ small enough so that the L^p norm of the Field V be uniformly contractive:

$$\forall t, \quad \int_t^{t+\tau} ||V(s,.)||_{Lip(D,\mathbb{R}^N)} dt \;\leq\; k(V) < 1 \tag{2.13}$$

We define classically the recursive sequence of mappings $T_t^n(x)$:

$$T_t^{n+1}(x) = x + \int_0^t V(s, T_s^n(x)) ds \tag{2.14}$$

with $T_t^0(x) = x$; we easily get:

$$||T_t^{n+1}(x) - T_t^n(x)||_{\mathbb{R}^N} \leq k(V)\, Max_{0\leq t\leq\tau} ||T_t^n(x) \;-\; T_t^{n-1}(x)||_{\mathbb{R}^N} \tag{2.15}$$

from which we derive that

$$||T^{n+p} - T^n||_{L^\infty(Q,\mathbb{R}^N)} \;\leq\; ||T^1 - T^0||_{L^\infty(Q,\mathbb{R}^N)} \frac{k(V)^n}{1-k(V)} \tag{2.16}$$

$$||T_t^1(x) - T_t^0(x)||_{L^\infty(Q,\mathbb{R}^N)} \leq \int_0^T ||V(t)||_{L^\infty(Q)} dt \tag{2.17}$$

We derive the uniform convergence on Q of the sequence of mappings to a mapping denoted $T_t(.)$. We verify now that it is effectively the flow mapping of the field V. For this we form the term $A_n(t,x) = T_t^{n+1}(x) - [x + \int_0^t V(s, T_s(x)) ds]$ and we have

$$\forall t, x \ ||A_{n+1}(x)|| \leq \int_0^t ||V(s, T_s^n(x)) - V(s, T_s(x))||ds \tag{2.18}$$
$$\leq k(V)||T_t^n(.) - T_t(.)||_{L^\infty(Q)} \to 0 \ , n \to \infty$$

so that, in the limit, we have that $T_t^n(x) - [x + \int_0^t V(s, T_s(x))ds]$, that is $T_t(x) = T_t(V)(x)$ in the sense that the map solves the functional equation:

$$T_t(V)(x) = x + \int_0^t V(s, T_s(V)(x))ds \tag{2.19}$$

The shape evolution analysis and the so-called *Shape differential equation* were introduced through the field perturbations of a subset Ω of D in the form $\Omega_t(V) = T_t(V)(\Omega)$. In a general setting, a boundary value problem is solved in the domain Ω_t (whose solution $y(\Omega)$ lies in some functional space "attached" to the set Ω_t) and the shape differential equation consists in taking the field $V(t)$ as an explicit function of $u, \nabla u, \ldots$. This problem (which will be formulated in the following sections) is a non linear problem on the unknown V. It has been solved in the strong formulation , i.e. in $\mathcal{E}^p(D)$ so that the flow mapping T_t is defined. The idea of the present *relaxation* is to consider the evolution of the characteristic function χ_{Ω_t} for non lipschitz-continuous fields.

3. First Variation of the Flow Mapping with respect to the Vector Field

3.1 The Tranversal field Z

The main question is now to determine the derivative, if it exists, of the mapping $V \to T_t(V)$. Given a second vector field W we consider, for any real number s, the flow mapping

$$\mathbf{T_s^t} = \mathbf{T_t(V + sW) o T_t(V)^{-1}} \tag{3.1}$$

For any fixed t, the transformation $\mathbf{T_s^t}$ maps D onto$D$ and the domain $\Omega_t(V)$ onto $\Omega_t(V + sW)$.

It follows that it is the flow mapping of a velocity vector field$\mathbf{Z^t}$. Of course $\mathbf{Z^t}$ is a non autonomous vector field and is defined as being the velocity field associated to the transformation $x \to \mathbf{T_s^t(x)}$, i.e.

$$\mathbf{Z^t(s, x)} = [\frac{\mathbf{d}}{\mathbf{ds}}(\mathbf{T_s^t(x)})]_{\mathbf{s=0}} \ \mathbf{o} \ \mathbf{T_s^t(x)^{-1}} \tag{3.2}$$

It turns out that when the fields V and W are smooth enough the field $\mathbf{Z^t(s, x)}$ exists and is characterized as the solution of a linear dynamical system associated to V and W.

Theorem 3.1. *We consider two non autonomous vector fields V and W given in E and we set*

$$Z(t,x) = \mathbf{Z^t(0,x)} \tag{3.3}$$

then the non autonomous vector field Z solves the following first order dynamical system in the cylindrical evolution domain $Q = [0,\tau] \times D$

$$Z(0,x) = 0 \tag{3.4}$$

$$\frac{d}{dt}Z(t,x)+DZ(t,x).V(t,x)-DV(t,x).Z(t,x) = W(t,x),\ for(t,x) \in [0,\tau]\times D \tag{3.5}$$

Proof: we write the integral characterization of the two transformations $T_t(V+sW)$ and $T_t(V)$, computing the difference and using the first order Taylor expansion of the field V(s,.) (here we use more then $V(s,.)$ in $Lip(D,\mathbb{R}^N)$ as it would simply derive from the linear space E):

$$\frac{(T_t(V+sW)-T_t(V))(x)}{s} = \tag{3.6}$$

$$\int_0^t DV(r,T_r(V)(x)+d_r[T_r(V+sW)(x)-T_r(V)(x)])$$

$$\frac{T_r(V+sW)(x)-T_r(V)(x)}{s}ds + \int_0^t W(r,T_r(V+sW)(x))dr$$

for each s the function $\frac{(T_t(V+sW)-T_t(V))(x)}{s}$ is, as a function of (t,x) , an element of $C^o([0,\tau],Lip(D,\mathbb{R}^N))$. From Gronwall classical lemma we have

$$||\frac{(T_t(V+sW)-T_t(V))(x)}{s}||_{C^o([0,\tau],Lip(D,\mathbb{R}^N))} \leq M \tag{3.7}$$

one can find a sequence s_n converging to zero such that,for example,

$$\frac{(T_t(V+s_nW)-T_t(V))(x)}{s_n}$$

converges to an element $S(t,x)$ in $L^\infty(Q)$-weak*. This element solves the following integral equation:

$$S(t,x) = \int_0^t DV(r,T_r(V)(x)).S(r,x)ds + \int_0^t W(r,T_r(V)(x))dr \tag{3.8}$$

finally we have $S(t,x) = Z(t,T_t(V)(x))$,and when V is smooth enough we get $t \to T_t(V)^{-1}(x)$ is in $C^1([0,\tau])$, so that $Z(t,x) = S(t,T_t(V)^{-1}(x))$ is differentiable with respect to t in$L^\infty(Q)$ and the derivative $Z_t(t,x)$ solves the dynamical system in the following integral sense:

$$Z(t,x) = -\int_0^t [DZ(r,T_r(V)(x)).V(r)]dr - \tag{3.9}$$

$$\int_0^t [DV(r,T_r(V)(x)).Z(r)]dr + \int_0^t W(r,T_r(V))(x)dr \tag{3.10}$$

In fact we show here that the dynamical system, for smooth enough data V, W, always has a solution Z, called the variational solution. Under smoothness properties (which will imply a boundary condition $Z(t,.) = 0$ on $[0,\tau] \times \partial D$) the linear first order dynamical system shall have a unique solution, so that the associated operator H will be an isomorphism, and the adjoint problem associated to the adjoint operator $H*$ will be well posed in a suitable functional space. Nevertheless it turns out to be more appropriate to study the variable $S(t)$ rather than $Z(t)$ and we derive easily that $S(.)$ is solution of the dynamical system:

$$S(0) = 0 \tag{3.11}$$

$$S_t(t) = DV(t)oT_t(V).S(t) + W(t)oT_t(V) \tag{3.12}$$

$$S(0,.) = 0 \quad \text{on } [0,\tau] \times \partial D \tag{3.13}$$

The operator $A(t)$ defined by:

$$A(t) = DV(t)oT_t(V) \tag{3.14}$$

is bounded in $H_0^1(D)$, as soon as the field is smooth enough, here $V \in E$ are regular, the preceding problem has a unique solution S in the linear space $C^1([0,\tau], H_0^1(D, \mathbb{R}^N))$. To derive the solution Z of the previous dynamical system we simply use the inverse transformation $Z(t) = S(t)oT_t(V)^{-1}$ so that, if V is smooth enough, the transformation $T_t(V)^{-1}$ maps the linear space $C^1([0,\tau], H_0^1(D, \mathbb{R}^N))$ onto itself and we obtain the existence and uniqueness of a solution Z in that space. The regularity required for the two fields V and W is difficult to obtain if one thinks of extremizing the Action $A_\alpha(V)$. We could look to $A(t)$ as being an unbounded operator in $H_0^1(D)$, but the question is then to know whether its domain is dense. The answer seems to be negative as an element φ in that domain should be zero in the neighborhood of points where $DV(t)$ is not locally Lipschitzian.

Remark 3.1. In fact the field Z can be characterized in terms of the Lie brackets of the two fields V and Z as follows.

$$\frac{d}{dt}Z + [Z,V] = W \tag{3.15}$$

3.2 Derivatives of the Flow Transformation

Theorem 3.2. *We have the following derivative:*

$$\frac{\partial}{\partial s}(T_t(V+sW))|_{s=0} = Z(t)oT_t(V)^{-1} \tag{3.16}$$

In order to handle this derivative, we have first to consider the analogous derivative for the term $T_t(V)^{-1}$

Theorem 3.3. *Let E be given by*

$$E(V)(t) = DT_t(V)^{-1}oT_t(V)^{-1} \tag{3.17}$$

then wc have the following derivative:

$$\frac{\partial}{\partial s}(T_t(V+sW)^{-1})|_{s=0} = -E(V)(t).Z(t) \tag{3.18}$$

4. Derivative of an Integral over the Evolution Tube with Respect to the velocity Field.

Let F be an element of $W^{1,1}(D)$ and consider a functional in the following form.

$$I(V) = \int_0^T (\int_{\Omega_t(V)} F(x)dx\ )dt \tag{4.1}$$

The question is now to compute the Gateaux derivative

$$I'(V,W) = [\frac{d}{ds}I(V+sW)\ ]_{s=0} \tag{4.2}$$

The functional can be written as

$$I(V+sW) = \int_0^T [\int_{\Omega_t(V+sW)} Fdx]dt$$

and, using the change of variable $x = \mathbf{T_s^t}(\mathbf{y})$, we obtain

$$I(V+sW) = \int_0^T [\int_{\Omega_t(V)} F\ o\ \mathbf{T_s^t}\ \mathbf{det}(\mathbf{DT_s^t})\mathbf{dx}\ ]\mathbf{dt} \tag{4.3}$$

so that the Gateaux derivative of I is easily obtained in terms of the derivative with respect to s of the transformation $\mathbf{T_s^t}$, that is in terms of the field $\mathbf{Z^t(s,.)}$ previously defined.

$$I'(V,W) = \int_0^T [\int_{\Omega_t(V)} \nabla F.\mathbf{Z^t(0,y)} + \mathbf{Fdiv(Z^t(0,y))dy}]\ \mathbf{dt} \tag{4.4}$$

setting $Z(t,y) = \mathbf{Z^t(0,y)}$ we have

$$I'(V,W) = \int_0^\tau [\int_{\Omega_t(V)} div(F.Z(t,y)dy]\ dt \tag{4.5}$$

When the boundary of $\Omega_t(V)$ is smooth enough we get

$$I'(V,W) = \int_0^\tau [\int_{\Gamma_t(V)} F\ Z(t,y).n_t(y)dy\ ]dt \tag{4.6}$$

The field Z depends on V and W through the dynamical system (3.1); in order to compute the Gateaux derivative $I'(V,W)$ in terms of the direction field V we have to introduce an adjoint problem. It is interesting to notice that we have no control problem; still, just to solve a variation problem, we have to introduce an adjoint problem. We can write

$$I'(V,W) = \int_0^\tau < -\nabla[\chi_{\Omega_t}(V)].F\ ,\ Z(t,.) > dt \tag{4.7}$$

where $\chi_{\Omega_t}(V)$ is the characteristic function of the moving domain and $<,>$ denotes the bilinear form pairing between the measures and the continuous functions over D . Let us denote by H the operator associated with the problem (4.1), i.e.

$$H.\phi = \phi_t + D\phi.V - DV.\phi \tag{4.8}$$

We denote by H^* the adjoint operator in the following sense.

$$\int_0^\tau < H^*.\phi\,,\psi > dt = \int_0^\tau < \phi\,,H.\psi > dt \tag{4.9}$$

and we define P as the solution of the adjoint problem

$$H^*.P = -\nabla[\chi_{\Omega_t}(V)].F \quad in[0,\tau] \times D \tag{4.10}$$

$$P = 0\ on\ [0,\tau] \times \delta D \tag{4.11}$$

$$p(\tau) = 0\, on\ D \tag{4.12}$$

and then we get

$$I'(V,W) = \int_0^\tau \int_D PW dxdt \tag{4.13}$$

The optimality of the functional I would imply the following necessary condition:

$$P = 0\ ,\ in\ [0,\tau] \times D \tag{4.14}$$

4.1 The adjoint problem

We turn now to the interpretation of the adjoint problem.

$$H^*.P = -p_t - (divV)\,p - Dp.V - (DV)^*.p \tag{4.15}$$

An interesting situation is when solving the adjoint equation in the form

$$H^*.P = -p_t - (divV)\,p - Dp.V - (DV)^*.p = \nabla g, \quad p(\tau) = 0 \tag{4.16}$$

the solution p can be found itself in gradient form. Setting $p = \nabla q$, the adjoint problem takes the following form:

$$-\,p_t - (divV)\,p - \nabla(< q, V >) = \nabla g, \quad p(\tau) = 0 \tag{4.17}$$

When $V(t)$ is a divergence free field we get

$$\nabla\{\ q_t + \ < q, V > \ + g\ \} = 0, \quad \nabla q(\tau) = 0 \tag{4.18}$$

and there exists a function $f(t)$ and a constant q_1 such that

$$q_t + \ < q, V > \ + g \ = \ f(t), \quad q(\tau) = q_1 \tag{4.19}$$

Of course the previous computation is formal, in the sense that we don't know yet whether the operator H defines an isomorphism between adequate linear spaces, so that the adjoint problem is well defined. Nevertheless, by construction, for each W one can get a solution Z to the equation $H.Z = W$ and from the classical theory of differential equations we know that for each W the solution Z is unique. The aim is then to find the good space in which the operator H realizes an isomorphism.

5. Non cylindrical Large Evolution of an Elastic Domain

An elastic material at rest, occupying a domain Ω, is charged by a load f. The domain moves at a speed V and we consider the displacement field $T_t(V) - I_d$ defined over D. We assume here a reversible behavior of the elastic material under consideration in order to simplify the writing: the strain energy developed in the material when forcing it from the old configuration Ω to the new one $T_t(V)(\Omega)$ is the same as the one developed in the inverse transformation. The main point here is that $T_t(V)^{-1}$ is defined on the moving domain Ω_t. In this first modeling we shall assume that the strain energy will be in the following form $||\epsilon(u)||^2$ with u given by:

$$u(V) = T_t(V)^{-1} - I_d \tag{5.1}$$

An important question here concerns the use of the linearized tensor $||\epsilon(u)||^2$. The underlying assumption is that the deformation tensor remains small during the displacement. Then we use the usual linear elasticity deformation tensor while we cannot identify the initial and "actual" configurations and this is the reason why we face non cylindrical evolution domain. We shall be able to give simple expressions for both kinetic and internal energies in the moving domain, and we will consider the Action, which is an integral over the non-cylindrical domain associated with the kinetic and elastic energies in the material. It takes the form of the following function

$$A(V) = \int_0^T A(\Omega_t)dt \tag{5.2}$$

with the density:

$$A(\Omega_t) = \int_{\Omega_t} |V(t)|^2 - |\epsilon(u(\Omega_t))|^2 \, dx - \int_{\Omega_t} f.u \, dx \tag{5.3}$$

We shall face to the derivative of the displacement u through the velocity vector field V, namely to the derivative

$$u'(V, W) = \frac{\partial}{\partial s} u(V + sW)|_{s=0}$$

The Action actually depends on the velocity vector field V and we shall be concerned with the extremality of $A(V)$ with respect to the vector field V. We shall consider the first variations with respect to the field V of the flow mapping $T_t(V)$, that is the first derivative with respect to the parameter s of the flow mapping $T_t(V + sW)$, derivative at $s = 0$. The derivative will involve the Lie brackets

$$[\, V(t) \, , \, W(t) \,] = DV(t).W(t) - DW(t).V(t) \tag{5.4}$$

5.1 Derivative of the Action

The derivative of the Action $A(V)$ is given by:

Theorem 5.1. *Let p be the solution of the following formal "adjoint problem":*

$$H^*.p = -\chi_{\Omega_t} E^* div\epsilon(u) + (E^*.\epsilon(u) + \frac{1}{2}(|V(t)|^2 - |\epsilon(u)|^2)I_d).\nabla\chi_{\Omega_t} \tag{5.5}$$

where $<,>$ stands for the bilinear duality form pairing between $D'(Q)$ and $D(Q)$, Q being the cylindrical evolution domain. With the final condition

$$p(\tau) = 0 \;\; in \;\; D \tag{5.6}$$

and the boundary condition

$$p = 0 \, on \, [0, \tau] \times D \tag{5.7}$$

The derivative of the Action is given by:

$$A'_o(V, W) = < \chi_{\Omega_t}.V - P, W > \tag{5.8}$$

In order to give a meaning to the previous adjoint problem we study the static part of the operator:

$$H^*_V.p = -(divV)\,p - Dp.V - (DV)^*.p \tag{5.9}$$

We have the following

$$< H^*_V.p, p > = - < (\frac{3}{2} I_d \, divV \; + \; DV\,).p \,,\, p > \tag{5.10}$$

5.2 Equations for the Free boundary

From the stationarity of that action we can deduce the equation resulting from the variational problem $A'(V, W) = 0$ for all W. Which is:

$$P = \chi_{\Omega_t}.V \tag{5.11}$$

that condition together with the equation of P leads to an equation for the field V itself, but also, through an integration by parts, to a new boundary condition verified by V on the lateral non cylindrical boundary of the evolution domain which is built from the boundary of the initial domain by the global flow transformation of the field V.

Theorem 5.2.

$$\frac{\partial}{\partial t} V + 2\epsilon(V)V \; + \; div(u)E(t)^* \; + div(V)V \; = \; 0 \tag{5.12}$$

where u is given by:

$$u(t)(.) = u(V) = T_t(V)^{-1} - I_d \tag{5.13}$$

and the boundary condition:

$$\frac{\partial}{\partial t} V + \frac{1}{2}(\; V.V - \epsilon(u)..\epsilon(u)\;)n_t + D(T_t(V)^{-1})^*.(\epsilon(u).n_t) = 0 \tag{5.14}$$

equation 5.12 can be rewritten as follows:

$$\frac{\partial}{\partial t} V + 2\epsilon(V)V + div(\; T_t(V)^{-1} - I_d)(DT_t(V)^*)^{-1} o T_t(V)^{-1} + div(V)V = 0. \tag{5.15}$$

6. Weak Convection of characteristic functions

6.1 An Uniqueness Result

Let a vector field $V \in L^2(0,\tau,L^2(D,R^N))$ with $divV(t,.) = 0$ be given in D and $V(t,.).n_{\partial D} = 0$ in $H^{-1/2}(\partial D)$.

For any smooth element $\phi \in C^\infty([0,\tau] \times D)$ we consider the first order operator $D_V.\phi = \frac{\partial}{\partial t}\phi + \nabla_x\phi.V \in L^2(0,\tau,L^2(D))$. We introduce the Hilbert space $\mathcal{H}_V$ defined as the completion of $C^\infty([0,\tau]\times D)$ for the following norm:

$$|\phi|_V = (\,|\phi|^2_{L^2(0,\tau,L^2(D))} + |D_V\phi|^2_{L^2(0,\tau,L^2(D))}\,)^{1/2}.$$

For any smooth ϕ, ψ we have:

$$\int_t^\tau \int_D D_V\phi\,\psi\,dxdt = \int_D \{\phi\psi(\tau) - \phi\psi(t)\}dx - \int_t^\tau \int_D \phi\,D_V\psi\,dxdt.$$

This identity gives a weak sense to pointwise values $\phi(t)$ as follows: we consider elements $\psi \in H^1(0,\tau,L^2(D)) \cap L^\infty(0,\tau,W^{1,\infty}(D))$, with $\psi(\tau) = 0$, so that $\nabla\psi \in L^\infty(0,\tau,L^\infty(D,R^N))$, $\nabla\psi.V \in L^2(0,\tau,L^2(D,R^N))$ then $D_V\psi \in L^2(0,\tau \times D)$ and for any t, $0 \le t < \tau$, the element $\phi(t)$ is weakly defined by:

$$\int_D \phi(t)\psi(t)\,dx = -\int_t^\tau \int_D D_V\phi\,\psi\,dxdt - \int_t^\tau \int_D \phi\,D_V\psi\,dxdt$$

Also from the first identity we get (taking $t = 0$), for all smooth ϕ with $\phi(0) = 0$:

$$\int_0^\tau \int_D \phi^2\,dtdx = 2\int_0^\tau \int_D (\tau - t)D_V\phi\,\phi\,dtdx$$

So that, by density, we derive

$$\forall\phi \in \mathcal{H}_V \text{ with } \phi(0) = 0,\ |\phi|_{L^2(0,\tau,L^2(D))} \le 2\tau\,|D_V\phi|_{L^2(0,\tau,L^2(D))} \qquad (6.1)$$

We consider the dynamical system

$$u(0) = \phi, \quad \frac{\partial}{\partial t}u(t) + < V(t), \nabla u(t) > = f \qquad (6.2)$$

with initial condition $\phi \in L^2(D)$ and right hand side $f \in L^2(0,\tau,\ L^2(D))$. If $u \in L^2(0,\tau,L^2(D))$ is a solution of (2), then u belongs to $\mathcal{H}_V$, so that $u(0) = 0$ makes sense: $\forall\psi \in H^1(0,\tau,L^2(D)) \cap L^\infty(0,\tau,W^{1,\infty}(D))$, with $\psi(\tau) = 0$,

$$\int_0^\tau \int_D f\,\psi\,dxdt = -\int_0^\tau \int_D u\,D_V\psi\,dxdt.$$

We derive the following uniqueness result:

Proposition 6.1. *For V given, $V \in L^2(0,\tau,L^2(D,R^N))$ with $divV(t,.) = 0$ in D and $V(t,).n_{\partial D} = 0$ in $H^{-1/2}(\partial D)$, the problem (2) has at most one solution in $\mathcal{H}_V$.*

Proof. Let u^i be two such solutions; then $u = u^2 - u^1$ solves the homogeneous problem. From the previous estimate we get $u = 0$.

6.2 The Galerkin Approximation

Proposition 6.2. *Let $V \in L^1(0,\tau,L^2(D,R^3))$ with*

$$divV \in L^2(0,\tau,L^2(D,R^3))$$

verifying the following uniform integrability condition:

$$\text{There exist } T_0 > 0,\ \rho < 1,\ s.t. \forall a \geq 0,\ \int_a^{a+T_0} ||V(t)||_{L^2(D,R^3)}dt \ \leq \rho < 1$$

(for shortness we consider $V \in L^p(0,\tau,L^2(D,R^3))$ with $p > 1$). We assume that the positive part of the divergence $divV = (divV)^+ - (divV)^-$, verifies

$$||(divV(t))^+||_{L^\infty(D,R^3)} \in L^1(0,\tau).$$

Then if $< V(t,.),n >= 0$ (as an element of $L^1(0,\tau,H^{-\frac{1}{2}}(\partial D))$,$f \in L^1(0,\tau,L^2(D))$ and initial condition $\phi \in L^2(D)$, there exists solutions,

$$u \in L^\infty(0,\tau,L^2(D)) \cap W^{1,p^*}(0,\tau,\ W^{-1,3}(D)) \subset C^0([0,\tau],W^{-\frac{1}{2},\frac{3}{2}}(D))$$

to the problem (2), where $\frac{1}{p} + \frac{1}{p^*} = 1$. Moreover there exists a constant M such that:

$$\forall \tau,\ ||u||_{L^\infty(0,\tau,L^2(D,R^N))} \leq M\{\ ||\phi||_{L^2(D)} + ||f||_{L^1(0,\tau,L^2(D))}\ \} \qquad (6.3)$$

$$(1 + \int_0^\tau (||(divV(s))^+||_{L^\infty(D,R^3)} + ||f(s)||_{L^2(D,R^3)})$$

$$\int_s^\tau (||(divV(\sigma))^+||_{L^\infty(D,R^3)} + ||f(\sigma)||_{L^2(D,R^3)})d\sigma)ds$$

If the field V is smoother,$V \in L^p(0,\tau,\ H^1(D,R^N))$, these solutions verify

$$u \in L^\infty(0,\tau,L^2(D)) \cap W^{1,p^*}(0,\tau,\ W^{-1,3}(D)) \subset C^0([0,\tau],W^{-\frac{1}{2},\frac{3}{2}}(D))$$

In both situations, if the initial condition is a characteristic function

$$\phi = \chi_{\Omega_0} \in L^2(D)$$

and if $f = 0$, then the unique solution $u \in H_V$ is itself a characteristic function:

$$a.e.(t,x), \quad u(t,x)\,(1-u(t,x)\,)=0 \quad \text{that is } u=\chi_{Q_V}$$

Where Q_V is a non cylindrical measurable set in $]0,\tau[\times D$. For a.e.t, we set

$$\Omega_t(V)=\{x\in D|\quad (t,x)\in Q_V\}.$$

If V is a divergence free field, $divV(t,x)=0$ for a.e. $t,\in]0,\tau[$, then the set $\Omega_t(V)$ verifies a.e.t, $meas(\Omega_t(V))=meas(\Omega_0)$.

Proof. Let us consider $V\in L^2(0,\tau,H_0^1(D))$ and a dense family $e_1,...e_m,...$ in$H_0^1(D)$ with each $e_i\in C_{comp}^\infty(D,R^3)$. Consider the approximated solution

$$u^m(t,x)=\Sigma_{i=1,...,m}\ u_i^m(t)\,e_i(x)$$

with $U^m=(u_1^m,...,u_m^m)$ solution of the following linear ordinary differential system:

$$\forall t, \int_D (\frac{\partial}{\partial t}U^m(t)+<V(t),\nabla U^m(t)>)\,e_j(x)\,dx=$$

$$\int_D f(t,x)e_j(x)\,dx,\ j=1,...,m$$

That is

$$\frac{\partial}{\partial t}U^m(t)+M^{-1}.A(t).U^m(t)=F(t) \tag{6.4}$$

where

$$M_{i,j}=\int_\Omega e_i(x)\,e_j(x)\,dx$$

$$A_{i,j}(t)=\int_D <V(t),\nabla e_i(x)>\ e_j(x)\,dx$$

The above is an ordinary linear differential systems possessing a global solution when $V\in L^p(0,\tau,L^2(D,R^N))$ for some p, $p>1$. By the classical energy estimate, as $\int_D<V(t),\nabla u^m(t)>u^m(t)\,dx=-\frac{1}{2}\int_D<u^m(t),u^m(t)>divV(t)\,dx$, a.e.t we obtain :

$$\forall \tau,\ \tau\le T,\ \|u^m(\tau)\|_{L^2(D)}^2\le\|u^m(0)\|_{L^2(\Omega)}^2$$

$$+\int_0^\tau\int_D<u^m(t,x),u^m(t,x)>(divV(t,x))^+\,dtdx$$

$$+2\int_0^\tau\int_D f(t,x)u(t,x)dtdx$$

Setting

$$\psi(t)=\|(divV(t,.))^+\|_{L^\infty(D,R^3)}$$

When $f=0$,

$$\frac{1}{2}\int_D u^m(t,x)^2dx\le\frac{1}{2}\int_D u^m(0,x)^2dx$$

$$+\frac{1}{2}\int_0^t \psi(s)\int_D \|u^m(t,x)\|^2\,dx$$

by Gronwall's lemma we get :

$$\int_D u^m(t,x)^2\,dx \le \int_D u^m(0,x)^2\,dx\ (1\ +\ \int_0^t \psi(s)exp\{\int_s^t \psi(\sigma)d\sigma\ \}\,ds\ )$$

By the choice of the initial conditions in the ordinary differential system we get

$$M>0, s.t.\ \forall \tau, \le T,\ \|u^m(\tau)\|_{L^2(D)}$$

$$\le M\ \|\phi\|_{L^2(D)}\ (1+\int_0^t \psi(s)exp\{\int_s^t \psi(\sigma)d\sigma\ \}\,ds)$$

When $\psi = 0$, we get

$$\forall \tau,\ \tau \le T,\ \|u^m(\tau)\|^2_{L^2(D)} \le \|u^m(0)\|^2_{L^2(\Omega)}$$

$$+2\int_0^\tau \int_D f(t,x)u(t,x)dtdx$$

In the general case, we use

$$\|u^m\| \le 1+ \|u^m\|^2$$

and we derive the following estimate :

$$\forall \tau,\ \tau \le T,\ \|u^m(\tau)\|^2_{L^2(D)} \le \|u^m(0)\|^2_{L^2(D,R^3)}$$

$$+\int_0^\tau \int_D <u^m(t,x),u^m(t,x)> (divV(t,x))^+\,dtdx$$

$$+2\int_0^\tau \int_D f(t,x)u(t,x)dtdx$$

$$\le \|u^m(0)\|^2_{L^2(D,R^3)} + \int_0^\tau \|f(t)\|_{L^2(D,R^3)}\,dt$$

$$+\int_0^\tau x <u^m(t,x),u^m(t,x)> (\psi(t) + \|f(t)\|_{L^2(D,R^3)}\ )\,dtdx$$

$$\le M\ (\|u_0\|^2_{L^2(D,R^3)} + \|f\|_{L^1(0,\tau,L^2(D,R^3))}\ )$$

$$+\int_0^t (\psi(t) + \|f(t)\|_{L^2(D,R^3)})\|u^m(s)\|^2_{L^2(D)}\,ds$$

From Gronwall's inequality we have:

$$\|u^m(\tau)\|^2_{L^2(D)} \le M(\|u_0\|^2_{L^2(D,R^3)} + \|f\|_{L^1(0,\tau.L^2(D,R^3))}\ )$$

$$\{\ 1+\int_0^t [\ (\psi(s) + \|f(s)\|_{L^2(D,R^3)})\int_s^t (\psi(\sigma) + \|f(\sigma)\|_{L^2(D,R^3)})d\sigma\ ]ds\ \}$$

In all cases u^m remains bounded in $L^\infty(0,\tau,L^2(D,R^N))$ and there exists an element u in that space and a subsequence, still denoted u^m, which weakly-* converges to u. In the limit u itself verifies the previous estimate from which the uniqueness follows. It can be verified that u solves the problem in a distribution sense, that is

$$\forall \phi \in H_0^1(0,\tau,L^2(D,R^3)) \cap L^2(0,\tau,H_0^1(D,R^3)),\ \phi(0)=0,$$

$$-\int_0^\tau \int_D u(\frac{\partial}{\partial t}\phi + div(\ \phi V)\ )dxdt = \int_D \phi(0)\, u_0\, dx + \int_0^\tau \int_D < f,\phi > dxdt.$$

When $V(t) \in L^2(D,R^3)$, the duality brackets $< \frac{\partial}{\partial t}u,\phi >$ are defined. If $\nabla\phi$ belongs to $L^\infty(D,R^3)$, this is verified, for example, when $\phi \in H_0^3(D)$ so that u_t is identified to an element of the dual space $H^{-3}(D)$. When $V(t) \in H^1(D,R^3)$ we get $a.e.t$, $u(t,.) \in L^2(D)$, $V(t,.) \in L^6(D)$ then $\nabla\phi(t)$ should be in $L^3(D,R^3)$,that is $\phi(t) \in W_0^{1,3}(D)$ and then, for $a.e.t$, the element u_t is in the dual space $W^{-1,\frac{3}{2}}(D)$ while $u_t \in L^{p^*}(0,\tau,\ W^{-1,\frac{3}{2}}(D))$ and $u \in W^{1,p^*}(0,\tau,\ W^{-1,\frac{3}{2}}(D))$. Then we have $u \in L^\infty(0,\tau,L^2(D)) \cap$

$$\cap W^{1,p^*}(0,\tau\, W^{-1,\frac{3}{2}}(D)) \subset L^2(0,\tau,W^{0,\frac{3}{2}}(D)) \cap W^{1,p^*}(0,\tau,\ W^{-1,\frac{3}{2}}(D))$$

$$\subset C^0([0,\tau],W^{-\frac{1}{2},\frac{3}{2}}(D))$$

When the initial data is a characteristic function, $u_0 = \chi_{\Omega_0}$, we verify that u^2 is also a solution and, by uniqueness, $u^2 = u\ a.e.$. We introduce $u_0^n \to u_0$ in $L^2(D,R^3)$ with $u_0^n \in C^\infty(D,R^3)$ and $u_0^n(x) \leq 1$. We also consider $V^n \to V$ in $L^2(0,\tau,L^2(D,R^3))$ with $V \in C^\infty$ and $div(V^n)^- \ \in L^\infty([0,\tau]\times \bar{D})$. The solution u^n associated to these data is obtained through the flow of V^n as follows:

$$u^n(t,x) = (u_0) o T_t(V^n)^{-1}(x).$$

As a consequence, $u^n \in C^\infty$, then $(u^n)^2$ is classically defined and, when $f=0$, it is obviously a solution to the equation associated to the initial condition u_0^2 and the field V^n. Now as $(u_0^n)^2 \leq u_0^n \leq 1$ we get $(u^n)^2 \leq u^n \leq 1$, then we can find a subsequence such that u^n is weakly converging to some element v, while $(u^n)^2$ is weakly converging to an element w, weakly in $L^2(0,\tau,L^2(D,R^3))$.

$$\forall \phi \in H_0^1(0,\tau,L^2(D,R^3)) \cap L^2(0,\tau,H_0^1(D,R^3)),\ \ \phi(0)=0,$$

$$-\int_0^\tau \int_D < u^n, \frac{\partial}{\partial t}\phi + \nabla\phi.V > dxdt = \int_D \phi(0)\, u_0^n\, dx$$

and

$$-\int_0^\tau \int_D < (u^n)^2, \frac{\partial}{\partial t}\phi + \nabla\phi.V > dxdt = \int_D \phi(0)\, (u_0^n)^2\, dx$$

In the limit we get

$$\forall \phi \in H_0^1(0,\tau,L^2(D,R^3)) \cap L^2(0,\tau,H_0^1(D,R^3)),$$

$$-\int_0^T \int_D < v, \frac{\partial}{\partial t}\phi + \nabla\phi.V > dxdt = \int_D \phi(0)\, u_0\, dx$$

and

$$-\int_0^T \int_D < w, \frac{\partial}{\partial t}\phi + \nabla\phi.V > dxdt = \int_D \phi(0)\, (u_0)^2\, dx$$

then we derive that $v = u$ and w is the solution associated with the initial condition $u_0^2 = \chi_{\Omega_0}$ and the field V. As $u = \chi_{\Omega_0}$ is a characteristic function, it can be verified that the convergence of u^n to $u = \chi_{\Omega_0}$ is strong : we verify the behavior of the norm

$$\lim_{n\to\infty} \int_0^T \int_D (u^n)^2\, dxdt \leq lim_{n\to\infty}, \int_0^T \int_D (u^n)\, dxdt = \int_0^T \int_D (u^2)\, dxdt$$

(As $u^2 = u = \chi_{\Omega_0}$). Then $(u^n)^2$ itself is strongly converging to w, and as a consequence we get $w = u^2 = \chi_{\Omega_0}$.

Proposition 6.3. *Let $V^n \to V$ in $L^2(0,\tau,L^2(D,R^3))$ with $divV^n(t,.) = divV(t,.) = 0$,*

$$\chi_{Q_{V^n}} \to \chi_{Q_{V^n}} \text{ in } L^2(0,\tau,L^2(D))$$

7. Variational Principle in Euler Problem

We consider a bounded domain D with lipschitzian boundary and the following Euler equation on a given domain Ω_0 in D and V_0, f defined on D:

$$\frac{\partial}{\partial t}V + DV(t).V(t) + \nabla p = f \text{ in } Q_V, \quad V(0) = V_0 \text{ on } \Omega_0 \tag{7.1}$$

Q_V is the non cylindrical evolution domain built by the field V. That is, in weak form,

$$\frac{\partial}{\partial t}\chi_{Q_V} + \nabla\chi_{Q_V}.V = 0, \quad \chi_{Q_V}(0) = \chi_{\Omega_0} \tag{7.2}$$

The Euler incompressible flow is then the solution u, p of the two previous coupled equations. We consider the functional

$$J(V) = \int_0^T \int_{Q_V} ||V(t,x)||^2\, dx + \sigma p(V) \tag{7.3}$$

where $p(V)$ could be the time-space perimeter of the non cylindrical evolution domain Q_V built by V , initiated at Ω_0 and relative to the cylinder $[0,\tau] \times D$. We consider the extremalization of $J(V)$ over the set of vector fields V verifying the initial condition $V(0) = V_0$ on $\Omega(0) = \Omega_0$. We will give a sense to the necessary condition associated to such extremality (assuming first the vector field "smooth enough"). Then we shall give an existence result to the optimization problem

$$Min\{\ J(V)\mid V\in\mathcal{E}\ \} \tag{7.4}$$

We shall replace the time space perimeter by a more physically realistic term and we consider a two fluids problem. The two fluids functional: in the moving domain Ω_t we assume the density $\rho_i = 1+a$ while in the exterior domain $\Omega_t^c = \bar{D} - \Omega_t$, the density is $\rho_e = a$. The one fluid configuration will correspond to $a = 0$. Then we assume $a \geq 0$.

$$J(V) = \int_I\int_D [(a+\chi_V)\ 1/2\ |V(t,x)|^2\ - \chi_V g(t,x)\ -\ f(t,x).V(t,x)\ ]\,dxdt \tag{7.5}$$

where χ_V (that we shall denote by χ when no confusion is possible) is the solution to the following problem:

$$\chi(0) = \chi_{\Omega_0},\ \frac{\partial}{\partial t}\chi + \nabla\chi.V = 0,\ \ \chi = \chi^2 \tag{7.6}$$

$$J'(V,W) = \int_I\int_D (\ \chi'\,(1/2|V|^2\ - g)\ +\ [(a+\chi)V - f]W\ )dxdt$$

where

$$\chi'(0) = 0,\ \frac{\partial}{\partial t}\chi' + \nabla\chi'.V = -\nabla\chi.W \tag{7.7}$$

We introduce the adjoint problem:

$$\lambda(\tau) = 0,\ -\frac{\partial}{\partial t}\lambda\ -\ \nabla\lambda.V = 1/2\ |V|^2\ - g \tag{7.8}$$

$$J'(V,W) = \int_I\int_D ((-\frac{\partial}{\partial t}\lambda - \ \nabla\lambda.V)\ \chi'\ \ +[\ (a+\chi)V - f\,]W)dxdt$$

$$= \int_I\int_D ((\frac{\partial}{\partial t}\chi' +\ \nabla\chi'.V)\ \lambda\ \ +[\ (a+\chi)V - f\,]W)dxdt$$

$$= \int_I\int_D ((-\nabla\chi.W)\ \lambda\ \ +[\ (a+\chi)V - f\,]W)dxdt$$

But as $\int_D ((-\nabla\chi.W)\ \lambda\, dx = \int_D\ \chi.W\,(\nabla\lambda)dx$ we get

$$J'(V,W) = \int_I\int_D [\chi\,\nabla\lambda\ +\ (a+\chi)V - f\,]\,W dxdt.$$

The necessary condition for extremality of J is : there exists a distribution π such that:

$$\chi_V\ \nabla\lambda\ +\ (a+\chi_V)V\ -\ f\ =\ -\nabla\pi \tag{7.9}$$

We set $\Lambda = \chi_V\ \nabla\lambda$ and we have the following condition:

$$\Lambda = f\, - (a+\chi_V)V\ -\ \nabla\pi$$

Moreover Λ solves the following problem:

Proposition 7.1. *The variable Λ solves the backward problem*

$$\Lambda(\tau)=0, \quad -\frac{\partial}{\partial t}\Lambda - D\Lambda.V - D^{*}V.\Lambda = \chi\nabla(\ 1/2|V|^2\ ) \tag{7.10}$$

Proof: from (5) we obtain

$$-\frac{\partial}{\partial t}\chi\ \ \nabla\lambda - \nabla\chi.V\ \ \nabla\lambda = 0 \tag{7.11}$$

Also from the definition of Λ we get

$$\frac{\partial}{\partial t}\lambda = -\frac{\partial}{\partial t}\chi\nabla\lambda + \chi\nabla(\frac{\partial}{\partial t}\lambda)$$

But from the equation verified by λ we get:

$$-\nabla(\frac{\partial}{\partial t}\lambda) - \nabla(\nabla\lambda.V) = 1/2\ \nabla(|V|^2) - \nabla g$$

i.e.

$$-\chi\ \nabla(\frac{\partial}{\partial t}\lambda) - \chi(D^{*}(\nabla\lambda).V\ +\ D^{*}V.\nabla\lambda) = 1/2\chi\nabla(|V|^2) - \chi\nabla g \tag{7.12}$$

By adding (7) and (8) we get:

$$-\chi\ \nabla(\frac{\partial}{\partial t}\lambda) - \frac{\partial}{\partial t}\chi\ \ \nabla\lambda - \nabla\chi.V\ \ \nabla\lambda - \chi D^{*}(\nabla\lambda).V + D^{*}V.(\chi\nabla\lambda)$$

$$= 1/2\chi\nabla(|V|^2) - \chi\nabla g$$

Notice that

$$[\nabla\chi.V\ \ \nabla\lambda + \chi D^{*}(\nabla\lambda).V]_i = \frac{\partial}{\partial x_j}\chi V_j\ \frac{\partial}{\partial x_i}\lambda + \chi\frac{\partial}{\partial x_i}(\frac{\partial}{\partial x_j}\lambda)V_j$$

$$= (\frac{\partial}{\partial x_j}\chi\frac{\partial}{\partial x_i}\lambda + \chi\frac{\partial}{\partial x_j}(\frac{\partial}{\partial x_i}\lambda))V_j$$

$$= \frac{\partial}{\partial x_j}(\chi\frac{\partial}{\partial x_i}\lambda)V_j = \frac{\partial}{\partial x_j}(\Lambda_i)\,V_j = (D\Lambda.V)_i$$

and then $\nabla\chi.V\ \ \nabla\lambda + \chi D^{*}(\nabla\lambda).V = D\Lambda V$.The proposition is proved. Inserting the necessary condition in (6) we obtain:

$$-\frac{\partial}{\partial t}(f - (a+\chi_V)V - \nabla\pi) - D(f - (a+\chi_V)V - \nabla\pi).V$$

$$- D^{*}V.(f - (a+\chi_V)V - \nabla\pi) = \chi\nabla(\ 1/2|V|^2 - g)$$

The above can be rewritten as follows:

$$\frac{\partial}{\partial t}(\,(a+\chi_v)V\,)+D(\,(a+\chi_v)V\,).V\;+\;\frac{\partial}{\partial t}(\nabla\pi-f)$$
$$+D(\,\nabla\pi).V\;+\;D^*V.(\,\nabla\pi)$$
$$-(\,D(f).V\;+\;D^*V.f)$$
$$=\,-\,D^*V.(\,(a+\chi_v)V\,)\;+\chi\,\nabla(\,1/2|V|^2\;-g\,)$$
$$=\;-a\,D^*V.V\;-\chi\,\nabla g\;=-a/2\,\nabla(\,|V|^2\,)-\chi\,\nabla g.$$

That is, as $D(\,\nabla\pi).V\;+\;D^*V.(\,\nabla\pi)\;=\nabla(\,\nabla\pi.V\,)$,

$$\frac{\partial}{\partial t}(\,(a+\chi_v)V\,)+D(\,(a+\chi_v)V\,).V\,)$$
$$\nabla(\,\frac{\partial}{\partial t}(\pi)\;+\nabla\pi.V\;+a/2\,|V|^2\,)$$
$$=\frac{\partial}{\partial t}f\;+\;(\,D(f).V\;+\;D^*V.f)\;-\chi\,\nabla g$$
$$=(\,D^*(f).V\;+\;D^*V.f)\;+(\,D(f)\;-\;D^*f).V\chi\nabla g$$
$$=\;\nabla(\,f.V\,)\;+\;(\,D(f)\;-\;D^*f).V\;-\chi\nabla g$$

then we get V as a solution to the problem

$$\frac{\partial}{\partial t}(\rho_V\,V\,)+D(\,\rho_V\,V\,).V\;)+\;\nabla P \tag{7.13}$$
$$=\;\frac{\partial}{\partial t}f\;+\;(D(f)\;-\;D^*f\,).V\;-\chi\nabla g$$

where the density is

$$\rho_V=(a+\chi_v)$$

and the pressure is given by

$$P=\;\frac{\partial}{\partial t}\pi\;\;+\nabla\pi.V\;+a/2\,|V|^2\;\;-f.V$$

If we assume that $\sigma(f)\;=\;Df\;-\;D^*f$ is zero, then we have $curl f\;=\;0$ (as $\sigma..\sigma=|curl f|^2$) and f derives from a potential, for example in the following form:

$$f(t,x)=\int_0^t\;\nabla F(\sigma,x)\,d\sigma.$$

Then V solves the following Euler equation

$$\frac{\partial}{\partial t}(\rho_V\,V\,)+D(\,\rho_V\,V\,).V\;\;+\;\nabla P \tag{7.14}$$
$$=\;\nabla F\;\;-\;\;\chi\nabla g.$$

In fact "F is of no use in the functional" as any additive gradient term can be "absorbed" by the pressure term P as follows:

$$\frac{\partial}{\partial t}(\,\rho_V\,V\,)+D(\,\rho_V\,V\,).V\;\;+\;\nabla P=\;g\;\nabla\chi \tag{7.15}$$

With

$$P=\;\frac{\partial}{\partial t}\pi\;\;+\nabla\pi.V\;\;+a/2\,|V|^2\;\;-f.V+\;\chi\;g-F$$

8. Differentiability

8.1 Smooth solutions

8.1.1 Bounded Velocity Field. Let W be a given element in $L^\infty(D, R^N)$ with $divW = 0$. We consider the unbounded operator A_W in the Hilbert space $H = L^2(D)$, with dense domain $D_W = H_0^1(D)$ and defined by

$$A_W.\phi = W.\nabla\phi \tag{8.1}$$

It can easily be verified that the adjoint unbounded operator verifies

$$A_W^* = -A_W.$$

Proposition 8.1. *The unbounded operator A_W is the infinitesimal generator of a semigroup of contractions in $H = L^2(D)$.*

Proof. As $H_0^1(D) \subset L^5(D)$, with $N = 3$, and as $\frac{1}{2} + \frac{1}{5} + \frac{3}{10} = 1$, let W_n be a sequence in $W^{1,\infty}(D, R^N)$ with $divW_n \to 0$ in $L^{\frac{5}{3}(D)}$ and converging to W in $L^{\frac{10}{3}}(D, R^N)$. We get:

$$\int_D A_W.\phi \; \phi \, dx = \int_D W.\nabla\phi \; \phi \, dx = lim\{ \int_D W_n.\nabla\phi\phi \, dx \} =$$

$$- lim\{ \int_D (\, divW_n \;\; (\phi)^2 \;\; + (W_n.\nabla\phi) \; \phi \,)dx \; \}$$

from which in the limit we deduce first that $\int_D (W.\nabla\phi) \; \phi \, dx = 0$ and then that the operator A_W is dissipative: $\int_D (A_W.\phi) \; \phi \, dx = 0$.

We consider now the evolution hyperbolic problem associated to any element

$$V \in W^{1,\infty}(0, \tau, L^\infty(D, R^N))$$

with $div\, V(t,.) = 0$ *a.e.t.* We consider the unbounded operator $A(t)$ in the Hilbert space $H = L^2(D)$ defined by $A(t).\phi = V(t).\nabla\phi$ with dense domain $D = H_0^1(D)$ which is independent on t. The Triplet { A(.),H,D } is then a CD-system in the sense of KATO [51](page 9) as we shall verify the following stability condition: for any times $t_1 < .... < t_k$,

$$|| R(t_k, \lambda)...R(t_2, \lambda).R(t_1, \lambda) ||, \leq M(\lambda - \beta)^{-1}$$

where the resolvent $R(t, \lambda) = (\lambda \, I_d \; + \; A(t)\,)^{-1}$ exists for any $t > \beta$. The stability condition is obviously verified from the contraction property of each operator $A(t_k)$. We obtain the following result (KATO [51] thm1.2 page 11)

Proposition 8.2. *Let $V \in W^{1,\infty}(0, \tau, L^\infty(D, R^N))$, $f \in Lip(0, \tau, L^2(D))$, and $\phi \in H_0^1(D)$, there exists a unique solution*

$$u \in C([0, \tau], H_0^1(D)) \cap C^1([0, \tau], L^2(D))$$

to the evolution problem (2).

8.1.2 Unbounded Velocity Field. We applied the theory directly in the Hilbert space $L^2(D)$ and we obtained that $V(t,.) \in L^\infty(D)$ was enough to describe the semigroup.

From the Sobolev embedding inequalities we have $H^1(D) \subset L^p(D)$ for any $p \leq \frac{2N}{N-1}$. Let V be a given element in $L^q(D, R^N)$ with $q > 2 + 2\frac{N-1}{N+1}$ and $divV = 0$. Let us observe that in dimension 3 the following inclusion holds: $H^1(D) \subset L^6(D)$ then as soon as $V(t,.) \in L^3(D)$ we get $\forall\, \phi \in H^1(D),\ \ \phi V(t,.).\nabla\phi \in L^1(D)$. The semigroup is also dissipative as well as his adjoint on the Banach space $H = L^3(D)$, while the dense domain is $\mathcal{D} = \mathcal{H}_\zeta^\infty(\mathcal{D})$. Then we have the

Proposition 8.3. *Let $V \in W^{1,\infty}([0,\tau], L^3(D, R^N))$ with $divV(t,.) = 0$.*

i) Let $f = 0$ and the initial data $\phi \in H^1_o(D)$; there exists a unique solution in $C^0([0,\tau], H^1_o(D)) \cap C^1([0,\tau], L^{\frac{6}{5}}(D))$ to the problem (2).
ii)For $f \in W^{1,\infty}([0,\tau], L^{\frac{6}{5}}(D))$ the dynamical system (2) has a unique solution in

$$C^0([0,\tau], H^1_o(D)) \cap C^1([0,\tau], L^{\frac{6}{5}}(D)).$$

8.2 Fields in $L^4(0,\tau, H^1_0(D, R^N))$

We consider a divergence free field V in $L^2(0,\tau, L^2(D, R^N))$. We consider

$$J(V) = MIN_{\zeta \in L^\infty(0,\tau,L^\infty(D))}\ MAX_{\phi \in \mathcal{H}_V}\ \ \mathcal{L}(\zeta, \phi)$$

with

$$\mathcal{L}(\zeta,\phi) = \int_0^\tau \int_D \{\ 1/2\zeta\,|V|^2\ -\ g\zeta\ +\ \zeta(-\frac{\partial}{\partial t}\phi - \nabla\phi.V)\ \}dxdt + \int_{\Omega_0} \phi(0)dx$$

or, for any integer $m \geq 1$:

$$\mathcal{L}^m(\zeta,\phi) = \int_0^\tau \int_D \{\ 1/2\zeta^m\,|V|^2 - g\zeta + \zeta(-\frac{\partial}{\partial t}\phi - \nabla\phi.V)\ \}dxdt + \int_{\Omega_0} \phi(0)dx$$

The Lagrangian $\mathcal{L}$ is concave-convex on $L^\infty(0,\tau, L^\infty(D)) \times \mathcal{H}_V$. Saddle points (χ, λ) are characterized by :

$$\frac{\partial}{\partial t}\chi + \nabla\chi.V = 0,\ \chi(0) = \chi_{\Omega_0}$$

$$\frac{\partial}{\partial t}\lambda + \nabla\lambda.V\ = 1/2\,|V|^2\ -\ g\,, \qquad \lambda(\tau) = 0$$

or

$$\frac{\partial}{\partial t}\lambda^m + \nabla\lambda^m.V\ = m/2\ \ \chi_V\,|V|^2\ -\ g\,,\ \ \lambda^m(\tau) = 0$$

If $V \in L^4(0,\tau, L^4(D, R^3))$ we get $1/2\,\chi_V\,|V|^2\ -\ g \in L^2(0,\tau, L^2(D))$, then the uncoupled system has a unique solution (χ_V, λ) in $L^2(0,\tau, L^2(D)) \times \mathcal{H}_V$.

If $V \in L^2(0, \tau, L^4(D, R^3))$ the solution λ is not necessarily unique (as the uniqueness result does not applies) then the saddle points are not unique, nevertheless the functional J in form of such min max as a well known Gateaux derivative. Then we have

$$J'(V, W) = \int_0^\tau \int_D \chi_V < (V - \nabla\lambda), W > dxdt$$

Notice that if V was a minimum of J with $V \in l^2$ we would get the necessary optimality condition in the form $\chi_V (V - \nabla\lambda) = \nabla\pi$. The differentiability of J at any divergence free vector field $V \in L^2(0, \tau, L^2(D, R^N))$ depends on the well posedness of the equation in λ with right hand side in $L^1(0, \tau, L^1(D))$.

9. Existence results

For any positive constants $\sigma \le 0$ and $\nu \le 0$ we shall consider the minimization associated to the following functional

$$J_{\sigma,\nu}(V) = J(V) + \sigma \int_0^\tau \|\nabla(\chi_V(t))\|_{M^1(D)} dt + \nu \int_0^\tau \int_D DV..DV dxdt \quad (9.1)$$

In the sequel, with $a > 0$ we shall consider the two situations associated with $\sigma + \nu > 0$ and $\sigma \nu = 0$. When ν is zero the term σ will play a surface tension role at the dynamical interface, while the second case should be consider as a mathematical regularization: as in the non usual variational interpretation developed in the previous section $\nu > 0$ does not lead to the usual viscosity term (i.e. does not lead to the Navier Stoke equations). Let

$$E = \{V \in L^2(0, \tau, L^2(D, R^N)) : div V(t) = 0$$

$$\text{a.e.t, } V(t).n_{\partial D} = 0 \text{ in } H^{-1/2}(\partial D)\}$$

Theorem 9.1. *Assuming $\sigma.\eta > 0$, there exists $V \in E$ such that $\forall W \in E$ we have $J_{\sigma,\nu}(V) \le J_{\sigma,\nu}(W)$.*

Proof: Let us consider a minimizing sequence V_n of $J_{\sigma,\nu}$ in E. Then V_n is bounded in $L^2(0, \tau, L^2(D, R^N))$, we consider a subsequence weakly converging to V.

Lemma 9.1. *Let the sequence V_n be bounded in E, then the sequence χ_{V_n} is bounded in $L^2(0, \tau, H^{-1}(D))$. More precisely, for all $V \in E$ we have*

$$\|\frac{\partial}{\partial t}\chi\|_{L^2(0,\tau,H^{-1}(D))} \le \|V\|_{L^2(0,\tau,L^2(D))} \quad (9.2)$$

Proof. $\forall\phi \in L^2(0,\tau,H_0^1(D))$ we have

$$\int_0^\tau < \frac{\partial}{\partial t}\chi, \phi >_{H^{-1}(D)\times H_0^1(D)} dt = \int_0^\tau \int_D \chi\, V.\nabla\phi \, dxdt$$

Then

$$|\int_0^\tau < \frac{\partial}{\partial t}\chi, \phi >_{H^{-1}(D)\times H_0^1(D)} dt| \leq ||\chi V||_{L^2(I\times D)} \; ||\phi||_{L^2(I,H^{-1}(D))}$$

The sequence χ_{V_n} is itself obviously bounded in $L^2(0,\tau,L^2(D))$ and as the continuous inclusion mapping $L^2(D) \to H^{-\epsilon}(D)$ is compact, $\forall\epsilon > 0$, from a classical result we get the strong convergence of the sequence:

$$\chi_{V_n} \to \chi \text{ strongly in } L^2(0,\tau,H^{-\epsilon}(D))$$

but also χ_{V_n} is $L^2(0,\tau,L^2(D))$-weakly convergent to the same element χ which verifies: $0 \leq \chi \leq 1$ *a.e.*.Let us assume now that $\nu > 0$, then we also have the weak-$L^2(0,\tau,H_0^1(D))$ convergence of V_n to V. This enables us to obtain in the limit in the weak version of the equation of χ_{V_n} :

$$\forall n, \int_0^\tau \int_D \chi_{V_n}\phi dxdt = \int_{\Omega_0} \phi(0,x)dx + \int_0^\tau \int_D (\tau - t)\chi_{V_n} V_n \, dxdt$$

which gives in the limit:

$$\int_0^\tau \int_D \chi\phi \, dxdt = \int_{\Omega_0} \phi(0,x)dx + \int_0^\tau \int_D (\tau - t)\chi V \, dxdt$$

that is:

$$\chi(0) = \chi_{\Omega_0}, \quad \frac{\partial}{\partial t}\chi + \nabla\chi = 0.$$

The problem has a unique solution , so $\chi = \chi_V = \chi^2$. Then also χ_{V_N} converges to χ_V strongly in $L^2(0,\tau,L^2(D))$. It follows that the limit V realizes the minimum of the functional $J_{0,\nu}$ over the linear space E. We turn to the case $\sigma > 0$ which requires first to extend the classical compactness inclusion result to the situation on which the space $L^2(0,\tau,)$ is replaced by $L^2(0,\tau,)$. Now the convergence $V_n \to V$ holds weakly in $L^2(0,\tau,L^2(D,R^N))$, which together with the weak L^2-convergence of χ_{V_n} is not enough for deriving $\chi = \chi^2$ and for passing to the limit in the previous weak equations. From $\sigma > 0$ we get the boundedness of χ_{V_n} in $L^2(0,\tau,BV(D))$. Now the boundedness of $\frac{\partial}{\partial t}\chi_{V_n}$ in $L^2(0,\tau,H^{-1}(D)$ still holds; then, with the next proposition, we obtain the strong convergence $\chi_{V_n} \to \chi_V$ strongly in $L^2(0,\tau,L^1(D))$, and then strongly in $L^2(0,\tau,L^2(D))$ (as we handle characteristic functions). In the sequel we shall use the compact injection of $BV(D)$ in $L^1(D)$. As $L^1(D)$ is not included in $H^{-1}(D)$ (but is included in $H^{-2}(D)$), we shall use the following result:

Proposition 9.1. *Let f_n be a bounded sequence in $L^2(0,\tau,BV(D))$ such that $\frac{\partial}{\partial t}f_n$ is bounded in $L^2(0,\tau,H^{-2}(D))$. Then there exists a subsequence strongly convergent in $L^2(0,\tau,L^1(D))$.*

We adapt to the present situation the proof of J.L.Lions (in the R.Temam's version).

Lemma 9.2. $\forall\eta>0$, *there exists a constant* c_η *with* $\forall\phi\in BV(D)$,

$$||\phi||_{L^1(D)}\le\eta||\phi||_{BV(D)}+c_\eta||\phi||_{H^{-2}(D)}$$

Proof of the lemma: assume that it is wrong. Then, $\forall\eta>0$, there exists $\phi_n\in BV(D)$ and $c_n\to\infty$ such that

$$||\phi_n||_{L^1(D)}\ge\eta||\phi_n||_{BV(D)}+c_n||\phi_n||_{H^{-2}(D)}$$

We introduce $\psi_n=\phi_n/||\phi_n||_{BV(D)}$, and we obtain:

$$||\psi_n||_{L^1(D)}\ge\eta+c_n||\psi_n||_{H^{-2}(D)}\ge\eta$$

But also $||\psi_n||_{L^1(D)}\le c||\psi_n||_{BV(D)}=c$, for some constant c. Then: $||\psi_n||_{H^{-2}(D)}\to 0$. But as $||\psi_n||_{BV(D)}=1$, there exists a subsequence strongly convergent in $L^1(D)\subset H^{-2}(D)$, which turns to be strongly convergent to zero. This is a contradiction with $||\psi_n||_{L^1(D))}\ge\eta$.

Proof of the proposition: from the lemma, $\forall\eta>0$, there exists a constant $d_n$such that

$$\forall f\in L^2(0,\tau,BV(D)),$$

$$||f||_{L^2(0,\tau,L^1(D))}\le\eta||f||_{L^2(0,\tau,BV(D))}+d_n||f||_{L^2(0,\tau,H^{-2}(D))}$$

Given $\epsilon>0$, as $||f_n||_{l^2(0,\tau,BV(D))}\le M$, we shall get;

$$||f_n||_{L^2(0,\tau,L^1(D))}\le 1/2\epsilon+d_n\,||f_n||_{L^2(0,\tau,H^{-2}(D))}$$

if we chose η such that $\eta M\le 1/2\,\epsilon$. At that point the conclusion will derive if we establish strong convergence to zero of f_n in $L^2(0,\tau,H^{-2}(D))$. Now, as$L^1(D)\subset H^{-2}(D)$, we get $f_n\in H^1(0,\tau,H^{-2}(D))\subset C^0([0,\tau],H^{-2}(D))$ so that by use of Lebesgue dominated convergence theorem it will be sufficient to prove the pointwise convergence of $f_n(t)$ strongly to zero in $H^{-2}(D)$. We shall prove it for $t=0$. We have $f_n(0)=a_n+b_n$, with

$$a_n=1/s\int_0^s f_n(t)dt,\quad b_n=-1/s\int_0^s(s-t)f_n'(t)dt$$

If $\epsilon>0$ is given we chose s such that

$$||b_n||_{H^{-2}(D)}\le\int_0^s\,||f_n'(t)||_{H^{-2}(D)}\,dt$$

Finally we observe that $a_n \to 0$ weakly in $BV(D)$, then strongly in $H^{-2}(D)$. Applying that proposition to $f_n = \chi_{V_n}$, where we have chosen a minimizing sequence $V_n \in E$ for the functional $J_{\sigma,0}$, we get $\chi_{V_n} \to \chi$ strongly in$L^2(0,\tau,L^1(D))$, then strongly in $L^2(0,\tau,L^2(D))$ to a characteristic function χ. Then the conclusion derives as in the previous case. Notice that we could also have worked out the proof with $H^{-1}(D)$ instead of $H^{-2}(D)$ by the following argument: as χ_{V_n}, associated to the minimizing sequence, converges strongly in $L^2(0,\tau,L^1(D))$, being a characteristic function it also converges in $L^2(0,\tau,L^2(D))$. Now that linear space is indeed a subspace of $L^2(0,\tau,H^{-1}(D))$.

10. Existence Results Under Capacitary Constraints

A classical problem in optimal design is to prove the existence of minimizers for shape functionals. We have a special interest for functionals defined using the solution of some variational equation. Let $B \subseteq R^N$ be an open ball . On the space $H_0^1(B)$ we consider the norm $||u|| = (\int_B | \nabla u |)^{1/2}$. Let be given a smooth symmetrical matrix $A \in M_{n\times n}(C^1(\overline{B}))$, $A = A^*$, and:

$$\alpha I_d \leq A \leq \beta I_d$$

where $0 < \alpha < \beta$ are the coercivity and the continuity coefficients. We define the associated operator $\mathcal{A} : H_0^1(B) \to H^{-1}(B)$:

$$\mathcal{A} = div A.\nabla \tag{10.1}$$

The right hand side f of the equation associated to this operator is given in $H^{-1}(B)$. For an open set Ω, $\Omega \subseteq B$, we consider the Dirichlet problem in Ω:

$$u \in H_0^1(\Omega) \ , \quad -\mathcal{A}u_\Omega = f \tag{10.2}$$

in the variational sense, i.e.:

$$\int_\Omega < A.\nabla u_\Omega, \nabla\phi > dx =< f_{|\Omega}, \phi >_{H^{-1}(\Omega)\times H_0^1(\Omega)} \quad \forall\phi \in \mathcal{D}(\Omega) \tag{10.3}$$

with $\mathcal{D}(\Omega) = C_0^\infty(\Omega)$, $f_{|\Omega}$ denoting the restriction of the distribution f to the open set Ω. Because $H_0^1(\Omega) = cl_{H_0^1(B)}(\mathcal{D}(\Omega))$, then (10.3) has a unique solution $u_\Omega \in H_0^1(\Omega)$, which we can extend as zero on $B \setminus \Omega$, to u_Ω^0, and $u_\Omega^0 \in H_0^1(B)$,$||u_\Omega^0||_{H_0^1(B)} = ||u_\Omega||_{H_0^1(\Omega)}$. When we consider the solution of (10.2), we will implicitly take its extension u_Ω^0. One can ask to minimize the functional:

$$J(\Omega) = \frac{1}{2}\int_B (u_\Omega - g)^2 dx$$

where $g \in L^2(B)$ is given. One way to prove the existence of extremal domains is to find compact sets in some topology on the space of domains, and to prove that the map:

$$\Omega \longrightarrow J(\Omega)$$

is continuous. A simple remark is that if we consider topologies too strong on the space of domains, we can get the continuity without difficulties, but the compact sets are trivial. An interesting topology is the one associated to the BV norm of the characteristic functions of the domain. For that topology the family of open sets is not closed and then we should consider weaker formulations of the problem. We can refer to a result by V.Sverak in two dimensions, where using the Hausdorff topology he obtained the continuity on the compact family of open sets whose complementary have a fixed maximal number of connected components. We study the N-dimensional case, and we obtain a similar result for more general elliptic problem, but our compact classes of opens are not so simple to describe geometrically. We will use classes of domains which satisfy some capacity density conditions.

10.1 Preliminaries about capacity and domains convergence

On the family of open subsets of B , we define the Hausdorff complementary topology, denoted H^c, given by the metric:

$$d(\Omega_1, \Omega_2) = \max(\sup_{x \in B \setminus \Omega_1} \inf_{y \in B \setminus \Omega_2} \|x - y\|, \sup_{x \in B \setminus \Omega_2} \inf_{y \in B \setminus \Omega_1} \|x - y\|) \quad (10.4)$$

d is a metric, and the family of all open subsets of B is compact. Moreover we have the following results:

Proposition 10.1. *If $\Omega_n \xrightarrow{H^c} \Omega$ then for all compact K, $K \subseteq \Omega$, there exists $N_K \in N$, $\forall n \geq N_K$ we have $K \subseteq \Omega_n$.*

Proposition 10.2. *If $\Omega_n \xrightarrow{H^c} \Omega$, and $x \in \partial\Omega$, then there exists a sequence $\{x_n\}_{n \in N}$, $x_n \in \partial\Omega_n$, and $x_n \longrightarrow x$.*

The problem which we will study is to find a family $\mathcal{O}$ of open subsets of B, which will be compact for the H^c topology and with the following continuity property:

$$\Omega_n \in \mathcal{O}, \Omega_n \xrightarrow{H^c} \Omega \quad \text{implies} \quad \Omega \in \mathcal{O}, \; u_{\Omega_n} \xrightarrow{H_0^1(B)} u_\Omega$$

Generally, this assertion is not true, and the main difficulty is that the limit of u_{Ω_n} is not the solution of (10.2) on Ω. To overcome this, we'll impose some constraints on the family of Ω_n in terms of capacity.

Definition 10.1. *The exterior capacity (or capacity) is defined as: -for compact sets in R^N:*

$$C(K) = \inf\{\|\varphi\|_{1,2}^2 \mid \varphi \in C_0^\infty(R^N), \varphi \geq 1 \quad on K\}$$

-for open sets $G \subseteq R!!^N$:

$$C(G) = \sup\{C(K) \mid K \subseteq G, K \, compact\}$$

-for arbitrary sets $E \subseteq R^N$:

$$C(E) = \inf\{C(G) \mid G \supseteq E, G \quad open\}$$

We will say that a property holds quasi everywhere (q.e.) if it holds in the complement of a set of zero capacity.

Definition 10.2. *For $r > 0$ and a compact $K \subseteq R^N$, the condenser capacity of K in the ball $B(x,r)$ is:*

$$C(K \cap B(x,r), B(x,2r))$$

$$= \inf\{\int_{B(x,2r)} |\nabla\varphi|^2 \, dV \mid \varphi \in C_0^\infty(B(x,2r)), \varphi \geq 1 \quad on \quad K \cap B(x,r)\}$$

10.2 Limiting process in the equation

In this section we shall prove that if $\Omega_n \xrightarrow{H^c} \Omega$, there exists a subsequence of solutions u_{Ω_n} weakly converging in $H_0^1(B)$ to some u satisfying the equation$-Au = f$ in Ω. A simple remark is that $u_\Omega$is the $H_0^1(\Omega)$ orthogonal projection of u. The next paragraph will be dedicated to prove that under capacitary conditions we get $u = u_\Omega$. In order to use the weak compactness of the unity ball in $H_0^1(B)$ we have the following result:

Proposition 10.3. *There exists a constant $M = M(\alpha, N, f)$, such that for all $\Omega \subseteq B$ we have:*

$$\|u_\Omega\| \leq M$$

where u_Ω is the solution of (10.3) in Ω.

Proof. Let $< \dots >$ denotes the duality bilinear form pairing between $H^{-1}(\Omega)$ and $H_0^1(\Omega)$. Because u_Ω is the solution of the equation in Ω, we have:

$$\int_\Omega < A.\nabla u_\Omega, \nabla\phi > dx =< f, \phi > \quad \forall\phi \in \mathcal{D}(\Omega) \tag{10.5}$$

Because $H_0^1(\Omega) = cl_{H_0^1(B)}(\mathcal{D}(\Omega))$ we can take in particular $\phi = u_\Omega$. Hence:

$$\int_{\Omega} < A.\nabla u_{\Omega}, \nabla u_{\Omega} > dx =< f, u_{\Omega} >$$

Using the coercivity property of A and the Cauchy-Schwartz inequality, there exists a constant M, such that:

$$\|u_{\Omega}\|_{H_0^1(B)} \leq \|f\|_{H^{-1}(\Omega)} \frac{1}{\alpha} = M$$

and so the proof is finished.

Let$\{\Omega_n\}_{n\in N}$ be a sequence of open sets of B, and $\Omega_n \xrightarrow{H^c} \Omega$. From the previous boundedness we derive the following convergence of solutions.

Proposition 10.4. *There exists a subsequence of* $\{\Omega_n\}_{n\in N}$*, which we still denote* $\{\Omega_n\}_{n\in N}$ *,such that:*

$$u_{\Omega_n} \overset{H_0^1(B)}{\rightharpoonup} u$$

and

$$-Au_{|\Omega} = f_{|\Omega} \ \ in \ \ \mathcal{D}'(\Omega) \times \mathcal{D}(\Omega) \tag{10.6}$$

Proof. Using the statement of the proposition 10.3, we have:

$$\|u_{\Omega_n}\|_{H_0^1(B)} \leq M \ \ , \forall n \in N$$

Because of the weak compactness of the closed unity ball, there exists a subsequence of $\{\Omega_n\}_{n\in N}$ still denoted $\{\Omega_n\}_{n\in N}$, and an element $u \in H_0^1(B)$, such that:

$$u_{\Omega_n} \overset{H_0^1(B)}{\rightharpoonup} u$$

We'll prove that u satisfies (10.6). Let $\varphi \in \mathcal{D}(\Omega)$. Then $supp\varphi = K \subseteq \Omega$, and using the properties of the Hausdorff convergence there exists $N_K \in N$, such that $\forall n \geq N_K$ we have $K \subseteq \Omega_n$. So, we can write:

$$\int_{\Omega_n} < A.\nabla u_{\Omega_n}, \nabla \phi > dx =< f, \phi >_{H^{-1}(\Omega_n)\times H_0^1(\Omega_n)}$$

or equivalently:

$$\int_{B} < A.\nabla u_{\Omega_n}, \nabla \phi > dx =< f, \phi >_{H^{-1}(B)\times H_0^1(B)}$$

Letting $n \to \infty$, and using the weak convergence $u_{\Omega_n} \overset{H_0^1(B)}{\rightharpoonup} u$ we obtain:

$$\int_{\Omega} < A.\nabla u_{\Omega}, \nabla \phi > dx =< f, \phi >$$

As$\varphi \in \mathcal{D}(\Omega)$ was arbitrary chosen we have the statement of the proposition.

10.3 Continuity under capacitary constraint

In view to preserve the Dirichlet condition in the limiting terms Ω and u, we are obliged to restrict our study to some smaller classes of domains. We should be careful that the relative capacity of the complementary of the domains near the boundary should not vanish. In order to handle this point, we introduce the following concepts related to the local capacity of the complementary near the boundary points.

Definition 10.3. *We say that an open set Ω has the (r,c) capacity density condition if*

$$\forall x \in \partial\Omega, \quad \frac{C(\Omega^c \cap B(x,r), B(x,2r))}{C(B(x,r), B(x,2r))} \geq c \tag{10.7}$$

Definition 10.4. *For $r < 1$ we define the following family of open subsets of B:*

$$\mathcal{O}_{c,r}(B) = \{\Omega \subseteq B \mid \forall r_0,\ 0 < r_0 < r\ ,\ \Omega \text{ has the } (r_0, c) \text{ capacity density condition}\}$$

Definition 10.5. *Let x be a point on $\partial\Omega$.The set Ω^c is thick at x if locally the complementary ofΩ has "enough " capacity, precisely if we have:*

$$\int_0^1 \left(\frac{C(\Omega^c \cap B(x,r), B(x,2r))}{C(B(x,r), B(x,2r))}\right)\frac{dr}{r} = \infty$$

Remark 10.1. If Ω has the (r,c)-capacity density condition, the complementary is thick at any point of the boundary.

Our main continuity result can be expressed as follows:

Theorem 10.1. *Let $\{\Omega_n\}_{n\in N}$ be a sequence in $\mathcal{O}_{c,r}(B)$, which converges in the H^c topology to an open Ω. Then $u_{\Omega_n} \overset{H_0^1(B)}{\longrightarrow} u_\Omega$.*

Proof. In a first step we consider that $f \in H^s(B)$,$s > N/2 - 2$.

Lemma 10.1. *Let v be an A-harmonic function (e.g. weak solution for the homogeneous version of (10.2), on the open set Ω. Then v can be redefined in a set of measure zero, so that it becomes continuous on Ω.*

We remark that the continuous representative from lemma 10.1 is in fact a quasi continuous $H_0^1(\Omega)$ representative. Indeed, let $v_1 = v$a.e. and v_1 continuous on Ω. We want to show that v_1 is a quasicontinuous representative of v. There exists a quasicontinuous representative v_2 of v, equal to v a.e. Sov_1 is continuous, v_2 is quasicontinuous and $v_1 = v_2$a.e.. We get$v_1 = v_2$ q.e.

Lemma 10.2. *Let Ω belongs to$\mathcal{O}_{c,r}(B)$. If $\theta \in H^1(\Omega) \cap C(\Omega)$, and if h is$A$-harmonic function inΩ with $h - \theta \in H_0^1(\Omega)$, then*

$$\lim_{x \to x_0} h(x) = \theta(x_0)$$

for any $x_0 \in \partial\Omega$.

As a remark, the fact that Ω belongs to$\mathcal{O}_{c,r}(B)$ hides the thickness to any point of its boundary, which is necessary in the proof of the lemma. Returning to the theorem 10.1 it will be sufficient to prove the continuity for a subsequence of $\{\Omega_n\}_{n\in N}$. Because of the proposition 10.4 there exists a subsequence of $\{\Omega_n\}_{n \subset N}$, which we'll still denote $\{\Omega_n\}_{n\in N}$, such that $u_{\Omega_n} \overset{H_0^1(B)}{\rightharpoonup} u$, and u satisfies the equation on Ω. We shall prove that in our hypothesis $u \in H_0^1(\Omega)$, which will imply that $u = u_\Omega$. For that it is sufficient to prove $u = 0$ q.e. on $B \setminus \Omega$ where u is a quasicontinuous representative. From the Banach-Saks theorem there exists a sequence of averages:

$$\psi_n = \sum_{k=n}^{N_n} \alpha_k^n u_{\Omega_n}$$

with

$$0 \leq \alpha_k^n \leq 1\,, \quad \sum_{k=n}^{N_n} \alpha_k^n = 1$$

such that

$$\psi_n \overset{H_0^1(B)}{\longrightarrow} u$$

Because of the strong convergence of ψ_n to u in$H_0^1(B)$, we have that :

$$\psi_n(x) \longrightarrow u(x) \quad q.e. \quad on \ B$$

for a subsequence of $\{\psi_n\}$ which we still denote $\{\psi_n\}$. Let G_0 be the set of zero capacity on which $\psi_n(x)$ does not converge to $u(x)$. Let $x \in B \setminus (\Omega \cup G_0)$, and $\varepsilon > 0$ arbitrary. We'll prove that $| u(x) | < \varepsilon$. We have:

$$| u(x) | \leq | u(x) - \psi_n(x) | + | \psi_n(x) |$$

We'll consider $n > N_{\varepsilon,x}$ such that

$$| u(x) - \psi_n(x) | < \frac{\varepsilon}{2}$$

Let consider u_B the solution of (10.2) on B. From the smoothness of B and the regularity of f we have that u_B is continuous on $\overline{B}$. Subtracting the corresponding equations, we obtain:

$$A(u_B - u_{\Omega_n}) = 0 \ \ in \ \ \mathcal{D}'(\Omega_n) \times \mathcal{D}(\Omega_n) \tag{10.8}$$

So $u_B - u_{\Omega_n}$ is$\mathcal{A}$-harmonic in Ω_n, and continuous on Ω_n from lemma 10.1. Moreover we get the continuity of u_{Ω_n} on the closure $\overline{\Omega_n}$, and the values of u_{Ω_n} on the boundary are zero. We use lemma 10.2 in the following way: θ is the restriction to Ω_n of the function u_B and $h_n = u_{B|\Omega_n} - u_{\Omega_n}$. Now $v - \theta = u_{\Omega_n}$ which belongs to $H_0^1(\Omega_n)$. From the continuity of $u_B$we obtain that the continuous extension of $u_{B|\Omega_n} - u_{\Omega_n}$ is equal to u_B on the boundary of Ω_n, and sothe extention of u_{Ω_n} to the boundary is zero. We obtain that ifh_n is δ - holderian on $\partial\Omega_n$ then it is δ_1 -holderian on allΩ_n. We have that u_B is$s+2-N/2$- holderian on B (with$s+2-N/2>0$) , with a constant M, because of the hypothesis on f, and the smoothness of B. Finally we get

$$\forall x,y \in \partial\Omega_n \quad | h_n(x) - h_n(y) |=| u_B(x) - u_B(y) |\leq M \mid x-y \mid^{s-N/2+2}$$

So, there exists $\delta_1 = \delta_1(N,\beta/\alpha,c)$, and

$$M_{1,n} = 80Mr^{-2}\max(1,(diam(\Omega_n))^2 \leq 80Mr^{-2}\max(1,(diam(B))^2 = M_1$$

such that:

$$| h_n(x) - h_n(y) |\leq M_{1,n} \mid x-y \mid^{\delta_1} \leq M_1 \mid x-y \mid^{\delta_1} \quad \forall x,y \in \Omega_n$$

By a simple argument we obtain that this inequality holds in B and hence $\forall x,y \in B$ we have:

$$| u_{\Omega_n}(x) - u_{\Omega_n}(y) |\leq| h_n(x) - h_n(y) | + | u_B(x) - u_B(y) |\leq$$

$$\leq M_1 \mid x-y \mid^{\delta_1} + M \mid x-y \mid \leq M_2 \mid x-y \mid^{\delta_2}$$

Let's chose $R>0$, such that $M_2R^{\delta_2} < \varepsilon/2$. Because of the H^c convergence of Ω_n to Ω $\exists n_R \in N$, such that $\forall n \geq n_R$ we have $(B\setminus\Omega_n)\cap B(x,R) \neq \emptyset$. Let's take $x_n \in (B\setminus\Omega_n)\cap B(x,R)$. We have:

$$| u_{\Omega_n}(x) |=| u_{\Omega_n}(x) - u_{\Omega_n}(x_n) |\leq M_2 \mid x - x_n \mid^{\delta_2} \leq M_2R^{\delta_2} \leq \frac{\varepsilon}{2}$$

because $u_{\Omega_n}(x_n) = 0$. So

$$| \psi_n(x) |=| \sum_{k=n}^{N_n} \alpha_k^n u_{\Omega_n}(x) |\leq \sum_{k=n}^{N_n} \alpha_k^n \frac{\varepsilon}{2} = \frac{\varepsilon}{2}, \ \forall n > n_R$$

Finally we obtain $| u(x) |\leq \varepsilon$. Because ε was taken arbitrarily we have $u(x) = 0$ q.e. on $B\setminus\Omega$, which implies that $u = u_\Omega$. The strong convergence of u_{Ω_n} to u_Ω is now immediate, because of the convergence of the A-norms of u_{Ω_n} to the A-norm of u_Ω, and the fact that the A-norm is equivalent to the $H_0^1(B)$-norm, i.e. :

$$\int_B < A.\nabla u_{\Omega_n}, \nabla u_{\Omega_n} > dx =< f, u_{\Omega_n} > \longrightarrow$$

$$\longrightarrow < f, u_\Omega > = \int_B < A.\nabla u_\Omega, \nabla u_\Omega > dx.$$

In the next step we shall prove that the continuity result is still valid for $f \in H^{-1}(B)$. The main idea is to use the continuous dependence of the solution u_Ω on f, which is uniformly in Ω. Indeed, let$\Omega \subseteq B$, and $f, g \in H^{-1}(B)$. Then, by a simple subtraction of the equations, we get:

$$\int_B | \nabla(u_{\Omega,f} - u_{\Omega,g}) |^2 \, dx = \int_B < (f-g)(u_{\Omega,f} - u_{\Omega,g}) > dx$$

and applying the Cauchy inequality we obtain:

$$\|u_{\Omega,f} - u_{\Omega,g}\|^2_{H^1_0(B)} \leq \|f-g\|_{H^{-1}(B)} \|u_{\Omega,f} - u_{\Omega,g}\|_{H^1_0(B)}$$

Finally we get:

$$\|u_{\Omega,f} - u_{\Omega,g}\|_{H^1_0(B)} \leq \|f-g\|_{H^{-1}(B)}$$

So,let $f \in H^{-1}(B)$ and $f_\varepsilon = f * \rho_\varepsilon$, where ρ_ε is a mollifier. Making $\varepsilon \to \infty$ we have $\|f_\varepsilon - f\|_{H^{-1}(B)} \to 0$. Let $Let\{\Omega_n\}_{n\in N} \subseteq \mathcal{O}_{c,r}(B)$, be a sequence which converges in the H^c topology to an open Ω. Then we have:

$$u_{\Omega_n, f_\varepsilon} \overset{H^1_0(B)}{\longrightarrow} u_{\Omega, f_\varepsilon}$$

because $f_\varepsilon \in H^s(B)$ and

$$u_{\Omega_n, f_\varepsilon} \overset{H^1_0(B)}{\longrightarrow} u_{\Omega_n, f}$$

uniformly on Ω_n, from the previous considerations. Let$\delta > 0$. We get:

$$\|u_{\Omega_n,f} - u_{\Omega,f}\|_{H^1_0(B)} \leq \|u_{\Omega_n,f} - u_{\Omega_n,f_\varepsilon}\|_{H^1_0(B)} + \|u_{\Omega_n,f_\varepsilon} - u_{\Omega,f_\varepsilon}\|_{H^1_0(B)}$$
$$+ \|u_{\Omega,f_\varepsilon} - u_{\Omega,f}\|_{H^1_0(B)}$$

We chose ε small enough such that $\|f_\varepsilon - f\|_{H^{-1}(B)} < \delta/4$ and for ε fixed we chosen$_{\varepsilon,\delta}$such that $\forall n > n_{\varepsilon,\delta}$ we have

$$\|u_{\Omega_n,f_\varepsilon} - u_{\Omega,f_\varepsilon}\|_{H^1_0(B)} < \frac{\delta}{2}$$

Finally we get

$$\|u_{\Omega_n,f} - u_{\Omega,f}\|_{H^1_0(B)} \leq \delta \quad \forall n > n_{\varepsilon,\delta}$$

As δ was arbitrarily chosen, the proof is finished.

In the case $N = 2$ we can obtain the result of Sverak.

Theorem 10.2. *Let $N = 2$ and l a positive integer. We define the set*

$$\mathcal{O}_l = \{\Omega \subseteq B \mid \#(B \setminus \Omega) \leq l\}$$

Then the set $\mathcal{O}_l$ is compact in the H^c-topology and the map:

$$\mathcal{O}_l \ni \Omega \longrightarrow u_\Omega \in H^1_0(B)$$

is continuous

By # we have denoted the number of connected components.

10.4 Continuity under flat cone condition

We shall study some classes of domains which have a simpler geometrical description , and satisfy the capacity density condition with some constants c and r.

Definition 10.6. *Let* $x \in R^N$, $1 > \omega > 0$, $h > 0$, *and* $\nu, \xi \in R^N$ *with* $\|\nu\| = \|\xi\| = 1$. *We define the "flat cone":*

$$C(x, \omega, h, \xi, \nu) = \{y \in R^N \mid 0 \le (\overline{xy}, \xi) \le \omega\|\overline{xy}\|,\ (\overline{xy}, \nu) = 0,\ \|\overline{xy}\| \le h\}$$

$C(x, \omega, h, \xi, \nu)$ lies in the hyperplane which contains x and has the normal ν, ω is its opening,h is its length and ξ its symmetry axis. That cone is called flat, for it is contained in a hyperplane. In particular its N-dimensional Lebesgue measure is zero.

Definition 10.7. *Let* $1 > \omega > 0$ *and* $h > 0$. *We say that an open set* Ω, $\Omega \subset B$, *satisfies the* (ω, h)*- flat cone condition* (ω, h)*-(f.c.c.), if* $\forall x \in \partial\Omega$, $\exists \xi_x, \nu_x \in R^N$ *unitary vectors, such that*

$$C(x, \omega, h, \xi_x, \nu_x) \subseteq \Omega^c$$

Let's denote by $C(\omega, h, B)$ the family of open subsets of B which satisfy the (ω, h)-f.c.c.

Remark 10.2. We can compare the f.c.c. with the uniform cone condition, for which is already known the continuity for the Dirichlet problem, and we observe that the f.c.c. is much less restrictive in two directions. We demand that the cone property arises in a point and not in a neighborhood, and the dimension of our cone is smaller (the cone has a zero measure), which has consequences on the possible flatness of the complementary. No local uniformity is required on the cone direction.

Proposition 10.5.

$$C(\omega, h, B) \subseteq \mathcal{O}_{c,r}(B)$$

Proof. It is immediately with $r = h$and

$$c = \frac{C(C(0, \omega, h, e_1, e_2) \cap \overline{B}(0, h), B(0, 2h))}{C(\overline{B}(0, h), B(0, 2h))}$$

and using the properties of capacity on translation and similarity.

Remark 10.3. Intuitively for N=2, the flat cone is a segment. The condition of connection implies the existence of a continuous curve, and hence the relation between f.c.c. and connection can be deduced .

Proposition 10.6. *The family*$C(\omega, h, B)$ *is compact in the* H^c*-topology.*

Proof. It is sufficient to prove that $C(\omega,h,B)$is closed. For that, let a sequence $\{\Omega_n\}_{n\in N} \subseteq C(\omega,h,B)$and $\Omega_n \xrightarrow{H^c} \Omega$. We shall prove that $\Omega \in C(\omega,h,B)$. Let $x \in \partial\Omega$, and then it exists a sequence of points , $x_n \in \Omega_n$, and $x_n \to x$. As $\Omega_n \in C(\omega,h,B)$, there exists a flat cone $\mathcal{C}(x_n,\omega,h,\xi_n,\nu_n) \subseteq \Omega_n^c$. It is easy to see that there exists subsequences $\{\xi_{n_k}\}$ and $\{\nu_{n_k}\}$ of$\{\xi_n\}$ and $\{\nu_n\}$ respectively, such that$\xi_{n_k} \to \xi$ and $\nu_{n_k} \to \nu$.Because of the properties of the H^c-topology we obtain that$\mathcal{C}(x,\omega,h,\xi,\nu) \subseteq \Omega^c$.Finallywe obtain that $C(\omega,h,B)$ is closed and hence compact.

Theorem 10.3. *If $\{\Omega_n\}_{n\in N}$ is a sequence of open sets, $\Omega_n \in C(\omega,h,B)$, $\forall n \in N$and $\Omega_n \xrightarrow{H^c} \Omega$, then $u_{\Omega_n} \xrightarrow{H_0^1(B)} u_\Omega$.*

Example 10.1. .

An example of a family which satisfies the flat cone condition is the family of open subsets of B obtained by "perforation". Let $0 < r$ and$0 < \mu < 1$. Let us consider

$$B_{r,\mu} = \{\Omega \mid \Omega = B\backslash\overline{\cup_{i\in I} B_0(x_i,r_i,\eta_i)}\ \ \forall i \in I,\ r_i \geq r > 0,\ x_i \in \mu B,\ \eta_i \in R^N\}$$

where I is an arbitrary set, $B_0(x_i,r_i,\eta_i)$ being the open R^{N-1}-ball, centered in x_i, of radius r_i, which lies in the normal plane to η_i:

$$B_0(x_i,r_i,\eta_i) = \{x \in R^N \mid < x - x_i, \eta_i >= 0,\ \|x - x_i\| < r_i\}$$

. Let $\Omega \in B_{r,\mu}$ and $y \in \partial\Omega$. Then there exists a sequence $y_n \in B_0(x_n,r,\eta_n) \subseteq \Omega^c$ such that $y_n \to y$. But, from the compactness of the set $\overline{B}$ and of the unity ball in R^N, we can assume that $x_n \to x$ and $\eta_n \to \eta$, and so $B_0(x,r,\eta) \subseteq \Omega^c$, but also $y \in B_0(x,r,\eta)$, and then the f.c.c.property follows.

10.5 Existence results for extremal domains

We shall prove that the set $\mathcal{O}_{c,r}(B)$ is compact in the H^c-topology, and because of the continuity of the shape functionals which are defined on the solutions we shall obtain some existence results.

Theorem 10.4. *The family $\mathcal{O}_{c,r}(B)$ defined in the definition 10.4 is compact.*

Proof. It is sufficient to prove that:

$$\forall c > 0,\ \ \forall r,\ 0 < r < 1\ \ cl_{H^c}\mathcal{O}_{c,r}(B) = \mathcal{O}_{c,r}(B)$$

. Let's take$\{\Omega_n\}_{n\in N} \subseteq \mathcal{O}_{c,r}(B)$ and $\Omega_n \xrightarrow{H^c} \Omega$. We shall denote by $K_n = \overline{B} \setminus \Omega_n$, and $K_\varepsilon = cl_{R^N}(\cup_{x\in K} B(x,\varepsilon)$. Let $x \in \partial\Omega$. We will prove in the point x, the capacity density condition for Ω. Because of the H^c convergence, $\forall\varepsilon > 0$, $\exists n_\varepsilon$, $\forall n > n_\varepsilon$ we have $K_n \subseteq K_\varepsilon$. For $\varepsilon > 0$ given, we consider $n_{\varepsilon/2}$, and $x_n \in \partial\Omega_n$ such that $\|x - x_n\| < \varepsilon/2$. We'll denote τ the translation of the vector $x_n - x$. As $\tau B(x,r) = B(x_n,r)$, and the condenser capacity being invariant on the translation of the two arguments, we get

$$C(K_\varepsilon \cap B(x,r_0), B(x,2r_0)) = C(\tau K_\varepsilon \cap B(x_n,r_0), B(x_n,2r_0))$$

Because $K_n \subseteq K_{\varepsilon/2}$, and $\|x - x_n\| < \varepsilon/2$ we have:

$$\tau K_\varepsilon \supseteq K_n$$

Then, from the monotonicity of the first argument we get

$$C(K_\varepsilon \cap B(x,r_0), B(x,2r_0)) \geq C(K_n \cap B(x_n,r_0), B(x_n,2r_0))$$

Using the capacity density condition for Ω_n we have:

$$C(K_\varepsilon \cap B(x,r_0), B(x,2r_0)) \geq c\, C(B(x,r_0), B(x,2r_0))$$

Making $\varepsilon \to 0$, and using the continuity of the capacity on decreasing sequences of compacts, and the fact that:

$$\bigcap_{\varepsilon>0} K_\varepsilon = K$$

we obtain the capacity density condition for Ω in x, and finally $\Omega \in \mathcal{O}_{c,r}(B)$.

A first example of extremal domain derives directly from theorem 10.4

Example 10.2. .

For any open $\Omega \subseteq B$ we can note the first eigenvalue of the Laplacian by

$$\lambda(\Omega) = \inf_{\varphi \in H_0^1(\Omega),\, \varphi \neq 0} \frac{\int_\Omega |\nabla\varphi|^2\, dx}{\int_\Omega |\varphi|^2\, dx} \tag{10.9}$$

Let $E \subseteq B$ be a given open set. Then the maximum

$$\max_\Omega \{\lambda(\Omega) \mid E \subseteq \Omega \subseteq B,\ \Omega \in \mathcal{O}_{c,r}(B)\} \tag{10.10}$$

is reached. In some sense the optimal solution is an $\mathcal{O}_{c,r}(B)$ approximation of the set E. For the proof of this result we have only to remark that the map $\Omega \to \lambda(\Omega)$ is upper semicontinuous in the H^c topology, and the admissible family of domains is compact. We give now the following general existence theorem.

Theorem 10.5. *If h is continuously defined from $H_0^1(B)$ in R then $J(\Omega) = h(u_\Omega)$ is continuously defined form $\mathcal{O}_{c,r}(B)$ in R and attains its extremal values on that set.*

Example 10.3. .

The hypothesis in theorem10.5 concerning the function h is verified in the example we gave in the introduction. We shall study a more realistic shape functional for which we can conclude the existence of the extremal domain even if the assumptions in theorem 10.5 concerning h fails. Let $\alpha > 0$ and

$$J(\Omega) = \frac{1}{2}\int_{\Omega} |\nabla u_{\Omega} - \bar{z}|^2 \, dx + \alpha \frac{1}{\int_{\Omega} |u_{\Omega}|^2 \, dx} \tag{10.11}$$

where $\bar{z}$ is given in $L^2(B; R^N)$. We remark that $\nabla u_{\Omega} = 0$ a.e. on$\partial\Omega$. Indeed, there exists a sequence $\{\varphi_n\}$ of elements of $\mathcal{D}(\Omega)$, such that $\varphi_n \overset{H_0^1(B)}{\longrightarrow} u_{\Omega}$. So $\nabla\varphi_n = 0$ a.e. on $\partial\Omega$ and as $\nabla\varphi_{n_k} \to \nabla u_{\Omega}$ a.e. we get $\nabla u_{\Omega} = 0$ a.e. on $\partial\Omega$. Effectively the functional can be rewritten as

$$J(\Omega) = \frac{1}{2}\int_{B} |\nabla u_{\Omega} - \bar{z}|^2 \, dx + \alpha \frac{1}{\int_{B} |u_{\Omega}|^2 \, dx} - \frac{1}{2}\int_{B\setminus\Omega} |\bar{z}|^2 \, dx \tag{10.12}$$

and the last term it is not of the form $h(u_{\Omega})$ for some h, but simply in the form $h(\Omega)$. It turns that J is not continuous for the H^c topology, but is lower semicontinuous for that topology, and then can be minimized. More precisely we have the following result.

Proposition 10.7. *Let μ be a positive finite measure on B. The application $\Omega \to \mu(\Omega)$ is lower semicontinuous with respect to the H^c topology.*

Proof. Let $\{\Omega_n\}$ be a sequence of open sets in B, $K_n = \bar{B} \setminus \Omega_n$, $K = \bar{B} \setminus \Omega$, and

$$\Omega_n \overset{H^c}{\longrightarrow} \Omega$$

Given $\varepsilon > 0$, we have $K_\varepsilon \supseteq K_n$, when $n > n_\varepsilon$. Then$\mu(K_\varepsilon) \geq \mu(K_n)$ when $n > n_\varepsilon$. So, $\forall\varepsilon > 0$

$$\mu(K_\varepsilon) \geq \overline{\lim_{n\to\infty}} \, \mu(K_n)$$

ButK_ε is a monotonic decreasing sequence and then

$$\mu(K_\varepsilon) \longrightarrow \mu(K) \quad \text{as } \varepsilon \longrightarrow 0$$

and the result follows.

Let's formulate the minimization problem. D is a given bounded domain in R^N, f is given in $H^{-1}(D)$, B being a ball containing D. Let μ be a positive measure on D, a a positive constant such that $0 < a < \mu(D) < +\infty$. We consider the equation:

$$-\, div(A.\nabla u_{\Omega}) = f \quad \text{in } \Omega \quad \text{and} \quad u = 0 \text{ on } \partial\Omega \tag{10.13}$$

For $J(\Omega)$ defined in (10.11) let's consider the following problem:

$$\text{Min}\{J(\Omega) \mid \Omega \in \mathcal{O}_{c,r}(B), \;\; \Omega \subseteq D, \; \mu(\Omega) \leq a\} \tag{10.14}$$

Theorem 10.6. *For any constants $c > 0$, and r, $0 < r < 1$ the problem (10.13)-(10.14) has at least a solution.*

Proof. For the proof it is sufficient to notice that in (10.12), the first term is continuous via the setting of theorem 10.5. The second term is lower semi-continuous from proposition 10.7 (with the measure of density $|\bar{z}|^2$). For the end we just recall that if $\Omega_n \subseteq D$ and $\Omega_n \xrightarrow{H^c} \Omega$, then $\Omega \subseteq D$. As $\alpha > 0$, a minimizing sequence can not converge to $\emptyset$. In fact for any admissible domain Ω_0 and for any optimal solution we have: $\int_\Omega |u_\Omega|^2 \, dx \geq \frac{\alpha}{J(\Omega_0)}$.

11. Geometry via the Oriented Distance function

11.1 Introduction

The *oriented boundary* (resp. *signed* or *algebraic*) distance function b_Ω of a subset Ω of $\mathbb{R}^N$ with nonempty boundary $\partial\Omega$, is particularly interesting since it provides a global access to fine local geometric properties of the set and its boundary. For C^2-domains it has been used to construct a fairly complete intrinsic tangential differential calculus on C^2 submanifolds of $\mathbb{R}^N$ of codimension one without using local bases and Christoffel symbols. This has led to an original application in the *theory of shells*. But there is deeper implications in *Differential Geometry* and many classical results can be reformulated in that framework. Among the many interesting properties of b_Ω is the fact that, for $C^{1,1}$ or smoother domains, the regularity of the boundary is completely characterized by the regularity of b_Ω in a neighborhood of $\partial\Omega$. Since b_Ω and its derivatives contain all the information on the geometry of Ω and its boundary, the question of characterizing domains which are less smooth naturally arises. It is natural to retain topologies on subsets or equivalence classes of subsets of $\mathbb{R}^N$ which correspond to some degree of differentiability of b_Ω in some neighborhood of the boundary of Ω. The region of interest lies between $W^{1,p}$ and $C^{1,1}$, that is from no assumption on the domain Ω to domains with a $C^{1,1}$ boundary. In the last case the elements of the matrix $D^2 b_\Omega$ of second order derivatives belong to L^∞ in a neighborhood of the boundary $\partial\Omega$ and the eigenvalues of this matrix at a point x are 0 and the $N-1$ principal curvatures of the corresponding level set of b_Ω through this point. In general the singularities of the gradient of b_Ω can be classified in two categories: the *skeleton* away from the boundary and the *set of cracks* which is contained in the boundary. In this chapter we consider the larger family of sets of *Locally Bounded Curvature*, that is of bounded curvature in every bounded open subset of $\mathbb{R}^N$. This local property is completely equivalent to the local bounded curvature in a neighborhood of the boundary. It will be shown that it contains the important family of *sets of positive reach* of H. Federer which correspond to sets for which the square of the distance function d_Ω belongs to $C^{1,1}$ in some neighborhood of their boundary. For compactness theorems the global boundedness condition is relaxed to a local one in the neighborhood of the boundary, which essentially yields the same results. This parallels the smooth case where the information on the smoothness of the boundary lies in a small neighborhood around it. In §2 we recall some basic definitions and results on topologies generated by distance functions. In §3 we introduce and characterize the class of sets of *locally bounded curvature* and show that they include convex and locally convex sets and sets with positive reach. We give a general classification of the singularities of the gradient. We also consider domains for which b_Ω belongs to $W^{2,p}$, $1 \leq p < \infty$ in a neighborhood of the boundary.§4 contains the compactness theorems for sets of locally bounded curvature. Finally §5 studies the continuity of the solution of the generic homogeneous Dirichlet problem in the class of sets for which b_Ω is uniformly locally $W^{2,p}$ and discusses its relationship with other types of continuity results.

11.2 Topologies generated by distance functions

In this section we recall a number of results. They make use of various types of distance functions. Recall that the distance function associated with a non-empty subset A of $\mathbb{R}^N$ ($N \geq 1$, a finite integer) is defined as

$$d_A(x) \min_{y \in A} |y - x|, \quad \forall x \in \mathbb{R}^N .$$

Consider subsets Ω of a bounded open *hold-all* D in $\mathbb{R}^N$ with a Lipschitzian boundary. The results can be extended to unbounded sets by locally restricting to bounded open subsets and using local spaces.

Introduce the following families

$$C_d(\overline{D}) := \{d_\Omega \,:\, \Omega \subset \overline{D},\, \Omega \neq \emptyset\} \tag{11.1}$$

$$C_d^c(D) := \{d_{C\Omega} \,:\, \Omega \text{ open subset of } D\} = \{d_{C\Omega} \,:\, \Omega \text{ open in } D,\, C\Omega \neq \emptyset\} \tag{11.2}$$

since $\overline{D} \neq \mathbb{R}^N$ and

$$C_b(\overline{D}) := \{b_\Omega \,:\, \Omega \subset \overline{D},\, \partial\Omega \neq \emptyset\}, \tag{11.3}$$

where $d_{C\Omega}$ is the *complementary distance function* associated with the set $C\Omega = \{\S \in \mathbb{R}^N \,:\, \curvearrowleft \notin \Omega\}$ and b_Ω is the *oriented boundary distance function* defined as

$$b_\Omega(x) := d_\Omega(x) - d_{C\Omega}(x). \tag{11.4}$$

There is a one-to-one correspondence between the functions d_Ω and b_Ω and the equivalence classes $[\Omega]_d = \{\Omega' \,:\, \overline{\Omega'} = \overline{\Omega}\}$ and $[\Omega]_b = \{\Omega' \,:\, \overline{\Omega'} = \overline{\Omega},\, \partial\Omega' = \partial\Omega\}$ respectively. Note that in general there is no open representative in the equivalence class $[\Omega]_b$, but if it does it is unique and equal to $\int \Omega$. The above families are compact in the space $C^0(\overline{D})$ of continuous functions in $\overline{D}$ and induce complete metric topologies on the associated equivalence classes of sets

$$\rho_H([\Omega_2],[\Omega_1]) = \|d_{\Omega_2} - d_{\Omega_1}\|_{C^0(\overline{D})} \tag{11.5}$$

$$\rho_{H^c}([\Omega_2],[\Omega_1]) = \|d_{C\Omega_\in} - d_{C\Omega_\infty}\|_{C^0(\overline{D})} \tag{11.6}$$

$$\rho_b([\Omega_2],[\Omega_1]) = \|b_{\Omega_2} - b_{\Omega_1}\|_{C^0(\overline{D})}. \tag{11.7}$$

We shall speak of *uniform (or Hausdorff) topologies*. Since the functions d_Ω, $d_{C\Omega}$ and b_Ω are uniformly Lipschitz of constant equal to one, we always have the following pointwise inequalities

$$|\nabla d_\Omega(x)| \le 1, \quad |\nabla d_{C\Omega}(x)| \le 1, \quad \text{and} \quad |\nabla b_\Omega(x)| \le 1, \quad \text{a.e. in } D.$$

So d_Ω, $d_{C\Omega}$ and b_Ω are elements of $W^{1,p}(D)$ for all p, $1 \le p \le \infty$. The sets $C_d(\overline{D})$,$C_d^c(D)$ and $C_b(\overline{D})$ are closed in $W^{1,p}(D)$ for all$p$, $1 \le p < \infty$. Therefore the $W^{1,p}$-norms also induce complete metric topologies on the associated equivalence classes of sets. The characteristic functions of the interior $\int \Omega$, closure $\overline{\Omega}$ and boundary$\partial\Omega$ and the complement $C\Omega$ of Ω are directly related to the gradient ofb_Ω and its positive and negative parts. For almost all $x \in \mathbb{R}^K$,

$$\chi_{\partial\Omega}(x) = 1 - |\nabla b_\Omega(x)| \tag{11.8}$$

$$\chi_{\overline{\Omega}}(x) = 1 - |\nabla d_\Omega(x)| = 1 - |\nabla b_\Omega^+(x)| \tag{11.9}$$

$$\chi_{\overline{C\Omega}}(x) = 1 - |\nabla d_{C\Omega}(x)| = 1 - |\nabla b_\Omega^-(x)| \tag{11.10}$$

$$\chi_{\int\Omega}(x) = |\nabla d_{C\Omega}(x)| = |\nabla b_\Omega^-(x)|, \tag{11.11}$$

since

$$b_\Omega^+(x) = \max\{b_\Omega(x), 0\} = d_\Omega(x) \tag{11.12}$$

$$b_\Omega^-(x) = \max\{-b_\Omega(x), 0\} = d_{C\Omega}(x). \tag{11.13}$$

The maps

$$b_\Omega \mapsto \chi_{\partial\Omega} \,:\, W^{1,p}(D) \to L^p(D) \tag{11.14}$$

$$b_\Omega \mapsto \chi_{\overline{\Omega}} \,:\, W^{1,p}(D) \to L^p(D) \tag{11.15}$$

$$b_\Omega \mapsto \chi_{\overline{C\Omega}} \,:\, W^{1,p}(D) \to L^p(D) \tag{11.16}$$

$$d_{C\Omega} \mapsto \chi_{\int\Omega} \,:\, W^{1,p}(D) \to L^p(D) \tag{11.17}$$

are all uniformly Lipschitz continuous for all p, $1 \le p < \infty$. In view of the above properties the set

$$C_b^0(\overline{D}) = \{b_\Omega \,:\, \Omega \subset \overline{D},\ \partial\Omega \ne \emptyset \text{ and } m(\partial\Omega) = 0\} \tag{11.18}$$

is closed in $W^{1,p}(D)$ for all p, $1 \le p < \infty$ (m, the N-dimensional Lebesgue measure). When $\nabla b_\Omega(x)$ exists at a point $x \notin \partial\Omega$, $|\nabla b_\Omega(x)| = 1$; when it exists at a point $x \in \partial\Omega$, then $\nabla b_\Omega(x) = 0$ except possibly on a subset of $\partial\Omega$

of zero N-dimensional Lebesgue measure. This does not contradict the fact that for a smooth domain $\nabla b_\Omega(x)$ exists and that its norm is equal to one in all points of $\partial\Omega$ since its Lebesque measure is precisely equal to zero. Of course when Ω is such that $b_\Omega = d_{\partial\Omega}$ in $\mathbb{R}^N$, then if $\nabla b_\Omega(x)$ exists at a point $x \in \partial\Omega$, $\nabla b_\Omega(x) = 0$. The identity $b_\Omega = d_{\partial\Omega}$ is verified for submanifolds of codimension greater or equal to two. In general the set

$$\partial^*\Omega = \{x \in \partial\Omega \,:\, \nabla b_\Omega(x) \text{ exists and } \nabla b_\Omega(x) \neq 0\}.$$

has zero Lebesgue measure and is to be compared with the *reduced boundary* in the theory of Caccioppoli sets.

11.2.1 Smoothness of boundary, curvatures, skeletons, convexity.

Smooth domains. In addition to providing new topologies on families of subsets of an open hold-all D, the function b_Ω captures all the geometric properties of the set and its boundary. The smoothness of the boundary is directly related to the smoothness of the distance function in a neighborhood of the boundary. Given an open subset V of $\mathbb{R}^N$ and an integer $k \geq 0$, $C^k(V)$ will be the space of k-times continuously differentiable functions on V; for $0 < \lambda \leq 1$, $C^{0,\lambda}(V)$ will be the space of locally λ-Hölderian continuous functions. $C^{k,\lambda}(V)$ will be the space of k-times differentiable functions on V whose partial derivatives up to order k belong to $C^{0,\lambda}(V)$ By convention we set $C^{k,0} = C^k$.

Note 11.1. Given $h > 0$ and a nonempty subset A in $\mathbb{R}^N$, the *open* (resp. *closed*)*h-tubular neighbourhood* of A is the set

$$U_h(A) := \{x \in \mathbb{R}^N : d_A(x) < h\} \text{ (resp. } A_h := \{x \in \mathbb{R}^N : d_A(x) \leq h\}). \tag{11.19}$$

When ∂A is not empty, then$U_h(\partial A)$ and $(\partial A)_h$ respectively coincide with

$$\{x \in \mathbb{R}^N : |b_A(x)| < h\} \text{ and } \{x \in \mathbb{R}^N : |b_A(x)| \leq h\}. \tag{11.20}$$

The *set of projections* of x onto $\bar{A}$ is the set

$$\Pi_A(x) := \{p \in \bar{A} \,:\, d_A(x) = |p - x|\}. \tag{11.21}$$

When ∂A is not empty, then $\Pi_{\partial A}(x)$ coincides with

$$\{p \in \partial A \,:\, |b_A(x)| = |p - x|\}. \tag{11.22}$$

It is important to observe that in general $(\overline{A})_h = A_h$ and $\overline{U_h(A)} \subset A_h$, but that $\overline{U_h(A)}$ is not necessarily equal to A_h.

Theorem 11.1. *Let Ω be an open subset of $\mathbb{R}^N$ such that $\partial\Omega \neq \emptyset$ and let $x \in \partial\Omega$.*

(i) Let $k \geq 1$ and $0 \leq \lambda \leq 1$. If there exists a neighborhood $U(x)$ of x such that $b_\Omega \in C^{k,\lambda}(U(x))$, then $\partial\Omega \cap U(x)$ lies in an $(N-1)$-submanifold of class $C^{k,\lambda}$ in $\mathbb{R}^N$.

(ii) Let $k \geq 2$ and $0 \leq \lambda \leq 1$, or $(k, \lambda) = (1, 1)$. If there exists a neighborhood $U(x)$ of x such that $\partial\Omega \cap U(x)$ lies in an $(N-1)$-submanifold of class $C^{k,\lambda}$ in $\mathbb{R}^N$, then there exists another neighborhood $U'(x)$ of x such that $b_\Omega \in C^{k,\lambda}(U'(x))$.

The converse of part (i) is generally not true for boundaries which are $C^{1,1-\varepsilon}, 0 < \varepsilon < 1$. Counterexamples can be constructed which show that ∇b_Ω has points of discontinuity in any neighborhood of $\partial\Omega$. This theorem gives an analytical description of the smoothness of submanifolds of codimension one which are $C^{1,1}$ or smoother. Recall that in the smooth case the matrix $D^2 b_\Omega$ is directly related to the curvatures of the boundary of $\partial\Omega$. The eigenvalues of the Hessian matrix on each level set of b_Ω are the $N-1$ principal curvatures of that level set and 0 since $\nabla b(x)$ is an eigenvector. It turns out that the *second fundamental form* associated with the submanifold $\Gamma = \partial\Omega$ of $\mathbb{R}^N$ coincides with the matrix $D^2 b$. Consider a local C^2 representation of $\partial\Omega$ in a neighborhood $U(x)$ of $x \in \partial\Omega$

$$\xi' = (\xi_1, \dots, \xi_{N-1}) \mapsto \phi(\xi') \,:\, A \subset \mathbb{R}^{N-1} \to \Gamma_x = \phi(A) \subset \partial\Omega \subset \mathbb{R}^N$$

with the usual assumptions and $\Gamma_x = \Gamma \cap U(x)$. Define the tangent vectors $(\mathbf{a}_1, \dots \mathbf{a}_{N-1})$ and the unit normal vector $\mathbf{a}_N$

$$\mathbf{a}_\alpha = \frac{\partial \phi}{\partial \xi_\alpha}, \quad \alpha = 1, \dots, N-1, \quad \mathbf{a}_N = -\nabla b_\Omega.$$

The choice of a_N coincides with the inner unit normal to Ω in order to make the principal curvatures of the unit ball all positive. We easily obtain the elements of the *second fundamental form* in term of $D^2 b_\Omega$

$$b_{\alpha\beta} = b_{\beta\alpha} := -\mathbf{a}_\alpha \cdot \frac{\partial}{\partial \beta}\mathbf{a}_N = \mathbf{a}_\alpha \cdot \frac{\partial}{\partial \beta}\nabla b_\Omega = \mathbf{a}_\alpha \cdot D^2 b_\Omega \, \mathbf{a}_\beta$$

where the greek indices ranges from 1 to $N-1$ and the roman indices from 1 to N. In view of the fact that $D^2 b \, \nabla b = 0$, for all $\{\xi_i\}_{i=1}^N$ and $\boldsymbol{\xi}$ of the form $\sum_{i=1}^N \xi_i \, \mathbf{a}_i$

$$D^2 b_\Omega \, \boldsymbol{\xi} = \sum_{i=1}^{N} \xi_i D^2 b_\Omega \, \mathbf{a}_i = \sum_{\alpha=1}^{N-1} \xi_\alpha D^2 b_\Omega \, \mathbf{a}_\alpha$$

and

$$\forall \boldsymbol{\zeta}, \boldsymbol{\xi}, \qquad b_{\alpha\beta} \, \xi_\alpha \, \zeta_\beta = D^2 b_\Omega \, \boldsymbol{\xi} \cdot \boldsymbol{\zeta}$$

and the second fundamental form coincides with the bilinear form generated by $D^2 b_\Omega$. It can also be shown that the *third fundamental form* coincides with the bilinear form generated by $(D^2 b)^2$ and that the mean curvature is equal to $\frac{1}{N-1} \Delta b_\Omega$.

11.2.2 Sets of bounded curvature and convex sets. Outside of the boundary the singularities of the vector field ∇b^2_Ω coincide with the points x where the projection $p_{\partial\Omega}(x)$ on $\partial\Omega$ is not unique. Whenever one exists the other exists and

$$p_{\partial\Omega}(x) = x - b_\Omega(x)\,\nabla b_\Omega(x) = x - \frac{1}{2}\nabla b^2_\Omega(x).$$

In general for an arbitrary set Ω the singularities can be classified as follows

$$\mathrm{Sk}(\Omega) := \{x \in \mathbb{R}^{\ltimes} \;:\; \nabla^{\nvDash}_{\Omega}(\curvearrowleft)\ \nexists\} \tag{11.23}$$

$$\mathrm{C}(\Omega) := \{x \in \mathbb{R}^{\ltimes} \;:\; \nabla^{\nvDash}_{\Omega}(\curvearrowleft)\exists,\ \nabla_{\Omega}(\curvearrowleft)\ \nexists\}. \tag{11.24}$$

The set $\mathrm{Sk}(\Omega)$ is the *total skeleton* of Ω. The second set is necessarily a subset of the boundary $\partial\Omega$. In summary

$$\mathrm{Sing}\,\nabla b_\Omega := \{x \in \mathbb{R}^{\ltimes} \;:\; \nabla_{\Omega}(\curvearrowleft)\ \nexists\} = \mathrm{Sk}(\Omega) \cup \mathrm{C}(\Omega)$$

$$\mathrm{Sk}(\Omega) \cap \partial\Omega = \emptyset, \quad \mathrm{C}(\Omega) \subset \partial\Omega,$$

where $m(\mathrm{Sing}\nabla b_\Omega) = m(\mathrm{Sk}(\Omega)) = m(\mathrm{C}(\Omega)) = 0$. Similarly

$$\mathrm{Sing}\nabla d_{C\Omega} := \{x \in \mathbb{R}^{\ltimes} \;:\; \nabla_{C\Omega}(\curvearrowleft)\ \nexists\} = \mathrm{Sk}_{\int}(\Omega) \cup \mathrm{C}_{\int}(\Omega)$$

where

$$\mathrm{Sk}_{\int}(\Omega) := \{x \in \mathbb{R}^{\ltimes} \;:\; \nabla^{\nvDash}_{C\Omega}(\curvearrowleft)\ \nexists\} \subset \int \Omega$$

$$\mathrm{C}_{\int}(\Omega) := \{x \in \mathbb{R}^{\ltimes} \;:\; \nabla^{\nvDash}_{C\Omega}(\curvearrowleft)\exists,\ \nabla_{C\Omega}(\curvearrowleft)\ \nexists\} \subset \partial\Omega$$

and $m(\mathrm{Sing}\nabla d_{C\Omega}) = m(\mathrm{Sk}_{\int}(\Omega)) = m(\mathrm{C}_{\int}(\Omega)) = 0$. Also

$$\mathrm{Sing}\nabla d_\Omega := \{x \in \mathbb{R}^{\ltimes} \;:\; \nabla_{\Omega}(\curvearrowleft)\ \nexists\} = \mathrm{Sk}_{\mathrm{ext}}(\Omega) \cup \mathrm{C}_{\mathrm{Ext}}(\Omega)$$

where

$$\mathrm{Sk}_{\mathrm{Ext}}(\Omega) := \{x \in \mathbb{R}^{\ltimes} \;:\; \nabla^{\nvDash}_{\Omega}(\curvearrowleft)\ \nexists\} \subset \int C\Omega$$

$$\mathrm{C}_{Ext}(\Omega) := \{x \in \mathbb{R}^{\ltimes} \;:\; \nabla^{\nvDash}_{\Omega}(\curvearrowleft)\exists,\ \nabla_{\Omega}(\curvearrowleft)\ \nexists\} \subset \partial\Omega,$$

and $m(\mathrm{Sing}\nabla d_\Omega) = m(\mathrm{Sk}_{Ext}(\Omega)) = m(\mathrm{C}_{Ext}(\Omega)) = 0$.

Clearly

$$\mathrm{Sk}(\Omega) = \mathrm{Sk}_{Int}(\Omega) \cup \mathrm{Sk}_{Ext}(\Omega) \subset \mathbb{R}^{\ltimes} \backslash \partial\Omega \quad \mathrm{C}(\Omega) \subset \mathrm{C}_{\mathrm{I}\ltimes\approx}(\Omega) \cup \mathrm{C}_{\mathrm{E}\curvearrowleft\approx} \subset \partial\Omega.$$

Notice that the set of cracks $C(\Omega)$ is a subset of the boundary $\partial\Omega$ with zero N-dimensional Lebesgue measure while the boundary itself may have non zero measure. Hence it would be quite natural to ask under what circumstances $C(\Omega)$ and $\mathrm{Sk}(\Omega)$ have finite $(N-1)$-dimensional Hausdorff measures.

Given a fixed open *hold-all* D in $\mathbb{R}^{\ltimes}$, attention is focused on subsets Ω of $\overline{D}$ such that

$$\nabla b_\Omega \ (\text{resp. } \nabla d_\Omega, \ \nabla d_{C\Omega}) \ \in BV(D)^N \qquad (11.25)$$

that is, such that the elements of the Hessian matrix of the function b_Ω (resp. d_Ω, $d_{C\Omega}$) are bounded measures on D

$$D^2 b_\Omega \ (\text{resp. } D^2 d_\Omega, \ D^2 d_{C\Omega}) \ \in M^1(D)^{N \times N}. \qquad (11.26)$$

$BV(D)$ is the space of functions in $L^1(D)$ for which ∇f belongs to $M^1(D)^N$ and $M^1(D)$ is the space of bounded measures on D.

Definition 11.1. *Let Ω be a subset of $\overline{D}$ for some open hold-all D of $\mathbb{R}^\kappa$.*

(i) If $\partial\Omega \neq \emptyset$, we say that Ω is of bounded curvature *in D if*

$$\nabla b_\Omega \in BV(D)^N. \qquad (11.27)$$

(ii) If $\Omega \neq \emptyset$ (resp. $C\Omega \neq \emptyset$), we say that Ω is of bounded exterior *(resp. interior) curvature in D if*

$$\nabla d_\Omega \in BV(D)^N (\text{resp. } \nabla d_{C\Omega} \in BV(D)^N). \qquad (11.28)$$

In this paper we consider sets with the local property.

Definition 11.2. *Let Ω be a subset of $\mathbb{R}^\kappa$.*

(i) The set Ω, $\partial\Omega \neq \emptyset$, is said to be locally of bounded curvature *if*

$$\nabla b_\Omega \in BV_{\text{loc}}(\mathbb{R}^\kappa)^N. \qquad (11.29)$$

(ii) The set Ω, $\Omega \neq \emptyset$ (resp. $C\Omega \neq \emptyset$), is said to be locally of bounded exterior *(resp. interior) curvature if*

$$\nabla d_\Omega \in BV_{\text{loc}}(\mathbb{R}^\kappa)^N \ (\text{resp. } c_\Omega \in BV_{\text{loc}}(\mathbb{R}^\kappa)^N). \qquad (11.30)$$

Theorem 11.2. *(i) Given a subset Ω of $\mathbb{R}^\kappa$ such that $\Omega \neq \emptyset$,*

$$\overline{\Omega} \text{ is convex iff } d_\Omega \text{ is convex.}$$

(ii) Given a subset Ω of $\mathbb{R}^\kappa$ such that $\partial\Omega \neq \emptyset$,

$$\overline{\Omega} \text{ is convex iff } b_{\overline{\Omega}} \text{ is convex.}$$

(iii) In both cases when $\overline{\Omega}$ is convex, then $\nabla b_{\overline{\Omega}}$ and ∇d_Ω belong to $BV_{\text{loc}}(\mathbb{R}^n)^N$.

There is also another interesting result which says that the convex envelope of the distance function of a set is the distance function of its closed convex hull. We believe that this result is new.

Theorem 11.3. *Let A be a nonempty subset of $\mathbb{R}^\kappa$,*

(i) The Fenchel transform

$$d_A^*(x^*) = \sup_{x \in \mathbf{R}^{\mathrm{K}}} x^* \cdot x - d_A(x)$$

of d_A is given by

$$d_A^*(x^*) = \sigma_A(x^*) + I_{\overline{B(0,1)}}(x^*) \tag{11.31}$$

where $\sigma_A(x^*)$ is the support function of A and

$$I_{\overline{B(0,1)}}(x^*) = \{ +\infty \quad x \notin \overline{B(0,1)} 0 \quad x \in \overline{B(0,1)}$$

is the indicator function of the closed unit ball $\overline{B(0,1)}$ at the origin.

(ii) The Fenchel transform of d_A^* is given by

$$d_A^{**}(x^{**}) = d_{\mathrm{co}\, A}(x^{**}) = d_{\overline{\mathrm{co}}\, A}(x^{**}). \tag{11.32}$$

In particular $d_{\mathrm{co}\, A}(x)$ is the convex envelope of $d_A(x)$.

Proof. (i) From the definition

$$\begin{aligned}
d_A^*(x^*) &= \sup_{x \in \mathbf{R}^{\mathrm{K}}} x^* \cdot x - d_A(x) \\
&= \sup_{x \in \mathbf{R}^{\mathrm{K}}} \left[x^* \cdot x - \inf_{p \in A} |x - p| \right] \\
&= \sup_{x \in \mathbf{R}^{\mathrm{K}}} \sup_{p \in A} x^* \cdot x - |x - p| \\
&= \sup_{p \in A} \sup_{x \in \mathbf{R}^{\mathrm{K}}} x^* \cdot x - |x - p| \\
&= \sup_{p \in A} \sup_{x \in \mathbf{R}^{\mathrm{K}}} x^* \cdot p + x^* \cdot (x - p) - |x - p| \\
&= \sup_{p \in A} x^* \cdot p + \sup_{x \in \mathbf{R}^{\mathrm{K}}} \left[x^* \cdot (x - p) - |x - p| \right] \\
&= \sup_{p \in A} x^* \cdot p + \sup_{x \in \mathbf{R}^{\mathrm{K}}} \left[x^* \cdot x - |x| \right] \\
&= \sigma_A(x^*) + I_{\overline{B(0,1)}}(x^*)
\end{aligned}$$

(ii) We now use the property that $\sigma_A = \sigma_{\mathrm{co}\, A}$

$$\begin{aligned}
d_A^{**}(x^{**}) &= \sup_{x^* \in \mathbf{R}^{\mathrm{K}}} x^{**} \cdot x^* - d_A^*(x^*) \\
&= \sup_{x^* \in \mathbf{R}^{\mathrm{K}}} x^{**} \cdot x^* - \sigma_A(x^*) - I_{\overline{B(0,1)}}(x^*) \\
&= \sup_{|x^*| \le 1} x^{**} \cdot x^* - \sigma_A(x^*) \\
&= \sup_{|x^*| \le 1} \left[x^{**} \cdot x^* - \sup_{x \in \mathrm{co}\, A} x^* \cdot x \right]
\end{aligned}$$

$$= \sup_{|x^*|\leq 1} \inf_{x \in \mathrm{co}\, A} (x^{**} - x) \cdot x^*$$

$$\leq \sup_{|x^*|\leq 1} (x^{**} - p(x^{**})) \cdot x^* \leq |x^{**} - p(x^{**})| = d_{\mathrm{co}\, A}(x^{**}).$$

But in fact we have a saddle point. If $x^{**} \in \overline{\mathrm{co}}\, A$, pick $x^* = 0$ and

$$d_A^{**}(x^{**}) = \sup_{|x^*|\leq 1} \inf_{x \in \mathrm{co}\, A} (x^{**} - x) \cdot x^*$$

$$\geq 0 = d_{\mathrm{co}\, A}(x^{**}).$$

For $x^{**} \notin \overline{\mathrm{co}}\, A$ rewrite the functional as

$$(x^{**} - x) \cdot x^* = (x^{**} - p(x^{**})) \cdot x^* + (p(x^{**}) - x) \cdot x^*.$$

Since $p(x^{**})$ is the minimizing point of the functional $|x - x^{**}|^2$ over the closed convex set $\overline{\mathrm{co}}\, A$, it is completely characterized by the variational inequality

$$(p(x^{**}) - x^{**}) \cdot (x - p(x^{**})) \geq 0, \quad \forall x \in \overline{\mathrm{co}}\, A. \tag{11.33}$$

By choosing the following special x^*

$$x^* = \frac{x^{**} - p(x^{**})}{|x^{**} - p(x^{**})|}$$

we get

$$d_A^{**}(x^{**}) = \sup_{|x^*|\leq 1} \inf_{x \in \mathrm{co}\, A} (x^{**} - x) \cdot x^*$$

$$\geq |x^{**} - p(x^{**})| + (p(x^{**}) - x) \cdot \frac{x^{**} - p(x^{**})}{|x^{**} - p(x^{**})|}$$

$$\geq |x^{**} - p(x^{**})| = d_{\mathrm{co}\, A}(x^{**})$$

in view of 11.33.

Therefore

$$\sup_{|x^*|\leq 1} \inf_{x \in \mathrm{co}\, A} (x^{**} - x) \cdot x^* = d_{\mathrm{co}\, A}(x^{**}) = \inf_{x \in \mathrm{co}\, A} \sup_{|x^*|\leq 1} (x^{**} - x) \cdot x^*.$$

Finally, from the previous theorem, $d_{\mathrm{co}\, A}$ is a convex function which coincides with the convex envelope since it is the bidual of d_A.

Therefore in view of the convexity of the distance function of a closed convex set, nonempty sets with convex closure are locally of bounded curvature. When $\overline{\Omega}$ is closed and convex and $\partial\overline{\Omega}$ is of class C^2, then for any $X \in \partial\overline{\Omega}$, there exists a strictly convex neighborhood $N(X)$ of X such that

$$b := b_{\overline{\Omega}} \in C^2(N(X)) \tag{11.34}$$

$$\forall x \in N(X),\ \forall \xi \in \mathbb{R}^N, \quad D^2 b(x)\xi \cdot \xi \geq 0, \tag{11.35}$$

or since $D^2 b(x)\, \nabla b(x) = 0$ in $N(X)$

$$\forall x \in N(X), \forall \xi \in \mathbb{R}^{\kappa} \text{ such that } \xi \cdot \nabla(\frown) = \nvdash, \quad \mathbb{D}^{\nvdash}(\frown)\xi \cdot \xi \geq \nvdash, \tag{11.36}$$

This is related to the notion of *elliptic boundary* in shell theory:

$$\exists c > 0, \forall x \in \partial\Omega, \forall \xi \in \mathbb{R}^{\kappa} \text{ such that } \xi \cdot \nabla(\frown) = \nvdash, \quad \mathbb{D}^{\nvdash}(\frown)\xi \cdot \xi \geq |\xi|^{\nvdash}.$$

All this motivates the introduction of the following natural notions.

Definition 11.3. *Let Ω be a closed subset of $\mathbb{R}^{\kappa}$ such that $\partial\Omega \neq \emptyset$.*

(i) *The set Ω is locally convex*(resp. locally strictly convex) if for each $X \in \partial\Omega$ there exists a strictly convex neighborhood $N(X)$ of X such that

$$b_\Omega \text{ is convex (resp. strictly convex) in } N(X).$$

(ii) The set Ω is *semiconvex* if

$$\exists \alpha \geq 0, \quad b_\Omega(x) + \alpha|x|^2 \text{ is convex in } \mathbb{R}^{\kappa}.$$

(iii) The set Ω is *locally semiconvex* if for each $X \in \partial\Omega$ there exists a strictly convex neighborhood $N(X)$ of X and

$$\exists \alpha \geq 0, \quad b_\Omega(x) + \alpha|x|^2 \text{ is convex in } N(X).$$

Remark 11.1. When Ω has a compact C^2 boundary, $D^2 b_\Omega$ is bounded in a bounded neighborhood of $\partial\Omega$ and Ω is necessarily locally semiconvex. When Ω is a locally semiconvex set, for each $X \in \partial\Omega$ there exists a strictly convex neighborhood $N(X)$ of X such that $\nabla b_\Omega \in BV(N(X))^N$. If, in addition, $\partial\Omega$ is compact, then there exists $h > 0$ such that $\nabla b_\Omega \in BV(U_h(\partial\Omega))^N$. □

Remark 11.2. If we fix a constant $\beta > 0$ and consider all the subsets of D which are semiconvex with constant $0 \leq \alpha \leq \beta$, then this set is closed for the uniform and the $W^{1,p}$, $1 \leq p < \infty$, topologies. □

It turns out that the local properties of Definition 11.2 only need to hold in some neighborhood of the boundary of the set.

We say that a function f defined on an open subset U of $\mathbb{R}^{\kappa}$ is *locally convex* if for each $x \in U$ there exists an open ball $B(x, \rho)$ of radius $\rho > 0$ in U where$f$ is convex.

Lemma 11.1. *Given a subset A, $A \neq \emptyset$, of $\mathbb{R}^{\kappa}$ and $h > 0$, the function*

$$k(x) := \{\, |\, x|^2 - 2\,h\,d_A(x), \quad d_A(x) \geq h|x|^2 - d_A^2(x) - h^2, \quad d_A(x) < h$$

is convex in $\mathbb{R}^{\kappa}$. In particular

$$k_{A,h}(x) := \frac{|x|^2}{2\,h} - d_A(x)$$

is locally convex in $\mathbb{R}^{\kappa} \backslash \mathbb{A}_{\approx}$ and

$$|x|^2 - d_A^2(x)$$

is locally convex in $U_h(A)$.

Proof. For all $p \in A$ define the convex function

$$\ell_p(x) = \{ |\, x - p| - h, \quad |x-p| \geq h0, \quad |x - p| < h.$$

Since ℓ_p is nonnegative, then ℓ_p^2 is convex and

$$\ell_p^2(x) = \{ |\, x|^2 - 2\,h\,|x-p| + |p|^2 + h^2 - 2\,x \cdot p, \quad |x-p| \geq h0, \quad |x-p| < h.$$

By subtracting the constant term $|p|^2 + h^2$ and the linear term $-2\,x \cdot p$ from ℓ_p^2, we get the new convex function

$$m_p(x) = \{ |\, x|^2 - 2\,h\,|x-p|, \quad |x-p| \geq h|x|^2 - |p-x|^2 - h^2, \quad |x-p| < h.$$

Then the function

$$k(x) := \sup_{p \in A} m_p(x)$$

is finite for each $x \in \mathbb{R}^N$, convex in x, and

$$k(x) = \{ |\, x|^2 - 2\,h\,d_A(x), \quad d_A(x) \geq h|x|^2 - d_A^2(x) - h^2, \quad d_A(x) < h.$$

If x is such that $d_A(x) \geq h$, then for all $p \in A$, $|x-p| \geq d_A(x) \geq h$ and $k(x) = |x|^2 - 2\,h\,d_A(x)$. If $d_A(x) < h$, then there exists $p \in \overline{A}$ such that $|x - p| < h$ and

$$\inf_{p \in A\, |x-p|<h} |x - p| = d_A(x).$$

Then either for all $p \in A$, $|x - p| < h$ and $k(x) = |x|^2 - h^2 - d_A^2(x)$ or there exists $p \in \overline{A}$ such that $|x - p| \geq h$,

$$\inf_{p \in A\, |x-p| \geq h} |p - x| \geq h \quad \Rightarrow$$

$$|x|^2 - 2\,h \inf_{p \in A\, |x-p| \geq h} |x-p| \leq |x|^2 - 2\,h^2 \leq |x|^2 - h^2 - d_A^2(x)$$

and $k(x) = |x|^2 - d_A^2(x) - h^2$. We recover the result of P.L. Lions by observing that the restriction of k to$\mathbb{R}^N \setminus A_{\approx}$ is locally convex. In addition the function $|x|^2 - d_A^2(x)$ is locally convex in $U_h(A)$.

This lemma has far reaching consequences.

Theorem 11.4. *Let A be a nonempty subset of $\mathbb{R}^N$.*

(i) The function $f_A(x) = \frac{1}{2}\,[|x|^2 - d_A^2(x)]$ is convex in $\mathbb{R}^N$ and $\nabla d_A^2 \in BV_{\text{loc}}(\mathbb{R}^N)^N$. For all x and y in $\mathbb{R}^N$

$$\forall p \in \Pi_A(x), \quad \frac{1}{2}[|y|^2 - d_A^2(y)] \geq \frac{1}{2}[|x|^2 - d_A^2(x)] + p \cdot (y - x) \quad (11.37)$$

or equivalently

$$\forall p \in \Pi_A(x), \quad d_A^2(y) - d_A^2(x) - 2\,(x-p) \cdot (y-x) \leq |x - y|^2. \quad (11.38)$$

The projection $p_A(x)$ onto $\overline{A}$ is the gradient of f_A whenever one or the other exists and

$$p_A(x) = \frac{1}{2}\nabla[|x|^2 - d_A^2(x)]. \tag{11.39}$$

For all x and y in $\mathbb{R}^N$

$$\forall p(x) \in \Pi_A(x), \forall p(y) \in \Pi_A(y), \quad (p(y) - p(x)) \cdot (y - x) \geq 0. \tag{11.40}$$

The function $d_A^2(x)$ is the difference of two convex functions

$$d_A^2(x) = |x|^2 - [|x|^2 - d_A^2(x)]. \tag{11.41}$$

(ii) *$\nabla d_A \in BV_{\text{loc}}(\mathbb{R}^N \backslash \overline{A})^N$. More precisely for all $x \in \mathbb{R}^N \backslash \overline{A}$ there exists $\rho > 0$, $0 < 3\rho < d_A(x)$, such that $k_{A,\rho}$ is convex in $B(x, 2\rho)$ and hence $\nabla d_A \in BV(B(x,\rho))^N$.*

(iii) *A is locally of bounded exterior curvature if and only if*

$$\forall x \in \partial A, \exists \rho > 0 \text{ such that } \nabla d_A \in BV(B(x,\rho))^N. \tag{11.42}$$

(iv) *Given $h > 0$ and a nonempty bounded open subset V of $\mathbb{R}^N$, there exist constants $c_0(V) > 0$ and $c_1(V) > 0$ such that*

$$\|D^2 d_A\|_{M^1(V)} \leq \|D^2 d_A\|_{M^1(U_h(p_A(V)))} + c_0(V) + \frac{c_1(V)}{h}. \tag{11.43}$$

(v) $$\exists c \geq 0, \quad f_c(x) = c\,|x|^2 - d_A(x) \text{ is convex in } \mathbb{R}^N$$

if and only if

$$\exists h > 0, \exists c \geq 0, \quad f_c(x) = c\,|x|^2 - d_A(x)$$

is locally convex in $U_h(A)$.

(vi) *Given a subset A of $\mathbb{R}^N$ such that $\emptyset \neq \overline{A} \neq \mathbb{R}^N$,*

$$\nexists c \geq 0, \quad f_c(x) = c\,|x|^2 - d_A(x) \text{ be convex in } \mathbb{R}^N.$$

It is important to emphasize that local bounded exterior curvature is equivalent to the same property in an arbitrarily small neighborhood of the boundary of A. Finally it shows that for non trivial sets A, $\emptyset \neq \overline{A} \neq \mathbb{R}^N$, $-d_A$ is never convex or semiconvex even if $\overline{A}$ is convex. Theorem 11.4 has its analogue for b_Ω since $b_\Omega^2 = d_{\partial\Omega}^2$.

Theorem 11.5. *Let Ω, $\partial\Omega \neq \emptyset$, be a subset of $\mathbb{R}^N$.*

(i) *The function $f_{\partial\Omega}(x) = \frac{1}{2}[|x|^2 - b_\Omega^2(x)]$ is convex in $\mathbb{R}^N$ and $\nabla b_\Omega^2 \in BV_{\text{loc}}(\mathbb{R}^N)^N$. For all x and y in $\mathbb{R}^N$*

$$\forall p \in \Pi_{\partial\Omega}(x), \quad \frac{1}{2}[|y|^2 - b_\Omega^2(y)] \geq \frac{1}{2}[|x|^2 - b_\Omega^2(x)] + p \cdot (y - x) \tag{11.44}$$

or equivalently

$$\forall p \in \Pi_{\partial\Omega}(x), \quad b_\Omega^2(y) - b_\Omega^2(x) - 2\,(x - p)\cdot(y - x) \le |x - y|^2. \quad (11.45)$$

The projection $p_{\partial\Omega}(x)$ onto $\partial\Omega$ is the gradient of $f_{\partial\Omega}$ whenever one or the other exists and

$$p_{\partial\Omega}(x) = \frac{1}{2}\nabla[|x|^2 - b_\Omega^2(x)]. \quad (11.46)$$

For all x and y in $\mathbb{R}^N$

$$\forall p(x) \in \Pi_{\partial\Omega}(x), \forall p(y) \in \Pi_{\partial\Omega}(y), \quad (p(y) - p(x))\cdot(y - x) \ge 0. \quad (11.47)$$

The function $b_\Omega^2(x)$ is the difference of two convex functions

$$b_\Omega^2(x) = |x|^2 - [|x|^2 - b_\Omega^2(x)]. \quad (11.48)$$

(ii) $\nabla b_\Omega \in BV_{\text{loc}}(\mathbb{R}^N \backslash \partial\Omega)^N$. More precisely for all $x \in \mathbb{R}^N \backslash \partial\Omega$ there exists $\rho > 0$, $0 < 3\rho < d_{\partial\Omega}(x)$, such that

$$b_\Omega = k_{C\Omega,\rho} - k_{\Omega,\rho} \quad (11.49)$$

where $k_{C\Omega,\rho}$ and $k_{\Omega,\rho}$ are convex in $B(x, 2\rho)$ and hence

$$\nabla b_\Omega \in BV(B(x,\rho))^N.$$

(iii) Ω is locally of bounded curvature if and only if

$$\forall x \in \partial\Omega, \exists \rho > 0 \text{ such that } \nabla b_\Omega \in BV(B(x,\rho))^N. \quad (11.50)$$

(iv) Given $h > 0$ and a nonempty bounded open subset V of $\mathbb{R}^N$, there exist constants $c_0(V) > 0$ and $c_1(V) > 0$ such that

$$\|D^2 b_\Omega\|_{M^1(V)} \le \|D^2 b_\Omega\|_{M^1(U_h(p_{\partial\Omega}(V)))} + c_0(V) + \frac{c_1(V)}{h}. \quad (11.51)$$

Proof. Proof of Theorem 11.4 (i) From Lemma 11.1 for all $h > 0$, $|x|^2 - \nabla d_A^2$ is locally convex. Therefore letting h go to infinity $|x|^2 - \nabla d_A^2$ is convex in $\mathbb{R}^N$ and hence $\nabla d_A^2 \in \text{BV}_{\text{loc}}(\mathbb{R}^N)^N$. Inequality 11.37 is verified when d_A is C^2. In general it follows directly from the inequality

$$\forall x \text{ and } y \in \mathbb{R}^N, \forall p(x) \in \Pi_A(x), \quad d_A(y) \le |p(x) - y|$$

since

$$d_A^2(y) - |y|^2 \le |p(x) - x + x - y|^2 - |y|^2$$

$$d_A^2(y) - |y|^2 \le |p(x) - x|^2 + |x - y|^2 + 2\,(p(x) - x)\cdot(x - y) - |y|^2$$

$$d_A^2(y) - |y|^2 \le |p(x) - x|^2 - |x|^2 + 2\,(p(x) - x)\cdot(x - y) + |x - y|^2 + |x|^2 - |y|^2$$

$$-2 f_A(y) \le -2 f_A(x) + 2\,p(x)\cdot(x - y) - 2\,x\cdot(x - y) + |x - y|^2 + |x|^2 - |y|^2$$

$$-2 f_A(y) \le -2 f_A(x) - 2\,p(x)\cdot(y - x) \;\Rightarrow\; f_A(y) \ge f_A(x) + 2\,p(x)\cdot(y - x).$$

Inequality 11.38 is 11.37 rewritten. Inequality 11.40 follows by adding inequality 11.37 to the same inequality with x and y permuted.

(ii) For any point x in $\mathbb{R}^N \backslash \overline{A}$, there exists ρ, $0 < 3\,\rho < d_A(x)$ such that the open ball $B(x, 2\,\rho)$ is contained in $\mathbb{R}^N \backslash A_\rho$ where $k_{A,\rho}$ is locally convex by Lemma 11.1. Hence $k_{A,\rho}$ is convex in $B(x, 2\,\rho)$ and $\nabla k_{A,\rho}$ and a fortiori∇d_A belong to $\mathrm{BV}\,(B(x,\rho))^N$.

(iii)For a set of locally bounded exterior curvature property 11.2.1 is verified by definition. Conversely, any open subset $V \subset\subset \mathbb{R}^N$ can be covered by a finite number of open balls of the form $B(x,\rho)$ and it is sufficient to establish that for each x there exists $\rho > 0$ such that$\nabla d_A \in \mathrm{BV}\,(B(x,\rho))^N$. This is true in $\int A$ where $\nabla d_A = 0$ and in∂A by assumption. It is also true for any point x in $\mathbb{R}^N \backslash \overline{A}$ by part (i).Therefore ∇d_A belongs to$\mathrm{BV}_{\mathrm{loc}}(\mathbb{R}^N)^N$ and A is of locally bounded curvature.

(iv)By definition for all $x \in V\backslash[p_A(V)]_{h/2}$, $d_A(x) = d_{p_A(V)}(x) > h/2$, $V\backslash[p_A(V)]_{h/2} \subset V\backslash A_{h/2}$, and

$$V \subset U_h(p_A(V)) \cup V\backslash U_h(p_A(V)) \subset$$

$$U_h(p_A(V)) \cup V\backslash[p_A(V)]_{h/2} = U_h(p_A(V)) \cup V\backslash A_{h/2}.$$

So it is sufficient to estimate the second term on the right-hand side of the following inequality

$$\|D^2 d_A\|_{M^1(V)} \leq \|D^2 d_A\|_{M^1(U_h(p_A(V)))} + \|D^2 d_A\|_{M^1(V\backslash A_{h/2})}.$$

By Lemma 11.1 the function $k = k_{A,h/2}$ is locally convex in $\mathbb{R}^N \backslash \mathbb{A}_{\approx/\not\models}$.The set $\overline{V\backslash A_{h/2}}$ is compact. For each $x \in \mathbb{R}^N \backslash \mathbb{A}_{\approx/\not\models}$,there exists $r_x > 0$ such that

$$B_x := \{y \in \mathbb{R}^N \;:\; |\frown - \frown| < \smallsetminus\frown\} \subset \mathbb{R}^N \backslash \mathbb{A}_{\approx/\not\models}$$

and the family $\{B_x\}$ is an open cover of $\overline{V\backslash A_{h/2}}$. There exists a finite subcover $B_n = B_{x_n}$, $1 \leq n \leq m$, of $\overline{V\backslash A_{h/2}}$ and it is sufficient to prove the result on each B_n. Define for fixed i, j, and $\xi = (\xi_1, \ldots, \xi_N)$, $|\xi| = 1$, the functional

$$L_\xi(\varphi) = \sum_{i,j=1}^{N} L_{ij}(\varphi), \quad L_{ij}(\varphi) = \int_{\mathbb{R}^N} k\,\partial^2_{ij}\varphi\,\xi_i\xi_j\,dx, \quad \forall\varphi \in C_c^\infty(\mathbb{R}^N).$$

Since k is convex in B_n, $L_\xi(\varphi) \geq 0$ for all $\varphi \geq 0$ in$C_c^\infty(B_n)$ and theL_ξ and the L_{ij}'s are Radon measures on $C_c(B_n)$. The constant of continuity can be evaluated as through the following construction. Letρ be a smooth function with compact support such that

$$\rho = 1 \text{ in } \overline{V}, \quad 0 \leq \rho \leq 1 \text{ in } \mathbb{R}^N.$$

Then the continuity constant can be chosen as $c_\xi = L_\xi(\rho)$. Since ρ has compact support

$$L_\xi(\rho) = \sum_{i,j=1}^{N} \int_{\mathbf{R}^N} k\, \partial_{ij}^2 \rho\, \xi_i \xi_j \, dx = - \sum_{i,j=1}^{N} \int_{\mathbf{R}^N} \partial_i k\, \partial_j \rho\, \xi_i \xi_j \, dx.$$

Moreover

$$\partial_i k = \frac{4}{h} x_i - \partial_i d_A \quad \Rightarrow \quad |\nabla k| \leq \frac{4}{h} c_V + 1$$

where

$$c_V = \sup_{x \in V} |x|$$

and finally

$$|L_\xi(\rho)| \leq (\frac{4}{h} c_V + 1) \int_{\mathbf{R}^N} |\nabla \rho| \, dx$$

where the constant is of the form $c_0(V) + c_1(V)/h$. Thus for all n and all pairs ij

$$\sup_{\varphi \in C_c(B_n)} |L_\xi(\varphi)| \leq 3\,(c_0(V) + \frac{c_1(V)}{h})\, \|\varphi\|_{C(\overline{B}_n)}$$

and necessarily

$$\sup_{\varphi \in C_c(V \setminus A_{h/2})} |L_\xi(\varphi)| \leq 3\,(c_0(V) + \frac{c_1(V)}{h})\, \|\varphi\|_{C(V \setminus A_{h/2})},$$

where the constants are independent of A.

(v) If f_c is convex in$\mathbb{R}^N$, it is locally convex in $U_h(A)$. Conversely assume that there exists $h > 0$ and $c \geq 0$ such that f_c is locally convex in $U_h(A)$, From part (i)

$$\frac{1}{h} |x|^2 - d_A(x)$$

is locally convex in $\mathbb{R}^N \setminus A_{h/2}$. Therefore for $\bar{c} = \max\{c, 1/h\}$ the function $f_{\bar{c}}$ is locally convex in $\mathbb{R}^N$ and hence convex.

(vi)Assume the existence of a $c \geq 0$ for wich f_c is convex in $\mathbb{R}^N$. For each $y \in \bar{A}$ the function

$$x \mapsto F_c(x, y) = c\,|x - y|^2 - d_A(x)$$

is also convex since it differs from $f_c(x)$ by the linear term

$$x \mapsto c\,(|y|^2 - 2\,x \cdot y).$$

Since $\emptyset \neq \overline{A} \neq \mathbb{R}^N$, there exist $x \in C\overline{A}$ and $p \in \bar{A}$ such that

$$0 < d_A(x) = |x - p| \leq \frac{1}{2c}.$$

For any $t > 0$, define$x_t = p - t\,(x - p)$ and $\lambda = t/(1 + t) \in]0, 1[$ and observe that

$$x_\lambda := \lambda\, x + (1 - \lambda)\, x_t = p \text{ and } F_c(x_\lambda, p) = 0.$$

But

$$\begin{aligned}
&\lambda F_c(x,p) + (1-\lambda) F_c(x_t,p) \\
&= \frac{t}{1+t} [c\,|x-p|^2 - d_A(x)] + \frac{1}{1+t} [c\,|x_t-p|^2 - d_A(x_t)] \\
&\le \frac{t}{1+t} [c\,d_A(x)^2 - d_A(x) + c\,t d_A(x)^2] \\
&\le \frac{t}{1+t} d_A(x) [(1+t)\,c\,d_A(x) - 1] \\
&\le \frac{t}{1+t} d_A(x) [(1+t)\frac{1}{2} - 1] = \frac{d_A(x)\,t}{1+t} \frac{t-1}{2},
\end{aligned}$$

since by construction $d_A(x) > 0$ and $c\,d_A(x) < 1/2$. Therefore for t, $0 < t < 1$, the above quantity is strictly negative and we have constructed two points x and x_t and a λ, $0 < \lambda < 1$, such that

$$F_c(\lambda\, x + (1-\lambda)\, x_t, p) = 0 > \lambda\, F_c(x,p) + (1-\lambda)\, F_c(x_t,p).$$

This contradicts the convexity of the function $x \mapsto F_c(x,p)$ and a fortiori of the function f_c.

Proof. Proof of Theorem 11.5 (i) Apply Theorem 11.4 to $d_{\partial\Omega} = |b_\Omega|$ and use the fact that $b_\Omega = d_\Omega$ in $\mathbb{R}^K \backslash \overline{\Omega}$ and $b_\Omega = d_{C\Omega}$ in $\mathbb{R}^K \backslash \overline{C\Omega}$ to conclude that $\nabla b_\Omega \in \mathrm{BV}_{\mathrm{loc}}(C\partial\Omega)^N$ since

$$C\partial\Omega = C[\overline{\Omega} \cap \overline{C\Omega}] = C\overline{\Omega} \cup C\overline{C\Omega}$$

(ii) and (iii) are obvious.

All C^2 domains with a compact boundary belong to all the categories of Definitions 11.1 and 11.2. The norms $\|D^2 b_\Omega\|_{M^1(U_h(\partial\Omega))}$,$\|D^2 d_{C\Omega}\|_{M^1(U_h(C\Omega))}$,and $\|D^2 d_\Omega\|_{M^1(U_h(\Omega))}$are all decreasing as h goes to zero.The limit is particularly interesting since it singles out the behaviour of the gradient in a shrinking neighborhood of the boundary $\partial\Omega$.

Exercise 11.1. If $\Omega \subset R^N$ has a compact boundary which is a C^2 $(N-1)$-submanifold of $\mathbb{R}^N$, then

$$\lim_{h\searrow 0} \|D^2 b_\Omega\|_{M^1(U_h(\partial\Omega))} = 0.$$

Exercise 11.2. Let $\Omega = \{x_i\}_{i=1}^I$ be I distinct points in R^N. Then $\partial\Omega = \Omega$ and

$$\lim_{h\searrow 0} \|D^2 b_\Omega\|_{M^1(U_h(\partial\Omega))} = \{2\ I - 1, N = 10, N \ge 2.$$

Exercise 11.3. Let Ω be a line in R^N of length $L > 0$, then $\partial\Omega = \Omega$ and

$$\lim_{h\searrow 0} \|D^2 b_\Omega\|_{M^1(U_h(\partial\Omega))} = \{2\ L, N = 20, N \ge 3.$$

Exercise 11.4. Let $N = 2$. For the finite square and the ball of finite radius

$$\lim_{h \searrow 0} \|\Delta d_\Omega\|_{M^1(U_h(\partial\Omega))} = \mathcal{H}^\infty(\partial\Omega),$$

where $\mathcal{H}^\infty$ is the one-dimensional Hausdorff measure.

Also by using $U_h(\Omega)$ andΔb_Ω, we can extract information about the skeleton ofΩ.

Exercise 11.5. Let Ω be the unit square in R^2, then

$$\lim_{h \searrow 0} \|\Delta b_\Omega\|_{M^1(U_h(\Omega))} = \frac{\sqrt{2}}{2}\mathcal{H}^\infty(\mathrm{Sk}(\Omega)),$$

where $\mathcal{H}^\infty$ is the one-dimensional Hausdorff measure and $\mathrm{Sk}(\Omega) = \mathrm{Sk}_{\mathrm{int}}(\Omega)$ is the skeleton of Ω made up of the two interior diagonals . It seems that in general

$$\lim_{h \searrow 0} \|\Delta b_\Omega\|_{M^1(U_h(\Omega))} = \int_{\mathrm{Sk}(\Omega)} |[\nabla b_\Omega] \cdot n| \, d\mathcal{H}^1$$

where $[\nabla b_\Omega]$ is the jump in ∇b_Ω and n is the unit normal to $\mathrm{Sk}(\Omega)$ (if it exists!).

11.3 Federer's sets of positive reach and curvature measures

For submanifolds of codimension larger than one $b_\Omega \geq 0$ (or equivalently $b_\Omega = d_{\partial\Omega}$) and the gradient has a discontinuity along $\partial\Omega$. In that case it is natural to go the square $d^2_{\partial\Omega} = b^2_\Omega$ of the function to relate its smoothness in a neighborhood to the smoothness of $\partial\Omega$. This is directly connected with the general concept of sets with positive reach as introduced by H. Federer .

Definition 11.4. *A non-empty set $A \subset \mathbb{R}^N$ is said to have* positive reach *if there exists $h > 0$ such that$\Pi_A(x) = \{p_A(x)\}$ is a singleton for every $x \in U_h(A)$. The maximum h for which the property holds is called the* reach *of A and denoted* reach (A).

This class also contains all nonempty convex sets since reach $(A) = +\infty$. A quite impressing result was to make sense of the classical Steiner-Minkowski formula for that class of sets. He also showed that for all r, $0 < r < h$, the boundaries of the tubular neighborhoods A_r are $C^{1,1}$ submanifolds of codimension one in $\mathbb{R}^N$. The next theorem summarizes several characterizations of sets with positive reach. Note that condition (vii) is a global condition on the smoothness of d^2_A in the tubular neighborhood $U_h(A)$ which parallels the ones of Theorem 11.1.

Theorem 11.6. *Given a nonempty subset A of $\mathbb{R}^N$, the following conditions are equivalent.*

(i) $\exists h > 0$*such that* d_A (resp. b_A) belongs to $C^{1,1}_{\mathrm{loc}}(U_h(A)\backslash\bar{A})$.

(ii) $\exists h > 0$such that d_A (resp. b_A) belongs to $C^1(U_h(A)\backslash\bar{A})$.
(iii) $\exists h > 0$ such that $\forall x \in U_h(A)\backslash\bar{A}$, $\Pi_A(x)$ is a singleton.
(iv) $\exists h > 0$ such that $\forall x \in U_h(A)$, $\Pi_A(x)$ is a singleton.
(v) A has positive reach, that is reach $(A) > 0$.
(vi) $\exists h > 0$such that p_A belongs to $C^{1,0}_{\text{loc}}(U_h(A))$.
(vii) $\exists h > 0$such that d_A^2 belongs to $C^{1,1}_{\text{loc}}(U_h(A))$.

Proof. The elements of the proof can be found in H. Federer works.
(i) $\Rightarrow$ (ii) $\Rightarrow$(iii)$\Rightarrow$ (iv) $\Rightarrow$ (v)are obvious.
(v) $\Rightarrow$ (vi)For each $x \in U_h(A)$, $\Pi_A(x) = \{p(x)\}$ is a singleton and for all $t \geq 0$ such that $t\,d_A(x) < h$,

$$d_A(p(x) + t(x - p(x))) \leq t\,|x - p(x)| = t\,d_A(x) < h.$$

By assumption for all $t \geq 0$, $t\,d_A(x) < h$, the projection of $p(x) + t(x - p(x))$ onto $\bar{A}$ is unique and equal to $p(x)$. Otherwise we could reduce $d_A(x)$ leading to a contradiction. Hence for all $t \geq 0$, $t\,d_A(x) < h$,

$$d_A(p(x) + t(x - p(x))) = t\,d_A(x).$$

For all $a \in \bar{A}$, $y \in U_h(A)$ and $t \geq 0$ such that $t\,d_A(y) < h$

$$|a - (p(x) + t(x - p(x)))|^2 \geq d_A(p(x) + t(x - p(x)))^2 = t^2 d_A(x)^2$$

$$|a - (p(x)|^2 + t^2|x - p(x))|^2 + 2t\,(a - (p(x))\cdot(p(x) - x) \geq$$

$$t^2 d_A(x)^(a - (p(x))\cdot(p(x) - x) \geq -\frac{1}{2t}\,|a - p(x)|^2.$$

So for any y_1 and y_2 in $\mathbb{R}^{\kappa}$ and $t \geq 0$ such that $t\,d_A(y_1) < t$ and $t\,d_A(y_2) < t$

$$(p(y_2) - (p(y_1))\cdot(p(y_1) - y_1) \geq -\frac{1}{2t}\,|p(y_2) - (p(y_1)|^2$$

$$(p(y_1) - (p(y_2))\cdot(p(y_2) - y_2) \geq -\frac{1}{2t}\,|p(y_1) - (p(y_2)|^2$$

$$(p(y_2) - (p(y_1))\cdot(y_2 - y_1) \geq \frac{t-1}{t}\,|p(y_2) - (p(y_1)|^2.$$

For any $x \in U_h(A)$, there exists $\rho > 0$ such that $d_A(x) + \rho < h$. Let $t = h/(d_A(x) + \rho)$ which is strictly greater than one. For all $y \in B(x,\rho)$, $d_A(y) < d_A(x) + \rho$ and $t\,d_A(y) < h$. Therefore for all y_1 and y_2 in $B(x,\rho)$

$$|p(y_2) - (p(y_1)| \leq \frac{h}{h - (d_A(x) + \rho)}|y_2 - y_1|.$$

(vi) $\Rightarrow$ (vii)¿From the identity $p_A(x) = x - \frac{1}{2}\nabla d_A^2(x)$.

(vii) $\Rightarrow$ (i)For each $U_h(A)\backslash\bar{A}$, there exists ρ, $0 < \rho$, such that $d_A(x) + \rho < h$. Therefore for all $y \in B(x,\rho)$, $h > d_A(x) + \rho > d_A(y) \geq d_A(x) - \rho > 0$, and $d_A^2 \in C^{1.1}(B(x,\rho))$. For any y_1 and y_2 in $B(x,\rho)$

$$\nabla d_A(y_2) - \nabla d_A(y_1) = \frac{1}{d_A(y_2)}[d_A(y_2)\nabla d_A(y_2) - d_A(y_2)\nabla d_A(y_1)]$$

$$= \frac{1}{2\,d_A(y_2)}[\nabla d_A^2(y_2) - \nabla d_A^2(y_1)]$$

$$+ \frac{1}{d_A(y_2)}\,[d_A(y_2) - d_A(y_1)]\nabla d_A(y_1).$$

So from the proof of (v) $\Rightarrow$
(vi) and $d_A(y_2) \geq d_A(x) - \rho > 0$

$$|\nabla d_A(y_2) - \nabla d_A(y_1)| \leq \frac{1}{2\,(d_A(x) - \rho)}\,\frac{2h - (d_A(x) + \rho)}{h - (d_A(x) + \rho)}|y_2 - y_1|$$

$$+ \frac{1}{d_A(x) - \rho}\,|y_2 - y_1|$$

and $\nabla d_A \in C^{0,1}(B(x,\rho))$, $d_A \in C^{1,1}(B(x,\rho))$ and $d_A \in C^{1,1}_{\text{loc}}(U_h(A)\backslash \bar{A})$.

One of the elements in the construction of curvature measures is the uniform convergence of d_{A_r} to d_A as r goes to zero.

Theorem 11.7. *Let A be a nonempty subset of $\mathbb{R}^N$.*

(i) As r goes to zero, $d_{A_r} \to d_A$ in $C(\mathbb{R}^N)$.
(ii) Assume that, in addition, there exist $h > 0$ and R, $0 < R < h$, such that $d_A^2 \in C^{1,1}(U_h(A))$ and $A_R \neq \mathbb{R}^N$. For r, $0 < r \leq R$, $\partial A_r \neq \emptyset$, $\partial \bar{A} \neq \emptyset$,

$$b_{A_r} = b_{\bar{A}} - r \text{ and } \nabla b_{A_r} = \nabla b_{\bar{A}} \text{ in } U_h(A),$$

and as r goes to zero $b_{A_r} \to b_{\bar{A}}$ in $C(\mathbb{R}^N)$. For $p \geq 1$ as r goes to zero

$$b_{A_r} \to b_{\bar{A}} \quad d_{A_r} \to d_{\bar{A}} \quad d_{CA_\triangledown} \to d_{C\bar{A}} \quad d_{\partial A_r} \to d_{\partial \bar{A}} \text{ in } W^{1,p}(U_h(A)).$$

Moreover $\bar{A}$ is of bounded curvature, finite perimeter, and $\partial\bar{A}$ of zero volume

$$\nabla b_{\bar{A}} \in BV_{\text{loc}}(\mathbb{R}^N)^N, \quad \chi_{\bar{A}} \in BV_{\text{loc}}(\mathbb{R}^N), \quad \gtrdot(\partial\bar{A}) = \nvdash. \tag{11.52}$$

The proof of the theorem requires the following two lemmas.

Lemma 11.2. *Assume that A is a nonempty subset of $\mathbb{R}^N$ and $r > 0$. Then*

$$0 \leq d_A(x) - d_{A_r}(x) \leq r \quad \text{in } \mathbb{R}^N \tag{11.53}$$

and if $A_r \neq \mathbb{R}^N$

$$r \leq d_{CA_\triangledown}(x) + d_A(x) \quad \text{in } \mathbb{R}^N$$

$$d_{CA_\triangledown}(x) \geq d_{\bar{A}}(x) + r \quad \text{in } \bar{A}.$$

Lemma 11.3. *Let A be a nonempty subset of $\mathbb{R}^{\kappa}$. Assume that there exists $h > 0$ such that $d_A^2 \in C^{1,1}(U_h(A))$, and that r is a real number such that $0 < r < h$.*

(i) For all $x \in U_h(A)\backslash\bar{A}$

$$y = p_A(x) + r\nabla d_A(x) \in \partial A_r, \quad d_A(y) = r \text{ and } p_A(y) = p_A(x).$$

If $A_r \neq \mathbb{R}^{\kappa}$,

$$\emptyset \neq \partial A_r \subset \{q : d_A(q) = r\}$$

$$d_A(x) = \inf_{q \in \partial A_r} |p_A(q) - x| = \inf_{d_A(q)=r} |p_A(q) - x|.$$

(ii) If $A_r \neq \mathbb{R}^{\kappa}$, then $\partial A_r \neq \emptyset, \partial\bar{A} \neq \emptyset$

$$d_{A_r}(x) = [d_A(x) - r]^+ \quad \text{in } U_h(A)$$

$$d_{CA_\nabla}(x) = [r - d_A(x)]^+ \quad \text{in } \mathbb{R}^{\kappa} \backslash \bar{A}$$

$$d_{CA_\nabla}(x) = d_{C\bar{A}}(x) + r \quad \text{in } \bar{A}$$

and

$$b_{A_r}(x) = b_{\bar{A}}(x) - r \quad \text{in } U_h(A) \tag{11.54}$$

and

$$b_{\bar{A}}(x) - b_{A_r}(x) = \{ d_A(x) - d_{A_r}(x), \quad h \leq d_A(x) r, \quad 0 \leq d_A(x) < h.$$

Proof. Proof of Lemma 11.2 (i) *First inequality.* For all q such that $d_A(q) = r$ and $p \in \Pi_A(q)$

$$d_A(x) \leq |p' - x| \leq |p' - q| + |q - x| = r + \inf_{d_A(q)=r} |q - x| \leq r + d_{A_r}(x).$$

Also for $r > 0$, $\bar{A} \subset A_r$ and $d_{A_r}(x) \leq d_A(x)$.

Second inequality. For all y such that $d_A(y) > r$ and all $p \in \bar{A}$

$$r \leq |y - p| \leq |y - x| + |x - p|$$

$$r \leq \inf_{d_A(y)>r} |y - x| + \inf_{p \in A} |x - p| = d_{CA_\nabla}(x) + d_A(x).$$

Third inequality. By definition for $r > 0$,

$$CA_\nabla = \{\dagger : \lceil_A(\dagger) > \nabla\} \quad \Rightarrow \quad \partial CA_\nabla \subset \{\dagger : \lceil_A(\dagger) = \nabla\}.$$

So for $x \in \bar{A}$

$$d_{CA_\nabla}(x) = d_{\partial CA_\nabla}(x) \geq \inf_{d_A(y)=r} |y - x|.$$

In particular

$$\exists y_r \in \partial CA_\nabla \quad \lceil_{\partial CA_\nabla}(\S) = |\S - \dagger_\nabla| = \nabla$$

$$\exists p \in \partial\bar{A} \quad |x - y_r| = |x - p| + |p - y_r|$$

since $d_A(q_0) = d_A(x) = 0$ and $d_A(q_1) = d_A(y_r) = r > 0$ for $q_t = x + t\,[y_r - x]$ and $t \mapsto d_A(q_t)$ is continuous. In view of $d_A(y_r) = r$

$$\begin{aligned} d_{CA_\nabla}(x) &= |x - y_r| = |x - p| + |p - y_r| \\ &\geq \inf_{p\in\partial\bar{A}} |x - p| + \inf_{p\in\partial\bar{A}} |p - y_r| \\ &\geq d_{C\bar{A}}(x) + d_A(y_r) = d_{C\bar{A}}(x) + r. \end{aligned}$$

Proof. Proof of Lemma 11.3 (i) In the region $U_h(A)\backslash\bar{A}$ the gradient ∇d_A is locally Lipschitz. For any point x, $0 < d_A(x) < h$, consider the flow

$$\frac{dy}{dt}(t) = \nabla d_A(y(t)), \quad y(0) = x.$$

There exists a unique local solution through x. Moreover

$$\frac{d}{dt} d_A(y((t)) = \nabla d_A(y(t)) \cdot \frac{dy}{dt}(t) = 1 \quad \Rightarrow \quad d_A(y((t)) = d_A(x) + t.$$

Therefore the solution exists and is unique for s, $0 \leq s < h - d_A(x)$. For any r, $d_A(x) < r < h$, let $y = y(r - d_A(x))$ and define

$$x(s) = p_A(y) + (s + d_A(x))\,\nabla d_A(y), \quad 0 \leq s \leq r - d_A(x).$$

It is readily seen that $d_A(x(s)) \leq s + d_A(x)$ and that

$$\begin{aligned} & d_A(y) - d_A(x(s)) \leq |y - x(s)| = r - (s + d_A(x)) \\ \Rightarrow \quad & d_A(x(s)) \geq s + d_A(x) \quad \Rightarrow \quad d_A(x(s)) = s + d_A(x) \\ \Rightarrow \quad & p_A(x(s)) = p_A(y) \text{ and } \nabla d_A(x(s)) = \nabla d_A(y). \end{aligned}$$

As a result

$$\frac{dx}{ds}(s) = \nabla d_A(y) = \nabla d_A(x(s)), \quad x(r - d_A(x)) = y = y(r - d_A(x))$$

and by uniqueness of solution $y(s) = x(s)$, $0 \leq s \leq r - d_A(x)$. In particular $x(0) = y(0) = x$

$$p_A(y) = p_A(y(0)) = p_A(x) \text{ and } \nabla p_A(y) = \nabla p_A(y(0)) = \nabla p_A(x)$$

and finally

$$\begin{aligned} y(s) &= p_A(x) + (s + d_A(x))\,\nabla d_A(x) = x + s\,\nabla d_A(x) \\ &= p_A(x) + (s + d_A(x))\,\nabla d_A(x). \end{aligned}$$

Hence for all r, $d_A(x) < r < h$,

$$y = p_A(x) + r\,\nabla d_A(x), \quad d_A(y) = r, \text{ and } p_A(y) = p_A(x).$$

(ii) *First identity.*¿From the first inequality in Lemma 11.2. $d_A(x) - d_{A_r}(x) \leq r$. Conversely for all $x \in \mathbb{R}^{\kappa}$, $d_A(x) \leq r$, $d_{A_r}(x) = 0$. From part (i) for $r < d_A(x) < h$, there exists $y = p_A(x) + (r/d_A(x))(x - p_A(x)) \in \partial A_r$ such that $d_A(q) = r$. Therefore

$$d_{A_r}(x) \leq |q - x| = d_A(x) - r.$$

Second identity. From the second inequality in Lemma 11.2, $r \leq d_{CA_\nabla}(x) + d_A(x)$.For x such that $d_A(x) > r$, $d_{CA_\nabla}(x) = 0$. For x such that $0 < d_A(x) \leq r$ always from part (i), $y = p_A(x) + r\,\nabla d_A(x) \in \partial A_r$,$d_A(y) = r$ and

$$d_{CA_\nabla}(x) = d_{\partial A_r}(x) \leq |q - x| = r - d_A(x).$$

Therefore $d_{CA_\nabla}(x) = [r - d_A(x)]^+$ in $\mathbb{R}^{\kappa} \setminus \bar{A}$.

Third identity. From the third inequality in Lemma 11.2, $d_{CA_\nabla}(x) \geq d_{C\bar{A}}(x) + r$ in $\bar{A}$. For $0 < r < h$, ∂A_r is $C^{1,1}$, $\partial A_r \neq \emptyset$, $\emptyset \neq A_r \neq \mathbb{R}^{\kappa}$, and $\emptyset \neq \bar{A} \subset A_r \neq \mathbb{R}^{\kappa}$, and $C\bar{A} \neq \emptyset$ and $\partial\bar{A} \neq \emptyset$. For each $x \in \bar{A}$, there exists a $p \in \partial\bar{A}$ such that $d_{C\bar{A}}(x) = |p - x|$.For $p \in \partial\bar{A}$ there exists a sequence $\{y_n\}$, $0 < d_A(y_n) < h$, such that $y_n \to p$. Consider the sequence

$$q_n = p_A(y_n) + r\,\nabla d_A(y_n).$$

From part (i) $q_n \in \partial A_r$, $p_A(q_n) = p_A(y_n)$. But p_A is Lipschitzian

$$|p_A(y_n) - p_A(p)| \leq c\,|y_n - p|$$

$$\Rightarrow \quad |q_n - p| \leq |q_n - p_A(y_n)| + |p_A(y_n) - p| \leq r + c\,|y_n - p|$$

and $\{q_n\}$ is bounded in ∂A_r. So there exists a q, $d_A(q) = r$, and a subsequence, still indexed by n, such that $q_n \to q$. By continuity of p_A in $U_h(A)$

$$p_A(q_n) \to p_A(q) \quad p_A(y_n) \to p_A(p) = p$$

$$\Rightarrow \quad \forall p \in \partial\bar{A},\ \exists q \in \partial A_r \text{ such that } p_A(q) = p.$$

Hence

$$d_{CA_\nabla}(x) = d_{\partial A_r}(x) \leq |q - x| \leq |q - p_A(q) + p - x|$$
$$\leq |q - p_A(q)| + |p - x| \leq r + |p - x| \leq r + d_{C\bar{A}}(x).$$

Fourth identity. From the first three.

Fifth identity. Outside $U_h(A)$, $d_A(x) \geq h > r$, $d_{C\bar{A}} = d_{A_r} = 0$ and $b_{\bar{A}}(x) - b_{A_r}(x) = d_A(x) - d_{A_r}(x)$.

Proof. Proof of Theorem 11.7

(i) From the first inequality in Lemma 11.2.

(ii) Clearly $\emptyset \neq \bar{A} \subset A_r \subset A_R \neq \mathbb{R}^\kappa$ implies $\mathbb{R}^\kappa \neq C\mathcal{A} \supset CA_\nabla \supset CA_R \neq \emptyset$ and $\partial A_r \neq \emptyset$ and $\partial \bar{A} \neq \emptyset$. The next identities and the convergence in $C(\mathbb{R}^\kappa)$ follow from Lemma 11.3

(ii) and Lemma 11.2. The convergence of b_{A_r} to $b_{\bar{A}}$ in $U_h(A)$ is a consequence of the fact that $\nabla b_{A_r} = \nabla b_{\bar{A}}$ a.e. in $U_h(A)$. Convergence in $W^{1,p}(U_h(A))$ of b_{A_r} also implies the convergence of $b_{A_r}^+$, $b_{A_r}^-$ and $|b_{A_r}|$ to $b_{\bar{A}}^+$, $b_{\bar{A}}^-$ and $|b_{\bar{A}}|$. Finally for $0 < r < h$ the boundary ∂A_r is $C^{1,1}$ and by Theorem 11.1 hence $b_{A_r} \in C_{loc}^{1,1}(U(\partial A_r))$ in some neighborhood $U(\partial A_r)$ of ∂A_r. By Theorem 11.5 $\nabla b_{A_r} \in \mathrm{BV}_{loc}(U_h(A))^N$ and since $\nabla b_{\bar{A}} = \nabla b_{A_r}$ in $U_h(A)$, $\nabla b_{\bar{A}} \in \mathrm{BV}_{loc}(U_h(A))$ and always by Theorem 11.5 $\nabla b_{\bar{A}} \in \mathrm{BV}_{loc}(\mathbb{R}^\kappa)^N$. Also $\emptyset \neq \bar{A} \subset A_r$ and since ∂A_r is $C^{1,1}$, A_r and a fortiori $\bar{A}$ are locally of finite perimeter. Moreover in $U_h(A)$

$$0 = \chi_{\partial A_r} = 1 - |\nabla b_{A_r}| = 1 - |\nabla b_{\bar{A}}| = \chi_{\partial \bar{A}} \quad \Rightarrow \quad m(\partial \bar{A}) = 0.$$

Recall that for a set A with positive reach, $0 \leq r < \mathrm{reach}\,(A)$ and a bounded Borel set Q in $\mathbb{R}^\kappa$

$$\int_{A_r \cap p_A^{-1}(Q)} dx = \sum_{i=0}^{\dot{N}} r^{N-i}\, \alpha(N-i)\, \Phi_i(A,Q)$$

where $\alpha(j)$ is the j dimensional measure of the unit ball in $\mathbb{R}^j$ and $\Phi_0(A,\cdot)$, $\Phi_1(A,\cdot)$, ..., $\Phi_N(A,\cdot)$ are the *Curvature measures* of H. Federer . They are related to the coefficients of the z-polynomial $\det[I + z\, D^2 b_{\bar{A}}(x)]$. For instance when Ω is a subset of $\mathbb{R}^\kappa$ such that $\partial\Omega \neq \emptyset$ and $b_\Omega \in C^{1,1}(U_h(\partial\Omega))$ for some $h > 0$. Then for all r, $0 < r < h$, and $\phi \in \mathcal{D}^0(\mathbb{R}^\kappa)$

$$\int_{\Omega_r} \phi \circ p_\Omega\, dx = \int_\Omega \phi \circ p_\Omega\, dx + \sum_{i=0}^{N-1} r^{N-i} \frac{1}{N-i} \int_{\partial\Omega} \lambda_i\, \phi\, d\Gamma$$

$$\int_{(\partial\Omega)_r} \phi \circ p_{\partial\Omega}\, dx = \sum_{i=0}^{N-1} r^{N-i} \frac{1 - (-1)^{N-i}}{N-i} \int_{\partial\Omega} \lambda_i\, \phi\, d\Gamma$$

where the $\lambda_i(x)$'s are the coefficients of the $(N-1)$-th order polynomial

$$\det[I + z\, D^2 b_\Omega(x)] = \sum_{i=0}^{N-1} \lambda_i(x)\, z^{N-i-1}, \quad |z| < h.$$

11.4 The $W^{2,p}$-case

In general, sets of locally bounded curvature do not have a boundary with zero Lebesgue measure, as can be seen from the following example.

Exercise 11.6. Let B be the open unit ball centered in 0 of $\mathbb{R}^N$ and define

$$\Omega = \{x \in B \, : \, x \text{ with rational coordinates}\}.$$

Then $\partial\Omega = \overline{B}$, $b_\Omega = d_B$ and for all $h > 0$

$$\nabla b_\Omega \in BV(U_h(\partial\Omega))^2$$

and

$$< \Delta b_\Omega, \varphi > \int_{\partial B} \varphi \, dx + \int_{CB} \frac{1}{|x|} \varphi \, dx.$$

The Lebesgue measure of the boundary is zero for all Lipschitian subsets Ω of D. It is also true for the following family of subsets of D.

Theorem 11.8. *Let D be a bounded open subset of $\mathbb{R}^N$ and let $\Omega \subset D$ be such that $\partial\Omega \neq \emptyset$. Assume that there exist $h > 0$ and $p \geq 1$ such that*

$$\int_{U_h(\partial\Omega)} |D^2 b_\Omega|^p \, dx < \infty. \tag{11.55}$$

Then

$$|\nabla b_\Omega|^2 = 1 \quad \textit{and} \quad \chi_{\partial\Omega} = 0 \quad \textit{a.e. in } \mathbb{R}^N \quad \textit{and} \quad >(\partial\Omega) = \nvdash. \tag{11.56}$$

Moreover when $p > N$, then $W^{2,p}(U_h(\partial\Omega)) \subset C^{1,\lambda}(U_h(\partial\Omega))$ for all λ, $0 < \lambda \leq 1 - N/p$ and the boundary of $\partial\Omega$ is Hölderian of class $C^{1,\lambda}$,$b_\Omega^2 = d_{\partial\Omega}^2 \in C^{1,1}_{\text{loc}}(U_h(\partial\Omega))$ and $\partial\Omega$ has positive reach.

Proof. (i) Consider the function $|\nabla b_\Omega|^2$. Since $|\nabla b_\Omega| \leq 1$, then $\nabla b_\Omega \in W^{1,p}(U_h(\partial\Omega))^N \cap L^\infty(U_h(\partial\Omega))^N$ and $|\nabla b_\Omega|^2 \in W^{1,p}(U_h(\partial\Omega)) \cap L^\infty(U_h(\partial\Omega))$.

But for almost all x, we know that $\nabla b_\Omega(x)$ is differentiable,

$$|\nabla b_\Omega(x)| = \{0\,, \quad x \in \partial\Omega 1, \quad x \notin \partial\Omega \tag{11.57}$$

and

$$|\nabla b_\Omega(x)| = 1 - \chi_{\partial\Omega}(x).$$

Necessarily

$$\nabla(|\nabla b_\Omega|^2)(x) = 0 \quad \text{a.e. in } U_h(\partial\Omega)$$

Since $\partial\Omega$ is compact, there exists a finite sequence of distinct points $x_i \in \partial\Omega$, $i \in I$, such that

$$\partial\Omega \subset \bigcup_{i \in I} B(x_i, h).$$

Define the following partition of the set I of indices

$$I_0 = \{i \in \{1,\dots,n\} : \exists x \in B(x_i,h),\ \nabla b_\Omega(x) = 0\}$$

$$I_1 = \{i \in \{1,\dots,n\} : \exists x \in B(x_i,h),\ \nabla b_\Omega(x) \exists \text{ and } |\nabla b_\Omega(x)| = 1\}.$$

Therefore for all $i \in I_0$, $\nabla b_\Omega(x) = 0$ in$B(x_i,h)$ and since b_Ω is Lipschizian and $b_\Omega(x_i) = 0$, then $b_\Omega = 0$ in $B(x_i,h)$ and $B(x_i,h) \subset \partial\Omega$. Similarly for all $i \in I_1$, $|\nabla b_\Omega(x)| = 1$ in $B(x_i,h)$. As a result $\partial\Omega$ can be partioned into two compact parts

$$\partial\Omega = (\partial\Omega)^0 \cup (\partial\Omega)^1$$

since

$$(\partial\Omega)^0 \subset V^0 := \bigcup_{i\in I_0} B(x_i,h) \quad (\partial\Omega)^1 \subset V^1 := \bigcup_{i\in I_1} B(x_i,h)$$

and

$$V^0 \cap V^1 = \emptyset.$$

But for $i \in I_0$

$$B(x_i,h) \subset \partial\Omega \quad \Rightarrow \quad \bigcup_{i\in I_0} B(x_i,h) \subset \partial\Omega$$

and

$$V^0 \subset V^0 \cap \partial\Omega = (\partial\Omega)^0 \subset V^0 \quad \Rightarrow \quad (\partial\Omega)^0 = V^0.$$

Therefore V^0 is both open and closed. It cannot be $\mathbb{R}^\kappa$ since Ω is bounded. So $V^0 = \emptyset$, $|\nabla b(x)| = 1$ in $U_h(\partial\Omega)$, and $\chi_{\partial\Omega} = 0$. This proves that $m(\partial\Omega) = 0$.(ii) In view of 11.55, $b_\Omega \in W^{2,p}(U_h(\partial\Omega))$ for $p > N$. Given ε, $0 < \varepsilon < h$, there exists $\rho_{\varepsilon,h} \in \mathcal{D}(\mathbb{R}^\kappa)$ such that

$$\rho_{\varepsilon,h} = \{\, 1\ , x \in \overline{U_{h-\varepsilon}(\partial\Omega)} 0, x \in \overline{U_h(\partial\Omega)\backslash U_{h-\frac{\varepsilon}{2}}(\partial\Omega)}.$$

As a result

$$\rho_{\varepsilon,h}\, b_\Omega \in W_0^{2,p}(U_h(\partial\Omega))$$

and

$$\begin{aligned} &\rho_{\varepsilon,h} b_\Omega \in C^{1,\lambda}(\,\overline{U_h(\partial\Omega)}\,), \quad 0 < \lambda \le 1 - N/p \\ \Rightarrow \quad & b_\Omega \in C^{1,\lambda}(\,\overline{U_{h-\varepsilon}(\partial\Omega)}\,), \quad 0 < \lambda \le 1 - N/p \\ \Rightarrow \quad & b_\Omega \in C^{1,\lambda}(U_h(\partial\Omega)), \quad 0 < \lambda \le 1 - N/p \end{aligned}$$

since ε can be made arbitrarily small. But then b_Ω and $b_\Omega^2 = d_{\partial\Omega}^2 \in C^1(U_h(\partial\Omega))$. Therefore $\Pi_{\partial\Omega}(x)$ is a singleton for each $x \in U_h(\partial\Omega)$ and from Theorem 11.6 $b_\Omega^2 = d_{\partial\Omega}^2 \in C^{1,1}_{\text{loc}}(U_h(\partial\Omega))$.

Remark 11.3. In dimension $N = 2$ the condition $b_\Omega \in W^{2,p}(N(\partial\Omega))$ for some bounded neighborhood $N(\partial\Omega)$ of $\partial\Omega$ is equivalent to $\Delta b_\Omega \in L^p(N(\partial\Omega))$. Recall that $m(\partial\Omega) = 0$, $|\nabla b_\Omega(x)|^2 = 1$, and $D^2 b_\Omega(x)\, \nabla b_\Omega(x) = 0$. Hence

$$\partial^2_{12} b_\Omega(x) = \partial^2_{21} b_\Omega(x) = -\partial_1 b_\Omega(x)\, \partial_2 b_\Omega(x)\, \Delta b_\Omega(x)$$

$$\partial^2_{11} b_\Omega(x) = (\partial_2 b_\Omega(x))^2\, \Delta b_\Omega(x) \partial^2_{22} b_\Omega(x) = (\partial_1 b_\Omega(x))^2\, \Delta b_\Omega(x).$$

□

11.5 Compactness theorems

In Shape Optimization, compactness theorems are used to establish the existence of optimal domains. The global boundedness condition on the generalized curvatures of the level sets of A is relaxed to a local condition in a neighborhood of the boundary of the set A. We emphasize the differences and similarities by separately giving versions for d_A, $d_{C\Omega}$ and b_Ω.

11.5.1 $W^{1,p}$-topology. For the $W^{1,p}(D)$-topology we recall the following compactness theorem .

Theorem 11.9. *Let D be a fixed bounded open Lipschitzian hold-all in $\mathbb{R}^N$. Let $\{A_n\}$, $\emptyset \neq A_n$, be a sequence of subsets of $\overline{D}$ such that*

$$\exists c > 0, \forall n \geq 1, \quad \|D^2 d_{A_n}\|_{M^1(D)} \leq c. \tag{11.58}$$

Then there exist a subsequence $\{A_{n_k}\}$ and a set A, $A \neq \emptyset$, such that $\nabla d_A \in BV(D)^N$ and

$$d_{A_{n_k}} \to d_A \text{ in } W^{1,p}(D)\text{-strong}$$

for all p, $1 \leq p < \infty$. Moreover for all $\varphi \in \mathcal{D}'(\mathcal{D})$

$$\lim_{n\to\infty} \langle \partial_{ij} d_{A_{n_k}}, \varphi\rangle = \langle \partial_{ij} d_A, \varphi\rangle, \quad 1 \leq i,j \leq N,$$

and

$$\|D^2 d_A\|_{M^1(D)} \leq c.$$

The global condition 11.58 on the fixed set D is now relaxed to a local one in a neighborhood of the boundary of each set of the sequence.

Theorem 11.10. *Let D be a fixed bounded open hold-all in $\mathbb{R}^N$. Let $\{A_n\}$,$\emptyset \neq A_n$, be a sequence of subsets of $\overline{D}$. Assume that there exist $h > 0$ and $c > 0$ such that*

$$\forall n, \quad \|D^2 d_{A_n}\|_{M^1(U_h(\partial A_n))} \leq c. \tag{11.59}$$

Then there exist a subsequence$\{A_{n_k}\}$ and a subset A, $\emptyset \neq A$, of $\overline{D}$ such that $\nabla d_A \in BV_{\text{loc}}(\mathbb{R}^N)^N$ and for all p, $1 \leq p < \infty$,

$$d_{A_{n_k}} \to d_A \text{ in } W^{1,p}_{\text{loc}}(\mathbb{R}^N)$$

$$\|D^2 d_A\|_{M^1(U_h(A))} \leq c$$

and $\chi_{\overline{A}} \in BV_{\text{loc}}(\mathbb{R}^N)$.

Remark 11.4. Condition 11.59 in $U_h(\partial A_n)$ implies the same condition on $U_h(A_n)$ since

$$U_h(A_n) = \int A_n \cup U_h(\partial A_n)$$

and $\nabla d_{A_n} = 0$ in $\int A_n$ imply that

$$\|\nabla d_{A_n}\|_{M^1(U_h(A_n))} \leq \|\nabla d_{A_n}\|_{M^1(\int A_n)} + |\nabla d_{A_n}\|_{M^1(U_h(\partial A_n))} \leq 0 + c.$$

The converse is also true since$U_h(\partial A_n) \subset U_h(A_n)$.

This compactness theorem is more interesting than the previous one. The sets are still contained in $\overline{D}$, but the boundedness condition is given in a tubular neighborhood which varies with each set. Its proof follows from the first theorem and Theorem 11.4 (iv).

Proof. FromRemark 11.4 we can assume that 11.59 holds in $U_h(A_n)$.

(i) for all

$$r > R := \left(\frac{N}{2N+2}\right)^{\frac{1}{2}} \operatorname{diam}\overline{D}$$

$\overline{U_r(D)}$ has a Lipschitzian boundary. Fix the set $U_{2R}(D)$. By assumption

$$\forall n \geq 1, \quad A_n \subset U_{2R}(A_n) \subset U_{2R}(D)$$

and by Theorem 11.4

(iii)

$$\|D^2 d_{A_n}\|_{M^1(U_{2R}(D))} \leq \|D^2 d_{A_n}\|_{M^1(U_h(A_n))} + c_0(U_{2R}(D)) + \frac{c_1(U_{2R}(D))}{h}$$

$$\leq c(R, D) := c + c_0(U_{2R}(D)) + \frac{c_1(U_{2R}(D))}{h}.$$

From Theorem 11.9 there exists $A \subset \overline{D}$ such that $A \neq \emptyset$ and $\nabla d_A \in \mathrm{BV}\,(U_{2R}(D))^N$ and a subsequence of $\{A_n\}$, still indexed by n, such that

$$d_{A_n} \to d_A \text{ in } W^{1,p}(U_{2R}(D))$$

for all $p \geq 1$. Moreover

$$\|D^2 d_A\|_{M^1(U_{2R}(D))} \leq c(R, D).$$

By Theorem 11.4 (ii) we can repeat the exercise for any bounded open set V by working in a set $U_{R'}(D)$ for a sufficiently large $R' > R$ such that $V \subset U_{R'}(D)$. As a result the subsequence of d_{A_n} converges to d_A in $W^{1,p}_{\mathrm{loc}}(\mathbb{R}^\kappa)$, $\nabla d_A \in \mathrm{BV}_{\mathrm{loc}}(\mathbb{R}^\kappa)^{\mathrm{N}}$ and $\chi_{\overline{A}} \in \mathrm{BV}_{\mathrm{loc}}(\mathbb{R}^\kappa)$.

(ii) As for the last estimate in $U_h(A)$, for all $\varepsilon > 0$, $0 < \varepsilon < h$, there exists $N > 0$ such that for all $n \geq N$

$$d_{A_n}(x) \leq d_A(x) + \varepsilon \quad \Rightarrow \quad U_{h-\varepsilon}(A) \subset U_h(A_n).$$

For each $\Phi \in \mathcal{D}(\mathcal{U}_((\mathcal{A}))^{\mathcal{N}\times\mathcal{N}}$, $\operatorname{supp}\Phi \subset U_h(A)$. Since $\overline{U_h(A)}$ is compact there exists $\varepsilon = \varepsilon(\Phi) > 0$ such that

$$\operatorname{supp}\Phi \subset U_{h-\varepsilon}(A).$$

Therefore for $n \geq N$, $\operatorname{supp}\Phi \subset U_{h-\varepsilon}(A) \subset U_h(A_n)$ and

$$\int_{U_h(A)} \nabla d_{A_n} \longrightarrow \operatorname{div}\Phi\, dx = \int_{U_{h-\varepsilon}(A)} \nabla d_{A_n} \longrightarrow \operatorname{div}\Phi\, dx =$$

$$= \int_{U_h(A_n)} \nabla d_{A_n} \longrightarrow \operatorname{div} \Phi \, dx$$

$$\left| \int_{U_h(A)} \nabla d_{A_n} \longrightarrow \operatorname{div} \Phi \, dx \right| \leq \|D^2 d_{A_n}\|_{M^1(U_h(A_n))} \, \|\Phi\|_{C^0(U_h(A_n))}$$

$$\leq c \, \|\Phi\|_{C^0(U_{h-\epsilon}(A))} = c \, \|\Phi\|_{C^0(U_h(A))}$$

and for all $n \geq N$ we have$\|D^2 d_{A_n}\|_{M^1(U_h(A))} \leq c$. This implies that

$$\|D^2 d_A\|_{M^1(U_h(A))} \leq \liminf \|D^2 d_{A_n}\|_{M^1(U_h(A))} \leq c.$$

Theorem 11.10 is to be compared with a similar result of H. Federer for sets of positive reach.

Theorem 11.11. *Let D be a fixed bounded open hold-all in $\mathbb{R}^{\kappa}$. Let $\{A_n\}$, $\emptyset \neq A_n$, be a sequence of subsets of $\overline{D}$. Assume that there exists $h > 0$ such that*

$$\forall n, \quad d^2_{A_n} \in C^{1,1}(U_h(A_n)). \tag{11.60}$$

Then there exist a subsequence $\{A_{n_k}\}$ and $A \subset \overline{D}$, $A \neq \emptyset$, such that $d^2_A \in C^{1,1}(U_h(A))$ and

$$d^2_{A_{n_k}} \to d^2_A \text{ in } C^1(U_h(A)).$$

11.5.2 $W^{1,p}$-complementary topology. When $\overline{D}$ is compact, $C^c_d(D)$ is compact for the uniform topology and closed in the $W^{1,p}(D)$-topology ($1 \leq p < \infty$) and we have the analogue of the two compactness theorems of the previous section.

Theorem 11.12. *Let D be a fixed bounded open Lipschitzian* (resp. bounded open) hold-all in $\mathbb{R}^{\kappa}$ .Let $\{\Omega_n\}$ be a sequence of open subsets of D. Assume that there exists $c > 0$ such that

$$\forall n, \quad \|D^2 d_{C\Omega_\backslash}\|_{M^1(D)} \text{ (resp. } \|D^2 d_{C\Omega_\backslash}\|_{M^1(U_h(C\Omega_\backslash))}) \leq c. \tag{11.61}$$

Then there exist a subsequence $\{\Omega_{n_k}\}$ and an open subset Ω of D such that $\nabla d_{C\Omega} \in \mathrm{BV}(\mathbb{R}^{\kappa})^{\mathrm{N}}$ and for all p, $1 \leq p < \infty$,

$$d_{C\Omega_{\backslash_\|}} \to d_{C\Omega} \text{ in } W^{1,p}(\mathbb{R}^{\kappa}).$$

Moreover $\|D^2 d_{C\Omega}\|_{M^1(\mathbb{R}^{\kappa})}$ (resp. $\|D^2 d_{C\Omega}\|_{M^1(U_h(C\Omega))}) \leq c$,

$$\chi_\Omega \in \mathrm{BV}(\mathbb{R}^{\kappa}),$$

and for all $\varphi \in \mathcal{D}'(\mathbb{R}^{\kappa})^{\mathrm{N}\times\mathrm{N}}$ (resp. $\mathcal{D}'(\mathcal{U}_((C\Omega))^{\mathcal{N}\times\mathcal{N}}$)

$$< D^2 d_{C\Omega_\backslash}, \varphi > \to < D^2 d_{C\Omega}, \varphi > .$$

Proof. (i) (Global condition on D) By assumption D is bounded, $\Omega_n \subset D$, and hence $\emptyset \neq CD \subset C\Omega_\backslash$ and $\emptyset \neq \partial D \subset C\Omega_\backslash \cap D$. Moreover

$$C\Omega_\backslash = [C\Omega_\backslash \cap \overline{D}] \cup [C\Omega_\backslash \cap C\overline{D}] \subset [C\Omega_\backslash \cap \overline{D}] \cup [C\Omega_\backslash \cap CD] \subset C\Omega_\backslash$$

$$\Rightarrow \quad d_{C\Omega_\backslash} = d_{[C\Omega_\backslash \cap \overline{D}] \cup CD} = \min\{d_{C\Omega_\backslash \cap \overline{D}}, d_{CD}\}$$

$$\Rightarrow \quad d_{C\Omega_\backslash} = d_{C\Omega_\backslash \cap \overline{D}} \text{ on } \overline{D},$$

since $C\Omega_\backslash \supset CD \Rightarrow C\Omega_\backslash \cap \overline{D} \supset \partial D \Rightarrow \lceil_{C\Omega_\backslash \cap \overline{D}} \leq \lceil_{\partial D}$ and

$$\forall x \in D, \quad d_{C\Omega_\backslash} = \min\{d_{C\Omega_\backslash \cap \overline{D}}, d_{\partial D}\} = d_{C\Omega_\backslash \cap \overline{D}}.$$

Setting $A_n = C\Omega_\backslash \cap \overline{D}$ we are back to the case of Theorem 11.9. There exist a subsequence, still indexed by n, and a set $A \subset \overline{D}$, $\nabla d_A \in BV(D)^N$ such that for all p, $1 \leq p < \infty$,

$$d_{C\Omega_\backslash \cap \overline{D}} = d_{A_n} \to d_A \text{ in } W^{1,p}(D)$$

$$\Rightarrow d_{C\Omega_\backslash} = \min\{d_{A_n}, d_{CD}\} \to \min\{d_A, d_{CD}\} = d_{A \cup CD} \text{ in } W^{1,p}(\mathbb{R}^\kappa)$$

since $d_{C\Omega_\backslash} = d_{A \cup CD} = 0$ on CD. Moreover since $d_{C\Omega_\backslash \cap \overline{D}} \leq d_{\partial D}$ on $\overline{D}$, $d_A \leq d_{\partial D}$ in $\overline{D}$ and $\partial D \subset \overline{A}$. Let

$$\Omega := C(\overline{A} \cup CD) = C\overline{A} \cup D \subset D$$

be the limit open set and by definition $d_{C\Omega_\backslash} \to d_{C\Omega}$ in $W^{1,p}(\mathbb{R}^\kappa)$ and since D has a Lipschitzian boundary and

$$d_{C\Omega} = d_{C\Omega \cap \overline{D}} = d_A \text{ in } D \text{ and } 0 \text{ in } CD$$

$$\nabla d_{C\Omega} = \nabla d_A \in BV(D)^N \quad \Rightarrow \quad \nabla d_{C\Omega} \in BV(\mathbb{R}^\kappa)^N.$$

In particular

$$\|D^2 d_{C\Omega}\|_{M^1(D)} \leq \liminf \|D^2 d_{C\Omega_\backslash}\|_{M^1(D)} \leq c.$$

(ii) (Local condition) We proceed as in the proof of Theorem 11.10 and work in the bounded open Lipschitzian domain $U_{2R}(D)$

$$\Omega_n \subset U_{2R}(\Omega_n) \subset U_{2R}(D)$$

and

$$\|D^2 d_{C\Omega_\backslash}\|_{M^1(U_{2R}(D))} \leq c + c_0(U_{2R}(D)) + \frac{c_1(U_{2R}(D))}{h}.$$

From part (i) there exist a subsequence, still denoted by n, and a set $\Omega \subset D$, $\nabla d_{C\Omega} \in BV(\mathbb{R}^\kappa)^N$ such that

$$d_{C\Omega_\backslash} \to d_{C\Omega} \text{ in } W^{1,p}(\mathbb{R}^\kappa) \text{ and } \|\mathbb{D}^\kappa c\Omega\|_{M^\kappa(U_\approx(C\Omega))} \leq \cdot$$

11.5.3 $W^{1,p}$-oriented distance topology. We combine the compactness theorems and the new local ones for the function b_Ω.

Theorem 11.13. *Let D be a fixed bounded open Lipschitzian* (resp. bounded open)hold-all in $\mathbb{R}^N$. Let $\{\Omega_n\}$ be a sequence of subsets of D such that $\partial\Omega_n \neq \emptyset$. Assume that there exists $c > 0$ such that

$$\forall n, \quad \|D^2 b_{\Omega_n}\|_{M^1(D)} \text{ (resp. } \|D^2 b_{\Omega_n}\|_{M^1(U_h(\partial\Omega_n))}) \leq c. \tag{11.62}$$

Then there exist a subsequence$\{\Omega_{n_k}\}$ and a subset Ω of D such that $\partial\Omega \neq \emptyset$, $\nabla b_\Omega \in \mathrm{BV}(D)^N$ (resp. $\nabla b_\Omega \in \mathrm{BV}_{\mathrm{loc}}(\mathbb{R}^N)^N$), and for all p, $1 \leq p < \infty$,

$$b_{\Omega_{n_k}} \to b_\Omega \text{ in } W^{1,p}(D) \text{ (resp. } W^{1,p}_{\mathrm{loc}}(\mathbb{R}^N)).$$

$$\text{Moreover } \|D^2 b_\Omega\|_{M^1(D)} \text{ (resp. } \|D^2 b_\Omega\|_{M^1(U_h(\partial\Omega))}) \leq c,$$

$$\chi_{\partial\Omega} \in \mathrm{BV}(D)\text{(resp. } \mathrm{BV}_{\mathrm{loc}}(\mathbb{R}^N)),$$

and for all $\varphi \in \mathcal{D}'(\mathcal{D})^{N\times N}$ (resp. $\mathcal{D}'(\mathbb{R}^N)^{N\times N}$)

$$< D^2 b_{\Omega_n}, \varphi > \to < D^2 b_\Omega, \varphi > .$$

Proof. (i) (Global condition on D) Same proof as Theorem 11.9.

(ii) As in the proof of Theorem 11.10 we work in the bounded open Lipschitzian domain $U_{2R}(D)$ where

$$\forall n \geq 1, \quad \Omega_n \subset U_{2R}(\Omega_n) \subset U_{2R}(D)$$

and by Theorem 11.5

(iii)

$$\|D^2 b_{\Omega_n}\|_{M^1(U_{2R}(D))} \leq \|D^2 b_{\Omega_n}\|_{M^1(U_h(\partial\Omega_n))} + c_0(U_{2R}(D)) + \frac{c_1(U_{2R}(D))}{h}.$$

From part (i) there exists Ω such that $\partial\Omega \neq \emptyset$, $\nabla b_\Omega \in \mathrm{BV}(U_{2R}(D))^N$, and a subsequence of $\{\Omega_n\}$, still indexed by n, such that for all $p \geq 1$

$$b_{\Omega_n} \to b_\Omega \text{ in } W^{1,p}(U_{2R}(D)).$$

By Theorem 11.5 (ii) we can repeat the exercise for any bounded open set V by working in a set $U_{R'}(D)$ for a sufficiently large $R' > R$ such that $V \subset U_{R'}(D)$. As a result b_{Ω_n} converges to b_Ω in $W^{1,p}_{\mathrm{loc}}(\mathbb{R}^N)$, $\nabla b_\Omega \in \mathrm{BV}_{\mathrm{loc}}(\mathbb{R}^N)^N$, and $\chi_{\partial\Omega} \in \mathrm{BV}_{\mathrm{loc}}(\mathbb{R}^N)$. For the estimate of$D^2 b_\Omega$ in $U_h(\partial\Omega)$ we proceed as in the proof of Theorem 11.10.

First we can complete the conclusions of Theorem 11.11 for sets of positive reach.

Corollary 11.1. *Let D be a fixed bounded open hold-all in $\mathbb{R}^N$. Let $\{A_n\}$, $\emptyset \neq A_n$, be a sequence of subsets of $\overline{D}$. Assume that there exists $h > 0$ such that*

$$\forall n, \quad d^2_{A_n} \in C^{1,1}(U_h(A_n)).$$

Then there exist a subsequence $\{A_{n_k}\}$ and $A \subset \overline{D}$, $\partial \bar{A} \neq \emptyset$, such that $d^2_A \in C^{1,1}(U_h(A))$, $\nabla b_{\bar{A}} \in BV_{\text{loc}}(\mathbb{R}^N)^N$, and for all p, $1 \leq p < \infty$,

$$b_{\bar{A}_{n_k}} \to b_{\bar{A}} \quad d_{A_{n_k}} \to d_A \quad d_{\complement \bar{A}_{n_k}} \to d_{\complement \bar{A}} \quad b_{\partial \bar{A}_{n_k}} \to b_{\partial \bar{A}} \textit{ in } W^{1,p}_{\text{loc}}(\mathbb{R}^N).$$

Proof. From Theorems 11.13 and 11.11.

Corollary 11.2. *Let D be a fixed bounded open Lipschitzian hold-all in $\mathbb{R}^N$. Let $\{\Omega_n\}$ be a sequence of subsets of D such that $\partial\Omega_n \neq \emptyset$. Assume that there exist $h > 0$, $p \geq 1$ and $c > 0$ such that*

$$\forall n, \quad \int_{U_h(\partial\Omega_n)} |D^2 b_{\Omega_n}(x)|^p \, dx \leq c. \tag{11.63}$$

Then, in addition to the conclusions of Theorem 11.12, $m(\partial\Omega) = m(\partial\Omega_n) = 0$ and

$$\int_{U_h(\partial\Omega)} |D^2 b_\Omega(x)|^p \, dx \leq c. \tag{11.64}$$

Moreover for $p > N$ the sets Ω_n's and the limit set Ω have Hölderian boundaries of class $C^{1,\lambda}$ for all λ, $0 < \lambda \leq 1 - N/p$, $b^2_\Omega = d^2_{\partial\Omega} \in C^{1,1}_{\text{loc}}(U_h(\partial\Omega))$ and $\partial\Omega$ has positive reach.

Proof. The first part of the corollary is a direct consequence of the Theorem 11.8. Since D is bounded, condition11.63 for some $p \geq 1$ implies the same condition for $p = 1$. Hence condition 11.61 in Theorem 11.12 is verified and its conclusions follow. The properties of the boundaries follow from Theorem 11.8

11.6 A continuity of the solution of the Dirichlet boundary value problem

Assume that

$$d_{\complement\Omega_n} \to d_{\complement\Omega} \text{ in } C^0(D) \tag{11.65}$$

for a sequence $\{\Omega_n\}$ of open subsets of D and an open subset Ω of D (here Ω can possibly be empty). Associate with each n the solution y_n of the homogeneous Dirichlet problem

$$\exists y_n = y(\Omega_n) \in H^1_0(\Omega_n), \quad \forall \varphi \in H^1_0(\Omega_n) \tag{11.66}$$

$$\textstyle\int_{\Omega_n} \nabla y_n \cdot \nabla\varphi - f\,\varphi \, dx = 0. \tag{11.67}$$

Introduce for any open subset Ω of D the closed linear subspace

$$H_0^1(\Omega; D) = \overline{\mathcal{D}(\Omega; \mathcal{D})}^{H^1}$$

of $H_0^1(D)$ where$\mathcal{D}(\Omega;\mathcal{D}) = \{\varphi \in \mathcal{D}(\mathcal{D}) \; : \; \mathrm{supp}\varphi \subset \Omega\}$.

As a consequence

$$\{\varphi_{|\Omega} \; : \; \varphi \in H_0^1(\Omega; D)\} = H_0^1(\Omega). \tag{11.68}$$

This defines a unique extension by zero in $H_0^1(D)$ of each element y_n of $H_0^1(\Omega_n)$. For simplicity this extension will also be denoted y_n.The sequence of extensions by zero of the solutions y_n's to problem 11.67 is uniformly bounded in $H_0^1(D)$. Hence there exists $y^* \in H_0^1(D)$ and a bounded subsequence, still indexed by n, such that

$$y_n \rightharpoonup y^* \text{ in } H_0^1(D)\text{-weak}.$$

If $y(\Omega)$ is the solution of the homogeneous Dirichlet problem on Ω

$$\exists y = y(\Omega) \in H_0^1(\Omega), \quad \forall \varphi \in H_0^1(\Omega) \tag{11.69}$$

$$\textstyle\int_\Omega \nabla y \cdot \nabla \varphi \, dx - f \varphi \, dx = 0, \tag{11.70}$$

can we conclude that $y^*_{|\Omega} = y(\Omega)$? In view of 11.65 if the open domain Ω is non-empty it has the *compactivorous property*:

$$\forall K \text{ compact } \subset \Omega, \exists N > 0, \forall n > N, \quad K \subset \Omega_n.$$

Therefore for each $\varphi \in \mathcal{D}(\Omega)$, there exists $N > 0$ such that

$$\forall n > N, \quad \varphi \in \mathcal{D}(\Omega_\backslash).$$

This property is sufficient to show that y^* verifies the variational equations 11.69.For each $\varphi \in \mathcal{D}(\Omega)$, its support $K := \mathrm{supp}\varphi$ is compact in Ω. As a result of the uniform complementarity convergence of $d_{C\Omega_\backslash}$

$$\exists N > 0, \forall n \geq N, \quad K \subset \Omega_n \quad \Rightarrow \varphi \in \mathcal{D}(\Omega_\backslash).$$

Then for $n \geq N$

$$0 = \int_{\Omega_n} \nabla y_n \cdot \nabla \varphi - f \varphi \, dx = \int_D \nabla y_n \cdot \nabla \varphi - f \varphi \, dx$$

$$\text{converges to } 0 = \int_D \nabla y^* \cdot \nabla \varphi - f \varphi \, dx = \int_\Omega \nabla y^* \cdot \nabla \varphi - f \varphi \, dx$$

and by density of $\mathcal{D}(\Omega)$ in $H_0^1(\Omega)$

$$\exists y^* \in H_0^1(D), \quad \forall \varphi \in H_0^1(\Omega)$$

$$\int_\Omega \nabla y^* \cdot \nabla \varphi - f \varphi \, dx = 0.$$

It remains to find under what conditions $y^* \in H_0^1(\Omega)$. The result is true in dimension $N = 1$. It follows from the fact that $H_0^1(\Omega_n) \subset C^0(\overline{\Omega_n})$ for all $n \geq 1$. In dimensions N higher than 1, the above result is no longer true and some additional assumptions are required on the sets Ω_n.

For all $n \geq 1$, assume that $\partial\Omega_n \neq \emptyset$. Now since $y_n \in H_0^1(D)$ and $d_{C\Omega_\setminus} \in W^{1,\infty}(D)$, then

$$y_n \, d_{\Omega_n} = 0 \quad \text{in } H_0^1(D).$$

The sequence $\{b_{\Omega_n}\}$has a convergent subsequence and there exists $\Omega^*, \partial\Omega^* \neq \emptyset$, such that

$$b_{\Omega_n} \rightharpoonup b_{\Omega^*} \quad \text{in } H^1(D)\text{-weak}.$$

As a result

$$d_{\Omega_n} \to d_{\Omega^*} \quad \text{in } H^1(D)\text{-weak and } C^0(\overline{D})\text{-strong}$$

$$d_{C\Omega_\setminus} \to d_{C\Omega^*} \quad \text{in } H^1(D)\text{-weak and } C^0(\overline{D})\text{-strong}$$

$$d_{\partial\Omega_n} \to d_{\partial\Omega^*} \quad \text{in } H^1(D)\text{-weak and } C^0(\overline{D})\text{-strong}.$$

In particular $\overline{C\Omega^*} = \overline{C\Omega}$ which means that

$$\Omega = \int \Omega^* \quad \text{and} \quad \partial\Omega \subset \partial\Omega^*.$$

Moreover

$$0 = y_n \, d_{\Omega_n} \to y^* \, d_{\Omega^*} \quad \text{in } H_0^1(D)\text{-weak}$$

$$y^* \, d_{\Omega^*} = 0 \text{ q.e. in } D$$

$$y^* = 0 \quad \text{q.e. in } D \backslash \overline{\Omega^*}.$$

For the family of sets characterized by Theorem 11.8 we have the desired continuity since the sequence of sets Ω_n and its limit Ω^* all have $C^{1,\lambda}$, $0 < \lambda \leq 1 - N/p$ boundaries and necessarily

$$y^*_{|\partial\Omega^*} = 0.$$

Theorem 11.14. *Let $p > N$ be given. Let $\{\Omega_n\}$, $\partial\Omega_n \neq \emptyset$, be a sequence of open subsets of D such that*

$$d_{C\Omega_\setminus} \to d_{C\Omega} \quad \textit{in } C^0(D)\textit{-strong} \tag{11.71}$$

for some open subset Ω (possibly empty) of D. Assume that there exist $c > 0$ and $h > 0$ such that

$$\forall n, \quad \int_{U_h(\partial\Omega_n)} |D^2 b_{\Omega_n}(x)|^p \, dx < c. \tag{11.72}$$

Then the domains $y(\Omega_n)$and $\Omega = \int \Omega^*$ are all of class $C^{1,\lambda}$, $0 < \lambda \leq N/p$, and for the solutions $y(\Omega_n)$ of the Dirichlet problem 11.67

$$y(\Omega_n) \to y(\Omega) \qquad \textit{in } H_0^1(D) \tag{11.73}$$

where $y(\Omega)$ *is the solution of problem 11.69 in* $H_0^1(\Omega; D)$ *or*

$$y(\Omega)_{|\Omega} \in H_0^1(\Omega) \tag{11.74}$$

is the solution of the homogeneous Dirichlet problem 11.69 in the domain Ω.

Proof. In view of 11.72 from Theorem 11.4 and Corollary 11.2, there exists a subsequence of $\{b_{\Omega_n}\}$, still indexed by n, and a set $\Omega^* \subset D$, $\partial\Omega^* \neq \emptyset$, such that

$$b_{\Omega_n} \to b_{\Omega^*} \quad \text{in } W^{1,p}(D) - \text{strong}$$

and the sets Ω^* and $\{\Omega_n\}$'s all have Hölderian boundaries of class $C^{1,\lambda}$, $0 < \lambda \leq N/p$. in particular

$$d_{C\Omega_\backslash} = b_{\Omega_n}^- \to b_{\Omega^*}^- = d_{C\Omega^*} \quad \text{in } C^0(D)$$

and

$$d_{C\Omega^*} = d_{C\Omega} \quad \Longrightarrow \quad \overline{C\Omega^*} = \overline{C\Omega} \quad \Longrightarrow \quad \Omega = \int \Omega^*.$$

Therefore Ω^* is of class $C^{1,\lambda}$. We conclude that

$$y^* = 0 \quad \text{on } \partial\Omega^*$$

and $y^* \in H_0^1(\Omega)$. Hence y^* is the solution of 11.69. To complete the proof we show that the whole sequence converges to $y(\Omega) \in H_0^1(\Omega)$ in the $H_0^1(D)$-strong topology. This is readily seen by noting that

$$\int_D |\nabla y_n|^2\,dx = \int_D \chi_{\Omega_n} f\, y_n\,dx \to \int_D \chi_\Omega f\, y^*\,dx = \int_D |\nabla y^*|^2\,dx.$$

Hence

$$\int_D |\nabla y_n - \nabla y^*|^2\,dx = \int_D |\nabla y_n|^2\,dx + \int_D |\nabla y^*|^2\,dx - 2\int_D \nabla y_n \cdot \nabla y^*\,dx$$

and the right-hand side converges to zero as n goes to infinity.

12. Derivative in a Fluid-Structure problem

12.1 Introduction

The problem arose from the industry. In the nineties, the "Aerospatial Inc." was interested in the stabilization of satellites. The observed instabilities were created by the combustion of the gas - the tank of a satellite represents more than a third of its total weight-. The tank is conceived to store enough gas to allow the satellite to manoeuver for years. Its weight diminishes significantly, and thus brings short structural instabilities that make up the satellite. In

1992, Clariond-Zolésio conducted a numerical study of the phenomenon and calculated numerically the first eigenvalues of the system. It then appears that, if we want to have a "control" on the spectrum, then we need to give information on the shape of the satellite which is sensitive to certain vibration eigenfunctions to which the structure is submitted; we then can hope to find the shape that remains the most insensitive (the less excited) to these modes. In that direction, we are here interested in the shape derivative of the first eigenvalue of the system. It is obvious that this control cannot have a meaning unless it acts on the eigenvalues that are significant to the system, and at which the first one may not play a major role. Nevertheless, this study constitutes an interesting approach for the study of the stability of the dynamical problem.

12.2 Definitions and existence results

The fluid is supposed to be irrotational and incompressible. Its speed V is then the gradient of a potential function: $V = \nabla\varphi$. The equations are stated on the manifold Γ. All the tangential differential operators are expressed via the intrinsic geometry tools such as the oriented distance function b_Ω and the projection mapping p. The tangential gradient of a function $\varphi \in H^1(\Gamma)$ is defined by $\nabla_\Gamma(\varphi) = \nabla(\varphi op)_{|\Gamma}$, the tangential divergence is given by $div_\Gamma\varphi = div(\varphi op)_{|\Gamma}$, the same way the Laplace Beltrami operator is defined by $\Delta_\Gamma(\varphi) = \Delta(\varphi op)_{|\Gamma}$. The membrane is modeled by linear elasticity (i.e. we consider small deformations of the structure) so that its elastic energy is the L^2-norm square of the tangential deformation tensor $\varepsilon_\Gamma = \frac{1}{2}\{D_\Gamma + {}^*D_\Gamma$ where for $u \in H^1(\Gamma; I\!R^N)$, we have ${}^*D_\Gamma u = (\nabla_\Gamma(u_1), ...\nabla_\Gamma(u_N))$. The tangential deformation tensor is defined by $\varepsilon_\Gamma(u) = \varepsilon(uop)_{|\Gamma}$. The integration by parts formulas on a manifold indicates that the elastic energy is very close to a vectorial version of the Laplace Beltrami operator. The point of view is the speed method by help of which we perturb the domain - which is a manifold here - using a speed vector field V whose flow $T_t(V)$ is linked to V by the following relation: $V = \frac{\partial T_t}{\partial t} oT_t^{-1}$. Then the manifold Γ is perturbed into $\Gamma_t = T_t(\Gamma)$ and if we denote by u_t the solution of the problem on Γ_t, its shape tangential derivative is defined by $u'_\Gamma = \dot{u}_{|\Gamma} - D_\Gamma u.V$ where $\dot{u} = \frac{\partial(u_t oT_t)}{\partial t}_{t=0}$ is the classical material derivative. Furthermore, the tangential derivative is a term used for the derivative with respect to a smooth manifold of solutions of tangential boundary differential value problems, conversely to the shape boundary derivative which is an appropriate term for the derivative of solutions of tangential differential operators - that is with respect to a manifold with no boundary -. This notion of shape derivative turns out to be well adapted as, according to a general structure theorem it only depends on the normal component of the speed field V and on its component $< V, \nu >$ where ν is the normal field to $\partial\Gamma$ outgoing to Γ.

12.2.1 The dynamical problem. In the forthcoming, τ will denote the time parameter. The dynamical problem of the coupling fluid-membrane is formulated from the Action extremality. The action A is the difference between the internal energy of the system and its kinetic energy. The internal energy is itself the sum of the potential energy V_g and the work of external forces W_{ext}:

$A = \int_0^T \{\mathcal{E}_i - \mathcal{E}_k\} d\tau$ where $\mathcal{E}_i = \int_\Gamma (E_d - W_{ext} + V_g) d\Gamma$. The sign minus in face of W_{ext} is because the pressure brings some energy to the system. Then the action becomes $A = \int_0^T [(E_d - W_{ext} + V_g) - (E_{k,\Gamma} + E_{k,\Omega})]\, d\tau$ where T denotes an arbitrary final time, E_d is the deformation energy associated with Γ, W_{ext} is the work of external forces (here the gravity), V_g is the potential energy associated with the fluid, $E_{k,\Gamma}$ is the kinetic energy associated with Γ and $E_{k,\Omega}$ is the kinetic energy of the fluid. The action A is extremal at the equilibria, not unique. This condition being translated by classical extremality conditions from which derives the strong formulation of the coupled problem whose solution is (u, φ), the vector displacement of the membrane and φ the potential of the fluid.

Let us denote by $u = v + w.n$ the vector displacement with $v(x)n(x) = 0$. $u.n$ is an element of the space $L^2(0, T, H^1_{0,*}(\Gamma))$ where $H^1_{0\,*}(\Gamma)$ is the Sobolev space $H^1_0(\Gamma)$ of functions u which satisfy $\int_\Gamma u.n\, d\Gamma = 0$. Following the generalized Bernoulli unsteady linearized condition we have $p - p_0 = -\rho_0(\varphi_\tau + gu.k)$. The expression of the action A becomes then:

$$A(u, \varphi) = \int_0^T \int_\Gamma (\frac{1}{2}\varepsilon_\Gamma(u)..\varepsilon_\Gamma(u) + \rho_0(\varphi_\tau + gu.k)w) d\Gamma d\tau -$$

$$+ \int_0^T \int_\Omega \rho_0 gz dx\ \tau - \rho_m \int_0^T \int_\Gamma \frac{1}{2}|u_\tau|^2 d\Gamma d\tau - \rho_0 \int_0^T \int_\Omega \frac{1}{2} |\nabla\varphi|^2 dx d\tau. \tag{12.1}$$

where ρ_m is the specific weight of the surface Γ and ρ_0 the volume density of the fluid. We introduce

$$\mathcal{S} = (L^2(0, T, H^1_0(\Gamma; T\Gamma) \times H^1_{0,*}(\Gamma; T\Gamma)), L^2(0, T, H^1(\Omega)))$$

The optimality conditions leads to:

$$A(u, \tilde{u}, \varphi, \tilde{\varphi}) = 0,\ \forall (\tilde{u}_\Gamma, \tilde{u}.n, \tilde{\varphi}) \in \mathcal{S} \tag{12.2}$$

or explicitly,

$$\int_0^T \int_\Gamma (\varepsilon_\Gamma(u)..\varepsilon_\Gamma(\tilde{u}) + \rho_0(\varphi_\tau \tilde{w} + g\tilde{u}.kw + gu.k\tilde{w}) + \rho_0 \tilde{\varphi}_\tau w) d\Gamma \tag{12.3}$$

$$-\rho_m \int_0^T \int_\Gamma \tilde{u}_\tau.u_\tau d\Gamma\ d\tau - \rho_0 \int_0^T \int_\Omega \nabla\varphi \nabla\tilde{\varphi} dx\ d\tau = 0, \qquad \forall (\tilde{u}, \tilde{\varphi}) \in \mathcal{S}$$

and after integrations by parts, with

$$(\tilde{u},\tilde{\varphi}) \in \mathcal{D}(0,T) \times \mathcal{D}(0,T,H_0^1(\Gamma;T\Gamma) \times H_{0,*}^1(\Gamma;T\Gamma)), L^2(0,T,H^1(\Omega)))$$

can be rewritten as follows

$$A(u,\tilde{u},\varphi,\tilde{\varphi})=$$
$$= \int_0^T \int_\Gamma (\varepsilon_\Gamma(u)..\varepsilon_\Gamma(\tilde{u}) + \rho_0(\varphi_\tau \tilde{w} + g\tilde{u}.kw + gu.k\tilde{w}) - \rho_0\tilde{\varphi}w_\tau)d\Gamma\, d\tau +$$
$$+\rho_m \int_0^T \int_\Gamma \tilde{u}.u_{\tau\tau} d\Gamma d\tau + \rho_0 \int_0^T (\int_\Omega \Delta\varphi\tilde{\varphi}dx + \int_\Gamma \frac{\partial\varphi}{\partial n}\tilde{\varphi}d\Gamma)d\tau. \tag{12.4}$$

φ turns to be the unique solution of the problem

$$\begin{cases} -\Delta\varphi & = 0 \quad \text{in } \Omega \\ \dfrac{\partial\varphi}{\partial n} & = w_\tau \quad \Gamma \\ \dfrac{\partial\varphi}{\partial n} & = 0 \quad \Sigma \end{cases} \tag{12.5}$$

as soon as $w = u.n$ is taken such that $\int_\Gamma w\, d\Gamma = 0$ in order to get the uniqueness of such an element φ in $H_*^1(\Omega)$. In fact, this zero mean value condition is imposed, as the fluid is incompressible, using Stockes' formula:

$$\int_\Gamma < \nabla\varphi, n >_{\mathbb{R}^N} d\Gamma = \int_\Omega div\ (\nabla\varphi)dx = 0.$$

Then we obtained the equation satisfied by the fluid. The strong equation associated with the vector displacement u is the following: $\forall \tilde{u} \in L^2(0,T,H_{0\,*}^1(\Gamma))$,

$$\int_0^T \int_\Gamma \{< -\mathbf{div}(\varepsilon_\Gamma(u)) + H\varepsilon_\Gamma(u).n \tag{12.6}$$

$$+\rho_0(\varphi_\tau n + gu.n\,k + gu.k\,n) + \rho_m u_{\tau\tau}\,, \tilde{u} >\}\ d\Gamma d\tau = 0$$

which gives the following strong formulation on $[0,T] \times \Gamma$

$$-\mathbf{div}_\Gamma(\varepsilon_\Gamma(u)) + H\varepsilon_\Gamma(u).n + \rho_0(\varphi_\tau n + gu.n\,k + gu.k\,n) + \rho_m u_{\tau\tau} = 0. \tag{12.7}$$

When, for some of τ, we have $u_\tau = u_{\tau\tau} = 0$, we recover that $u = 0$ is the unique solution of equation (12.6).

Proposition 12.1. *The problem (5)-(7) with initial and boundary conditions has a unique solution* (u,φ) .

<u>Proof:</u> Use Galerkin's approximation. .

12.2.2 The eigenvalue problem. At this step, we are looking for vibratory solutions, that is the eigen frequencies and eigenfunctions of the system fluid-membrane. Let us denote by $(\bar{u}, \bar{\varphi})$ the solution of the dynamical problem. We are looking for vibrating solutions:

$$u = \bar{u} + \alpha(x)\; exp\,(i\,\delta\,\tau) \tag{12.8}$$

where α and δ are any functions. The initial conditions of the problem ($\bar{u}(0) = \bar{u}_0 + \alpha(x)$ and $\bar{u}_\tau(0) = \bar{u}_1$) become $u(0) = \bar{u}_0 + \alpha(x)$ and $u_\tau(0) = \bar{u}_1 + i\delta\alpha$. The solutions of this form impose to α and δ to be the solutions of an eigenvalue problem we are going to characterize. In practice, we are more interested in α which is the physical interesting value. The solutions we are looking for are in the form:

$$\begin{cases} u(x,\tau) \text{ in the form } exp(\lambda\tau)u(x), \\ \varphi(x,\tau) \text{ in the form } exp(\lambda\tau)\varphi(x). \end{cases} \tag{12.9}$$

We obtain

$$\forall(\tilde{u}, \tilde{\varphi}) \in (H_0^1(\Gamma) \times H_{0,*}^1(\Gamma), H^1(\Omega)),$$

$$\int_\Gamma \{\varepsilon_\Gamma(u)..\varepsilon_\Gamma(\tilde{u}) + \rho_0(\lambda\varphi\tilde{w} + g\tilde{u}.kw + gu.k\tilde{w}) - \rho_0\tilde{\varphi}\lambda w + \rho_m\lambda^2\tilde{u}.u\}d\Gamma$$
$$+\rho_0\,(\int_\Omega \Delta\varphi\tilde{\varphi}dx - \rho_0\int_\Gamma \tilde{\varphi}\frac{\partial\varphi}{\partial n}d\Gamma) = 0. \tag{12.10}$$

where now, φ becomes the unique solution of the following problem:

$$\begin{cases} -\Delta\varphi & = 0 & ,\Omega \\ \dfrac{\partial\varphi}{\partial n} & = \lambda w & ,\Gamma \\ \dfrac{\partial\varphi}{\partial n} & = 0 & ,\Sigma \end{cases} \tag{12.11}$$

Then from this expression we get the strong formulation (τ is a unitary tangential vector, that is such that $< \tau, n >= 0$). We set $\varphi = \lambda\,\Phi$ in order to eliminate λ in the equation . Then λ is the solution of the problem:

$$\exists(u,\varphi), (u = v + w.n) \in (H_0^1(\Gamma) \times H_{0,*}^1(\Gamma), H^1(\Omega)) \text{ such that}$$

$$-\,\mathbf{div}_\Gamma(\varepsilon_\Gamma(w.n))n + \frac{H}{2} < \varepsilon_\Gamma(w\;n).n, n > \tag{12.12}$$
$$+2\rho_0 gwk.n + \rho_0\lambda^2.K(w) + \rho_m\lambda^2 w = 0$$

$$-\mathbf{div}_\Gamma(\varepsilon_\Gamma(v))\tau + \frac{H}{2} < \varepsilon_\Gamma(v).n, \tau >= -\lambda^2\rho_m\|v\|$$

where K is the pseudo differential operator defined as the trace of the operator K_* (K_* being such that $\Phi = K_*(w)$, that is $\Phi_{|\Gamma} = K(u.n)$). The unknown becomes the couple (u, Φ) where now Φ is the solution of the following problem:

$$\begin{cases} -\Delta\Phi & = 0 \quad \text{in} \quad \Omega \\ \dfrac{\partial\Phi}{\partial n} & = w \quad \text{on} \quad \Gamma \\ \dfrac{\partial\Phi}{\partial n} & = o \quad \text{on} \quad \Sigma \end{cases} \tag{12.13}$$

For simplicity, in the forthcoming we shall write $\Phi = K(u.n) = K(w)$. It is easy to verify that Φ is an element of $H^1_*(\Omega)$. We set $\lambda = i\eta$ as we are interested in the above vibratory solutions where η is a real number. The eigenvalue problem can be written

$$\text{Find} \qquad (u, \lambda) \in (H^1_0(\Gamma) \times H^1_{0,*}(\Gamma), H^1(\Omega)) \times I\!R^*_+ \tag{12.14}$$

$$\text{such that} \qquad Au = \eta^2 Bu.$$

where A and B are the two following matrices:

$$A = (-a_{11} \quad 00 \quad a_{22})$$

Where

$$a_{11} = \mathbf{div}_\Gamma(\varepsilon_\Gamma(v))\tau + \frac{H}{2} < \varepsilon_\Gamma(v).n, \tau >_{I\!R^N}$$

$$a_{22} = -\mathbf{div}_\Gamma(\varepsilon_\Gamma(w.n))n + \frac{H}{2} < \varepsilon_\Gamma(w\ n).n, n >_{I\!R^N} + 2\rho_0 g w k.n$$

$$B = (\rho_m \quad 00 \quad \rho_0 K(w) + \rho_m)$$

Lemma 12.1. *The first eigenvalue of the coupled problem fluid-membrane is defined using the classical Rayleigh quotient:*

$$\eta_1^2(\Gamma) = Min_{\{u \in (H^1_0(\Gamma;T\Gamma) \times H^1_{0,*}(\Gamma;T\Gamma))\}} \tag{12.15}$$

$$\{< Au, u >_{H^1 \times H^1} \times < Bu, u >^{-1}{}_{L^2(\Gamma) \times L^2(\Gamma)}\}$$

$$= Min_{\{u \in (H^1_0(\Gamma;T\Gamma) \times H^1_{0,*}(\Gamma;T\Gamma))\}}$$

$$\{\int_\Gamma \varepsilon_\Gamma(u)..\varepsilon_\Gamma(u) + \rho_0 g k.u\} \times \{\int_\Gamma \rho_m |u|^2 + \rho_0 K(w).w d\Gamma\}^{-1}$$

Proof. In order to show that η_1^2 is well defined, we have to show the following lemma

Lemma 12.2. $\forall f \in H^{-\frac{1}{2}}(\Gamma)$, $\exists!\, u \in (H_0^1(\Gamma;T\Gamma) \times H_{0,*}^1(\Gamma;T\Gamma)$, $(I_d + K)u = f\, on\, \Gamma$

Proof. The condition on Γ is equivalent to showing that there exists a unique $\Phi \in H^1(\Omega)$ - in fact $H_*^1(\Omega)$ - such that

$$\frac{\partial \Phi}{\partial n} + \Phi = f \text{ on } \Gamma. \tag{12.16}$$

For this we consider the following minimization problem and we show that there exists at most one solution:

$$Min_{\{\Phi \in H_*^1(\Omega)\}} \int_\Omega \frac{1}{2}|\nabla\Phi|^2 dx + \int_\Gamma \frac{1}{2}\Phi^2 d\Gamma - < f, \Phi >_{H^{-\frac{1}{2}} \times H^{\frac{1}{2}}_{0\,0}(\Gamma)} . \tag{12.17}$$

The functional $J(\Phi) = \int_\Omega \frac{1}{2}|\nabla\Phi|^2 dx + \int_\Gamma \frac{1}{2}\Phi^2 d\Gamma - < f, \Phi >_{H^{-\frac{1}{2}} \times H^{\frac{1}{2}}_{0\,0}(\Gamma)}$ is coercive on $H_*^1(\Omega)$. Furthermore, it is lower semi continuous so that the minimum is obviously reached over the space $H_*^1(\Omega)$ - and the uniqueness is easy - then the operator $(I_d + K)$ is invertible and $\eta_1^2(\Gamma)$ has a sense.

Lemma 12.3. *The operator K is self-adjoint.*

Proof. We denote by $\partial\Omega$ the boundary of Ω. We have $\partial\Omega = \Sigma \cup \Gamma$. And then Γ has a boundary made of two points where the vector displacement u (and the vector field V) is zero. In such a case, it is obvious that the dual space of $H_0^{\frac{1}{2}}(\Gamma)$ is not equal to the Sobolev space $H^{-\frac{1}{2}}(\Gamma)$.

The Sobolev space $H^{\frac{1}{2}}_{0\,0}(\Gamma)$ is: $H^{\frac{1}{2}}_{0\,0}(\Gamma) = \{ \Phi_{|\Gamma},\ \ \Phi \in H^{\frac{1}{2}}(\partial\Omega), \Phi = 0$ a.e. on Σ $\}$ and $(H^{\frac{1}{2}}_{0\,0}(\Gamma))' = H^{-\frac{1}{2}}(\Gamma)$. It is easy to verify that the operator

$$\begin{array}{rll} K: & H^{-\frac{1}{2}}(\Gamma) & \longrightarrow H^{\frac{1}{2}}_{0\,0}(\Gamma) \\ & u & \mapsto \Phi \end{array}$$

is in $\mathcal{L}(\ H^{-\frac{1}{2}}(\Gamma), H^{\frac{1}{2}}_{0\,0}(\Gamma)\ )$. For any ξ in $H^{-\frac{1}{2}}(\Gamma)$, let us define $\mathcal{Z} = K(\xi)$. Then we have

$$\begin{aligned} \int_\Omega \nabla\mathcal{Z}.\nabla\Phi dx &= \int_\Omega -\Delta\mathcal{Z}\Phi dx + \int_{\Gamma\cup\Sigma} \frac{\partial\mathcal{Z}}{\partial n}\, \Phi\, dl \\ &= \int_\Omega -\Delta\Phi.\mathcal{Z} dx + \int_{\Gamma\cup\Sigma} \frac{\partial\Phi}{\partial n}\, \mathcal{Z}\, dl \\ &= \int_{\Gamma\cup\Sigma} \frac{\partial\Phi}{\partial n}\, \mathcal{Z}\, dl \end{aligned}$$

but $\Delta\mathcal{Z} = \Delta\Phi = 0$ in Ω so that $< K(\xi), \Phi > = < \xi, K(\Phi) >_{H^{-\frac{1}{2}}(\Gamma) \times H^{\frac{1}{2}}_{0\,0}(\Gamma)}$ and K is self-adjoint.

12.3 The static case

In the forthcoming, t will denote the "perturbation's variable " which is associated with a flow transformation $T_t(V)$ by help of which the domain is perturbed: $T_t(V)(\Omega) = \Omega_t$. In this section, we are looking for the static equations of the problem in order to determine the shape tangential derivative of u_t, the solution of on the perturbed manifold Γ_t. In the dynamical case, the coupling was coming from Bernouilli's linearized equation - as the problem dealt with a succession of small displacements - and in the static case, $p - p_0 = -\rho_0 g u.k$ so that the variable φ does not appear anymore. In the static situation, the coupling does not exist anymore. This is an expected situation because in the static case, the speed of the fluid is zero and the membrane is naturally fixed. Then the strong equations are obtained via the dynamical equation for the vector u. For the stationary problem, at the equilibria we have:

$$-\mathbf{div}(\varepsilon_\Gamma(u)) + H\varepsilon_\Gamma(u).n + \rho_0\, g u.n\, k + \rho_0\, g u.k\, n = 0 \quad \text{on } \Gamma. \tag{12.18}$$

This equation is made of a membrane term and several "easy to differentiate " terms. Let us then concentrate on the differentiation of the central term; that is the intrinsic operator associated with a membrane: $\int_\Gamma \{< -\mathbf{div}(\varepsilon_\Gamma(u)) + H\varepsilon_\Gamma(u).n, \tilde{u} >\} d\Gamma = \int_\Gamma \varepsilon_\Gamma(u)..\varepsilon_\Gamma(\tilde{u}) d\Gamma \quad \forall \tilde{u} \in H^1(\Gamma; \mathbb{R}^N)$. In the next

section, we give a result of shape tangential derivative for the term corresponding to the membrane with Dirichlet conditions on the boundary $\partial\Gamma$ of Γ. Of course, in our context, the membrane being attached at its two extremities, the vector field V is zero on $\partial\Gamma$ so that the following given boundary condition fails. Anyway, we prefer to consider the general situation.

12.4 Shape derivative of the solution

In this section, we give the context of shape tangential derivative using intrinsic geometry tools. Then we compute that derivative using the method of the commutator. In Let D a hold all of $\mathbb{R}^N$ contain Ω an open set of $\mathbb{R}^N, \Gamma = \partial\Omega$ its boundary of class $C^2$and A is a L^∞ function. The problem is then equivalent to the minimization of the elastic energy of the system. It can be written in the following way:

$$Min_{\left\{u \in \frac{H^1(\Gamma)}{\mathcal{N}}\right\}} \frac{1}{2} \int_\Gamma \left\{ A\varepsilon_\Gamma(u)..\varepsilon_\Gamma(u) \right\} d\Gamma - \int_\Gamma \mathbf{f} u d\Gamma \tag{12.19}$$

where u is the vector displacement of a point of the membrane $\mathbf{f}$ is an exterior force which acts on the membrane, $\varepsilon_\Gamma = \frac{1}{2}(D_\Gamma + {}^*D_\Gamma)$ is the tangential deformation tensor and $\mathcal{N} = ker(\varepsilon_\Gamma)$.

Remark 12.1. in the special case of Ω included in $I\!R^3$, $\mathcal{N} = \Big\{Au + B, A =$ rotation matrix, $B =$ translation vector$\}$ i.e. $\mathcal{N} = \{$rigid body motion $\}$

In order to differentiate with respect to the domain the solution u, we need to define the transformation $T_t(V)$ which maps the domain Ω (and more precisely, its boundary Γ) into Ω_t(whose boundary is Γ_t). Classically, to any speed vector field $\mathbf{V} \in C^0([0,\tau[, C^k(\bar{D}, I\!R^N))$ $(k \geq 2)$, we can associate a transformation $T_t(\mathbf{V}) \in C^1([0,\tau[, C^k(\bar{D}, I\!R^N))$, which is the flow of $\mathbf{V}$. $T_t(\mathbf{V})$ and $\mathbf{V}$ are linked by the following equality: $V(t,x) = (\frac{\partial T_t}{\partial t}) o T_t^{-1}(x)$. Then we have the shape boundary derivative (or " tangential derivative") u'_Γ of u, solution of the equation (2).

Definition 12.1. *If B is any matrix, then we denote by S its symmetrized part $S(B) = \frac{1}{2}\{B +^* B\}$. Antisym$(B)$ denotes the antisymetric part of B : Antisym$(B) = \frac{1}{2}\{B -^* B\}$*

Theorem 12.1. *Let u_t be the solution of the problem with a right hand side $\mathbf{f}$ in $H^{\frac{3}{2}+\varepsilon}(D)$ and $A \in L^\infty(\Gamma)$. The shape tangential derivative u'_Γ of u_t exists in $L^2(\Gamma)$ and is the unique solution of the following equation:*

$$-\,\mathbf{div}_\Gamma(A\varepsilon_\Gamma(u'_\Gamma)) + HA\varepsilon_\Gamma(u'_\Gamma).n \tag{12.20}$$

$$= \mathbf{div}_\Gamma\Big[2A\varepsilon_\Gamma(u)\,(\frac{1}{2}HI_d - D^2b)V.n + AS(D_\Gamma u.\nabla_\Gamma(V.n)^* n)$$

$$+A\ Antisym(^*D_\Gamma u.D^2 b\ V.n) + A\varepsilon_\Gamma(u).n^*\nabla_\Gamma(V.n) + A'_\Gamma\varepsilon_\Gamma(u)\Big]$$

$$-H\ A\varepsilon_\Gamma(u).n - AS(D_\Gamma u.\nabla_\Gamma(V.n)\ ^*n).n - A'_\Gamma\varepsilon_\Gamma(u).n$$

$$+2\frac{\partial \mathbf{F}}{\partial n}V.n + 2H\mathbf{f}V.n on\ \ \Gamma$$

With the boundary condition:

$$u'_\Gamma = -D_\Gamma u.\nu\ V.\nu \quad on \quad \partial\Gamma$$

where $M(V.n) = \nabla_\Gamma(V.n).^*n - V.nD^2b$, H is the mean curvature on Γ and A'_Γ, the shape boundary derivative of A is $A'_\Gamma = \frac{\partial A}{\partial n}V.n$.

The shape tangential derivative (or shape Boundary derivative) u'_Γ of the solution u_t of the initial problem defined on Γ_t, is defined in the following way:

Definition 12.2. *Let u be a vector. If $\dot{u} = lim_{t\searrow 0}\{\frac{u_t o T_t - u}{t}\}$ in $H^1(\Gamma)$-norm exists then $u'_\Gamma = \dot{u} - D_\Gamma u.V$*

The material derivative $\dot{u}$ of the problem on Γ_t has to exist in order to have the shape boundary derivative u'_Γ well defined. Let $\Phi \in \mathcal{D}(\mathbb{R}^N)$ be such that its restriction to Γ is equal to φ, an element of $H^1(\Gamma)(\Phi_{|\Gamma} = \varphi)$. Let us denote by P the projection on Γ. We choose Φ such that $\Phi = \varphi o P$; Φ is then built as an extension of φ in a tubular neighborhood $\mathcal{U}$ of $\Gamma : \mathcal{U} = \bigcup_{-h \leq t \leq h} \Gamma_t$ where $\Gamma_t = \{x(t) \in \mathbb{R}^N / x(t) = X + t.\nabla b_\Omega(X), X \in \Gamma\}$. Such a construction gives to Φ the property of being constant along the normal vector field n to Γ at each point X of Γ. The so-called function b_Ω is the oriented distance function defined as follows:

$$b_\Omega(x) = \begin{cases} d_\Gamma(x) & if x \in \mathbb{R}^N \setminus \overline{\Omega}. \\ -d_\Gamma(x) & if x \in \Omega. \end{cases}$$

More precisely, b_Ω is C^k in a neighborhood of Γ if and only if the domain Ω is C^k itself. The gradient of b_Ω is an extension to that neighborhood of the normal vector field n defined on Γ. The last property of the oriented distance function which we are interested in is that the matrix $D^2 b$ (second order derivatives of b, Γ being considered of class C^2) contains the curvatures and the main directions of curvatures . Now, we have roughly defined the context and the tools that appear in the result for the shape boundary derivative of the solution to our problem. Here we face a problem: the Shape derivative of a function y_t (or of a vector) defined over the whole space $\mathbb{R}^N$ goes through the operators such that, for example, $(\nabla y)' = \dot{\nabla y_t} - D(\nabla y).V = \nabla y'$ where similarly, $y' = \dot{y} - \nabla y.V$. This property is very useful in order to characterize the shape derivative of a state. For tangential calculus, the shape boundary (or shape tangential) derivative has not got this property. So, this technique seems to be no more valid. Here the interest is that we find a relation between the shape and the shape boundary derivatives in order to recover this very useful property. The two following lemmas are crucial in order to derive the final result. The first one is an integration by parts lemma on a manifold which has a boundary and the second one deals with the differentiation through an integral defined on a smooth perturbed manifold.

Lemma 12.4. *Let Γ be a C^2 manifold whose boundary $\partial\Gamma$ is also C^2. y being any vector in $H^1(\Gamma; \mathbb{R}^N)$ and φ a regular function, we get:*

$$\int_\Gamma \varepsilon_\Gamma(y)..\varepsilon_\Gamma(\varphi) d\Gamma =$$

$$\int_\Gamma \{-\mathbf{div}_\Gamma(\varepsilon_\Gamma(y)) + H(\varepsilon_\Gamma(y)).n\} \varphi \, d\Gamma - \int_{\partial\Gamma} \varphi \varepsilon_\Gamma(y).\nu \, dl \tag{12.21}$$

where $\nu(X)$ is the normal field to $\partial\Gamma$ outgoing to Ω and contained in the tangent space to $\partial\Gamma$ at point X.

Lemma 12.5. *Let $F \in H^1(\bar{D})$,*

$$\frac{\partial}{\partial t}\Big\{\int_{\Gamma_t} F(t,x)d\Gamma_t\Big\}_{t=0} = \int_\Gamma (F'_\Gamma + HFV.n)d\Gamma = \int_\Gamma F' + (\frac{\partial F}{\partial n} + HF)V.n \quad d\Gamma \tag{12.22}$$

Lemma 12.6. $\frac{\partial}{\partial t}(p_t oT_t)_{|t=0} = V + b^* DV.\nabla b + \nabla(b < nop, Vop - V >)$

Proof. p_t being the projection mapping onto Γ_t and b_t the oriented distance function on Γ_t, we have:

$$\begin{aligned} p_t oT_t &= T_t - b_t oT_t.\nabla b_t oT_t \\ &= T_t - b_t oT_t.{}^* DT_t^{-1}\nabla(b_t oT_t) \end{aligned}$$

but we need the material derivative of b_t in order to differentiate this relation: $\frac{\partial}{\partial t}(p_t oT_t)_{|t=0} = V - \frac{\partial}{\partial t}(b_t oT_t)_{|t=0}.\nabla b + b^* DV.\nabla b - b.\nabla(\frac{\partial}{\partial t}(b_t oT_t)_{|t=0})$.

Lemma 12.7. $\frac{\partial}{\partial t}(b_t oT_t)_{|t=0} = - < nop, Vop - V >$

Proof. $b^2_{\Omega_t}(x) = Min_{y_t \in \Gamma_t}\|x - y_t\|^2 = Min_{y \in \Gamma}\|x - T_t(y)\|^2$ and we differentiate this expression $2b_\Omega(x)b'_\Omega(x) = 2 < x - y, -V(0,y) >$ with $y = p(x)$ then finally $b'_\Omega(x) = < \frac{x-p(x)}{b_\Omega(x)}, -V(0,p(x)) >$ and $\frac{x-p(x)}{b_\Omega(x)} = n(p(x)) = nop(x)$. Notice that $b'_\Gamma = 0$.

Lemma 12.8. *$DP = I_d - \nabla b.{}^*\nabla b - bD^2 b$ is the differential of the projection mapping p, $D(uop) = D(uopop) = D(uop).DP$*

Lemma 12.9. $\frac{\partial}{\partial t}(u_t op_t)op = u'_\Gamma op$.

Proof. For any function y defined in a neighborhood of Γ, $y'_\Gamma = (y')_{|\Gamma} + \frac{\partial y}{\partial n}$ then for a function y in the form yop, that is with zero normal derivative, $(yop)'_{|\Gamma} = y'_\Gamma$. And the same argument is still valid in the vectorial case.

Proposition 12.2. *u is a vector in $H^1(\Gamma, I\!R^N)$ and its transported, u_t is a vector in $H^1(\Gamma_t, I\!R^N)$. In a neighborhood $\mathcal{U}$ of Γ we have the following relation between the Shape and the Shape boundary derivatives:*

$$(u_t op_t)' = u'_\Gamma op + D(uop)op.m(V) + D(uop)op.V - D(uop).V + b\,D(uop)\,D^2 b.V \tag{12.23}$$

where $m(V) = b^* DV\nabla b + \nabla(b < nop, Vop - V >)$.

Remark 12.2. $m(V)$ vanishes on Γ and $[D(uop)op]_{|\Gamma}.V = [D(uop)]_{|\Gamma}.V$ so that we recover the relation $(u_t op_t)'_{|\Gamma} = u'_\Gamma$.

Proof. Using the definition, we have $(u_t op_t)' = \dot{\widehat{(u_t op_t)}} - D(uop).V$. Using the obvious property $p_t op_t = p_t$, we express the derivative of $(p_t op_t oT_t)$ at $t = 0$. This property of the projection mapping is essential in order to get the relation between the shape and the shape boundary derivatives. In fact, the tangential gradient of u is only defined on the manifold so that in the relation we need to get an expression of this gradient in a neighborhood of Γ. This is possible thanks to the above property.

$$\begin{aligned}\dot{\widehat{(u_t o p_t)}} &= \frac{\partial}{\partial t}(u_t o p_t o T_t)_{|t=0}\\ &= \frac{\partial}{\partial t}(u_t o p_t o p_t o T_t)_{|t=0}\\ &= \frac{\partial}{\partial t}(u_t o p_t)_{|t=0} op + D(uop)op.\frac{\partial}{\partial t}(p_t o T_t)_{|t=0}\\ &= \frac{\partial}{\partial t}(u_t o p_t)_{|t=0} op + (D_\Gamma u)op.\frac{\partial}{\partial t}(p_t o T_t)_{|t=0}\end{aligned}$$

and then using lemmas 3.5 and 3.6 we get:

$$\dot{\widehat{(u_t o p_t)}} = u'_\Gamma op + D(uop)op.(V + b^* DV.\nabla b + \nabla(b < nop, Vop - V >))$$

but $D(uopop) = D(uop).DP = D(uop).(I_d - \nabla b^* \nabla b - bD^2 b)$ and so we get for the shape derivative of $(u_t o p_t)$:

$$(u_t o p_t)' =$$

$$= u'_\Gamma op + D(uop)op.(V + b^* DV.\nabla b + \nabla(b < nop, Vop - V >)) - D(uop).V$$

$$= u'_\Gamma op + D(uop)op.m(V) + D(uop)op.V - D(uop)\,(I_d - \nabla b^* \nabla b - bD^2 b).V =$$

$$= u'_\Gamma op + D(uop)op.m(V) + D(uop)op.V - D(uop)\,(I_d - bD^2 b).V =$$

$$= u'_\Gamma op + D(uop)op.m(V) + D(uop)op.V - D(uop).V + b\,D(uop)\,D^2 b.V$$

as $D(uop).\nabla b = 0$ by construction. We have then proved the relation between the shape and the shape boundary derivatives in a neighborhood $\mathcal{U}$ of Γ. In order to go further into the computations, we need several lemmas.

Lemma 12.10. *E is any vector and α a scalar, we have:* $D_\Gamma(\alpha E) = \alpha D_\Gamma E + E^* \nabla_\Gamma(\alpha)$.

Lemma 12.11. *$\mathbf{a}$ and $\mathbf{b}$ are any vectors, we have* $\nabla(< \mathbf{a}, \mathbf{b} >= {}^*D\mathbf{a}.\mathbf{b} + {}^*D\mathbf{b}.\mathbf{a}$

Proof. We write the ith component of the gradient:

$$\begin{aligned}\partial_i < \mathbf{a}, \mathbf{b} > &= \partial_i(a_j bj)\\ &= \partial_i(a_j).bj + \partial_i(b_j).aj\\ &= {}^*D\mathbf{a}.\mathbf{b} + {}^*D\mathbf{b}.\mathbf{a}\end{aligned}$$

as soon as we notice that $[D(\mathbf{b})]_{ij} = \partial_j b_i$.

Lemma 12.12. *Let n_t be the normal vector on Γ_t the perturbed manifold. Its shape tangential derivative is given by:* $n'_\Gamma = -\nabla_\Gamma(V.n)$.

Proposition 12.3. *$\mathcal{U}$ is a neighborhood of Γ, E is any vector in $H^1(\mathcal{U})$ and E_Γ its tangential component, we have:* $-D^2 b.E_\Gamma = {}^*D_\Gamma E.n - \nabla_\Gamma(E.n)$.

Proof. It follows easily by writing $< E_\Gamma, n >= 0$ on Γ where E_Γ is a tangential vector. Then we differentiate this relation to have:

$$\begin{aligned}
\nabla_\Gamma(< E_\Gamma, n >) &= 0 \text{ on } \Gamma \\
&= {}^*D_\Gamma E_\Gamma.n + {}^*D_\Gamma n.E_\Gamma \\
&= {}^*D_\Gamma E_\Gamma.n + D^2b.E_\Gamma \text{ on } \Gamma \\
&= {}^*D_\Gamma E.n + {}^*D_\Gamma(E.nn).n + D^2b.E \\
&= {}^*D_\Gamma E.n - E.n^*D_\Gamma n.n + \nabla_\Gamma(E.n).{}^*nn + D^2b.E \\
&= {}^*D_\Gamma E.n + \nabla_\Gamma(E.n) + D^2b.E
\end{aligned}$$

so that the proposition 3.2 is proved.

Lemma 12.13. *A being any matrix and v any vector, we have* $[D(A.v)]_{i\,j} = [A.Dv]_{i\,j} + \partial_j(A_{i\,k})\, v_k$

Proof. We develop the differential:

$$\begin{aligned}
[D(A.v)]_{i\,j} &= \partial_j(A_{i\,k} v_k) \\
&= \partial_j(v_k)\ A_{i\,k} + \partial_j(A_{i\,k})\ v_k \\
&= [A.Dv]_{i\,j}\partial_j(A_{i\,k})\ v_k
\end{aligned}$$

As the tangential gradient $\nabla_\Gamma u$ of a function u only defined on Γ can be expressed as the restriction to Γ of the classical gradient of an extension, $\nabla_\Gamma u = (\nabla(uop))_{|\Gamma}$, similarly as the matrix ϵ_Γ is made of $\frac{1}{2}(D_\Gamma + {}^*D_\Gamma)$, we have:

$$\varepsilon_\Gamma(u) = \varepsilon(uop)_{|\Gamma}$$

The restriction to Γ of the differential of the vector $m(V)$ is given by

Lemma 12.14. $Dm_{|\Gamma} = -D^2b.V^*n + \nabla_\Gamma(V.n)\ {}^*n - n^*n.DV.n^*n$

Proof. We explicit the expression of the derivative of $m(V)$:

$$[Dm(V)]_{i\,j} = [D(b^*DV\nabla b + \nabla(b < nop, Vop - V >))]_{i\,j} =$$

$$= [D(b^*DV\nabla b)]_{i\,j} + [D(\nabla(b < nop, Vop - V >))]_{i\,j}$$

$$= \partial_j(b\ \partial_i V_k \partial_k b) + \partial_j(\partial_i(b < nop, Vop - V >)) =$$

$$= [Dm(V)]_{i\,j_{|\Gamma}} =$$

$$= \partial_j b\ \partial_i V_k\ \partial_k b + \partial_j(\ \partial_i b < nop, Vop - V > + b\ \partial_i < nop, Vop - V >) \text{ on } \Gamma$$

$$= {}^*DV.n^*n + \partial_i b\ \partial_j(< nop, Vop - V >) + \partial_j b\ \partial_i(< nop, Vop - V >) \text{ on} \Gamma$$

Using the lemma 3.7, we have $\partial_i(< nop, Vop - V >) = [{}^*D(nop).(Vop - V) + {}^*D(Vop - V).(nop)]_i$ and on Γ it remains $\partial_i(< nop, Vop - V >)_{|\Gamma} = [{}^*D(Vop - V).(nop)]_{i|\Gamma} = [{}^*D_\Gamma.n - {}^*DV.n]_i = -n^*n.{}^*DV.n$, so that

$$[Dm(V)]_{i\,j_{|\Gamma}} =$$

$$= {}^*DV.n^*n - n^*n.{}^*DV.n^*n - n^*n.DV.n^*n$$
$$= {}^*DV.n^*n - 2n^*n.\varepsilon(V).n^*n$$
$$= {}^*D_\Gamma V.n^*n - n^*n.DV.n^*n \text{ (by use of proposition 3.2)}$$
$$= -D^2b.V^*n + \nabla_\Gamma(V.n)\,{}^*n - n^*n.DV.n^*n$$

and so lemma 3.9 is proved. Then, using first lemma 3.1 in order to differentiate the integral over Γ_t and then proposition 3.1, we obtain the following weak equation for the shape boundary derivative u'_Γ:

$$\frac{\partial}{\partial t}\Big(\int_{\Gamma_t} \varepsilon_{\Gamma_t}(u)..\varepsilon_{\Gamma_t}(\varphi)d\Gamma_t\Big)_{|t=0} =$$

$$= \frac{\partial}{\partial t}\Big(\int_{\Gamma_t} \varepsilon(uop_t)..\varepsilon(\varphi op_t)d\Gamma_t\Big)_{|t=0} = \qquad (12.24)$$

$$= \int_\Gamma \Big[\varepsilon_\Gamma(u'_\Gamma)..\varepsilon_\Gamma(\varphi) + \varepsilon_\Gamma(\varphi)..D((D_\Gamma u)op.m(V)) +$$
$$+\varepsilon_\Gamma(\varphi)..D(\,b\,D(uop)\,D^2b.V) - \varepsilon_\Gamma(\varphi)..D(D(uop).V) +$$
$$+\varepsilon_\Gamma(u)..D((D_\Gamma\varphi)op.m(V)) + \varepsilon_\Gamma(u)..D(\,b\,D(\varphi op)\,D^2b.V) -$$
$$+\varepsilon_\Gamma(u)..D(D(\varphi op).V) + \Big\{\frac{\partial}{\partial n}(\varepsilon(uop)..\varepsilon(\varphi op)) + H\varepsilon_\Gamma(u)..\varepsilon_\Gamma(\varphi)\Big\}V.n\Big]d\Gamma$$

In dimension 2 , let φ be any vector and D^2b the hessian matrix of the oriented distance function b, then the matrix $D^2b.D_\Gamma(\varphi)$ is symmetric on Γ. Proof. For any vectors E and φ in $I\!R^N$, we denote by E_Γ and $E.n$ - resp - the tangential and normal components of the vector E. $D^2b\ D_\Gamma\varphi.E$ and ${}^*D_\Gamma(\varphi)\ D^2b.E$ are both tangential vectors (with no normal component). They both only act on the tangential component of the vector E as obviously $D_\Gamma\varphi n = 0$ and n is in the kernel of D^2b. Then we just have to show that the tangential components of the vectors $D^2b\ D_\Gamma\varphi.E$ and ${}^*D_\Gamma\varphi\ D^2b.E$ are equal which is obvious as soon as we notice that if we denote by τ the orthogonal vector to the normal n, such that (n,τ) form a basis of the space $I\!R^N$ (that is $E = \|E_\Gamma\|\tau + E.nn$ on Γ) , then on Γ we get:

$$\begin{aligned} < D^2b.D_\Gamma\varphi.E,\tau > &=< D_\Gamma\varphi.E, D^2.\tau > \\ &= \|E_\Gamma\| < D_\Gamma\varphi.\tau, D^2.\tau > \\ &=< D_\Gamma\varphi.\tau, D^2b.\|E_\Gamma\|\tau > \\ &=< D_\Gamma\varphi.\tau, D^2b.E_\Gamma > \\ &=< {}^*D_\Gamma\varphi\ D^2bE,\tau > \end{aligned}$$

At this step, in order to verify the structure's theorem, we have to express the terms

$$\int_\Gamma \varepsilon_\Gamma(\varphi)..\Big\{D((D_\Gamma u)op.m(V)) + D(D(uop)op.V) +$$

$$-D(D(uop).V) + D(b\,D(uop)\,D^2b.V\Big\}\,d\Gamma$$

and the symmetric term

$$\int_\Gamma \varepsilon_\Gamma(u)..\Big\{D((D_\Gamma\varphi)op..m(V)) + D(D(\varphi op)op.V) +$$

$$-D(D(\varphi op).V) + D(b\,D(\varphi op)\,D^2b.V\Big\}\,d\Gamma$$

which a priori do not only depend on the normal component of V. In fact, it does:

Lemma 12.15.

$$\begin{aligned}\textstyle\int_\Gamma \varepsilon_\Gamma(\varphi) \quad ..\Big\{&D((D_\Gamma u)op.m(V)) + D(D(uop)op.V) - \\ &D(D(uop).V) + D(b\,D(uop)\,D^2b.V\Big\}\,d\Gamma \\ =\ &\tfrac{1}{2}\textstyle\int_\Gamma D_\Gamma\varphi..\{n.^*\nabla_\Gamma(V.n)\ ^*D_\Gamma u\}\,d\Gamma\end{aligned}$$

Proof. Let us notice the following relation

Lemma 12.16. $\int_\Gamma \varepsilon_\Gamma(\varphi)..\Big\{D(D(uop)op.V) - D(D(uop).V)\Big\}\,d\Gamma = 0$

Proof. Using lemma 3.8 we have $D(D(uop)op.V)_{|\Gamma} = D_\Gamma u.V + \partial_j[D_\Gamma u]_k.V_k$ and similarly for $D(D(uop).V)_{|\Gamma} = D_\Gamma u.V + \partial_j[D_\Gamma u]_k.V_k$. So the lemma is proved. The initial term of lemma 3.10 consists of two parts; let us find a specific structure for each of them and then compare both forms for which we expect to have an expression depending only on $V.n$, following the structure's theorem.

Lemma 12.17.

$$\int_\Gamma \varepsilon_\Gamma(\varphi)..D((D_\Gamma u)op.m(V))d\Gamma$$

$$= \frac{1}{2}\int_\Gamma D_\Gamma\varphi..\Big\{n.^*\nabla_\Gamma(V.n)\ ^*D_\Gamma u - n^*VD^2b^*D_\Gamma u\ \Big\}d\Gamma$$

Proof. Using the lemma 1.27, we can develop the matrix $D((D_\Gamma u)op.m(V))$ on Γ:

$$\begin{aligned}D((D_\Gamma u)op.m(V))_{|\Gamma} &= (D_\Gamma u)op\,Dm(V) \ \text{ because } m_{|\Gamma} = 0 \\ &= (D_\Gamma u)op\ \ Dm(V) \\ &= D_\Gamma u\,(^*DV.n.^*n - 2n.^*n\epsilon(V)n.^*n) \quad (\text{ lemma 0.26}) \\ &= D_\Gamma u\ ^*D_\Gamma V.n.^*n\end{aligned}$$

but applying lemma 0.5 we can isolate the vector V and see the hessian matrix D^2b appear:

$$D((D_\Gamma u)op.m(V)) \quad = D_\Gamma u\,[-D^2b.V^*n + \nabla_\Gamma(V.n)^*n],$$

this matrix D^2b forces the expression to depend on the tangential component of V and not only on its normal component as we would expect to. But the other term $\int_\Gamma \varepsilon_\Gamma(\varphi)..D(\, b\, D(uop)\, D^2b.V)d\Gamma$ associated with this expression "kills" the disturbing term " in D^2b ". Then returning to the initial expression we have:

Lemma 12.18.

$$\int_\Gamma \varepsilon_\Gamma(\varphi)..D(\, b\, D(uop)\, D^2b.V)d\Gamma = \frac{1}{2}\int_\Gamma D_\Gamma(\varphi)..n^* V D^2b^* D_\Gamma u\; d\Gamma.$$

Proof. Let us develop the term $D(\, b\, D(uop)\, D^2b.V)$ on Γ:

$$\begin{aligned} D(\, b\, D(uop)\, D^2b.V) &= bD(\, D(uop)\, D^2b.V) + D(uop)\, D^2b.V^*\nabla b \\ &= D(uop)\, D^2b.V^*\nabla b \text{ on } \Gamma \\ &= D(uop)\, D^2b.V^* n \text{ on } \Gamma \end{aligned}$$

then we obtain

$$\begin{aligned} \textstyle\int_\Gamma \varepsilon_\Gamma(\varphi) .. D(\, b\, D(uop)\, D^2b.V)d\Gamma &= \textstyle\int_\Gamma \varepsilon_\Gamma(\varphi)..D(uop)\, D^2b.V^* n \\ &= \textstyle\frac{1}{2}\int_\Gamma {}^*D_\Gamma(\varphi)..\, D(uop)\, D^2b.V^* n\; d\Gamma \\ &= \textstyle\frac{1}{2}\int_\Gamma D_\Gamma(\varphi)..n^* V.D^2b.{}^*D_\Gamma u\; d\Gamma \end{aligned}$$

And for the similar term associated with u, we get the similar result. Finally, we have proved the lemma 0.31 and for the whole expression we get

$$\int_\Gamma \varepsilon_\Gamma(\varphi)..\Big\{D((D_\Gamma u)op.m(V)) + D(b\, D(uop)\, D^2b.V\Big\}$$

$$+\varepsilon_\Gamma(u)..\Big\{D((D_\Gamma\varphi)op..m(V)) + D(b\, D(uop)\, D^2b.V\Big\}\; d\Gamma$$

$$= \frac{1}{2}\int_\Gamma D_\Gamma\varphi..\{n.{}^*\nabla_\Gamma(V.n)\; {}^*D_\Gamma u\}\; d\Gamma$$

$$+\frac{1}{2}\int_\Gamma D_\Gamma u..\{n.{}^*\nabla_\Gamma(V.n)\; {}^*D_\Gamma\varphi\}d\Gamma$$

$$= \frac{1}{2}\int_\Gamma D_\Gamma\varphi..\Big\{n\,{}^*\nabla_\Gamma(V.n)\; .{}^*D_\Gamma u + \;{}^*D_\Gamma u.n^*\nabla_\Gamma(V.n)\Big\}\; d\Gamma$$

which is only depending on $V.n$. Concerning the term of normal derivative we have (" $\frac{\partial}{\partial n}\varepsilon(uop)$")..$\varepsilon(\varphi op)$ in the following sense:

Lemma 12.19. $\frac{\partial}{\partial n}\Big(\varepsilon_{j\,k}(uop)\Big)\varepsilon_{j\,k}(\varphi op) = -(D_\Gamma u D^2 b)..\varepsilon_\Gamma(\varphi).$

Lemma 12.20. *Let* ξ *be an element of* $H^1(\Gamma)$, *we have* $\dfrac{\partial}{\partial n}\varepsilon_{j\,k}(\xi op) = -\frac{1}{2}\Big(((D^2b.(\nabla_\Gamma\xi_j)op)_k + (D^2b.(\nabla_\Gamma\xi_k)op)_j\Big)$

Proof. $\frac{\partial}{\partial n}\varepsilon_{j\,k}(\xi op) =< \nabla(\varepsilon_{j\,k}(\xi op)), n >$ and so $(\nabla(\varepsilon_{j\,k}(\xi op)))_i = \partial_i(\varepsilon_{j\,k}(\xi op))$.

Lemma 12.21. *p being the projection mapping and n_i denoting the ith component of the normal vector n on Γ, we have* $\partial_i(DP)_{j\,l_{|\Gamma}} n_i = -\partial^2_{j\,l} b$

Proof. The differential DP of the projection mapping is $DP = Id - \nabla b.^*\nabla b - bD^2b$ then $\partial_i(DP)_{jl} = \partial_i(\delta_{jl} - \partial_j b\partial_l b - b\partial^2_{jl}b)$ and $\partial_i(DP)_{j\,l_{|\Gamma}} = 0 - \partial^2_{i\,j}b\partial_l b - \partial_j b\partial^2_{i\,l}b - \partial_i b\partial^2_{j\,l}b$, finally we get $\partial_i(DP)_{j\,l_{|\Gamma}} n_i = -\partial^2_{j\,l}b$. We have

$$\begin{aligned}\varepsilon_{j\,k}(\xi op) &= \tfrac{1}{2}\Big(\partial_j(\xi_k op) + \partial_k(\xi_j op)\Big)\\ &= \tfrac{1}{2}\Big((DP.(\nabla_\Gamma \xi_k)op)_j + (DP.(\nabla_\Gamma \xi_j)op)_k\Big)\end{aligned}$$

and

$$\begin{aligned}\partial_i\varepsilon_{j\,k}(\xi op) &= \tfrac{1}{2}\Big(\partial_i(DP.(\nabla_\Gamma \xi_k)op)_j + \partial_i(DP.(\nabla_\Gamma \xi_j)op)_k\Big)\\ &= \tfrac{1}{2}\Big\{(^*D(DP.(\nabla_\Gamma \xi_k)op))_{ij} + (^*D(DP.(\nabla_\Gamma \xi_j)op))_{ik}\Big\}\end{aligned}$$

but $(^*D(Am))_{i\,j} = (^*Dm.^*A)_{i\,j} + \partial_i(A_{jk})m_k$, then we have

$$\begin{aligned}&\partial_i\varepsilon_{j\,k}(\xi op)\, n_i\\ &= \tfrac{1}{2}\Big\{(^*D_\Gamma((\nabla_\Gamma \xi_k)op.DP))_{ij}\, n_i + (^*D_\Gamma((\nabla_\Gamma \xi_j)op.DP))_{ik}\, n_i\\ &\quad -\partial^2_{k\,l}b.(\nabla_\Gamma \xi_j)_l op - \partial^2_{j\,l}b.(\nabla_\Gamma \xi_k)_l op\Big\}.\end{aligned}$$

On the other hand, we have the very useful following result

Lemma 12.22. $(^*D_\Gamma(\nabla_\Gamma \xi_k)op.DP)_{i\,j}\, n_i = 0$

Proof.

$$\begin{aligned}(^*D_\Gamma(\nabla_\Gamma \xi_k)op.DP)_{i\,j}\, n_i &= \Big[^*D_\Gamma((\nabla_\Gamma \xi_k)op)\Big]_{i\,l}(DP)_{l\,j}\, n_i\\ &= \Big[^*D_\Gamma((\nabla_\Gamma \xi_k)op)\Big]_{i\,l} n_i(DP)_{l\,j}\\ &= (DP)_{l\,j}.(D_\Gamma((\nabla_\Gamma \xi_k)op))_{l\,i}\, n_i\\ &= (DP)_{l\,j}.(D_\Gamma((\nabla_\Gamma \xi_k)op.n)_l\\ &= (DP.D_\Gamma((\nabla_\Gamma \xi_k)op).n)_j\\ &= 0\end{aligned}$$

because $D_\Gamma(\nabla_\Gamma \xi_k)op.n = 0$ and in the same way we have

$$\begin{aligned}(^*D_\Gamma(\nabla_\Gamma \xi_j)op.DP)_{i\,k}\, n_i &= \Big[^*D_\Gamma((\nabla_\Gamma \xi_j)op)\Big]_{il}(DP)_{l\,k}\, n_i\\ &= \Big[^*D_\Gamma((\nabla_\Gamma \xi_k)op)\Big]_{l\,i} n_i(DP)_{l\,k}\\ &= (DP)_{l\,k}.\Big[D_\Gamma((\nabla_\Gamma \xi_j)op).n\Big]_l\\ &= (DP)_{kl}.\Big[D_\Gamma((\nabla_\Gamma \xi_j)op.n\Big]_l\\ &= \Big[DP.D_\Gamma((\nabla_\Gamma \xi_j)op).n\Big]_k\\ &= 0.\end{aligned}$$

then $\partial_i\varepsilon_{j\,k}(\xi op)\; n_i \;=\; -\frac{1}{2}\Big(\partial^2_{k\,l}b.(\nabla_\Gamma\xi_j)_l op + \partial^2_{jl}b.(\nabla_\Gamma\xi_k)_l op\Big)$ and finally $\frac{\partial}{\partial n}\varepsilon_{j\,k}(\xi op) = -\frac{1}{2}\Big((D^2b.(\nabla_\Gamma\xi_j)op)_k + (D^2b.(\nabla_\Gamma\xi_k)op)_j\Big)$ then returning to the initial expression, we get

$$\frac{\partial}{\partial n}\Big\{\varepsilon(\varphi op)..\varepsilon(uop)\Big\}$$

$$\begin{aligned}
&= \frac{\partial}{\partial n}\Big(\varepsilon_{jk}(uop)\varepsilon_{jk}(\varphi op)\Big)\\
&= \frac{\partial}{\partial n}\Big[\varepsilon_{j\,k}(uop)\Big]\varepsilon_{j\,k}(\varphi op) + \frac{\partial}{\partial n}\Big[\varepsilon_{j\,k}(\varphi op)\Big]\varepsilon_{j\,k}(uop)
\end{aligned}$$

and

$$\frac{\partial}{\partial n}\Big[\varepsilon_{j\,k}(uop)\Big]\varepsilon_{j\,k}(\varphi op)$$

$$\begin{aligned}
&= -\tfrac{1}{4}\Big[(D^2b.(\nabla_\Gamma u_j)op)_k + (D^2b.(\nabla_\Gamma u_k)op)_j\Big]\times\\
&\quad (\partial_j(\varphi_k op) + \partial_k(\varphi_j op))\\
&= -\tfrac{1}{4}\{< D^2b.\nabla_\Gamma u_j, \nabla_\Gamma\varphi_j > + < D^2b.\nabla_\Gamma u_k, \nabla_\Gamma\varphi_k >\}\\
&\quad -\tfrac{1}{4}\{(D^2b.(\nabla_\Gamma u_j)op)_k\partial_j(\varphi_k op) - (D^2b.(\nabla_\Gamma u_k)op)_j\}\\
&= -\tfrac{1}{2} < D^2b.\nabla_\Gamma u_j, \nabla_\Gamma\varphi_j > -\tfrac{1}{2}(D^2b.(\nabla_\Gamma u_j)op)_k\partial_j(\varphi_k op)\\
&= -\tfrac{1}{2} < D^2b.\nabla_\Gamma u_j, \partial_k(\varphi_j op) + \partial_j(\varphi_k op) >
\end{aligned}$$

but $(\varepsilon_\Gamma(\varphi))_{j\,k} = \frac{1}{2}\Big\{\partial_j(\varphi_k op) + \partial_k(\varphi_j op)\Big\}$ and

$$\begin{aligned}
&(D^2b.\nabla_\Gamma u_j)_k =\\
&= \partial^2_{k\,l}b.\partial_l(u_j op)\\
&= (D^2b.{}^*D_\Gamma u)_{k\,j}\\
&= (D_\Gamma u.D^2b)_{j\,k}
\end{aligned}$$

then the lemma is proved. And the equation becomes, $\forall\varphi \in \frac{H^1(\Gamma,\mathbb{R}^N)}{N}$,

$$\int_\Gamma \Big[\varepsilon_\Gamma(u'_\Gamma)..\varepsilon_\Gamma(\varphi) + \frac{1}{2}\int_\Gamma D_\Gamma\varphi..\Big\{n\,{}^*\nabla_\Gamma(V.n)\;.{}^*D_\Gamma u + {}^*D_\Gamma u.n{}^*\nabla_\Gamma(V.n)\Big\}\,d\Gamma$$

$$+\Big\{H\varepsilon_\Gamma(u)..\varepsilon_\Gamma(\varphi) - (D_\Gamma u D^2b)..\varepsilon_\Gamma(\varphi) - (D_\Gamma\varphi D^2b)..\varepsilon_\Gamma(u)\Big\}V.n\Big]\,d\Gamma = 0$$

whose strong formulation is the following one:

$$\begin{gathered}
-\mathbf{div}_\Gamma(\varepsilon_\Gamma(u'_\Gamma)) + H\varepsilon_\Gamma(u'_\Gamma).n =\\
\mathbf{div}_\Gamma(H\varepsilon_\Gamma(u)V.n) - H^2\varepsilon_\Gamma(u).nV.n - \mathbf{div}_\Gamma(S(D_\Gamma u.D^2b)V.n)\\
+\frac{1}{2}\,\mathbf{div}_\Gamma\Big\{n\,{}^*\nabla_\Gamma(V.n)\;.{}^*D_\Gamma u + {}^*D_\Gamma u.n{}^*\nabla_\Gamma(V.n)\Big\}\\
-\frac{1}{2}\,H\,\Big\{n\,{}^*\nabla_\Gamma(V.n)\;.{}^*D_\Gamma u + {}^*D_\Gamma u.n{}^*\nabla_\Gamma(V.n)\Big\}.n
\end{gathered}$$

$$+HS(D_\Gamma u.D^2b).nV.n - \mathbf{div}_\Gamma(\varepsilon_\Gamma(u)D^2bV.n) + H\varepsilon_\Gamma(u)D^2b.nV.n$$

where $S(D_\Gamma u.D^2b)$ denotes the symmetrized part of the matrix $D_\Gamma u.D^2b$, but

$S(D_\Gamma u.D^2b).n = \frac{1}{2}D^2b\,{}^*D_\Gamma u.n = D^2b\ \varepsilon_\Gamma(u).n$ and $\varepsilon_\Gamma(u)D^2b.n = 0$ as long as n belongs to the kernel of the hessian matrix of b, D^2b. So finally we get:

$$-\mathbf{div}_\Gamma(\varepsilon_\Gamma(u'_\Gamma)) + H\varepsilon_\Gamma(u'_\Gamma).n$$

$$= \mathbf{div}_\Gamma(H\varepsilon_\Gamma(u)V.n) - \mathbf{div}_\Gamma(\varepsilon_\Gamma(u)D^2bV.n) - \mathbf{div}_\Gamma(S(D_\Gamma u.D^2b)V.n)$$

$$+\frac{1}{2}\,\mathbf{div}_\Gamma\left\{n\,{}^*\nabla_\Gamma(V.n)\ .{}^*D_\Gamma u + {}^*D_\Gamma u.n{}^*\nabla_\Gamma(V.n)\right\}$$

$$-\frac{1}{2}\,H\left\{n\,{}^*\nabla_\Gamma(V.n)\ .{}^*D_\Gamma u\right\}.n + HD^2b\ \varepsilon_\Gamma(u).nV.n - H^2\varepsilon_\Gamma(u).nV.n.$$

12.4.1 Boundary Condition.

Lemma 12.23. *u_t is the solution of the coupled problem in Ω_t. Its shape tangential derivative u'_Γ verifies the following boundary condition on $\partial\Gamma$:*

$$u'_\Gamma = -D_\Gamma u.\nu\ V.\nu \qquad \text{on } \partial\Gamma, \tag{12.25}$$

where $\nu(X)$ is the normal to $\partial\Gamma$ at point X outgoing to Γ, contained in the tangent space to Γ at point X, $D_\Gamma u.\nu$ is the vector defined with the transposed matrix of $D_\Gamma u$ as: ${}^*D_\Gamma u.\nu = (\frac{\partial u_1}{\partial \nu}, ..., \frac{\partial u_N}{\partial \nu}) = (< \nabla_\Gamma u_1, \nu >, ..., < \nabla_\Gamma u_N, \nu >)$.

Proof. Let $\varphi \in \mathcal{D}(I\!R^N)$, $\int_{T_t(\partial\Gamma)} u_t\ \varphi\ dl_t = \int_{\partial\Gamma} u_t oT_t\ \varphi oT_t\ \theta(t)\ dl = 0$ We differentiate this expression and we obtain, at $t = 0$: $\int_{\partial\Gamma}\{\dot{u}\ \varphi\ \theta(0) + u\ \dot{\varphi}\ \theta(0) + u\ \varphi\ \theta'(0)\}\ dl = 0$ where the notation $\dot{u}$ stands for the material derivative of u_t, where $u = 0$ on $\partial\Gamma$ and where θ is the jacobian whose value at $t = 0$ is equal to 1 (it is then associated with the Identity transformation).

Then we get $\int_{\partial\Gamma}\{\dot{u}\ \varphi\}\ dl = 0$ and as $\dot{u} = u'_\Gamma + D_\Gamma u.V_\Gamma$ then we finally get: $\int_{\partial\Gamma}\{u'_\Gamma\ \varphi + < D_\Gamma u.V_\Gamma, \varphi >\}\ dl = 0$ but $u = 0$ on $\partial\Gamma$ then $D_\Gamma u.V_\Gamma = D_\Gamma u.\nu\ V_\Gamma.\nu$ on $\partial\Gamma$.

12.5 The Exterior Navier Stokes Problem

Let B denote a bounded domain in R^3 and Q its boundary. The fluid domain is the exterior $\Omega = R^3 - \bar{B}$. The speed vector field of the steady fluid U will be defined in Ω and be divergence free with sticking conditions on the boundary Q and given behavior at infinity. In practice this means that when the body B is empty we assume the fluid to have a constant and uniform flow speed $\mathbf{c} \in R^3$. The presence of the body B creates a "speed perturbation" $u = U - \mathbf{c}$ formally the fluid problem can be written as

$$DU - \nu\Delta U + \nabla P = -g\mathbf{z} \tag{12.26}$$

$$divU = 0 \tag{12.27}$$

$$U = 0 \text{ on } Q, \quad U(x) \to \mathbf{c}, \quad x \to \infty \tag{12.28}$$

In order to get homogeneous boundary conditions on the boundary Q as well as at infinity the use of the perturbation field u is not adequate as u would be search in the closed convex of admissible fields having the trace $\mathbf{c}$ on Q. For that reason we introduce the use of a function $\mathcal{C} \in C^{\infty}(\Omega \cup Q, R^3) \cap W^{1,\infty}(\Omega, R^3)$ having the following properties

$$\mathcal{C} \in W^{1,\infty}(\Omega, R^3), \;\; D\mathcal{C}\,(1+|x|^2) \in L^{\infty}(\Omega, R^{N^2}), \; div\mathcal{C} = 0, \tag{12.29}$$

$$\mathcal{C} = 0 \text{ on } Q, \;\; \mathcal{C}(x) \to \mathbf{c}, \; x \to \infty \tag{12.30}$$

Then we set

$$\mathcal{L} = \nu\, \Delta\mathcal{C} \; - \; D\mathcal{C}.\mathcal{C} \tag{12.31}$$

Considering the "$\mathcal{C}$-perturbation speed flow "$v = U - \mathcal{C}$ and the dynamic pressure $p = P + gz$, we get the equivalent problem

$$Dv.v \; - \; \nu\, \Delta v \; + \; D\mathcal{C}.v \; + \; Dv.\mathcal{C} \; + \; \nabla p \; = \; \mathcal{L} \tag{12.32}$$

$$divv \; = \; 0 \tag{12.33}$$

$$v \; = \; 0, \text{ on } Q, \quad v(x) \; \to 0 \; x \; \to \infty \tag{12.34}$$

13. The Outer Sobolev Space

We consider the following space

$$W(\Omega) = \{v \in L^2_{loc}(\Omega, R^3) \mid Dv \in L^2(\Omega, R^{N^2}), \\ (1+|x|^2)^{-\frac{1}{2}}\, v \in L^2(\Omega, R^3)\} \tag{13.1}$$

Proposition 13.1. *There exists a positive constant $M > 0$ such that*

$$\forall v \in W(\Omega), \;\; \int_{\Omega} Dv..Dv\, dx \geq \frac{1}{M^2} \int_{\Omega} \{(1+|x|^2)^{-\frac{1}{2}} |v(x)|\}^2 \, dx \tag{13.2}$$

Or equivalently

$$\forall v \in W(\Omega), \;\; M \; \|v\|_{H^m} \geq \|(1+|x|^2)^{-\frac{1}{2}}\, v\|_{L^2(\Omega, R^3)}$$

We set

$$W_0(\Omega) \; = \; \{ \; v \in W(\Omega) \mid \; v \; = \; 0 \text{ on } Q \; \} \tag{13.3}$$

equipped with the equivalent norm

$$\|v\|_{W(\Omega)} \; = \; (\int_{\Omega} Dv..Dv \; dx)^{\frac{1}{2}} \tag{13.4}$$

13.1 Existence Result

Then, setting

$$\nu_0 = INF\{M^2 \,\|(1+|x|^2)DC\|_{L^\infty(\Omega,R^9)} \mid C \quad \text{verifies the conditions (12.29),(12.30)} \,\} \tag{13.5}$$

we can derive the following result

Proposition 13.2. *Let $\nu > \nu_0$ $(\nu_0 \geq 0)$, then there exists (at least one) solution v to the outer problem.*

The proof uses a Brouwer result in finite dimensional Hilbert space. For that let us consider a dense family $\mu_1, \mu_2, ..., \mu_k, ...$ in the Hilbert space $W_0(\Omega)$. We assume that the support K_k of μ_k is an increasing sequence of compacts sets in Ω, H^m is the Hilbert space generated in $W(\Omega)$ by the m first elements and we consider the mapping $\Phi^m : H^m \longrightarrow H^m$ defined by

$$< \Phi^m(v^m), \theta^m >_{H^m} = \int_\Omega \{ < Dv^m.v^m, \theta^m > + \nu Dv^m..D\theta^m + \tag{13.6}$$

$$+ < DC.v^m + Dv^m.\mathcal{C}, \theta^m > - < \mathcal{L}, \theta^m > \} \, dx$$

Lemma 13.1. *Assume K^m is a compact set in Ω with smooth enough boundary and $v^m \in H_0^1(K_m \cap \Omega)$, then*

$$\forall \theta, \; div\, \theta = 0 \; in \; \Omega, int_\Omega < Dv^m.\theta, v^m > dx = 0 \tag{13.7}$$

Proof. Performing by part on (13.7)

$$2 \int_\Omega < Dv^m.\theta, v^m > dx = - \int_\Omega |v^m|^2 \, div\, \theta \, dx + \int_{\partial K_m} |v^m|^2 \; < \theta, n > d\Gamma$$

In the specific case where $\theta^m = v^m$, we have

$$< \Phi^m(v^m), v^m >_{H^m} = \int_\Omega \{\nu \, Dv^m..Dv^m + < DC.v^m, v^m > - < \mathcal{L}, v^m > \} \, dx$$

Using the Cauchy Schwartz inequality we obtain

$$< \Phi^m(v^m), v^m >_{H^m} \geq \nu \, \|v^m\|^2_{H^m} + \|(1+|x|^2)\, DC\|_{L^\infty(\Omega,\, R^3)}$$

$$\|(1+|x|^2)^{-\frac{1}{2}}\, v^m\|^2_{L^2(\Omega,\, R^3)} - \|(1+|x|^2)^{\frac{1}{2}}\, \mathcal{L}\|_{L^2(\Omega,\, R^3)} \, \|(1+|x|^2)^{-\frac{1}{2}}\, v^m\|_{L^2(\Omega,\, R^3)}$$

and, using the (13.2) inequality,

$$\begin{aligned} < \Phi^m(v^m), v^m >_{H^m} \;\; & \geq \;\; (\nu - M^2\, \|(1+|x|^2)\, DC\|_{L^\infty(\Omega,\, R^3)})\, \|v^m\|^2_{H^m} \\ & \quad - \|(1+|x|^2)^{\frac{1}{2}}\, \mathcal{L}\|_{L^2(\Omega,\, R^3)} \;\; \|v^m\|_{H^m} \end{aligned}$$

but as $\|v^m\|_{H^m} = \|v^m\|_{W(\Omega)}$, we get

$$< \Phi^m(v^m), v^m >_{H^m} \geq (M_1 \|v^m\|_{W(\Omega)} - M_2) \|v^m\|_{W(\Omega)} \tag{13.8}$$

where the constants M_1 and M_2 do not depend on m.

$$M_1 = \nu - M^2 \|(1 + |x|^2) DC\|_{L^\infty(\Omega, R^3)}$$

$$M_2 = \|(1 + |x|^2)^{\frac{1}{2}} C\|_{L^2(\Omega, R^3)}$$

A consequence of the classical Brouwer's fixed-point theorem is the following corollary

Corollary 13.1. *Let H be a finite-dimensional Hilbert space whose scalar product is denoted by $< .,. >_H$ and the corresponding norm by $\|.\|_H$. Let Φ be a continuous mapping from H into H with the following property*

$$< \Phi(f), f >_H \geq 0, \ \forall f \in H \ withVert f\|_H = \mu \tag{13.9}$$

Then, there exists an element f in H such that

$$\Phi(f) = 0 \ , \ \|f\|_H \leq \mu. \tag{13.10}$$

In our situation, as $\nu > \nu_0$ then, there exists an C such that $M_1 > 0$ and so, we are in the conditions of the corollary 13.1 with $\mu = M_2.(M_1)^{-1}$. Then, in the ball $\|v^m\|_{W(\Omega)} \leq \mu$, for each m there exists an element $v^m \in H^m$ such that$\Phi^m(v^m) = 0$. As μ is independent on m that sequence is bounded in $W(\Omega)$, then we consider a weakly converging subsequence to an element v in that ball of radius μ and it is classical to verify that this element v solves the problem. The main point is that, a priori, Ω being unbounded, that weak convergence in $W(\Omega)$ is not strong in $L^2(\Omega, R^3)$. But to verify that the limiting element solves the equation in weak sense it is enough to consider any element $\psi \in \mathcal{D}(\Omega)$, for which there exists an integer q such that $k \geq q$ implies that K = support of $\psi \subset K_k$. It derives that the convergence is strong in $L^2(K, R^3)$.

13.2 The coupling with a Potential Flow

The idea is now to introduce a tubular neighborhood$\mathcal{U}$ of the body boundary $\Gamma = Q$ having itself a boundary $\partial\mathcal{U} = \Sigma \cup \Gamma$. We assume that the surface Σ is well chosen so that in the outer domain $\Omega_P = R^3 \setminus (\mathcal{U} \cup \Sigma)$, the previous flow is irrotational

$$\text{curl } v = 0 \text{ in } \Omega_P \tag{13.11}$$

In practice this means that the curl part of the fluids appears through the sticking condition on the boundary $Q = \Gamma$ of the body and then is propagated in a wake which is contained in the neighborhood $\mathcal{U}$. By the Helmholtz decomposition theorem we know that the field v can always be decomposed in the form $v = \nabla\phi +$ curlψ. We have curl$v =$ curl curl$\psi = 0$ then, $v = \nabla \phi$. The continuity conditions on the interface Σ will be

$$v = \nabla\phi \text{ on } \Sigma \tag{13.12}$$

that condition will be split in the two following conditions

$$< v, n > = \frac{\partial}{\partial n}\phi \text{ on } \Sigma, \quad v_\Sigma = \nabla_\Gamma(K.\frac{\partial}{\partial n}\phi\,) \tag{13.13}$$

In the exterior domain Ω_P,the flow is potential, i.e. $v = \nabla\phi$ and the incompressibility condition turns to the harmonicity of that potential

$$\Delta\phi = 0 \text{ in } \Omega_P \tag{13.14}$$

In some sense we shall "project" the solution of the previous outer Navier Stokes problem on the elements which are expressed as gradients in the outer domain Ω_P. For that, purpose we introduce the following linear space

$$\mathcal{V} = \{\underline{\sqsubset} \in \mathcal{H}^\infty_{\text{loc}}(\Omega, \mathcal{R}^\mathcal{N})\,|\,,\ \underline{\sqsubset} = \prime \text{ on } \Gamma,\ \underline{\sqsubset}|_{\Omega_P} = \nabla\varphi, \tag{13.15}$$

$$\varphi \in W^2(\Omega_P)/R\ \ \}$$

Where the outer space W^2 is defined as

$$W^2(\Omega_P) = \{\ \ \varphi \in L^2_{loc}(\Omega_P)\ \ |\,(1+|x|^2)^{-1}\varphi \in L^2(\Omega_P), \tag{13.16}$$

$$(1+|x|^2)^{-\frac{1}{2}}\nabla\varphi \in L^2(\Omega_P, R^3),\ \ D^2\varphi \in L^2(\Omega_P, R^9)\ \ \}$$

Notice that the constant functions are in $W^2(\Omega_P)$, i.e. $R \subset W^2(\Omega_P)$.

We consider the closed subspace

$$\mathcal{V}^0 = \{\ u \in \mathcal{V}\ |\ divu = 0\ \text{ in } \Omega\}$$

Any element $vin\mathcal{V}^0$ verifies

$$\Delta\varphi = 0 \text{ in } \Omega_P,\ \ \nabla\varphi = w \text{ on}\Sigma \tag{13.17}$$

where w denotes the restriction to $\mathcal{U}$ of the element $v \in \mathcal{V}$. Equipped with the following norm the linear space $\mathcal{V}$ is a Banach space

$$||v||^2 = \int_{\Omega_P} D^2\varphi..D^2\varphi\,dx + \int_{\mathcal{U}} Dw..Dw\,dx \tag{13.18}$$

The main point to be verified is that $||v|| = 0$ implies $v = 0$. For that, let us remark that if $||v||$ is zero then each of the two integrals is zero, and from the second we conclude that w is constant through $\mathcal{U}$, but as $w = 0$ on the boundary Γ, then w itself is zero in$H^1(\mathcal{U})$. Then, with the second integral we obtain that $D(\nabla\varphi) = 0$ in the outer domain Ω_P and then the function$\nabla\varphi$ is constant in the outer domain, as $\nabla\varphi = w = 0$ on the smooth boundary Σ, we get that $\nabla\varphi$ itself is zero in the outer domain. Then φ itself is a function constant through Ω_P, which implies $\varphi = 0$ in the quotient space. The norm of$W^2(\Omega_P)/R$ is equivalent to the L^2 norm of $D^2\varphi$.

13.2.1 The Projected Problem. In order to use the previous Galerkin approximation we consider a dense family $E_1, ..., E_m, ...$ in the Hilbert space $\mathcal{V}$ previously defined. We shall assume that each of these elements are bounded on the interface Σ : $\forall i,\ E_i \in L^\infty(\Sigma, R^N)$. From the previous properties of the space$\mathcal{V}$ it turns out that elements E_i can be written as $E_i|_{\Omega_P} = K(e_i)$, where the elements e_i are in $H^1(\mathcal{U})$. We denote by $\mathcal{V}^{\Updownarrow}$ the subspace spanned by $E_1, ..., E_m$ and we consider the mapping $\phi^m : \mathcal{V}^{\Updownarrow} \rightarrow \mathcal{V}^{\Updownarrow}$ defined in weak form as follow :

$$\forall j \in [1, ..., m],\ \forall u^m \in \mathcal{V}^{\Updownarrow}$$

$$< \phi^m(u^m), E_j >= \int_{\mathcal{U}} \{< Dw^m.w^m, E_j > +\nu\, Dw^m..DE_j \tag{13.19}$$

$$+ < Dw^m.\mathcal{C} + D\mathcal{C}.w^m, E_j > - < \mathcal{L}, E_j > \}dx$$

$$+ \int_{\Omega_P} \{< D\nabla\varphi^m.\nabla\varphi^m, \nabla\psi_j > +\nu\, D^2\varphi^m..D^2\psi_j$$

$$+ < D^2\varphi^m.\mathcal{C} + D\mathcal{C}.\nabla\varphi^m, \nabla\psi_j > - < \mathcal{L}, \nabla\psi_| > \}\lceil\S = ,$$

As before, we can show that for each m, the following problem has at least one solution

$$\forall j \in [1, ..., m],\ u^m \in \mathcal{V}^{\Updownarrow}\ \int.\sqcup. :$$

$$< \phi^m(u^m), E_j >= 0 \tag{13.20}$$

and we derive the existence of a constant such that $||u^m|| \leq M$. Then, there exists a subsequence, still denoted u^m, which weakly converges to u in $\mathcal{V}$. The difficulty is that, the domain being unbounded, we have no compact inclusions and the problem is how to pass in the limit through the non linear term in the previous weak formulation of the problem. We bypass that point by using the integration by parts in the outer domain Ω_P and the harmonicity of the element φ^m associated to u^m. We only get integral terms on the smooth boundary Σ which permit to use compact injections on the compact surface Σ. Finally we obtain the following existence result.

Theorem 13.1. *Let Σ be a smooth compact surface in R^3 dividing the outer domain Ω in two disjoint open domains $\mathcal{U}$ and Ω_P, $\Omega = \Omega_P \cup \mathcal{U} \cup \Sigma$ with $\mathcal{U}$relatively compact and $\partial\mathcal{U} = \Sigma \cup \Gamma$.*
There exists (at least one) $u \in \mathcal{V}$ such that $\forall\theta \in \mathcal{V}$ we have

$$\int_{\mathcal{U}} \{< Du.u, \theta > +\nu\, Du..D\theta \tag{13.21}$$

$$+ < Du.\mathcal{C} + D\mathcal{C}.\sqcap, \theta > - < \mathcal{L}, \theta > \}\lceil\S$$

$$+ \int_{\Omega_P} \{< D\nabla\varphi.\nabla\varphi, \nabla\psi > +\nu\, D^2\varphi..D^2\psi + < D^2\varphi.\mathcal{C} + D\mathcal{C}.\nabla\varphi, \nabla\psi >$$

$$- < \mathcal{L}, \nabla\psi > \}\lceil\S = ,$$

where we denote

$$\nabla\varphi = u|_{\Omega_P}$$

$$\nabla\psi = \theta|_{\Omega_P}$$

We call this problem the "$\mathcal{V}$−projected problem", in the sense that its weak formulation is the same as that of the outer Navier-Stokes problem but with solution and test functions in the space $\mathcal{V}$ of vector fields which are gradients fields in the outer domain Ω_P. By construction the solutions u verify $curl\, u = 0$ in the outer domain Ω_P. The weak formulation of the "projected problem" can also be written as follows:

Proposition 13.3.

$$\forall \theta \in \{v \in H^1(\mathcal{U}) \mid divv = 0,\, v = 0 \ \ on \ \ \Gamma, \}$$

$$\int_{\mathcal{U}} \{< Du.u, \theta >) - \nu < \Delta u\,, \theta > + < Du.C + DC.u, \theta > - < \mathcal{L}, \theta >\}dx$$
$$- \int_{\Sigma} \{\nu K^*.[div_\Gamma[(Du.n - D^2\varphi.n)_\Gamma] + \frac{1}{2}|\nabla\varphi|^2 + < c, \nabla\varphi >\} < \theta, n > d\Sigma = 0 \tag{13.22}$$

Before proving this proposition, let us make the following integrations by parts in the space $\mathcal{V}^0$:

Lemma 13.2. *Let ϕ be given in $\mathcal{V}^0$, then $D^2\phi.n \in H^{-\frac{1}{2}}(\Sigma, R^3)$; moreover for $\psi \in \mathcal{V}$ we have :*

$$\int_{\Omega_p} D^2\phi..D^2\psi \; dx \; = \; - \int_{\Sigma} < D^2\phi.n, \nabla\psi > d\Sigma \tag{13.23}$$

where the integral stands for the duality bilinear form pairing between $H^{-\frac{1}{2}}(\Sigma, R^3)$ and $H^{\frac{1}{2}}(\Sigma, R^3)$. In this paper we shall always denote by integrals such duality forms.

Proof. As $div\, D^2\phi = \nabla(\Delta\phi) = 0$ in Ω_p, considering $\Omega_p^r = \{x \in \Omega_p \mid ||x|| \leq r\}$, the usual integration by parts leads to

$$\int_{\Omega_P^r} D^2\phi..D^2\psi \; dx \; = \; - \int_{\Sigma} < D^2\phi.n, \nabla\psi > d\Sigma$$
$$+ \int_{\{\;||x||=r\;\}} < D^2\phi.n, \nabla\psi > ds$$

from the density of $\mathcal{D}(\Omega, R^3)$ in $\mathcal{V}$ the second integral can be dropped.

Lemma 13.3. *Let $\phi \in \mathcal{V}$, $\psi \in \mathcal{V}^0$, then we have:*

$$\int_{\Omega_P} < D\nabla\phi.\nabla\phi,\; \nabla\psi > dx = - \int_{\Sigma} \frac{1}{2}|\nabla\phi|^2 \; \frac{\partial}{\partial n}\psi \; d\Sigma$$

Proof. It suffices to remark that $D\nabla\phi.\nabla\phi = \frac{1}{2}\nabla(|\nabla\phi|^2)$.

Lemma 13.4.

$$< Du.n, n > = < D^2\phi.n, n > \quad on \ \Sigma$$

Proof. Notice that $< Du.n, n > - < \epsilon(u).n, n >= 0$. From (13.75) it follows that the jump across Σ of $< Du.n, n >$ is zero; this jump is precisely the term under consideration.

Lemma 13.5.

$$\int_{\Omega_P} < D^2\phi.\mathcal{C}, \nabla\psi > dx = -\int_\Sigma < \nabla\phi, c > \frac{\partial\psi}{\partial n} d\Sigma$$

Proof.

$$\begin{aligned}\int_{\Omega_P} < D^2\phi.\mathcal{C}, \nabla\psi > dx &= \int_{\Omega_P} < D^2\phi.\mathbf{c}, \nabla\psi > dx \\ &= \int_{\Omega_P} < \nabla < \nabla\phi, c >, \nabla\psi > dx \\ &= -\int_{\Omega_P} < \nabla\phi, c > \Delta\psi\, dx - \int_\Sigma < \nabla\phi, c > \frac{\partial\psi}{\partial n} d\Sigma\end{aligned}$$

Proof. (of the proposition)

Applying the lemma 13.2 to (13.21) we get, as $\boldsymbol{div Du} = \boldsymbol{\Delta u}$ and $\theta = \nabla\psi$ on Σ,

$$\int_{\mathcal{U}} \nu Du..D\theta\, dx +, \int_{\Omega_p} \nu\, D^2\phi..D^2\psi\, dx$$

$$= \int_{\mathcal{U}} \nu < \mathbf{div u}, \theta > dx + \int_\Sigma \nu < Du.n - D^2\phi.n\, , \theta > d\Sigma$$

but $div u = 0$, with the lemma 13.3 we get:

$$\int_{\mathcal{U}} \{< Du.u, \theta >) + < Du.\mathcal{C} + D\mathcal{C}.u, \theta > - < \mathcal{L}, \theta > \}\lceil\S \qquad (13.24)$$

$$+ \int_\Sigma [\nu < (Du.n - D^2\varphi.n), \theta > +\{-\frac{1}{2}|\nabla\varphi|^2 + < c, \nabla\varphi >\} < \theta, n >] d\Sigma = 0$$

As $\theta = \psi$ on Σ and $\Delta\psi = 0$ in Ω_p we get

$$\theta_\Gamma = \nabla_\Gamma\psi = \nabla_\Gamma(, K.(< \theta, n >)) \text{ on } \Sigma$$

where $K \in \mathcal{L}(H^{-\frac{1}{2}}(\Sigma), H^{\frac{1}{2}}(\Sigma))$is the Neuman pseudo-differential map defined at section 13.11. Then it derives

$$\int_\Sigma \nu < (Du.n - D^2\varphi.n), \theta > d\Sigma =$$

$$\int_\Sigma \nu < (Du.n - D^2\varphi.n), n > < \theta, n > d\Sigma +$$

$$+ \int_\Sigma \nu < (Du.n - D^2\varphi.n), \nabla_\Gamma(K.(< \theta, n >) > d\Sigma$$

From Lemma 13.4 the first term of the right hand side is zero, then by tangential integration by parts we have:

$$\int_{\Sigma} \nu < (Du.n - D^2\varphi.n), \theta > d\Sigma =$$

$$-\nu \int_{\Sigma} K^*.div_{\Gamma}[\ (Du.n - D^2\varphi.n)] < \theta, n > d\Sigma$$

and (13.22) follows.

13.3 Strong Formulation of the Projected Problem

The formulation (13.22) leads to the following strong formulation

$$Du.u - \nu\Delta u + Du.C + DC.u + \nabla p = \mathcal{L} \text{ in } \mathcal{U} \tag{13.25}$$

and the boundary condition in weak form

$$\forall\theta \in \mathcal{V},\ \int_{\mathcal{U}} < -\nabla p, \theta > dx - \int_{\Sigma} \{\ \nu K^*.[\ div_{\Gamma}(\,(Du.n - D^2\varphi.n)_{\Gamma}\,)\ ] \tag{13.26}$$

$$-\frac{1}{2}|\nabla\varphi|^2 + + < c, \nabla\varphi > \}\ < \theta, n > d\Sigma = 0$$

that is

$$\int_{\Sigma} \{\ -p < \theta, n > - \nu K^*.[\ div_{\Gamma}(\,(Du.n)_{\Gamma} - (D^2\varphi.n)_{\Gamma})] \tag{13.27}$$

$$-\frac{1}{2}|\nabla\varphi|^2 + < c, \nabla\varphi > \}\ < \theta, n > d\Sigma = 0$$

from which we obtain that there exists a constant k such that

$$-p - \nu K.[\ div_{\Gamma}(\,(Du.n)_{\Gamma} - (D^2\varphi.n)_{\Gamma}\,)] - \frac{1}{2}|\nabla\varphi|^2 + < c, \nabla\varphi > = k \tag{13.28}$$

In fact the jump term $(Du.n - D^2\varphi.n)_{\Gamma}$ on Σverifies from (13.75)

$$(Du.n - D^2\varphi.n)_{\Gamma} = (\sigma(u).n + \epsilon(u).n - \epsilon(\nabla\varphi))_{\Gamma} = 2\sigma(u).n \text{ on } \Sigma \tag{13.29}$$

where $2\sigma(E) = DE - D^*E$ is the anti-symmetrical part of the Jacobian matrix DE, obviously for any potential field that term is zero : $\sigma(\nabla\varphi) = 0$. It can also be verified that the normal component of the jump term is zero ;

$$< (Du.n - D^2\varphi, n >= 2 < \sigma(u).n, n >= 0 \quad \text{on } \Sigma \tag{13.30}$$

(the condition on Σ simplifies to the following one

$$-p - 2\nu K.[\ div_{\Gamma}(\sigma(u).n] - \frac{1}{2}|\nabla\varphi|^2 + < c, \nabla\varphi > \ = k \text{ on } \Sigma \tag{13.31}$$

Finally the projected problem gives, for any smooth compact manifold Σ, divergence free solutions to the problem(13.25)-(13.31).

13.4 From the Projected problem to the Navier Stokes Flow

In this section we address the following question : under which extra condition does a solution u to the previous projected problem(13.25)-(13.31) verify the Navier Stokes equation? For this purpose we consider the following result

Proposition 13.4. *Let the smooth unbounded domain Ω be divided, as previously, as $\Omega = \Omega_P \cup \Sigma \cup \mathcal{U}$ where Σ is a smooth compact surface which divides Ω in two open subsets, $\mathcal{U}$ being bounded with two disjoint boundaries Γ and Σ. We consider two divergence free elements $v \in W(\Omega_P, R^3)$ and $w \in H^1(\mathcal{U}, \mathcal{R}^3)$, solutions of the Navier Stokes equation respectively in Ω_P and in $\mathcal{U}$ with $w = 0$ on Γ. We consider the element of $L^2_{loc}(\Omega, R^3)$ defined by : $u = w$ in Ω_P , $u = v$ in $\mathcal{U}$. Then u solves the Navier Stokes equation in Ω if and only if the following conditions hold:*

$$v = w \;\; on \;\; \Sigma \tag{13.32}$$

There exists a constant denoted by k such that, q and p being the pressures respectively in the flow occupying Ω_P and $\mathcal{U}$,

$$p = q + k \tag{13.33}$$

the jump across Σ of the following tangential component is zero

$$(\,\epsilon(v).n\,)_\Gamma \; - \; (\,\epsilon(w).n\,)_\Gamma \; = 0 \;\; on \;\; \Sigma. \tag{13.34}$$

that last condition can be written as

$$(\,\sigma(v).n\,)_\Gamma \; - \; (\,\sigma(w).n\,)_\Gamma \; = 0 \;\; \text{on} \;\; \Sigma.$$

In the specific situation where the outer solution v is a potential flow, we compare that last jump condition on the interface $\Sigma$with the previous condition (13.31) verified by the $\mathcal{V}$-projected flow we discover a strong similitude. We must recall that in the outer domain Ω_P for the potential flow, φ being in fact the "perturbation potential" (due to the presence of the divergence free function C) the Bernoulli condition leads to

$$q = \frac{1}{2}|\nabla\varphi|^2 - < c, \nabla\varphi > \tag{13.35}$$

Then (13.31), using (13.30) can be written as

$$2\,\nu\, K.(\;div_\Gamma[\,(\sigma(u).n)\,]\;) \;+ p - q = k \tag{13.36}$$

Of course conditions (13.33) and (13.34) imply(13.36), conversely it depends on the choice of the surface Σ, if that surface is chosen such that a solution u of the $\mathcal{V}$-projected problem verifies also (13.34),then it is a solution of the Navier Stokes flow as it verifies (13.33). If the solution u verifies only the pressure jump condition(13.33) on the interface Σ, then we obtain, since the linear operator K^* is an isomorphism, a condition on the tangential divergence of the jump

$$div_\Gamma(\ (\epsilon(u).n)_\Gamma\ -\ (D^2\varphi.n)_\Gamma\ \ ) = div_\Gamma(\ (\sigma(u).n)\ \ ) = 0$$

we use the tangential Helmholtz decomposition

Proposition 13.5. *There exist Φ and ψ such that $u = \nabla\Phi + curl\,\psi$ in $\mathcal{U}$.*

if we assume u smooth enough, then the decomposition is valid on the interface Σ and we get the condition

$$div_\Gamma(\,\sigma(curl\psi).n\,) = 0$$

and the divergence free field $u \in \mathcal{V}$ will be solution of the Navier Stokes if the last condition implies $\sigma(curl\psi).n = 0$.

13.5 Uniqueness of Solution for both Navier Stokes flow and $\mathcal{V}$-projected Problems

The present "philosophy" is not to obtain the Navier Stokes flow from the projected one, since this is in general impossible, as these two problems are different, but to recognize, by solving only the projected problem, numerically much easier, when the interface Σ is well chosen. If we assume the datum (velocity c, viscosity ν, gravity$g$, ...)such that these two problems have a unique solution then when the interface is well chosen (i.e. such that the neighborhood $\mathcal{U}$ ofΓ contains the support of ψ) both solution coincide. The uniqueness result for the Navier Stokes problem is related to regularity result on the solution u. Let $b(u,v,w) = \int_\Omega < Du.v, w > dx$ be the trilinear form which occurs in the weak formulation. Classically, when the domain $\Omega$is smooth and bounded we have $b(u,v,w) = -b(w,v,u) + \frac{1}{2}\int_{\partial\Omega} < u,w >< v,n > d\Gamma$, so that if $< v,n >= 0$ on the boundary we get $b(u,v,u) = 0$ for any element u.

13.6 The Weak Formulation of the $\mathcal{V}$-projected Problem

We shall solve the projected problem by a shell approximation in the domain$\mathcal{U}$. In order to perform the shell approximation in intrinsic geometry, we introduce some considerations concerning the weak problem in $\mathcal{U}$.

Proposition 13.6. *The weak formulation (13.21) of the $\mathcal{V}$-projected problem can be expressed without any integral on the outer domain but with the use of the self adjoint pseudo-differential operator $K \in \mathcal{L}(H^{-\frac{1}{2}}(\Sigma), H^{\frac{1}{2}}(\Sigma))$ introduced at(13.13)*

there exist $u \in \mathcal{V}_\Sigma$, such that $\forall\theta \in \{v \in H^1(\mathcal{U}, R^3), v = 0$ on $\Gamma,\ div v = 0\,\}$

$$\int_{\mathcal{U}}\{ < Du.u,\theta > +\nu\, Du..D\theta + < Du.C + DC.u,\theta > - < \mathcal{L},\theta > \}dx$$
$$- \int_{\Sigma}\{ \ [\frac{1}{2}(|u|^2 + < c,u >) + 2\nu\, div_{\Gamma}u] < \theta,n > \\ +2\nu < \epsilon(u).n,\ \nabla_{\Gamma}K.(\theta.n) > \}d\Sigma \tag{13.37}$$

The proof is done using by parts integration in the outer domain Ω_P.We simply give the main points.

$$\int_{\Omega_P} < D^2\varphi, \nabla\varphi\nabla\psi >$$

Proposition 13.7.

$$\epsilon(u).n = -(\,div_{\Gamma}\,(u_{\Gamma}\,) + H < u,n >)n + \nabla_{\Gamma}(< u,n >) - D^2b.u_{\Gamma} + \sigma(u).n \tag{13.38}$$

In view of the "shell treatment" of (13.37) it is convenient to introduce the linear space

$$\mathcal{V}_{\Sigma} = \{v \in H^1(\mathcal{U},R^3), divv = 0,\ v = 0 \text{ on} \Gamma,\ v_{\Gamma} = \nabla_{\Gamma}(\ K.(< v,n >))\ \tag{13.39}$$

Then (13.37) can be rewritten as

there exist $u \in \mathcal{V}_{\Sigma}$, such that $\forall\theta \in \mathcal{V}_{\Sigma}$

$$\int_{\mathcal{U}}\{ < Du.u,\theta > +\nu\, Du..D\theta + < Du.C + DC.u,\theta > - < \mathcal{L},\theta > \}dx$$
$$- \int_{\Sigma}\{ \ [\frac{1}{2}(|u|^2 + < c,u >) + 2\nu\, div_{\Gamma}u] < \theta,n > \\ +2\nu < \epsilon(u).n,\ \theta_{\Gamma} > \}d\Sigma = 0 \tag{13.40}$$

At that point we can return to the formulation of the problem in the physical flow $U = u + C$, and (13.40) leads to

$$\int_{\mathcal{U}}\{ < DU.U,\theta > +\nu\, DU..D\theta \ \}dx \tag{13.41}$$
$$+ \int_{\Sigma}\{ < DC.n,\theta > -[\frac{1}{2}(|U-c|^2 + < c,U > -|c|^2) - 2\nu\, div_{\Gamma}U] < \theta,n > \\ -2\nu < \epsilon(U-C).n,\ \theta_{\Gamma} > \}d\Sigma = 0$$

If the function C is taken equal to the constant field c in a neighborhood of the surface Σ, then the derivatives of C at Σ are null and we have the problem

$$\int_{\mathcal{U}}\{ < DU.U,\theta > +\nu\, DU..D\theta \ \}dx \tag{13.42}$$
$$- \int_{\Sigma}\{ \ [\frac{1}{2}(|U-c|^2 + 2 < c,U > -|c|^2) - 2\nu\, div_{\Gamma}U] < \theta,n > \\ -2\nu < \epsilon(U).n,\ \theta_{\Gamma} > \}d\Sigma = 0$$

and notice that $|U-c|^2 + 2 < c,U > -|c|^2 = |U|^2$

13.7 Shell Approach

We make use of oriented distance function. Γ is the boundary of a smooth compact domain Q in R^3, h is the given thickness of $\mathcal{U}$. The shell approach consists in looking for approximations of the flow field in the form $\sum_{i=0,..,d} \tilde{q}_i ob\ q_i op$ lying in the space $\mathcal{V}_\Sigma$. We have to consider three conditions. The condition $v = 0$ on Γ is taken in account with $\tilde{q}_i(0) = 0,\ i = 1,..,d$ and $q_0 = 0$. The divergence free condition can be considered at least by three main possibilities: the exact treatment, by building such functions q_i onΓ for a given choice of polynomial functions $\tilde{q}_i$. In the section, this construction is done for particular polynomial basis. This construction leads to very heavy computations. The second approach is to use a mixed formulation for the linearized version of the problem (7.1); this approach is developed in section. We concentrate here on the penalization method which is classically used in such analysis. The third condition is simplified first by the choice of given polynomial $\tilde{q}_i(z)$ such that $\tilde{q}_i(h) = 0, i = 0,..,d-1$ and $\tilde{q}_d(h) = 1$. Then only the function q_d is concerned by the third condition of the linear space $\mathcal{V}_\Sigma$. The function $q_d$will be searched in the finite dimensional functional space on the boundary Γ generated by a family $\mathbf{E}^1 oT_h, ..., \mathbf{E}^m oT_h$,where T_h is the flow mapping of the field ∇b,and each function $\mathbf{E}^j$ is defined on the surface Σ as follows : given any scalar function e^j on the surface Σwe set $\mathbf{E}^j = e^j \mathbf{n} + \nabla_\Gamma(K.e^j)$ Notice that the sticking condition on Γ implies, under this choice of the polynomial function $\tilde{q}_i$, that $q_0 = 0$. Concerning the other functions $q_i, i = 1, ..., d-1$, no such condition occurs, and they can be expanded on any family of functions E^i_j. The penalized approach of the problem (13.42) is the following : there exists $U_\rho \in \mathcal{V}_\Sigma + \mathcal{C}$, such that, $\forall \theta \in \mathcal{V}_\Sigma$

$$\int_{\mathcal{U}} \{ < DU_\rho.U_\rho, \theta > + \nu\, DU_\rho..D\theta \ \} dx \tag{13.43}$$

$$+ \frac{1}{\rho} \int_{\mathcal{U}} div(U_\rho)\, div(\theta)$$

$$- \int_\Sigma \{ \ [\frac{1}{2}|U_\rho|^2 - 2\nu\, div_\Gamma U_\rho] < \theta, n >$$

$$-2\nu < \epsilon(U_\rho).n,\ \theta_\Gamma > \ \} d\Sigma = 0$$

In the sequel we shall drop the index ρ and write $U$for U_ρ.

13.8 Shell Structures

13.8.1 Oriented Distance Function. We consider a bounded body in R^N occupying a set $\bar{Q}$ with boundary Γ piecewise smooth. The outer domain occupied by the fluid, denoted by $\Omega = R^N \setminus \bar{Q}$, is an unbounded open set inR^N with compact boundary Γ . We consider the oriented distance function to the body b_Q, lipschitz continuous over R^N by

$$b_Q(x) = -\min_{y\in Q}|x-y|, x\in \bar{Q}, \ = \min_{y\in\bar{\Omega}}|x-y|, \ x\in\Omega$$

We recall from chapter 4 the smoothness properties of the boundary Γ and of the function b (we shall denote $b$for $b_{\bar{Q}}$). More precisely, there exists a tubular neighborhood of Γ in R^N in the following form $U_h = \{x \mid |b(x)| \leq h\}$, for some $h > 0$ such that the following property holds

Proposition 13.8.

$$b \in C^{1,1}(U_h) \quad \textit{if and only if } \Gamma \textit{ is a} C^{1,1} \textit{ manifold} \tag{13.44}$$

Moreover the projection mapping p is continuously defined from U_h onto Γ

In the present boundary layer situation, we consider the tubular neighborhood in the fluid domain : $\mathcal{U}_\langle = \{\S \mid \prime < \lfloor(\S) < \langle\} \subset \mathcal{U}_\langle$

Definition 13.1. *A vector field u defined in the neighborhood $\mathcal{U}_\langle$ is said to be in Shell Form of order d if there exists a familyu$_i$ $i = 1, .., d$ of vector fields defined on the boundaryΓ, $u_i : \Gamma \longrightarrow R^N$ and a family of functions $\tilde{u}_i(z)$ such that*

$$u = \sum_{i=0}^{d} \tilde{u}_i \circ b \ u_i \circ p \tag{13.45}$$

p, the projection onto Γ is defined in $\mathcal{U}_\langle$ for h small enough and verifies the following properties

$$p = I - b\,\nabla b$$

$$\nabla b = \nabla b \circ p = n \circ p$$

We have also $Dp = I - \nabla b.\nabla b^* - bD^2 b = D^* p$ which can be written as $Dp = P \circ p - bD^2 b$ where $P = I - n.n^*$ is the projector onto the tangential linear space $T_{p(x)}(\partial\Omega)$
We notice that $Dp.\nabla b = 0$
Dp cannot be written in the form $A \circ p$. We could choose a first order approximation

$$\partial^2_{ij} b(x) \cong \partial^2_{ij} b(p(x)) + < x - p(x), \nabla \partial^2_{ij} b(p(x)) >$$

$$D^2 b \cong D^2 b \circ p + < I - p, D^3 b \circ p >$$

but $I-p = b\,\nabla b$ then $\partial^2_{ij} b(x) = (\partial^2_{ij} b) \circ p(x) + b(x) < (\nabla_k b)(x), (\partial^3_{ijk} b) \circ p(x) >$ and as $(\partial_k b)(x) = (\partial_k b) \circ p(x)$ in $\mathcal{U}_\langle$

$$D^2 b \cong D^2 b \circ p + b\ D^3 b \circ p.\nabla b$$

We set $D^3 b.\nabla b$ is a two order tensor defined by

$$(D^3 b.\nabla b)_{ij} = \sum_k \partial^3_{ijk} b.\partial_k b = -((D^2 b)_{ij})^2$$

More precisely, as we will see in lemma 13.6, in a neighborhood of Γ, D^2b is completely defined by $D^2b \circ p$ with

$$D^2b = \sum_{i=0}^{\infty} (-b)^i (D^2b \circ p)^{i+1} \tag{13.46}$$

In our model, we consider all the terms of its expansion and don't make any geometrical approximation

$$p(X) = X - b(X)\ \nabla b(X)$$

$$T_z(X) = X + z\ \nabla b(X) = X + z\ n \circ p(X)$$

$$DT_z = I + z\ D^2b$$

$$p \circ T_z = I_\Gamma$$

$$T_z \circ p = I_{\Gamma_z}$$

$$Dp = (I - n.n^*) \circ p(X) + \sum_{i=1}^{\infty} (-b(X) \mathrm{d}^2 b \circ p(X))^i \quad , \forall X \in \mathcal{U}_($$

Lemma 13.6. $\forall x \in \Gamma$, $\forall z$, $|z| < \rho^{-1}$; $\rho = spec(D^2b)$

$$D^2b \circ T_z(x) = \sum_{i=0}^{\infty} (-z)^i\ (D^2b(x))^{i+1} \tag{13.47}$$

$$= -\frac{1}{z} \sum_{i=1}^{\infty} (-zD^2b(x))^i \tag{13.48}$$

and

$$Dp \circ T_z(x) = (I - n.n^*)(x) + \sum_{i=1}^{\infty} (-z\ D^2b(x))^i \tag{13.49}$$

moreover $D^2b(x)$ and $D^2b \circ T_z(x)$ are simultaneously diagonalizable.

Proof. we have

$$D^2b = D^2b \circ T_z[I + zD^2b]$$

Now, let consider the following expression

$$A_n(z) = \sum_{i=0}^{n} (-zD^2b)^i$$

and denote by ρ the spectral radius of D^2b then, $\forall z, |z| < \rho^{-1}$, A_n converges uniformly (in z) to $[I + zD^2b]^{-1}$

$$A_n(z)\,[I + zD^2b] = [I + zD^2b]\,A_n(z) = I - (-zD^2b)^{n+1} \rightarrow I \text{ as } n \rightarrow \infty$$

Let V be an eigenvector of$D^2b(x)$, $D^2b(x).V = \lambda V$,with $\lambda \neq 0$.

$$(D^2b\,.\,A_n(z)).V = \lambda\left(\sum_{i=0}^{n}(-z\lambda)^i\right)V \to \frac{\lambda}{1+z\lambda}V mboxas\ n \to \infty$$

For $\lambda = 0$, obviously, the normal vector n is eigenvector of D^2b and $D^2b \circ T_z$.Then the symmetrical operators $D^2b \circ T_z$ and D^2b are simultaneously diagonalizable and the curvatures of the z level curves

$$\lambda_j \circ T_z = \frac{\lambda_j}{1+z\lambda_j}$$

$$P^{-1}.(D^2b \circ T_z).P = -\frac{1}{z}\sum_{i=1}^{\infty}(-z)^i(P^{-1}.D^2b.P)^i$$

$$\lambda_j \circ T_z = -\frac{1}{z}\sum_{i=1}^{\infty}(-z\lambda_j)^i$$

Now, as $(D^2b)^3 - H(D^2b)^2 + \kappa D^2b = 0$, we can compute $(D^2b)^n$ only depending on $D^2b$and $(D^2b)^2$, as follow

Theorem 13.2. $\forall n \geq 0$,
in the specific case where$\lambda_1 = \lambda_2 = \lambda$

$$(D^2b)^{n+1} = n\lambda^{n-1}(D^2b)^2 - (n-1)\lambda^n D^2b \tag{13.50}$$

and in the general case $(\lambda_1 \neq \lambda_2)$

$$(D^2b)^{n+1} = \frac{\lambda_1^n - \lambda_2^n}{\lambda_1 - \lambda_2}(D^2b)^2 - \lambda_1\lambda_2\frac{\lambda_1^{n-1} - \lambda_2^{n-1}}{\lambda_1 - \lambda_2}D^2b \tag{13.51}$$

Proof. From the Caley-Hamilton theorem, as the minimal polynomial of D^2b is $P(x) = x(x^2 - Hx + \kappa) = x^3 - Hx^2 + \kappa x$,

$$(D^2b)^3 - H(D^2b)^2 + \kappa D^2b = 0$$

then, there exist α_n, $\beta_n$and γ_n such that

$$(D^2b)^{n+1} = \alpha_{n+1}\ (D^2b)^2 + \beta_{n+1}\ D^2b + \gamma_{n+1}\ I$$

As the eigenvalue associated to the eigenvector ∇b is zero, $\gamma_{n+1} = 0$

$$(D^2b)^{n+1} = \alpha_{n+1}\ (D^2b)^2 + \beta_{n+1}\ D^2b.$$

Let us assume that the two non zero other eigenvalues $\lambda_1$and λ_2 of D^2b are separate; then

$$\lambda_1^{n+1} = \alpha_{n+1}\ \lambda_1^2 + \beta_{n+1}\ \lambda_1 \tag{13.52}$$

$$\lambda_2^{n+1} = \alpha_{n+1}\ \lambda_2^2 + \beta_{n+1}\ \lambda_2 \tag{13.53}$$

Multiplying (13.52) by λ_2, (13.53) by $-\lambda_1$ and adding, we obtain

$$\lambda_2\,\lambda_1^{n+1} - \lambda_1\,\lambda_2^{n+1} = \alpha_{n+1}\,(\lambda_1^2\,\lambda_2 - \lambda_2^2\,\lambda_1)$$

then,

$$\alpha_{n+1} = \frac{\lambda_1^n - \lambda_2^n}{\lambda_1 - \lambda_2}$$

Multiplying (13.52)by λ_2^2, (13.53) by$-\lambda_1^2$ and adding, we obtain

$$\lambda_2^2\,\lambda_1^{n+1} - \lambda_1^2\,\lambda_2^{n+1} = \beta_{n+1}\,(\lambda_1\,\lambda_2^2 - \lambda_2\,\lambda_1^2)$$

then,

$$\beta_{n+1} = -\lambda_1\,\lambda_2\,\frac{\lambda_1^{n-1} - \lambda_2^{n-1}}{\lambda_1 - \lambda_2}$$

The case $\lambda_1 = \lambda_2$ is obtain when λ_1 tends to λ_2

$$\lim_{\lambda_2\to\lambda_1} \alpha_{n+1} = n\,\lambda_1^{n-1} \quad \text{and} \quad \lim_{\lambda_2\to\lambda_1} \beta_{n+1} = -(n-1)\,\lambda_1^n$$

This and (13.47) permit us to give an useful expression of$D^2b\circ T_z$

Corollary 13.2. $\forall z \in R\setminus\left\{-\frac{1}{\lambda_1}, -\frac{1}{\lambda_2}\right\}$,*On* Γ; $\forall z$, $-\min_i \frac{1}{max(\lambda_i,0)} < z < \min_i \frac{1}{max(-\lambda_i,0)}$

$$D^2b\circ T_z = \frac{1}{j(z)}\{(1+zH)D^2b - z(D^2b)^2\} \tag{13.54}$$

(Denoting $j(z) = 1 + zH + z^2\kappa$).

Proof. From 13.51,

$$(D^2b)^{n+1} = \frac{\lambda_1^n - \lambda_2^n}{\lambda_1 - \lambda_2}(D^2b)^2 - \lambda_1\lambda_2\frac{\lambda_1^{n-1} - \lambda_2^{n-1}}{\lambda_1 - \lambda_2}D^2b$$

that we substitute in$D^2b\circ T_z$ given by 13.47.

$$\begin{aligned}
D^2b\circ T_z &= \sum_{i=0}^{\infty}(-z)^i\,(D^2b)^{i+1} \\
&= -\frac{D^2b}{\lambda_1 - \lambda_2}\left\{\lambda_2\sum_{i=0}^{\infty}(-z\lambda_1)^i - \lambda_1\sum_{i=0}^{\infty}(-z\lambda_2)^i\right\} \\
&\quad + \frac{(D^2b)^2}{\lambda_1 - \lambda_2}\left\{\sum_{i=0}^{\infty}(-z\lambda_1)^i - \sum_{i=0}^{\infty}(-z\lambda_2)^i\right\} \\
&= A\,D^2b + B\,(D^2b)^2
\end{aligned}$$

Taking account that

$$\forall x,\ |x| < 1,\ \sum_{i=1}^{\infty}(-x)^i = \frac{-x}{1+x}$$

which is verified by $(z\,\lambda_{\frac{1}{2}})$ for all $z, \|z| < (spec(D^2b))^{-1}$,and we can establish

$$\begin{aligned} A &= -\frac{1}{\lambda_1 - \lambda_2}\left\{\lambda_2 + \lambda_2\frac{-z\lambda_1}{1+z\lambda_1} - \lambda_1 - \lambda_1\frac{-z\lambda_2}{1+z\lambda_2}\right\} \\ &= \frac{(\lambda_1 - \lambda_2)(j - z^2\kappa)}{(\lambda_1 - \lambda_2)\,j} = \frac{1+zH}{j} \end{aligned}$$

$$\begin{aligned} \text{and} \quad B &= \frac{1}{\lambda_1 - \lambda_2}\left(\frac{-z\lambda_1}{1+z\lambda_1} + \frac{z\lambda_2}{1+z\lambda_2}\right) \\ &= \frac{-z\lambda_1 - z^2\lambda_1\,\lambda_2 + z\lambda_2 + z^2\lambda_1\,\lambda_2}{(\lambda_1\,\lambda_2)(1 + z(\lambda_1+\lambda_2) + z^2(\lambda_1\,\lambda_2))} = \frac{-z}{j} \end{aligned}$$

Then, we obtain an expression of $D^2b \circ T_z$ depending on$z$. Its validity domain is out of $j(z)$ zeros $-\frac{1}{\lambda_1}$and $-\frac{1}{\lambda_2}$
Effectively,

$$j(z_0) = 0 \equiv z_0 = \frac{-H \overset{+}{-} \sqrt{H^2 - 4\kappa}}{2\kappa}$$

That we can rewrite as $\sqrt{H^2 - 4\kappa} = |\lambda_1 - \lambda_2|$

$$\begin{aligned} z_0 &= \frac{-\lambda_1 - \lambda_2 \overset{+}{-} |\lambda_1 - \lambda_2|}{2\kappa} \\ &= -\frac{1}{2}\left\{\frac{1}{\lambda_1} + \frac{1}{\lambda_2} \overset{-}{+} sgn(\lambda_1\,\lambda_2)\left|\frac{1}{\lambda_1} - \frac{1}{\lambda_2}\right|\right\} \end{aligned}$$

$$\text{Finally} \quad z_0 = \begin{cases} -\max_i & \frac{1}{\lambda_i} \\ -\min_i & \frac{1}{\lambda_i} \end{cases}$$

Finally, the roots of $j(z)$are $-\max_i \frac{1}{\lambda_i}$ and $-\min_i \frac{1}{\lambda_i}$ i.e. $-\frac{1}{\lambda_1}$ and $-\frac{1}{\lambda_2}$. That we can see more simply noticing that

$$j(z) = 1 + z(\lambda_1 + \lambda_2) + z^2\lambda_1\lambda_2 = (1 + z\lambda_1)(1 + z\lambda_2)$$

From this corollary, we can derive the following one

Corollary 13.3. *On Γ; $\forall z,\ -\min_i \frac{1}{max(\lambda_i,0)} < z < \min_i \frac{1}{max(-\lambda_i,0)}$*

$$(D^2b)^2 \circ T_z = \frac{1}{j^2}\{z\kappa(2+zH)\,D^2b + (1 - z^2\kappa)\,(D^2b)^2\} \tag{13.55}$$

And as $Dp \circ T_z = I - n.n^* - zD^2b \circ T_z$ on Γ

Corollary 13.4. *On Γ; $\forall z$, $-\min_i \frac{1}{max(\lambda_i,0)} < z < \min_i \frac{1}{max(-\lambda_i,0)}$*

$$Dp \circ T_z = I - n.n^* - \frac{z(1+zH)}{j(z)} D^2 b + \frac{z^2}{j(z)} (D^2 b)^2 \tag{13.56}$$

As a consequence, we can also establish the following corollary

Corollary 13.5. *On Γ; $\forall z$, $-\min_i \frac{1}{max(\lambda_i,0)} < z < \min_i \frac{1}{max(-\lambda_i,0)}$*

$$\begin{aligned}(Dp)^2 \circ T_z &= I - n.n^* \\ &\quad - \frac{2z(1+zH)^2 + z^4 \kappa H}{j^2} D^2 b \\ &\quad + \frac{z^2(3 + 2zH + z^2\kappa)}{j^2} (D^2 b)^2\end{aligned}$$

Proof. Taking (13.56) to the power two and considering that

$$(D^2 b)^3 = H (D^2 b)^2 - \kappa D^2 b$$

and

$$(D^2 b)^4 = (H^2 - \kappa)(D^2 b)^2 - H\kappa D^2 b$$

obtained considering the case$n = 3$ in (13.51).

From (13.54) we can also establish the following properties of geometrical elements.

Lemma 13.7. *On Γ; $\forall z$, $-\min_i \frac{1}{max(\lambda_i,0)} < z < \min_i \frac{1}{max(-\lambda_i,0)}$*

i)

$$H \circ T_z = H - \frac{z}{j(z)} (zH\kappa + D^2 b..D^2 b) \tag{13.57}$$

ii)

$$H \circ T_z = \frac{H + 2z\kappa}{j(z)} = \frac{\partial}{\partial z} Log\ j(z) \tag{13.58}$$

Proof. *i)* Applying the *trace* operator to $D^2 b \circ T_z$ given by(13.54),

$$\begin{aligned}H \circ T_z &= trace(D^2 b \circ T_z) \\ &= \frac{1}{j(z)} \{(1+zH)H - zD^2 b..D^2 b\} \\ &= H - \frac{z}{j(z)} (zH\kappa + D^2 b..D^2 b)\end{aligned}$$

ii) $\lambda_i \circ T_z = \frac{\lambda_i}{1+z\lambda_i}$ so,

$$\begin{aligned}H \circ T_z &= \frac{\lambda_1}{1+z\lambda_1} + \frac{\lambda_2}{1+z\lambda_2} \\ &= \frac{\lambda_1 + \lambda_2 + 2z\lambda_1\lambda_2}{1 + z(\lambda_1+\lambda_2) + z^2\lambda_1\lambda_2} \\ &= \frac{H + 2z\kappa}{j(z)}\end{aligned}$$

Lemma 13.8.

$$\kappa \circ T_z = \frac{\kappa}{j(z)}$$

Proof.

$$\kappa \circ T_z = \frac{\lambda_1}{1 + z\lambda_1}$$

$$frac\lambda_2 1 + z\lambda_2$$

$$= \frac{\lambda_1\lambda_2}{1 + z(\lambda_1 + \lambda_2) + z^2\lambda_1\lambda_2}$$

$$= \frac{\kappa}{j(z)}$$

Remark 13.1. Comparing the two expressions of $H \circ T_z$ given by(13.57) and (13.58), we can verify that

$$\kappa = \frac{1}{2}(H^2 - D^2b..D^2b) \tag{13.59}$$

The penalized problem in that tubular boundary layer $\mathcal{U}_($ can be formulated on the level curves of the function b, see section 13.14.1, and we get

$$\int_0^h \int_\Gamma \{< DU_\rho oT_z.U_\rho oT_z, \theta oT_z > + \nu\, DU_\rho oT_z..D\theta oT_z\ \}[1 + Hz + z^2K]\, d\Gamma\, dz \tag{13.60}$$

$$+\frac{1}{\rho}\int_0^h \int_\Gamma (div(U_\rho))oT_z\, (div(\theta))oT_z\ [1 + zH + z^2\, K]\, d\Gamma\ dz$$

$$-\int_0^h \int_\Gamma \{\ [\frac{1}{2}(|U_\rho oT_z - c|^2 + < c, U_\rho oT_z >) + 2\nu\, (div_\Gamma U_\rho)oT_z]\ < \theta oT_z, noT_z >$$

$$+2\nu\ < \epsilon(U_\rho)oT_z.noT_z, \theta_\Gamma oT_z\ >\ \}[1 + zH + z^2\, K]\ d\Gamma\ dz$$

Where T_z is the flow mapping of the field ∇b:$T_z(x) = x + z\, n(x)$ maps the boundary Γ onto the surface$\Sigma = b^{-1}(h)$. We shall now investigate the weak formulation when the field U is in Shell form with an adequate family of polynomial functions . The functions U_ρ and $\theta \in \mathcal{V}_\Sigma$ should be taken in shell form in this weak formulation. As the divergence free condition is now treated by the penalization approach, these functions should only verify the sticking condition $\theta = 0$ on Γ and $\theta_\Gamma = \nabla_\Gamma K.(\theta.n)$ on the surface Σ. These two conditions shall be taken in account by the previous choice of polynomial functions $\tilde{q}_i$. We shall apply the chain rule formulae

$$(Du)oT_z = D(uoT_z).(DT_z)^{-1} \tag{13.61}$$

so that when u is in shell form, $u = \tilde{u}_i ob\ u_i op$ in$\mathcal{U}_($ we get $uoT_z = \tilde{u}_i(z)\ u_i$ on Γ. After several sections devoted to the properties of the various intrinsic operators involved in this question, we shall derive the final version of the penalized shell Navier Stokes flow.

13.9 The Linear Tangent Operator Du

In the weak form of the various problems we consider, we make use of the following integration by parts formula

Lemma 13.9.

$$\int_\Omega Du..Dv\,dx = \int_\Omega 2\epsilon(u)..\epsilon(v) + \nabla(divu)\;v\,dx - \int_{\partial\Omega} < D^*u.n, v > d\Gamma$$

Proof.

$$2\int_\Omega \epsilon(u)..\epsilon(v)\,dx \quad = \quad \int_\Omega \frac{1}{2}(\partial_i u_j + \partial_j u_i)(\partial_i v_j + \partial_j v_i)\,dx$$

$$= \quad \int_\Omega \partial_i u_j\;\partial_i v_j + \partial_j u_i$$

$$partial_j v_i\,dx$$

$$= \quad \int_\Omega Du..Dv - \partial_j\partial_i u_i\;v_j\,dx + \int_{\partial\Omega} \partial_j u_i\;v_j\;n_i\,d\Gamma$$

We need to decompose on any surface Σ the restrictions to the surfaceΣ of the differential operators such as Du, u being defined in a neighborhood of the surface Σ. We shall always denote by $D_\Gamma u$ the tangential differential operator on the surfaceΣ; we do here emphasize that this notation will hold when the surface is Γ itself or any smooth surface Σ. The following holds on Σ

$$Du|_\Sigma = D_\Gamma(u|_\Sigma) + (Du)|_\Sigma.n.n^* \tag{13.62}$$

where n stands here for the unitary normal field on Σ. The boundary Layer $\mathcal{U}_\langle$ has "two boundaries" (i.e its boundary has two connected components Γ and $\Sigma = b^{-1}(h)$). We shall consider the situation when the wake is neglected: then $Q$is the domain occupied by the solid domains . The situation in which the wake S assumed to be a bounded surface and in which Q stands for the volume occupied by the rigid body augmented of the wake S. In both situations when h is large enough the surface Σ is smooth and compact. On each level curve $b^{-1}(z)$ $0 \le z \le h$ the projector Dp onto the linear tangential space is given by

$$Dp \circ T_z = (I - \nabla b.\nabla^* b - b\;D^2 b) \circ T_z = I - n.n^* - z\;D^b \circ T_z \tag{13.63}$$

$$\int_\Omega < Du.\quad v \quad , w > dx = \int_\Omega \partial_j u_i v_j w_i\,dx$$

$$= \quad -\int_\Omega u_i v_j \partial_j w_i\,dx - \int_\Omega u_i \partial_j v_j w_i\,dx + \int_{\partial\Omega} u_i v_j w_i n_j\,d\Gamma$$

$$= \quad -\int_\Omega < Dw.v, u > dx - \int_\Omega < u, w > \;divv\,dx$$

$$+ \int_{\partial\Omega} < u, w >< v, n > d\Gamma$$

Such that, $\forall v$, $div\ v = 0$ in Ω,

$$\int_{\Omega} < Du.v, u > dx = \frac{1}{2}\int_{\partial\Omega} u^2 < v, n > d\Gamma$$

Lemma 13.10.

$$\int_{\partial\Omega} < Du.n, v > d\Gamma = \int_{\partial\Omega} \frac{\partial}{\partial n}(< u, v >) - < Dv.n, u > d\Gamma$$

Proof.

$$\nabla(< u, v >) = D^*u.v + D^*v.u$$

$$\begin{aligned}
\int_{\partial\Omega} < Du.n, v > d\Gamma &= \int_{\partial\Omega} < D^*u.v, n > d\Gamma \\
&= \int_{\partial\Omega} \frac{\partial}{\partial n}(< u, v >) - < D^*v.u, n > d\Gamma \\
&= \int_{\partial\Omega} \frac{\partial}{\partial n}(< u, v >) - < Dv.n, u > d\Gamma
\end{aligned}$$

From which we get

$$\int_{\partial\Omega} < Du.n, u > d\Gamma = \frac{1}{2}\int_{\partial\Omega} \frac{\partial}{\partial n}(|u|^2)\, d\Gamma$$

13.10 Weak formulation for the vector field Δu

Lemma 13.11.

$$\int_{\Omega} < \Delta u, v > dx = -\int_{\Omega} Du..Dv\, dx + \int_{\partial\Omega} < Du.n, v > d\Gamma \tag{13.64}$$

Proof.

$$\int_{\Omega} \partial^2_{jj} u_i\ v_i\, dx = -\int_{\Omega} \partial_j u_i\ \partial_j v_i\, dx + \int_{\partial\Omega} \partial_j u_i v_i n_j\, d\Gamma$$

13.10.1 The Deformation Tensor. $\epsilon(u)$ We classically denote by $\epsilon(u)$ the anti-symmetric part of the deformation tensor Du.

$$\epsilon(u)\hat{e}q\frac{1}{2}(Du + D^*u)$$

Lemma 13.12.

$$2\int_{\Omega} \epsilon(u)..\epsilon(v)\, dx = -\int_{\Omega} < \Delta u + \nabla(divu), v > dx + 2\int_{\partial\Omega} < \epsilon(u).n, v > d\Gamma$$

Proof.

$$\begin{aligned}
2\int_{\Omega} \epsilon(u)..\epsilon(v) \quad dx &= \frac{1}{2}\int_{\Omega}(\partial_j u_i + \partial_i u_j)(\partial_j v_i + \partial_i v_j)\, dx \\
&= \int_{\Omega} \partial_j u_i \partial_j v_i + \partial_j u_i \partial_i v_j \, dx \\
&= -\int_{\Omega} \partial^2_{jj} u_i v_i + \partial^2_{ij} u_i v_j \, dx + \int_{\partial\Omega} \partial_j u_i v_i n_j + \partial_j u_i v_j n_i \, d\Gamma \\
&= -\int_{\Omega} < \Delta u + \nabla(divu), v > \, dx \\
&\quad + \int_{\partial\Omega} < Du.n, v > + < D^* u.n, v > \, d\Gamma
\end{aligned}$$

Lemma 13.13.

$$\int_{\Omega} \epsilon(u)..\epsilon(v)\, dx = -\int_{\Omega} < div(\epsilon(u)), v > \, dx + \int_{\partial\Omega} < \epsilon(u).n, v > \, d\Gamma$$

Proof.

$$\begin{aligned}
\int_{\Omega} \epsilon(u)..\epsilon(v)\, dx &= \int_{\Omega} \frac{1}{2}(\epsilon(u))_{ij}(\partial_j v_i + \partial_i v_j)\, dx \\
&= -\frac{1}{2}\int_{\Omega} \partial_j(\epsilon(u))_{ij} v_i + \partial_i(\epsilon(u))_{ij} v_j \, dx \\
&\quad + \frac{1}{2}\int_{\partial\Omega} (\epsilon(u))_{ij} v_i n_j + (\epsilon(u))_{ij} v_j n_i \, d\Gamma
\end{aligned}$$

From lemma 13.12and 13.13,

$$2\ div(\epsilon(u)) = \Delta u + \nabla(divu)$$

Anti-symmetrical part $\sigma(u)$ Let denote $\sigma(u)$ the anti-symmetrical component of the linear operator Du

$$\sigma(u)\hat{e}q\frac{1}{2}(Du - D^* u)$$

Lemma 13.14. *i)*

$$2\int_{\Omega} \sigma(u)..\sigma(v)\, dx = \int_{\Omega} < -\Delta u + \nabla(divu), v > \, dx + 2\int_{\partial\Omega} < \sigma(u).n, v_{\Gamma} > \, d\Gamma$$

ii) $< curl(u), curl(v) >= 2\ \sigma(u)..\sigma(v)$

Proof.

$$\begin{aligned}
2\int_\Omega \sigma(u) \quad .. \quad \sigma(v)\,dx &= \frac{1}{2}\int_\Omega (\partial_j u_i - \partial_i u_j)(\partial_j v_i - \partial_i v_j)\,dx \\
&= \int_\Omega \partial_j u_i \partial_j v_i - \partial_j u_i \partial_i v_j \,dx \\
&= \int_\Omega -\partial^2_{jj} u_i v_i + \partial^2_{ij} u_i v_j \,dx + \int_{\partial\Omega} \partial_j u_i v_i n_j - \partial_j u_i v_j n_i \,d\Gamma \\
&= \int_\Omega < -\Delta u + \nabla(divu), v > dx \\
&\quad + \int_{\partial\Omega} < Du.n, v > - < D^* u.n, v > d\Gamma
\end{aligned}$$

Lemma 13.15. $\forall (u,v) \in H^1(\Omega; R^3)^2$,

$$\int_\Omega \sigma(u)..\sigma(v)\,dx = \int_\Omega \epsilon(u)..\epsilon(v) +$$

$$< \nabla(divu), v > dx - \int_{\partial\Omega} < D^* u.n, v > d\Gamma$$

Proof. From lemma13.12 and 13.14.

$$\sigma(u \circ p) = \frac{1}{2}(Du \circ p.Dp - Dp.D^* u \circ p)$$

13.11 The pseudo-differential operator K

$$\begin{aligned}
K : H^{-\frac{1}{2}}(\Sigma) &\longrightarrow H^{\frac{1}{2}}(\Sigma) \\
\mu &\longmapsto \gamma_\Sigma \phi
\end{aligned}$$

Where ϕ is the solution of the non-homogeneous outer Neumann problem

$$\Delta\phi = 0 \text{ in } \Omega$$

$$\frac{\partial\phi}{\partial n} = \mu \text{ on} Sigma = \partial\Omega$$

Proposition 13.9. *K is self adjoint.*

Proof. For all $(\mu,\theta) \in (H^{-\frac{1}{2}}(\Sigma))^2$, we define$(\phi,\pi) \in (H^{\frac{1}{2}}(\Sigma))^2$ as follow, $K.\mu = \gamma_\Sigma \phi$ and $K.\theta = \gamma_\Sigma \pi$

$$\begin{aligned}
\int_\Omega < \nabla\phi, \nabla\pi > dx &= \int_\Omega -\Delta\phi \,\pi\,dx + \int_\Sigma \pi \frac{\partial\phi}{\partial n}\,d\Gamma \\
&= \int_\Omega -\Delta\pi \,\phi\,dx + \int_\Sigma \frac{\partial\pi}{\partial n}\phi\,d\Gamma
\end{aligned}$$

Where $\Delta\phi = 0$ and $\Delta\pi = 0$in Ω so,

$$< K.\theta, \mu >_{H^{\frac{1}{2}}(\Sigma)\times H^{-\frac{1}{2}}(\Sigma)} = < \theta, K.\mu >_{H^{\frac{1}{2}}(\Sigma)\times H^{-\frac{1}{2}}(\Sigma)}$$

$$u_\Gamma = \nabla_\Gamma K.(< u,n >) + c_\Gamma + \nabla_\Gamma K.(- < c,n >)$$

$$u =< u,n > n + u_\Gamma$$

$$u = c + \nabla_\Gamma K.(< u - c,n >)+ < u - c,n > n$$

13.12 Coupling Navier-Stokes and Potential Flows

13.12.1 Weak Formulation in the Whole Domain Ω. Find $u \in H^1(\Omega; R^3)$, $div(u) = 0$, $u = 0$on Γ, $u = c$ on S

$$\begin{cases} \forall v \in H_0^1(\Omega; R^3), div(v) = 0, \\ \int_\Omega < Du.u, v > +2\nu\ \epsilon(u)..\epsilon(v)\, dx = 0 \end{cases} \tag{13.65}$$

Boundary Strengths onΓ: We propose to substitute to the sticking condition a boundary strength constraint f acting on the boundary of the fluid domain.

$$\int_\Omega < Du.u, v > +2\nu\ \epsilon(u)..\epsilon(v)\, dx = \int_{\partial\Omega} < f, v > d\Gamma \tag{13.66}$$

$\forall v \in H^1(\Omega; R^3)$, $div(v) = 0$

$$\int_{\partial\Omega} < f, v > d\Gamma = \int_\Omega < Du.u - \nu Deltau, v > dx + \int_{\partial\Omega} 2\nu\ < \epsilon(u).n, v > d\Gamma$$

For all $v, div(v) = 0$, there exist p in $L^2(\Omega)$, such that$Du.u - \nu\ \Delta u = -\nabla p$ in Ω,the boundary acting term can be rewrite as

$$\int_{\partial\Omega} < f, v > d\Gamma = \int_\Omega < -\nabla p, v > dx + \int_{\partial\Omega} 2\nu\ < \epsilon(u).n, v > d\Gamma$$

Performing by part and taking account that $div(v) = 0$,this expression becomes

$$\int_{\partial\Omega} < f, v > d\Gamma = \int_{\partial\Omega} < 2\nu epsilon(u).n - p\, n, v > d\Gamma$$

In consequence, the force applied by the fluid to the wall

$$f = 2\nu\ \epsilon(u).n - p\, n \tag{13.67}$$

and the resulting force

$$F = \int_{\partial\Omega} 2\nu\ \epsilon(u).n - p\, n\, d\Gamma$$

Its component T in the c leading velocity direction, called the drag, can be expressed as follow:

$$T =< F, -\frac{c}{|c|} >= \int_{\partial\Omega} < 2\nu\ \epsilon(u).n - pn, -\frac{c}{|c|} > d\Gamma$$

Proposition 13.10.

$$T = \frac{1}{|c|} 2\nu \int_\Omega \epsilon(u)..\epsilon(u)\, dx$$

Proof. Let us consider (13.66)

$$\forall v \in H^1(\Omega; R^3), div(v) = 0$$

and

$$u \text{solution of the problem}(13.65),$$

$$\int_\Omega < Du.u, v > +2\nu\ \epsilon(u)..\epsilon(v)\, dx = \int_{\partial\Omega} < f, v > d\Gamma$$

in the particular case where $v = u - c$, ($v = 0$ on S and $v = -c$on Γ), the drag force can be written as

$$T\,|c| = \int_\Omega < Du.u, u - c > +2\nu\ \epsilon(u)..\epsilon(u)\, dx$$

where

$$\int_\Omega < Du.u, u - c > dx = \int_S (\frac{1}{2}|u|^2 - < u, c >) < u, n > d\Gamma$$

$$\text{and } u = c \text{ on } S$$

$$= -\frac{1}{2}|c|^2\ < c, \int_S n\, d\Gamma >= 0$$

Which can also be derived from

$$\begin{aligned}\int_\Omega < Du.u, u - c > dx &= \int_\Omega < Dv.u, v > dx\\ &= \frac{1}{2}\int_\Gamma |v|^2 < u, n > d\Gamma = 0\end{aligned}$$

Multi-domain Navier Stokes Problem: Ω bounded open set of R^N, Σsurface, $\Omega \setminus \Sigma = \Omega_P \cup \mathcal{U}$
then $u \in H_0^1(\Omega)$ such that $v = u_{|\Omega_P}$ and $w = u_{|\mathcal{U}}$. In $\Omega \setminus \mathcal{U}$

$$Dv.v - \nu\Delta v + \nabla q = g \tag{13.68}$$

$$v = c \text{ on } \S \tag{13.69}$$

$$divv = 0 \tag{13.70}$$

In $\mathcal{U}$

$$Dw.w - \nu\Delta w + \nabla p = g \tag{13.71}$$

$$w = 0 \text{ on } \Gamma \tag{13.72}$$

$$divw = 0 \tag{13.73}$$

Let us assume that the normal vector n to Σ is oriented outward of $\mathcal{U}$.

Lemma 13.16.

$$u \in H^1(\Omega) \equiv v = w \quad on\ \Sigma$$

Lemma 13.17.

$$divu = 0 \quad in \quad \Omega \equiv \begin{cases} divv = 0 \quad in \quad \Omega_P \\ divw = 0 \quad in \quad \mathcal{U} \\ < v, n > = < w, n > \quad on\ \Sigma \end{cases}$$

Proof.

$$< divU, v >_{\mathcal{D}'(\Omega) \times \mathcal{D}(\Omega)} =$$

$$\int_{\mathcal{U}} div(u)\ v\,dx + \int_{\Omega \backslash \mathcal{U}} div(\nabla\phi) v\,dx + \int_{\Sigma} < u - \nabla\phi, n > d\Gamma$$

$divv = 0$ in Ω_P
$div(u) = 0$ in $\mathcal{U}$,
$\Delta\phi = 0$ in $\Omega \setminus \mathcal{U}$,
$u = \nabla\phi$ on Σ

Let us establish a few properties of this decomposition

13.12.2 S.N.C. for a global stationary Navier Stokes Flow.

Lemma 13.18. *Si* $div(U) = 0$ *in* Ω,

$$< \epsilon(U).n, n > = < DU.n, n > = -div_\Gamma(U_\Gamma) - H \quad < U, n > \quad on \quad \Sigma$$

Proof.

$$0 = \Delta\phi = \Delta_\Gamma\phi + H\frac{\partial\phi}{\partial n} + \frac{\partial^2\phi}{\partial n^2}$$

On one hand,

$$\begin{aligned} div_\Gamma U &= div_\Gamma(U_\Gamma + < U, n > n) \text{ and } \div_\Gamma (n) = trace(D^2 b) = H \\ &= div_\Gamma(U_\Gamma) + H \quad < U, n > \end{aligned}$$

and on the other hand,

$$\begin{aligned} div_\Gamma U &= divU - < DU.n, n > \\ &= divU - < \epsilon(U).n, n > \end{aligned}$$

Lemma 13.19. $\forall\tau, < \tau, n > = 0$,

$$< \epsilon(U)\ n, \tau > = < \nabla_\Gamma(< U, n >), \tau > - < D^2 b. < \sigma(U).n, \tau > \quad on \quad \Sigma$$

$$(\epsilon(U).n)_\Gamma = \nabla_\Gamma(< U, n >) - D^2 b.U_\Gamma + \sigma(U).n$$

Proof.

$$\nabla(< U, \nabla b >) = D^*U.\nabla b + D^2 b.U$$

where $D^*U = \epsilon(U) - \sigma(U)$ so,

$$< \nabla_\Gamma(< U, n >), \tau > = < \epsilon(U).n, \tau > - < \sigma(U).n, \tau > + < D^2 b.U_\Gamma, \tau >$$

Moreover, $\sigma(U)$ is antisymetric,

$$< \sigma(U).n, n > = - < n, \sigma(U).n > = 0$$

. $\sigma(U).n$is consequently only tangential. This permits to conclude.

From lemma 13.18 and13.19, we can establish

$$D^*u.n = -(div_\Gamma u_\Gamma + H < u, n >)\, n + \nabla_\Gamma(< u, n >) - D^2 b.u_\Gamma \quad (13.74)$$

and

$$\epsilon(u).n = -(div_\Gamma u_\Gamma + H < u, n >)\, n + \nabla_\Gamma(< u, n >) - D^2 b.u_\Gamma + \sigma(u).n \quad (13.75)$$

13.13 Potential case inΩ_P

In particular, in $\Omega_P$where $U = \nabla\phi$,

$$< D^2\phi.n, \tau > = < \nabla_\Gamma(\frac{\partial\phi}{\partial n}), \tau > - < D^2 b.\nabla_\Gamma\phi, \tau >$$

Et si $U = \nabla_\Gamma\phi$,

$$< D^2\phi.n, \tau > = - < D^2 b.\nabla_\Gamma\phi, \tau >$$

Finally, in the sliding case on Σ,

$$\epsilon(\nabla\phi).n = D^2\phi.n = -\Delta_\Gamma\phi\, n - D^2 b.\nabla_\Gamma\phi \text{ on } \Sigma$$

Proposition 13.11. *(U, P) is a Navier Stokes solution in all Ω if and only if (U, P) is a Navier Stokes solution in $\mathcal{U}$, $\Omega \setminus \mathcal{U}$ and*

$$\left.\begin{array}{l} p = -\frac{1}{2}|u|^2 \\ \sigma(u).n = 0 \end{array}\right\} \text{ on } \Sigma$$

Proof. For all V, $div(V) = 0$, Cas Ω

$$\begin{aligned}
\int_\Omega < DU.U, V > &+ 2\nu\, \epsilon(U)..\epsilon(V)\, dx = \int_\Omega < DU.U - \nu\Delta U, V > dx \\
&+2\nu \int_{\partial\Omega} < \epsilon(U).n, V > d\Gamma \quad, \forall V, div(V) = 0 \\
&= \int_\Omega < \nabla P, V > dx - 2\nu \int_{\partial\Omega} < \epsilon(U).n, V > d\Gamma \\
&= \int_{\partial\Omega} \{P - 2\nu < \epsilon(U).n, n >\} < V, n > \\
&\qquad -2\nu < \epsilon(U).n, V_\Gamma > d\Gamma
\end{aligned}$$

From the previous calculations, (U, P) is Navier Stokes solution in Ω if and only if

$$\int_{\Sigma} \text{ljump}P - 2\nu < \epsilon(U).n, n > \text{rjump}_{\Sigma} < V, n > -$$
$$2\nu\text{ljump} < \epsilon(U).n, V_{\Gamma} > \text{rjump}_{\Sigma}\, d\Gamma = 0 \quad ,\forall V$$

Where ljump.rjump$_{\Sigma}$ is the jump across Σ.

$$\begin{cases} \text{ljump}P - 2\nu < \epsilon(U).n, n > \text{rjump}_{\Sigma} = 0 \\ \text{ljump} < \epsilon(U).n, V_{\Gamma} > \text{rjump}_{\Sigma} = 0 \end{cases}$$

From the continuity of Uon Σ and lemma 13.18,ljump $< \epsilon(U).n, n >$ rjump$_{\Sigma}$ $= 0$. So, the pressure P is continuous across Σ and $p = -\frac{1}{2}|u|^2$ on Σ. In the same way, from lemma 13.19,ljump $< \epsilon(U).n, V_{\Gamma} >$ rjump$_{\Sigma}$ $=<$ $\sigma(u).n, V_{\Gamma} >$and so $(\sigma(u).n)_{\Gamma} = 0$ on Σ.

Proposition 13.12.

$$\int_{\mathcal{U}} < Du.u, V > +2\nu\, \epsilon(u)..\epsilon(V)\, dx + \int_{\Sigma} \{\frac{1}{2}|u|^2 - 2\nu < \epsilon(U).n, n >\} < V, n > +$$
$$-2\nu < \epsilon(U).n, V_{\Gamma} > \, d\Gamma = 0 \ ,\forall V$$

Proof. Cas $\Omega - U$ et U

$$0 = \int_{\Omega} < DU.U, V > +2\nu\, \epsilon(U)..\epsilon(V)\, dx$$

$$= \int_{\}\mathcal{U}} < DU.U, V > +2\nu\, \epsilon(U)..\epsilon(V)\, dx$$
$$+ \int_{\Omega\setminus\mathcal{U}} < DU.U, V > +2\nu$$

epsilon$(U)..\epsilon(V)\, dx$

$$= \int_{\mathcal{U}} < Du.u, V > +2\nu\, \epsilon(u)..\epsilon(V)\, dx$$
$$+ \int_{\Omega\setminus\mathcal{U}} < \nabla(\frac{1}{2}|\nabla\phi|^2), V > \, dx$$
$$-2\nu \int_{\Sigma} < \epsilon(\nabla\phi).n, V > \, d\Gamma$$
$$= \int_{\mathcal{U}} < Du.u, V > +2\nu\, \epsilon(u)..\epsilon(V)\, dx$$
$$+ \int_{\Sigma} \{\frac{1}{2}|u|^2 - 2\nu < \epsilon(U).n, n >\} < V, n >$$
$$-2\nu < \epsilon(U).n, V_{\Gamma} > \, d\Gamma$$

On Σ, $u =< u, n > n + \nabla_{\Gamma}K.(< u, n >) + c_{\Gamma} + \nabla_{\Gamma}K.(- < c, n >)$

13.14 Shell Representation

Suppose that all variables are of the following form

$$\begin{aligned} F: \ & \mathcal{U}_{(} \to R \\ & X \mapsto \tilde{f} \circ b(X)\ f \circ p(X) \end{aligned}$$

Where $\tilde{f} : [0, h] \to R$ and $f : \Gamma \to R$

$$\begin{aligned} U: \ & \mathcal{U}_{(} \to \mathcal{R}^{\ni} \\ & X \mapsto \tilde{u} \circ b(X)\ u \circ p(X) \end{aligned}$$

Where $\tilde{u} : [0, h] \to R$ and $u : \Gamma \to R^3$Gradient Operator ∇:

$$\nabla(\tilde{f} \circ b) = \tilde{f}' \circ b\ \nabla b \tag{13.76}$$

Lemma 13.20.

$$\nabla(f \circ p) = D^* p.(\nabla_\Gamma f) \circ p$$

Proof.

$$\begin{aligned} (\nabla(f \circ p))_i &= \partial_i (f \circ p) \\ &= \partial_k f \circ p\ \ \partial_i p_k \\ &= (\nabla f \circ p)_k\ (Dp)_{ki} \end{aligned}$$

and Dp is self adjoint.

Lemma 13.21.

$$\nabla(\tilde{f} \circ b\ f \circ p) = \tilde{f}' \circ bnablab\ f \circ p + \tilde{f} \circ b\ Dp.(\nabla_\Gamma f) \circ p$$

Proof.

$$\begin{aligned} \nabla(\tilde{f} \circ b\ f \circ p) &= \nabla(\tilde{f} \circ b)\ f \circ p + \tilde{f} \circ b\ \nabla(\ f \circ p) \\ &= \tilde{f}' \circ b\ \nabla b\ f \circ p + \tilde{f} \circ b\ Dp.(\nabla_\Gamma f) \circ p \end{aligned}$$

so that $\frac{\partial}{\partial n}(\tilde{f} \circ b\ f \circ p) = \tilde{f}' \circ b\ \ f \circ p$ and

$$\nabla(\tilde{f} \circ b\ f \circ p) \circ T_z = \tilde{f}'(z)\ f\ n + \tilde{f}(z)\ Dp \circ T_z . \nabla_\Gamma f\ , \qquad \text{on } \Gamma_0$$

Divergence Operator *div*:

Lemma 13.22.

$$div(u \circ p) = D_\Gamma u \circ p..D^* p$$

Proof.

$$\begin{aligned} div((u \circ p) \circ p) &= \partial_k (u_k \circ p) \circ p \\ &= \partial_l (u_k \circ p) \circ p\ \partial_k p_l \\ &= D(u \circ p) \circ p..D^* p \quad \text{et} \quad D^* p = Dp \end{aligned}$$

Lemma 13.23.

$$div(\tilde{u} \circ b\ u \circ p) = \tilde{u}' \circ b\ \ < u, n > \circ p + \tilde{u} \circ b\ D_\Gamma u \circ p..Dp$$

Proof.

$$\begin{aligned} div(\tilde{u} \circ b\ u \circ p) &= < \nabla(\tilde{u} \circ b), u \circ p > + \tilde{u} \circ b\ div(u \circ p) \\ &= < \tilde{u}' \circ b\ \nabla b, u \circ p > + \tilde{u} \circ b\ D_\Gamma u \circ p..Dp \\ &= \tilde{u}' \circ b\ \ < u, n > \circ p + \tilde{u} \circ b \mathrm{d}_\Gamma u \circ p..Dp \end{aligned}$$

$Dp \circ T_z = I - n.n^* - z\ D^2 b \circ T_z$ and $D_\Gamma u..(I - n.n^*) = div_\Gamma u$

$$\{div(\tilde{u} \circ b\ u \circ p)\} \circ T_z(x) =$$

$$\tilde{u}'(z) < u(x), n(x) > + \tilde{u}(z)\{div_\Gamma u(x) - z D_\Gamma u(x)..D^2 b \circ T_z(x)\}$$

$$\{div_\Gamma(\tilde{u} \circ b\ u \circ p)\} \circ T_z(x)$$

$$= \tilde{u}(z)\{div_\Gamma u(x) - z D_\Gamma u(x)..D^2 b \circ T_z(x)\}$$

Jacobian Operator DP:

$$D(u \circ p) = D_\Gamma u \circ p.Dp \tag{13.77}$$

Lemma 13.24.

$$D(\tilde{u} \circ b\ u \circ p) = \tilde{u}' \circ b(u.n^*) \circ p + \tilde{u} \circ b\ D_\Gamma u \circ p.Dp \tag{13.78}$$

Proof.

$$\begin{aligned} D(\tilde{u} \circ b\ u \circ p) &= u \circ p.(\nabla(\tilde{u} \circ b))^* + \tilde{u} \circ b\ D(u \circ p) \\ &= u \circ p.(\tilde{u}' \circ b\ \nabla b)^* + \tilde{u} \circ b\ D_\Gamma u \circ p.Dp \\ &= \tilde{u}' \circ b\ (u.n^*) \circ p + \tilde{u} \circ b\ D_\Gamma u \circ p.Dp \end{aligned}$$

Laplace Operator Δ: We know that for Γ and f smooth enough,

$$\Delta f = \Delta_\Gamma f + H\ \frac{\partial f}{\partial n} + \frac{\partial^2 f}{\partial n^2} \quad \text{on } \Gamma \tag{13.79}$$

Lemma 13.25.

$$\boldsymbol{\Delta} p = -\nabla(H\ b)$$

Proof.

$$\begin{aligned} \boldsymbol{\Delta} p &= \boldsymbol{\Delta}(I - b\ \nabla b) \\ &= -\boldsymbol{\Delta}(b\ \nabla b) \ \text{ and} \end{aligned}$$

$$vec\Delta(f\ u) = \Delta f\ u + 2Du.\nabla f + f\ \boldsymbol{\Delta} u$$

$$= -\Delta b\ \nabla b - 2D^2 b.\nabla b - b$$

$$vec\Delta(\nabla b) \ \text{ and } \ \Delta b = H$$

$$= -(H\ \nabla b + b\ \nabla H)$$

Lemma 13.26.

$$\Delta(\tilde{f} \circ b) = \tilde{f}'' \circ b + Htildef' \circ b \tag{13.80}$$

Proof.

$$\begin{aligned}
\Delta(\tilde{f} \circ b) &= \partial^2_{ii}(\tilde{f} \circ b) \\
&= \partial_i(\tilde{f}' \circ b\, \partial_i b) \\
&= \tilde{f}'' \circ b\, \partial_i b\, \partial_i b + \tilde{f}' \circ b \\
partial^2_{ii} b & \\
&= \tilde{f}'' \circ b < \nabla b, \nabla b > + \tilde{f}' \circ b\, \Delta b
\end{aligned}$$

Lemma 13.27.

$$\Delta(f \circ p) = D^2 f \circ p..(Dp)^2 - b\ < \nabla_\Gamma f \circ p, \nabla_\Gamma H >$$

Proof.

$$\begin{aligned}
\Delta(f \circ p) &= \partial^2_{ii}(f \circ p) \\
&= \partial_i(\partial_k f \circ p\, \partial_i p_k) \\
&= \partial^2_{kl} f \circ p\, \partial_i p_l\, \partial_i p_k + \partial_k f \circ p \partial^2_{ii} p_k \\
&= ((D^2 f \circ p)_{kl}(Dp)_{li})(Dp)_{ki} + (\nabla f \circ p)_k (\boldsymbol{\Delta} p)_k \\
&= D^2 f \circ p..(Dp)^2 + < \nabla_\Gamma f \circ p, \boldsymbol{\Delta} p >
\end{aligned}$$

And $(\boldsymbol{\Delta} p)_\Gamma = -b\, \nabla_\Gamma H$ from lemma 13.25.

Lemma 13.28.

$$\Delta(\tilde{f} \circ b\, f \circ p) = (\tilde{f}'' \circ b + H\, \tilde{f}' \circ b)\, f \circ p + \tilde{f} \circ b\, \{D^2 f \circ p..(Dp)^2 + < \nabla_\Gamma f \circ p, \boldsymbol{\Delta} p >\}$$

Proof.

$$\begin{aligned}
\Delta(\tilde{f} \quad &\circ \quad b\, f \circ p) = \\
&= \Delta(\tilde{f} \circ b)\, f \circ p + 2 < \nabla(\tilde{f} \circ b), \nabla(f \circ p) > + \tilde{f} \circ b\, \Delta(f \circ p) \\
&= (\tilde{f}'' \circ b + H\, \tilde{f}' \circ b)\, f \circ p + 2 < \tilde{f}' \circ b\, \nabla b, D^* p.(\nabla_\Gamma f) \circ p > + \\
&\quad \tilde{f} \circ b\, \{D^2 f \circ p..(Dp)^2 + < \nabla f \circ p, \boldsymbol{\Delta} p >\} \\
&= (\tilde{f}'' \circ b + H\, \tilde{f}' \circ b)\, f \circ p + 2\tilde{f}' \circ b < Dp.\nabla b, (\nabla_\Gamma f) \circ p > + \\
&\quad \tilde{f} \circ b\, \{D^2 f \circ p..(Dp)^2 + < \nabla f \circ p, \boldsymbol{\Delta} p >\}
\end{aligned}$$

Where $Dp.\nabla b = 0$

Lemma 13.29.

$$(\boldsymbol{\Delta}(u \circ p))_i = D^2 u_i..(Dp)^2 + (Du \circ p.\boldsymbol{\Delta} p)_i$$

Proof.

$$\begin{aligned}(\boldsymbol{\Delta}(u\circ p))_i &= \partial^2_{jj}(u_i\circ p)\\ &= \partial_j\{(\partial_k u_i)\circ p\ \partial_j p_k\}\\ &= \partial^2_{lk}u_i\ \partial_j p_l \partial_j p_k + (\partial_k u_i)\circ p\ \partial^2_{jj}p_k\\ &= D^2 u_i..(Dp)^2 + (Du\circ p.\boldsymbol{\Delta} p)_i\end{aligned}$$

Lemma 13.30.

$$\boldsymbol{\Delta}(\tilde{u}\circ b\ u\circ p) = (\tilde{u}''\circ b + H\tilde{u}'\circ b)\ u\circ p - b\ D_\Gamma u\circ p.\nabla_\Gamma H$$

Proof. $\boldsymbol{\Delta}(f\ u) = \Delta f\ u + 2Du.\nabla f + f$
$vec\Delta u$ and lemmas 13.80,13.77, 13.76 and 13.29.

Curl Operator *curl*:

$$A\wedge\wedge B\hat{e}q\sum_k A_{.k}\wedge B_{.k}$$

$$(A\wedge B)_{.j}\hat{e}qA_{.j}\wedge B_{.j}$$

$$curl(u\circ p) = \ D^*p\wedge\wedge\ D_\Gamma u\circ p$$

Lemma 13.31.

$$curl(\tilde{u}\circ b\ u\circ p) = \tilde{u}'\circ b\ (n\wedge u)\circ p + \tilde{u}\circ b\ D^*p\wedge\wedge D_\Gamma u\circ p$$

Proof.

$$\begin{aligned}curl(\tilde{u}\circ b\ u\circ p) &= \nabla(\tilde{u}\circ b)\wedge u\circ p + \tilde{u}\circ b\ curl(u\circ p)\\ &= \tilde{u}'\circ b\ \nabla b\wedge u\circ p + \tilde{u}\circ b\ D^*p\wedge\wedge D_\Gamma u\circ p\\ &= \tilde{u}'\circ b\ (n\wedge u)\circ p + \tilde{u}\circ b\ D^*p\wedge\wedge D_\Gamma u\circ p\end{aligned}$$

$$curl(b^i u_i\circ p) = b^i\ Dp\wedge D_\Gamma u_i\circ p + ib^{i-1}(n\wedge u_i)\circ p$$

13.14.1 Intrinsic Equations. We make use of the Federer's decomposition to compute every integrals over $\mathcal{U}$.x

$$\int_{\mathcal{U}_h} f(X)\,dX = \int_0^h\int_{\Gamma_h} f\,d\Gamma_z\,dz = \int_{\Gamma_0}\int_0^h f\circ T_z(x)\mathrm{j}_x(z)\,dz\,d\Gamma_x$$

Where $j_x(z) = 1 + H_x z + \kappa_x z^2$ when $N=3$Where $H = tr(D^2 b)$ is the mean curvature of Γ_o and κ its total or Gaussian curvature.Let us consider the previous Navier Stokes problem in the particular case where $\Sigma = \Gamma_h = \{x\in\Omega|b(x)=h\}$ $\mathcal{U} = \mathcal{U}_($, tubular neighborhood of Γ.

$$\mathcal{U}_h = \{x\in\Omega|0 < b(x) < h\}$$

13.15 Intrinsic Shell Form of $a(\theta; u, v)$

$$\begin{cases} u(X) = \sum_{i=0}^{d} \tilde{u_i} \circ b(X)\ u_i \circ p(X) \\ \theta(X) = \sum_{j=0}^{d} \tilde{\theta_j} \circ b(X) \\ theta_j \circ p(X) \\ v(X) = \sum_{k=0}^{d} \tilde{v_k} \circ b(X)\ v_k \circ p(X) \end{cases}$$

Where the function $\tilde{u_i},\ \tilde{\theta_j}$,
$tildev_k : [0, h] \to R$ supposed known (chosen before hand).

$$u_i : \Gamma \to R^3, \theta_j : \Gamma \to R^3, v_k : \Gamma \to R^3$$

From lemma 13.78, $\forall x \in \Gamma_0$,

$$\{Du\} \circ T_z(x) = \tilde{u_i}'(z)\ u_i(x).n^*(x) + \tilde{u_i}(z)\ D_\Gamma u_i(x).Dp \circ T_z(x)$$

Lemma 13.32. $\forall x \in \Gamma_0$,

$$\{Du\} \circ T_z(x) = \tilde{u_i}'(z)\ u_i(x).n^*(x) + \tilde{u_i}(z)\ D_\Gamma u_i(x).Dp \circ T_z(x)$$

Proof.

$$\begin{aligned} \{Du\} \circ T_z(x) &= \{D(\tilde{u_i} \circ b(X)\ u_i \circ p(X))\} \circ T_z(x) \\ &= \{u_i \circ p\ (\nabla(\tilde{u_i} \circ b))^* + \tilde{u_i} \circ b\ D(u_i \circ p)\} \circ T_z(x) \\ &= \{u_i \circ p\ \tilde{u_i}' \circ b\ \nabla^* b + \tilde{u_i} \circ b\ (D_\Gamma u_i) \circ p.Dp\} \circ T_z(x) \\ &= \tilde{u_i}'(z)\ u_i(x).n^*(x) + \tilde{u_i}(z)\ D_\Gamma u_i(x).Dp \circ T_z(x) \end{aligned}$$

In a similar manner,

$$\{Du\ n\} \circ T_z(x) = \tilde{u_i}'(z)\ u_i(x)$$

$$a(\theta; u, v) = a_0(u, v) + a_1(\theta; u, v)$$

$$\{Du..Dv\} \circ T_z$$

$$\begin{aligned} &= Du \circ T_z..Dv \circ T_z \\ &= \{\tilde{u_i}'(z)\ u_i.n^* + \tilde{u_i}(z) D_\Gamma u_i.Dp \circ T_z\}.. \\ &\qquad ..\{\tilde{v_j}'(z)\ v_j.n^* + \tilde{v_j}(z) D_\Gamma v_j.Dp \circ T_z\} \\ &= \tilde{u_i}'(z)\tilde{v_j}'(z)\ < u_i, v_j > \\ &\quad + \tilde{u_i}'(z)\tilde{v_j}(z)(u_i.n^*)..(D_\Gamma v_j.Dp \circ T_z) \\ &\quad + \tilde{u_i}(z)\tilde{v_j}'(z)\ (D_\Gamma u_i.Dp \circ T_z)..(v_j.n^*) \\ &\quad + \tilde{u_i}(z)\tilde{v_j}(z)(D_\Gamma u_i.Dp \circ T_z)..(D_\Gamma v_j.Dp \circ T_z) \\ &= \tilde{u_i}'(z)\tilde{v_j}'(z)\ < u_i, v_j > \\ &\quad + \tilde{u_i}(z)\tilde{v_j}(z)(D_\Gamma u_i.Dp \circ T_z)..(D_\Gamma v_j.Dp \circ T_z) \end{aligned}$$

$$a_0(u,v)$$

$$\begin{aligned}
&= \int_{\mathcal{U}_\ell} Du..Dv\,dX \\
&= \int_\Gamma \int_0^h \{Du..Dv\} \circ T_z(x) j_x(z)\,dz\,d\Gamma_x \\
&= \int_\Gamma \int_0^h \{\tilde{u_i}'(z)\; u_i(x).n^*(x) + \tilde{u_i}(z)\; D_\Gamma u_i(x).Dp \circ T_z(x)\}.. \\
&\quad \{\tilde{v_k}'(z)\; v_k(x).n^*(x) + \tilde{v_k}(z)\; D_\Gamma v_k(x).Dp \circ T_z(x)\} \;\; j_x(z)\,dz\,d\Gamma_x \\
&= \int_\Gamma \int_0^h \{(\tilde{u_i}'\tilde{v_k}')(z) < u_i, v_k > (x) + \\
&\quad (\tilde{u_i}\tilde{v_k})(z)\; (D_\Gamma u_i.(Dp)^2 \circ T_z)(x)..(D_\Gamma v_k)(x)\}\{1 + H_x z + \kappa_x z^2\}\,dz\,d\Gamma_x
\end{aligned}$$

Now, considering the expression of $(Dp)^2 \circ T_z$ given at lemma 13.5, $a_0(u,v)$ can be rewrite as follows,

$$\begin{aligned}
\int_{\mathcal{U}_\ell} Du..Dv\,dX &= \int_\Gamma A^0_{ij}(h) < u_i, v_k > \\
&\quad +(D_\Gamma u_i \;\; . \;\; \{A^1_{ij}(h)\,I + A^2_{ij}(h)\,D^2 b + A^3_{ij}(h)\,(D^2 b)^2\})..D_\Gamma v_k\;d\Gamma
\end{aligned}$$

Where the scalar functionsA_{ij} are given by

$$A^0_{ij}(h) = \int_0^h \tilde{u_i}'(z)\;\tilde{v_k}'(z)\;j(z)\,dz$$

$$A^1_{ij}(h) = \int_0^h \tilde{u_i}(z)\;\tilde{v_k}(z)\;j(z)\,dz$$

$$A^2_{ij}(h) = -\int_0^h \tilde{u_i}(z)\;\tilde{v_k}(z) frac2z(1+zH)^2 + z^4\kappa H j(z)\,dz$$

and

$$A^3_{ij}(h) = \int_0^h \tilde{u_i}(z)\;\tilde{v_k}(z)\;\frac{z^2(3+2zH+z^2\kappa)}{j(z)}\,dz$$

Remark 13.2. All these scalar functions on Γ can be computed as soon as the normal expansion functions $\tilde{u_i}$ and $\tilde{v_k}$ are chosen. See for example section for the Bernstein basis application.

$$\begin{aligned}\int_\Omega & < Du.u, v > dx = \\ &= \int_0^h \int_{\Gamma_z} \{ib^{i-1}u_i \circ p.\nabla^* b + b^i(D_\Gamma u_i) \circ p.Dp\} b^j u_j \circ p \; v \circ p \; d\Gamma_z \; dz \\ &= \int_0^h \int_\Gamma \{iz^{i-1}u_i.n^* + z^i(D_\Gamma u_i).(Dp \circ T_z)\} z^j u_j \; v \; \omega(z) \, d\Gamma \, dz \\ &= \text{.. as } \omega(z) = 1 + Hz + \kappa z^2\end{aligned}$$

$$a_1(\theta; u, v)$$

$$\begin{aligned}&= \int_{\mathcal{U}_\ell} < Du.\theta, v > dX \\ &= \int_\Gamma \int_0^h < \{(Du) \circ T_z(x)\}.\theta \circ T_z(x), v \circ T_z(x) > \; j_x(z) \, dz \, d\Gamma_x \\ &= \int_\Gamma \int_0^h < \{\tilde{u_i}'(z) \; u_i(x).n^*(x) + \tilde{u_i}(z) \; D_\Gamma u_i(x).Dp \circ T_z(x)\}.\theta_j(x) \; \tilde{\theta_j}(z), \\ &\qquad \tilde{v_k}(z) \; v_k(x) > j_x(z) \, dz \, d\Gamma_x \\ &= \int_\Gamma \int_0^h \{(\tilde{u_i}'\tilde{\theta_j}\tilde{v_k})(z) < u_i.n^*.\theta_j, v_k > (x) + \\ &\quad (\tilde{u_i}\tilde{\theta_j}\tilde{v_k})(z) < D_\Gamma u_i.Dp \circ T_z.\theta_j, v_k > (x)\} \; \{1 + H_x z + \kappa_x z^2\} \; , dz \, d\Gamma_x \\ &= \int_\Gamma \int_0^h \{(\tilde{u_i}'\tilde{\theta_j}\tilde{v_k})(z)(< \theta_j, n >< u_i, v_k >)(x) + \\ &\quad (\tilde{u_i}\tilde{\theta_j}\tilde{v_k})(z) < D_\Gamma u_i.Dp \circ T_z.\theta_j, v_k > (x)\}\{1 + H_x z + \kappa_x z^2\} \, dz \, d\Gamma_x\end{aligned}$$

Taking the form of $Dp \circ T_z$ given by 13.56 into account, this term can be written as

$$\begin{aligned}\int_{\mathcal{U}_\ell} & < Du.\theta, v > dx = \int_\Gamma B^0_{ijk}(h) < u_i, v_k >< \theta_j, n > \\ & + < D_\Gamma u_i.\{B^1_{ijk}(h) \, I + B^2_{ijk}(h) \, D^2 b + B^3_{ijk}(h) \, (D^2 b)^2\} .\theta_j, v_k > d\Gamma\end{aligned}$$

$$B^0_{ijk}(h) = \int_0^h \tilde{u_i}'(z) \, \tilde{\theta_j}(z) \, \tilde{v_k}(z) \, j(z) \, dz$$

$$B^1_{ijk}(h) = \int_0^h \tilde{u_i}(z) \, \tilde{\theta_j}(z) \, \tilde{v_k}(z) \, j(z) \, dz$$

$$B^2_{ijk}(h) = - \int_0^h \tilde{u_i}(z) \, \tilde{\theta_j}(z) \, \tilde{v_k}(z) \, z(1 + zH) \, dz$$

$$B^3_{ijk}(h) = \int_0^h \tilde{u_i}(z) \, \tilde{\theta_j}(z) \, \tilde{v_k}(z) \, z^2 \, dz$$

13.16 Intrinsic Shell Form of $b(u,q)$

As before, in $\mathcal{U}_{(}$,we decompose every functions as follows

$$\begin{cases} u(X) = \sum_{i=0}^{d} \tilde{u_i} \circ b(X)\ u_i \circ p(X) \\ q(X) = \sum_{j=0}^{e} \tilde{q_j} \circ b(X)\ q_j \circ p(X) \end{cases}$$

Where $\tilde{u_i}$ and $\tilde{q_j} : [0,h] \to R$ are given (basis functions before hand choose). And $u_i : \Gamma \to R^3$ are the unknowns. $q_j : \Gamma \to R$ so that the test functions $q : \mathcal{U} \to R$

$$u \circ T_z(x) = \tilde{u_i}(z)\ u_i(x)\ , forall x \in \Gamma$$

$$q \circ T_z(x) = \tilde{q_j}(z)\ q_j(x)$$

$$\begin{aligned} \{\nabla(\tilde{q_j} \circ b)\} \circ T_z(x) &= \{\tilde{q_j}' \circ b\ \nabla b\} \circ T_z(x) \\ &= \tilde{q_j}'(z) n(x) \end{aligned}$$

$$\begin{aligned} \{\nabla(q_j \circ p)\} \circ T_z(x) &= \{Dp.\nabla_\Gamma q_j \circ p\} \circ T_z(x) \\ &= Dp \circ T_z(x).\nabla_\Gamma q_j(x) \end{aligned}$$

Which permit us to establish

$$(\nabla q) \circ T_z(x) = \tilde{q_j}'(z) q_j(x)\ n(x) + \tilde{q_j}(z)\ Dp \circ T_z(x).\nabla_\Gamma q_j(x)\quad ,\ \forall x \in \Gamma$$

$$\begin{aligned} b(u,q) &= \int_{\mathcal{U}_{(}} < u, \nabla q > dx \\ &= \int_\Gamma \int_0^h < u \circ T_z(x), (\nabla q) \circ T_z(x) > \ j_x(z)\, dz\, d\Gamma_x \\ &= \int_\Gamma \int_0^h < \tilde{u_i}(z)\ u_i(x)\ ,\ \tilde{q_j}'(z) q_j(x)\ n(x) + \\ &\quad \tilde{q_j}(z)\ Dp \circ T_z(x).\nabla_\Gamma q_j(x) > \{1 + H_x z + \kappa_x z^2\}\, dz\, d\Gamma_x \\ &= \int_\Gamma \int_0^h \tilde{u_i}(z)\ \{\tilde{q_j}'(z) q_j(x) < u_i(x), n(x) > + \\ &\quad \tilde{q_j}(z) < u_i(x), Dp \circ T_z(x).\nabla_\Gamma q_j(x) >\}\ j(z)\, dz\, d\Gamma_x \end{aligned}$$

Let introduce the notation

$$B_{ij}^{klm}(h) = \int_0^h z^k\ \tilde{u_i}^{(l)}(z) tildeq_j{}^{(m)}(z)\, dz \qquad (13.81)$$

and considering $Dp \circ T_z$in power of z,

$$Dp \circ T_z = I - n.n^* + \sum_{k=1} (-z\ D^2 b)^k$$

$$b(u,q) = \int_\Gamma < u_i, n > q_j\{B_{ij}^{0,0,1} + H\ B_{ij}^{1,0,1} + \kappa\ B_{ij}^{2,0,1}\} +$$
$$+ \sum_{k=0} < u_i, (-D^2b)^k.\nabla_\Gamma q_j > \{B_{ij}^{k,0,0} + H\ B_{ij}^{k+1,0,0} + \kappa\ B_{ij}^{k+2,0,0}\}$$
$$+ o(h^p)\, d\Gamma$$

and with the expression of $Dp \circ T_z$ given by (13.56), this can be written as:

$$\begin{aligned}
\int_{U_(} < u, \nabla q > dx \quad = \quad & \int_\Gamma C_{ij}^0(h) < u_i, n > \\
& + C_{ij}^1(h) < u_i, \nabla_\Gamma q_j > \\
& - C_{ij}^2(h) < u_i, D^2b.\nabla_\Gamma q_j > \\
& + C_{ij}^3(h) < u_i, (D^2b)^2.\nabla_\Gamma q_j > d\Gamma
\end{aligned}$$

where

$$C_{ij}^0(h) = \int_o^h \tilde{u}_i(z)\, \tilde{q}_j{}'(z)\, j(z)\, dz$$
$$C_{ij}^1(h) = \int_o^h \tilde{u}_i(z)\, \tilde{q}_j(z)\, j(z)\, dz$$
$$C_{ij}^2(h) = \int_o^h \tilde{u}_i(z)\, \tilde{q}_j(z)\, z\,(1 + zH)\, dz$$
$$\text{and} \quad C_{ij}^3(h) = \int_o^h \tilde{u}_i(z)\, \tilde{q}_j(z)\, z^2\, dz$$

$$(Du) \circ T_z(x) = \tilde{u}_i{}'(z)\ u_i(x).n^*(x) + \tilde{u}_i(z)\ D_\Gamma u_i(x).Dp \circ T_z(x)$$
$$(divu) \circ T_z(x) =$$
$$\tilde{u}_i{}'(z) < u_i(x), n(x) > + \tilde{u}_i(z)\{div_\Gamma u_i(x) - zD_\Gamma u_i(x)..D^2b \circ T_z(x)\}$$
$$(div_\Gamma u) \circ T_z(x) = \tilde{u}_i(z)\{div_\Gamma u_i(x) - zD_\Gamma u_i(x)..D^2b \circ T_z(x)\}$$

in the same way,

$$\begin{aligned}
\int_{U_(} div(u)\, q\, dX \quad = \quad & \int_\Gamma \{D_{ij}^0(h)\ < u_i, n > + D_{ij}^1(h)\, div_\Gamma u_i \\
& - D_{ij}^2(h)\, D_\Gamma u_i..D^2b + D_{ij}^3(h)\, D_\Gamma u_i..(D^2b)^2\ \}\, q_j\, d\Gamma
\end{aligned}$$

where

$$D_{ij}^0(h) = \int_o^h \tilde{u}_i{}'(z)\tilde{q}_j(z) j(z)\, dz$$
$$D_{ij}^1(h) = C_{ij}^1(h) = \int_o^h \tilde{u}_i(z)\, \tilde{q}_j(z)\, j(z)\, dz$$
$$D_{ij}^2(h) = C_{ij}^2(h) = \int_o^h \tilde{u}_i(z)\, \tilde{q}_j(z)\, z\,(1 + zH)\, dz$$
$$\text{and} \quad D_{ij}^3(h) = C_{ij}^3(h) = \int_o^h \tilde{u}_i(z)\, \tilde{q}_j(z)\, z^2\, dz$$

13.17 Intrinsic Shell Form of $\epsilon..\epsilon$

$$\epsilon(u)..\epsilon(v) = Du..\epsilon(v)$$

From lemma 13.78

$$
\begin{aligned}
&2\ \{\epsilon(u)..\epsilon(v)\} \circ T_z \\
= \quad & Du \circ T_z..\{Dv + D^*v\} \circ T_z \\
= \quad & \{\tilde{u_i}'(z)\ u_i.n^* + \tilde{u_i}(z) D_\Gamma u_i.Dp \circ T_z\}.. \\
& \qquad \{\tilde{v_j}'(z)\ v_j.n^* + \tilde{v_j}(z) D_\Gamma v_j.Dp \circ T_z \\
& \qquad +\tilde{v_j}'(z)\ n.v_j^* + \tilde{v_j}(z) D^* p \circ T_z.D_\Gamma^* v_j\} \\
= \quad & \tilde{u_i}'(z)\tilde{v_j}'(z)\ (< u_i, v_j > + < u_i, n >< v_j, n >) \\
& + \tilde{u_i}'(z)\tilde{v_j}(z)(u_i.n^*)..(D_\Gamma v_j.Dp \circ T_z + D^* p \circ T_z.D_\Gamma^* v_j) \\
& + \tilde{u_i}(z)\tilde{v_j}'(z)\ (D_\Gamma u_i.Dp \circ T_z)..(v_j.n^* + n.v_j^*) \\
& + \tilde{u_i}(z)\tilde{v_j}(z)\ (D_\Gamma u_i.Dp \circ T_z)..(D^* p \circ T_z.D_\Gamma^* v_j) \\
= \quad & \tilde{u_i}'(z)\tilde{v_j}'(z)\ (< u_{i\Gamma}, v_{j\Gamma} > +2 < u_i, n >< v_j, n >) \\
& +\tilde{u_i}'(z)\tilde{v_j}(z)(u_i.n^*)..(D^* p \circ T_z.D_\Gamma^* v_j) \\
& + \tilde{u_i}(z)\tilde{v_j}'(z)(D_\Gamma u_i.Dp \circ T_z)..(n.v_j^*) \\
& + \tilde{u_i}(z)\tilde{v_j}(z)(D_\Gamma u_i.Dp \circ T_z)..(D^* p \circ T_z.D_\Gamma^* v_j)
\end{aligned}
$$

we rearrange the terms by the powers of z and introducing the notation,

$$C_{ij}^{klm}(h) = \int_0^h z^k\ \tilde{u_i}^{(l)}(z) tildev_j{}^{(m)}(z)\, dz \tag{13.82}$$

we can rewrite it as an integral on Γ only.

$$
\begin{aligned}
& \int_{\mathcal{U}_(} 2\epsilon(u)..\epsilon(v)\, dx \\
= \quad & \int_\Gamma (< u_i, v_j > + < u_i, n >< v_j, n >)(C_{ij}^{0,1,1} + H\ C_{ij}^{1,1,1} + \kappa\ C_{ij}^{2,1,1}) \\
& +(u_i.n^*)..\{(I - n.n^*).D_\Gamma^* v_j\}(C_{ij}^{0,1,0} + H\ C_{ij}^{1,1,0} + \kappa\ C_{ij}^{2,1,0}) \\
& + \sum_{k=1} (u_i.n^*)..\{(-D^2 b)^k.D_\Gamma^* v_j\}(C_{ij}^{k,1,0} + H\ C_{ij}^{k+1,1,0} + \kappa\ C_{ij}^{k+2,1,0}) \\
& +(v_j.n^*)..\{(I - n.n^*).D_\Gamma^* u_i\}(C_{ij}^{0,0,1} + HC_{ij}^{1,0,1} + \kappa\ C_{ij}^{2,0,1}) \\
& + \sum_{k=1} (v_j.n^*)..\{(-D^2 b)^k.D_\Gamma^* u_i\}(C_{ij}^{k,0,1} + HC_{ij}^{k+1,0,1} + \kappa\ C_{ij}^{k+2,0,1}) \\
& + \sum_{k,l=0} \{D_\Gamma u_i.(-D^2 b)^k\}..\{(-D^2 b)^l.D_\Gamma^* v_j\} \\
& \qquad (C_{ij}^{k+l,0,0} + HC_{ij}^{k+l+1,0,0} + \kappa C_{ij}^{k+l+2,0,0}) \\
& +o(h^e)\ d\Gamma
\end{aligned}
$$

Intrinsic Shell Form of $\sigma..\sigma$:

$$\sigma(u)..\sigma(v) = Du..\sigma(v)$$

$$2\ \{\sigma(u)..\sigma(v)\} \circ T_z$$

$$= \quad Du \circ T_z..\{Dv - D^*v\} \circ T_z$$

$Du \circ T_z$ given by lemma 13.78

$$\begin{aligned}
&= \{\tilde{u_i}'(z)\ u_i.n^* + \tilde{u_i}(z) D_\Gamma u_i.Dp \circ T_z\}.. \\
&\qquad \{\tilde{v_j}'(z)\ v_j.n^* + \tilde{v_j}(z) D_\Gamma v_j.Dp \circ T_z \\
&\qquad -\tilde{v_j}'(z)\ n.v_j^* - \tilde{v_j}(z) D^* p \circ T_z.D_\Gamma^* v_j\} \\
&= \tilde{u_i}'(z)\tilde{v_j}'(z)\ (< u_i, v_j > - < u_i, n >< v_j, n >) \\
&\quad + \tilde{u_i}'(z)\tilde{v_j}(z)(u_i.n^*)..(D_\Gamma v_j.Dp \circ T_z - D^* p \circ T_z.D_\Gamma^* v_j) \\
&\quad + \tilde{u_i}(z)\tilde{v_j}'(z)(D_\Gamma u_i.Dp \circ T_z)..(v_j.n^* - n.v_j^*) \\
&\quad + \tilde{u_i}(z)\tilde{v_j}(z)(D_\Gamma u_i.Dp \circ T_z)..(D^* p \circ T_z.D_\Gamma^* v_j)
\end{aligned}$$

ou $Dp.n = 0$,

$$\begin{aligned}
&= \tilde{u_i}'(z)\tilde{v_j}'(z)\ < u_{i\Gamma}, v_{j\Gamma} > \\
&\quad - \tilde{u_i}'(z)\tilde{v_j}(z)(u_i.n^*)..(D^* p \circ T_z.D_\Gamma^* v_j) \\
&\quad - \tilde{u_i}(z)\tilde{v_j}'(z)(v_j.n^*)..(D^* p \circ T_z.D_\Gamma^* u_i) \\
&\quad + \tilde{u_i}(z)\tilde{v_j}(z)(D_\Gamma u_i.Dp \circ T_z)..(D^* p \circ T_z.D_\Gamma^* v_j)
\end{aligned}$$

$$\int_{\mathcal{U}_(} 2\sigma(u)..\sigma(v)\, dx = \int_\Gamma \int_0^h 2\ \{\sigma(u)..\sigma(v)\} \circ T_z\ (1 + Hz + \kappa z^2)\, dz\, d\Gamma$$

If we rearrange the terms under power of z and identifying the C_{ij}^{klm} coefficients previously introduced in (13.82), we obtain an integral only on Γ.

$$\int_{\mathcal{U}_(} 2\sigma(u)..\sigma(v)\, dx$$

$$\begin{aligned}
= \quad & \int_\Gamma < u_{i\Gamma}, v_{j\Gamma} > (C_{ij}^{0,1,1} + H\ C_{ij}^{1,1,1} + \kappa\ C_{ij}^{2,1,1}) \\
& -(u_i.n^*)..\{(I - n.n^*).D_\Gamma^* v_j\}(C_{ij}^{0,1,0} + H\ C_{ij}^{1,1,0} + \kappa\ C_{ij}^{2,1,0}) \\
& -\sum_{k=1}(u_i.n^*)..\{(-D^2 b)^k.D_\Gamma^* v_j\}(C_{ij}^{k,1,0} + H\ C_{ij}^{k+1,1,0} + \kappa\ C_{ij}^{k+2,1,0}) \\
& -(v_j.n^*)..\{(I - n.n^*).D_\Gamma^* u_i\}(C_{ij}^{0,0,1} + H\ C_{ij}^{1,0,1} + \kappa\ C_{ij}^{2,0,1}) \\
& -\sum_{k=1}(v_j.n^*)..\{(-D^2 b)^k.D_\Gamma^* u_i\}(C_{ij}^{k,0,1} + H\ C_{ij}^{k+1,0,1} + \kappa\ C_{ij}^{k+2,0,1}) \\
& +\sum_{k,l=0}\{D_\Gamma u_i.(-D^2 b)^k\}..\{(-D^2 b)^l.D_\Gamma^* v_j\} \\
& \qquad (C_{ij}^{k+l,0,0} + H C_{ij}^{k+l+1,0,0} + \kappa\ C_{ij}^{k+l+2,0,0}) \\
& +o(h^e)\ d\Gamma
\end{aligned}$$

Lemma 13.33.

$$D_\Gamma u.(I - n.n^*) = D_\Gamma u$$

Proof.

$$D_\Gamma u = Du.(I - n.n^*)$$
$$D_\Gamma u.(I - n.n^*) = Du.(I - n.n^*)^2 = Du.(I - n.n^*) = D_\Gamma u$$

Intrinsic Form of the Free Divergence Penalty Term:

$$\int_{\mathcal{U}_\ell} div(u) div(v)\, dx =$$
$$= \int_\Gamma \int_0^h \{div(u)\} \circ T_z\ \{div(v)\} \circ T_z\ (1 + z\ H + z^2\ \kappa)\, dz\, d\Gamma$$
$$= \int_\Gamma \int_0^h \{\tilde{u_i}'(z) < u_i, n > + \tilde{u_i}(z)(div_\Gamma u_i - z D_\Gamma u_i..D^2 b \circ T_z)\}$$
$$\{\tilde{v_j}'(z) < v_j, n > + \tilde{v_j}(z)(div_\Gamma v_j - z D_\Gamma v_j..D^2 b \circ T_z)\}$$
$$(1 + z\ H + z^2\ \kappa)\ \ dz\, d\Gamma$$

substituting $D^2 b \circ T_z$ by its expansion given by(13.47) and taking account of lemma 13.33, we can rewrite the penalization term as an integral on Γ.

$$\int_{\mathcal{U}_\ell} div(u)\ div(v)\, dx$$
$$= \int_\Gamma \int_0^h \{\tilde{u_i}'(z) < u_i, n > + \tilde{u_i}(z) \sum_{k=0} z^k D_\Gamma u_i..(-D^2 b)^k\}$$
$$\{\tilde{v_j}'(z) < v_j, n > + \tilde{v_j}(z) \sum_{l=0} z^l D_\Gamma v_j..(-D^2 b)^l\}$$
$$(1 + z\ H + z^2\ \kappa)\, dz\, d\Gamma$$

and identifying the$C_{ij}^{klm}(h)$ (13.82), the divergence penalization term takes the following form

$$\int_{\mathcal{U}_\ell} div(u)\ div(v)\, dx$$
$$= \int_\Gamma \{ < u_i, n >< v_j, n > (C_{ij}^{0,1,1} + H\ C_{ij}^{1,1,1} + \kappa\ C_{ij}^{2,1,1})$$
$$+ < u_i, n > \sum_{l=0} D_\Gamma v_j..(-D^2 b)^l (C_{ij}^{l,1,0} + H\ C_{ij}^{l+1,1,0} + \kappa\ C_{ij}^{l+2,1,0})$$
$$+ < v_j, n > \sum_{k=0} D_\Gamma u_i..(-D^2 b)^k (C_{ij}^{k,0,1} + H C_{ij}^{k+1,0,1} + \kappa\ C_{ij}^{k+2,0,1})$$
$$+ \sum_{k,l=0} D_\Gamma u_i..(-D^2 b)^k\ D_\Gamma v_j..(-D^2 b)^l$$
$$(C_{ij}^{k+l,0,0} + H C_{ij}^{k+l+1,0,0} + \kappa\ C_{ij}^{k+l+2,0,0})\ \}\ d\Gamma$$

But a better way to compute the penalty term is to consider the expression of $Dp \circ T_z$ given by (13.56) in $div(u) \circ T_z$ given by.

$$div(u) \circ T_z$$

$$\begin{aligned} &= \tilde{u_i}'(z) < u_i, n > + \tilde{u_i}(z) D_\Gamma u_i..Dp \circ T_z \\ &= \tilde{u_i}'(z) < u_i, n > + \tilde{u_i}(z)\{div_\Gamma(u_i) - \\ &- \frac{z(1+zH)}{j(z)} D_\Gamma u_i..D^2b + \frac{z^2}{j(z)} D_\Gamma u_i..(D^2b)^2\} \end{aligned}$$

This permit us to give an exact expression with a finite number of terms.

$$\int_{\mathcal{U}_\ell} div(u)\, div(v)\, dx =$$

$$= \int_\Gamma \begin{bmatrix} < u_i, n > \\ div_\Gamma u_i \\ D_\Gamma u_i..D^2b \\ D_\Gamma u_i..(D^2b)^2 \end{bmatrix}^* .Div_{ij}(h). \begin{bmatrix} < v_j, n > \\ div_\Gamma v_j \\ D_\Gamma v_j..D^2b \\ D_\Gamma v_j..(D^2b)^2 \end{bmatrix} d\Gamma \qquad (13.83)$$

scalar matrix is given by

$$Div_{ij}(h)$$

$$= \int_0^h \begin{bmatrix} \tilde{u_i}'(z) \\ \tilde{u_i}(z) \\ -\tilde{u_i}(z)\, \frac{z(1+zH)}{j(z)} \\ \tilde{u_i}(z)\, \frac{z^2}{j(z)} \end{bmatrix} \cdot \begin{bmatrix} \tilde{v_j}'(z) \\ \tilde{v_j}(z) \\ -\tilde{v_j}(z)\, \frac{z(1+zH)}{j(z)} \\ \tilde{v_j}(z)\, \frac{z^2}{j(z)} \end{bmatrix}^* j(z)\, dz$$

$$= \int_0^h \begin{pmatrix} \tilde{u_i}'\tilde{v_j} & -(z+z^2H)\tilde{u_i}'\tilde{v_j} & z^2\tilde{u_i}'\tilde{v_j} & \\ \tilde{u_i}\tilde{v_j} & -(z+z^2H)\tilde{u_i}\tilde{v_j} & z^2\tilde{u_i}\tilde{v_j} & \\ \tilde{v_j}' & -(z+z^2H)\tilde{u_i}\tilde{v_j} & \frac{z^2(1+zH)^2}{j}\tilde{u_i}\tilde{v_j} & -\frac{z^3+z^4H}{j}\tilde{u_i}\tilde{v_j} \\ \tilde{u_i}\tilde{v_j} & -\frac{z^3+z^4H}{j}\tilde{u_i}\tilde{v_j} & \frac{z^4}{j}\tilde{u_i}\tilde{v_j} & \end{pmatrix} dz$$

13.18 The Fluid Shell Equation

All these preliminary computations permit us to establish the shell form of the Navier Stokes problem (13.43)

$$\begin{aligned}\int_\Gamma \quad & \nu \ \{A^0_{ij}(h) < u_i, v_k > + \\ & (D_\Gamma u_i.\{A^1_{ij}(h)\, I + A^2_{ij}(h)\, D^2 b + A^3_{ij}(h)\,(D^2 b)^2\})..D_\Gamma v_k\} \\ + \quad & B^0_{ijk}(h) < u_i, v_k >< \theta_j, n > \\ + \quad & < D_\Gamma u_i.\{B^1_{ijk}(h)\, I + B^2_{ijk}(h)\, D^2 b + B^3_{ijk}(h)\,(D^2 b)^2\}\,.\theta_j, v_k > \\ + \quad & \frac{1}{\rho}\{\}\, d\Gamma = 0\end{aligned}$$

$$A^0_{ij}(h) = \int_0^h \tilde{u_i}'(z)\ \tilde{v_k}'(z)\ j(z)\, dz$$

$$A^1_{ij}(h) = \int_0^h \tilde{u_i}(z)\ \tilde{v_k}(z)\ j(z)\, dz$$

$$A^2_{ij}(h) = -\int_0^h \tilde{u_i}(z)\ \tilde{v_k}(z) frac2z(1+zH)^2 + z^4 \kappa H j(z)\, dz$$

$$A^3_{ij}(h) = \int_0^h \tilde{u_i}(z)\ \tilde{v_k}(z)\ \frac{z^2(3+2zH+z^2\kappa)}{j(z)}\, dz$$

$$B^0_{ijk}(h) = \int_0^h \tilde{u_i}'(z)\,\tilde{\theta_j}(z)\,\tilde{v_k}(z)\, j(z)\, dz$$

$$B^1_{ijk}(h) = \int_0^h \tilde{u_i}(z)\,\tilde{\theta_j}(z)\,\tilde{v_k}(z)\, j(z)\, dz$$

$$B^2_{ijk}(h) = -\int_0^h \tilde{u_i}(z)\,\tilde{\theta_j}(z)\,\tilde{v_k}(z)\, z(1+zH)\, dz$$

$$B^3_{ijk}(h) = \int_0^h \tilde{u_i}(z)\,\tilde{\theta_j}(z)\,\tilde{v_k}(z)\, z^2\, dz$$

13.19 Intrinsic Equation

$$U = \tilde{u_i} \circ b \; u_i \circ p$$

$$\{DU\} \circ T_z = \tilde{u_i}'(z)\; u_i.n + \tilde{u_i}(z) \mathrm{d}_\Gamma u_i.Dp \circ T_z$$

Now, considering the geometrically exact form of $Dp \circ T_z$ given by

$$Dp \circ T_z = I - n.n^* - \frac{z(1+zH)}{j(z)} D^2 b + \frac{z^2}{j(z)} (D^2 b)^2$$

we get

$$\{DU\} \circ T_z =$$

$$= \tilde{u_i}'\; u_i.n^* + \tilde{u_i}\; D_\Gamma u_i - \tilde{u_i} \frac{z(1+zH)}{j(z)}\; D_\Gamma u_i.D^2 b + \tilde{u_i} \frac{z^2}{j(z)}\; D_\Gamma u_i.(D^2 b)^2$$

Now, introducing the two order tensor N which contain all the geometrical information of the boundary,

$$N = \int_a^b \begin{bmatrix} \tilde{u_i}'(z) \\ \tilde{u_i}(z) \\ -\tilde{u_i}(z)\; \frac{z(1+zH)}{j(z)} \\ \tilde{u_i}(z)\; \frac{z^2}{j(z)} \end{bmatrix} \cdot \begin{bmatrix} \tilde{v_j}'(z) \\ \tilde{v_j}(z) \\ -\tilde{v_j}(z)\; \frac{z(1+zH)}{j(z)} \\ \tilde{v_j}(z)\; \frac{z^2}{j(z)} \end{bmatrix}^* j(z)\, dz$$

$$= \int_a^b \begin{pmatrix} \tilde{u_i}'\tilde{v_j} & -(z+z^2H)\tilde{u_i}'\tilde{v_j} & z^2\tilde{u_i}'\tilde{v_j} & \\ \tilde{u_i}\tilde{v_j} & -(z+z^2H)\tilde{u_i}\tilde{v_j} & z^2\tilde{u_i}\tilde{v_j} & \\ \tilde{v_j}' & -(z+z^2H)\tilde{u_i}\tilde{v_j} & \frac{z^2}{j}\tilde{u_i}\tilde{v_j} & -\frac{z^3+z^4H}{j}\tilde{u_i}\tilde{v_j} \\ \tilde{u_i}\tilde{v_j} & -\frac{z^3+z^4H}{j}\tilde{u_i}\tilde{v_j} & \frac{z^4}{j}\tilde{u_i}\tilde{v_j} & \end{pmatrix} dz$$

Then, we can show the shell form of the elasticity general term

$$\int_{\mathcal{U}} Du..C..Dv \qquad dx = \int_\Gamma \int_a^b \{Du\} \circ T_z..C..\{Dv\} \circ T_z\, j(z)\, dz d\Gamma$$

$$= \int_\Gamma N..\{ \begin{bmatrix} u_i.n^* \\ D_\Gamma u_i \\ D_\Gamma u_i.D^2 b \\ D_\Gamma u_i.(D^2 b)^2 \end{bmatrix} ..C.. \begin{bmatrix} v_j.n^* \\ D_\Gamma v_j \\ D_\Gamma v_j.D^2 b \\ D_\Gamma v_j.(D^2 b)^2 \end{bmatrix}^* \}\, d\Gamma$$

Proposition 13.13.

$$\{DU\} \circ T_z = d_n\tilde{u_i} \,.\, d_s u_i$$
$$= d_{s,t} u_i \,.\, d_n\tilde{u_i}$$

$$(d_s v)_{k,l,1} = (v_j.n^*)_{kl}$$

$$(d_{s,t} u)_{m,i,j} = (d_s u)_{i,j,m}$$

$$\int_{\mathcal{U}} DU..C..DV\, dx = \int_{\Gamma} d_{s,t} u \; ...calC_{ab} \; ... \; d_s v \, d\Gamma$$

$$[C_{ab}]_{\alpha ijkl\beta} = \int_a^b d_n\tilde{u_\alpha}(z)\, d_n\tilde{v_\beta}(z)\, j(z)\, dz$$

13.20 Numerical approximation

13.20.1 The Non stationary Problem.

$$\int_{\mathcal{U}_\langle} \frac{1}{\delta t} < u_{n+1} - u_n, v > + < Du_{n+1}.u_n, v > + 2\nu Du_{n+1}..Dv\, dx + \int_\Gamma d\Gamma = 0$$

$$\int_{\mathcal{U}_\langle} < \frac{1}{\delta t} u_{n+1} + Du_{n+1}.u_n, v > + 2\nu Du_{n+1}..Dv\, dx + \int_\Gamma d\Gamma =$$
$$= \frac{1}{\delta t} \int_{\mathcal{U}_\langle} < u_n, v > dx$$

Particularization of the normal expansion functions: a wise choice of the normal expansion functions for the data in $\mathcal{U}_\langle$ greatly simplifies the boundary conditions in $z = 0$and $z = h$. In effect, if we consider for q the Bernstein basis

$$\tilde{q}_i(z) = \binom{e}{i} (\frac{z}{h})^i (1 - \frac{z}{h})^{e-i} \quad i = 0, .., e \;\; \text{where} dstyl \binom{e}{i} = \frac{e!}{i!\,(e-i)!}$$

we obtain $q = q_0 \circ p$ on Γ_0 and$q = q_e \circ p$ onΓ_h
Furthermore,

$$\frac{\partial q}{\partial n} \circ T_0(x) = \tilde{q_j}'(0)\; q_j(x) = \frac{e}{h}\{q_1(x) - q_0(x)\}$$

and

$$\frac{\partial q}{\partial n} \circ T_h(x) = \tilde{q_j}'(h)\; q_j(x) = \frac{e}{h}\{q_e(x) - q_{e-1}(x)\}$$

$$(Du\; n) \circ T_0(x) = \tilde{u_i}'(0)\; u_i(x) = \frac{e}{h}\{u_1(x) - u_0(x)\}$$

$$(Du\; n) \circ T_h(x) = \tilde{u_i}'(h)\; u_i(x) = \frac{e}{h}\{u_e(x) - u_{e-1}(x)\}$$

The computation of the C_{ij}^{klm}: when the basis function of the normal expansion functions are chosen, the coefficients C_{ij}^{klm} can be computed. And their expressions can easily be obtained by symbolic computation.

14. MinMax Shape Derivative

14.1 Notation and definitions

This chapter is concerned with a sensitivity analysis of an optimal shape control problem for the linearized stationary Navier-Stokes models. Using the regularity at the boundary via the extractor technique and the material derivative method we obtain a regularity result for the solutions of the Oseen equations. The flow is in an unbounded domain, so we reduce the problem to a bounded domain D by introducing an artificial boundary ∂D on which we set the Speed flow $u = u_\infty$, where u_∞ is a given constant field. We introduce the perturbation speed field U solution of (14.1) which vanishes on ∂D. The incompressibility will be (mainly for numerical purpose) treated by penalty approach. Also we introduce, in view of the numerical fixed point method a "linear-splitted" model of the stationary Navier-Stokes flow, as well as the linearized model.

In order to give sense to the shape density gradient of the several functional studied in this paper, we consider the regularity at the obstacle boundary for these problems. Under a main density assumption we establish, via the extractor technique, the $L^2(\Gamma, R^3)$ of the normal derivative terms $Du.n$ and $D\lambda.n$. Then we consider the pressure approximation due to that modeling. Then we are concerned with the calculus of shape gradient of three basic functional: the fineness involving the pressure and fineness involving pressure and viscosity, and this for several flows: the penalized linear-splitted model, the full non linear penalized Navier Stokes equations and the incompressible Navier-Stokes equations. For the first two functionals we give three different expressions of the shape gradient. The fully distributed one which only requires the "energy" regularity of the flow u and adjoint state λ and the two expressions of the density gradient g^e_m and g^e_c derived by different approaches based on a min-max formulation.

The main difficulty arises when considering functionals such as the fineness which involves unbounded observation operator (trace operators which are not defined in H^1). We introduce this question by first solving the usual scalar equation while the adjoint equation is introduced in an unusual way by transposition techniques. In the present fineness problem the difficulty is also in the fact that the unbounded operator acts through the pressure at the boundary and the shape derivative of the pressure field is not managed by classical methods.

The $min - max$ approach for the functionals governed by the linear-splitted problems leads to an unusual adjoint problem governed by non homogeneous boundary conditions at the body. This fact leads to a new formulation searching the "Lagrange multiplier" λ (the adjoint state of control) ranging in a closed convex set. The parameter differentiability technique of a $min - max$ requires the convex set $K(\Omega_s)$ to be independent of the parameter s. Again in this case the usual transportation techniques in the form

$\varphi A.\varphi \mapsto oT_s^{-1}$ fail. We are obliged to introduce a heavier technique (which is a parametrization). Finally the two families of expressions g_c^e and g_m^e are not directly comparable while they look similar in the numerical point of view.

Then we take care of the non uniqueness of the solutions for the Navier-Stokes flow governing the functional. Indeed we choose the robust control approach such as H^∞ approach. The non uniqueness of the solution may be considered as an "uncertainty" on the solution (as it is usual in control theory, where some parameters are not exactly known;) then we control the "worst case" that is $\mathcal{J}_\Gamma^{max}(\Omega)$ which is the maximum of the cost functional for v ranging among all the solutions. We derive an approximation schema and we also give the shape gradient for that functional.

14.2 The Navier Stokes problem

The flow is in an unbounded domain so we reduce the problem to a bounded domain D by introducing an artificial boundary ∂D on which we set the Speed flow $u = u_\infty$, where u_∞ is a given constant field. Hence we consider a bounded open set D in R^3 , that domain is assumed filled with a viscous flow having a constant uniform speed u_∞. That uniform speed is perturbed by the presence of a compact obstacle (or body) S which is chosen as a compact subset in D. Let Ω be the effective fluid domain : $\Omega = D \setminus S$ in which the viscous fluid has a steady speed vector u. That fluid is sticking on the obstacle , that is $u = 0$ on ∂S. For technical reasons we shall introduce an intermediary smooth set B verifying

$$S \subset B \subset D$$

We denote by Γ the boundary $\partial\Omega$ by ∂S the boundary of S , by ∂D the boundary of D and by ∂B the boundary of B. Hence $\Gamma = \partial D \bigcup \partial S$. Let n be the unitary normal on Γ outgoing to Ω . We note $(Du)^*$ by D^*u, the transpose matrix of Du and $\epsilon(u) = \frac{1}{2}(Du + D^*u)$ is the deformation tensor.

We define the following spaces:

$L_0^2(\Omega) = \{\varphi \in L^2(\Omega, R^3) / \int_\Omega \varphi \, dx = 0\}$

Let $q > 0$.

$H_q(D) = \{v \in H^q(D, R^3)\}$

$H_q^-(D) = \{v \in H^q(D, R^3) / divv \leq 0 \text{ in } D\}$

$H_q^{div} = \{v \in H^q(D, R^3) / divv = 0 \text{ in } D\}$

We will study the boundary smoothness of the solution of Navier-Stokes equations with Dirichlet boundary conditions with respect to the one of the domain Ω.

14.2.1 The flow generated by u_∞. Assume that $f \in L^2(D, R^3)$. Let α, β be two strictly positive real number. We denote by $C = C(\Omega) = (\lambda_1(\Omega))^{-\frac{1}{2}})$ the Poincare constant. $C_i(\Omega) = C_i$ $(i = 1, 2, ...\text{etc})$ is an norm which depends only with the domain Ω, deriving from the continuous injection of Hilbert space $H^m(\Omega)$ in $L^p(\Omega)$ space. Thus C_1 is the norm generated from the continuous inclusion of $H^1(\Omega)$ in $L^4(\Omega)$. The physical flow is an unbounded domain. We model that situation by taking a "large enough" domain D on which is imposed the speed "at infinity", then we consider

$$\begin{cases} -\alpha\Delta u + Du.u + \nabla p = f \text{ in } \Omega \\ divu = 0 \text{ in } \Omega \\ u = 0 \text{ on } \partial S \\ u = u_\infty \text{ on } \partial D \end{cases} \tag{14.1}$$

We introduce in view of the numerical fixed point method a "linear-splitted" model (Oseen model) of the stationary Navier-Stokes equations. In order to verify the incompressibility condition we would study its penalty case hence we consider this following problem: Let $v \in H_1^-(D)$.

$$\begin{cases} -\alpha\Delta u + Du.v - \beta\nabla(divu) = f \text{ in } \Omega \\ u = 0 \text{ on } \partial S \\ u = u_\infty \text{ on } \partial D \end{cases} \tag{14.2}$$

Proposition 14.1. *Let $R \in H^2(D, R^3)$ be the solution of*

$$\begin{cases} \Delta^2 R + \nabla q = 0 \text{ in } \Omega\setminus\overline{B} \\ divR = 0 \text{ in } \Omega\setminus\overline{B} \\ R = 0 \text{ in } B \\ R = \frac{\partial R}{\partial n} = 0 \text{ on } \partial B \\ R = u_\infty \text{ on } \partial D \end{cases} \tag{14.3}$$

Let $F = f + \alpha\Delta R - DR.v$, F is in $L^2(\Omega, R^3)$ There exists an unique solution u in $H^1(\Omega, R^3)$ to the problem (11.1). And we have

$$\|u\|_{H^1(\Omega,R^3)} \leq \frac{C^2+1}{\alpha}\|F\|_{L^2(\Omega,R^3)} + \|R\|_{H^1(\Omega,R^3)}$$

Proof. Let $U = \begin{cases} u - R \text{ in } \Omega\setminus\overline{B} \\ u \text{ in } \overline{B} \end{cases}$

The system (11.1) leads to the vector U which is a solution of

$$\begin{cases} -\alpha\Delta U + DU.v - \beta\nabla(divU) = F \text{ in } \Omega \\ U = 0 \text{ on } \Gamma \end{cases} \tag{14.4}$$

We consider the continuous and bilinear form on $H_0^1(\Omega, R^3)$ defined by

$$a(U,w) = \int_\Omega \alpha DU..Dw\, dx + \int_\Omega DUvw\, dx\ ; + \int_\Omega \beta\, divU divw\, dx$$

For $w = U$ we get

$$a(U,U) = \int_\Omega \alpha DU..DU\, dx + \int_\Omega DUvU\, dx + \int_{\Omega=} \beta\, (divU)^2\, dx$$

But as $divv \leq 0$ in Ω we have $\int_\Omega DUvU\, dx = -\frac{1}{2}\int_\Omega divv\, U^2 dx \geq 0$ so

$$a(U,U) = \int_\Omega \alpha DU..DU\, dx + \int_\Omega \beta\, (divU)^2\, dx - \frac{1}{2}\int_\Omega divv\, U^2 dx$$

From the Poincare inequality and as α and β are strictly positive we have

$$\|DU\|^2_{L^2(\Omega,R^3)} \geq \frac{1}{C^2+1}\|U\|^2_{H^1(\Omega,R^3)} \tag{14.5}$$

It follows that

$$a(U,U) \geq \frac{\alpha}{C^2+1}\|U\|^2_{H^1(\Omega,R^3)} \tag{14.6}$$

With the Lax-Milgram theorem we get an unique solution $U \in H^1_0(\Omega, R^3)$ such as

$$-\alpha\Delta U - \beta\nabla(divU) = F - DU.v$$

We have $a(U,U) = \int_\Omega FU dx$.

Hence from Cauchy-Schwarz inequality we get

$$a(U,U) \leq \|F\|_{L^2(\Omega,R^3)}\|U\|_{H^1(\Omega,R^3)}.$$

Therefore, from (14.6) we have

$$\|U\|_{H^1(\Omega,R^3)} \leq \frac{C^2+1}{\alpha}\|F\|_{L^2(\Omega,R^3)} \tag{14.7}$$

Finally we obtain

$$\|u\|_{H^1(\Omega,R^3)} \leq \frac{C^2+1}{\alpha}\|F\|_{L^2(\Omega,R^3)} + \|R\|_{H^1(\Omega,R^3)} \tag{14.8}$$

14.2.2 The Speed Method. We consider the smooth map $T_s(V(s,x))$, the flow of the velocity field V that generates perturbations of the initial domain Ω. We have $\Omega_s = T_s(V)(\Omega) = D \setminus \bar{S}_s$.

We have $V(s,x) = \frac{\partial}{\partial s}(T_s(s,x)) \circ T_s^{-1}(s,x))$. We denote by V V(0,x).

We perturb the boundary of S but not the one of D. We have $T_s(V)(D) = D$ and $T_s(V)(\partial D) = \partial D$. So we take the flow V such as V is null for all x which belong to ∂D.

Let $T > 0, V \in V^{ad} = \{V \in C^0([0,T[, C^2(D,R^3))/divV = 0$ in D, $SuppV \subset D\}$.

Definition 14.1. *Given a speed field V, as previously defined , $J(\Omega)$ is shape differentiable if*

(1) $J(\Omega)$ has an Eulerian semiderivative at Ω in the direction V, i.e if

$$dJ(\Omega,V) = lim_{s\to 0}\left[\frac{J(\Omega_s(V)) - J(\Omega)}{s}\right]$$
$$= \frac{dJ(\Omega_s)}{ds}|_{s=0}$$

exists and is finite.

(2) $V \mapsto dJ(\Omega,V)$ is linear and continuous over admissible vector fields. In the distribution sense we have

$$dJ(\Omega,V) =< G(\Omega), V >$$

$G(\Omega)$ is called the shape gradient of the design functional $J(\Omega)$ at Ω in the direction V.

The method of finding $G(\Omega)$ in this way is called the material derivative method or Speed method. We have no perturbation on the boundary ∂D so the following results are right for the case where u is null on Γ.

14.2.3 The extractor identity.

Proposition 14.2. *Assume that $v \in H^1(D,R^3)$ and $u \in H^2(\Omega,R^3)$ where $u = 0$ on ∂S. Given a speed field $V \in V^{ad}$. We have*

$$\int_{\partial S}(\alpha < Du.n >^2 + \beta < Du.n, n >^2) < V, n > \, d\Gamma =$$
$$\int_\Omega < 2(\alpha\Delta u + \beta\nabla(divu)), Du.V > dx +$$
$$+ \int_\Omega 2\alpha Du.DV..Du \, dx + \int_\Omega < 2\beta\nabla(divu), DV.u > dx +$$
$$+ \int_\Omega < D^*u.u, DV.v > dx + \int_\Omega divv < Du.V, u > dx$$

Proof. $SuppV \subset D$ so V is zero in a neighborhood of ∂D. The proof is in three steps : the extractor is computed by direct shape calculus and by change of variable and finally identifying the two different expressions we deduce the expression for the boundary terms.

The extractor , following (11.1), is defined as follows:

For $s > 0$ we consider the functional defined over $H_0^1(\Omega_s, R^3)$

$$I(u)(s) = \int_{\Omega_s} \alpha D(u\circ T_s^{-1})..D(u\circ T_s^{-1}) + \beta(div(u\circ T_s^{-1}))^2 +$$
$$+ < D(u\circ T_s^{-1}).v, u\circ T_s^{-1} > dX$$

and the extractor

$$\mathcal{E}(u)(V) = \left[\frac{\partial}{\partial s} I(u)(s)\right]_{s=0}$$

The first expression for the extractor. It is obtained by computing the derivative of an integral over a moving domain. We have

$$\mathcal{E}(u)(V) = \int_\Gamma (\alpha Du..Du+\beta\,(divu)^2) < V,n > d\Gamma - \int_\Omega 2\alpha Du..D(Du.V)dx +$$

$$-\int_\Omega 2\beta divu\; div(Du.V)\; dx + \int_\Omega divv < Du.V, u > dx$$

We first concentrate on the trilinear part of the extractor:

$I_1(u)(s) = \int_{\Omega_s} < D(u \circ T_s^{-1})v, u \circ T_s^{-1} > dX$
By shape calculus we get

$$\left[\frac{\partial}{\partial s} I_1(u)(s)\right]_{s=0} = -\int_\Omega < D(DuV)v, u > + < Du.v, Du.V > dx +$$

$$+\int_\Gamma < Duv, u >< V, n > d\Gamma$$

Then we have:

$$\left[\frac{\partial}{\partial s} I(s)\right]_{s=0} =$$

$$= \left[\frac{\partial}{\partial s} I_1(u)(s)\right]_{s=0} - \int_\Omega 2\,[\alpha\, D(Du.V)..Du \,+\, \beta div(Du.V)div(u)]\; dx$$

hence the extractor becomes

$$\mathcal{E}(u)(V) = \int_\Omega < 2(\alpha\Delta u+\beta\nabla(divu)), Du.V > dx + \int_\Omega divv < Du.V, u > dx- \tag{14.9}$$

$$\int_{\partial S} (\alpha < Du.n >^2 + \beta < Du.n, n >^2) < V, n > \; d\Gamma+$$

By Green' s theorem we have

$$-\int_\Omega < D(Du.V).v, u.dx = \int_\Omega divv < Du.V, u > \; dx+ \tag{14.10}$$

$$+\int_\Omega < Du.v, Du.V > dx \; - \int_\Gamma < Du.V, u >< v, n > d\Gamma$$

And

$$\int_\Omega Du..D(-Du.V)dx = \int_\Omega < \Delta u, Du.V > dx - \int_\Gamma < Du.n, Du.V > d\Gamma \tag{14.11}$$

$$\int_\Omega divu\, div(-Du.V)dx = \int_\Omega < \nabla(divu), Du.V > dx + \tag{14.12}$$

$$- \int_\Gamma divu < Du.V, n > d\Gamma$$

As $divu = div_\Gamma u + < Du.n, n >= div_\Gamma u + < \epsilon(u).n, n >$, the matrix $Du = D_\Gamma u + Du.n.n^*, u$ is constant on ∂S then we have $\mathrm{div}_\Gamma u = 0$ and $D_\Gamma u = 0$. Hence we get

$$\begin{cases} divu|_{\partial S} =< \epsilon(u).n, n > \\ Du|_{\partial S} = Du.n.n^* \end{cases} \tag{14.13}$$

And we have $u = 0$ on $\partial S, < V, n >= 0$ on ∂D. Thus we obtain the following equations

$$\int_\Gamma < Du.n, Du.V > d\Gamma = \int_{\partial S} < Du.n >^2 < V, n > \; d\Gamma \tag{14.14}$$

$$\int_\Gamma divu < Du.V, n > d\Gamma = \int_{\partial S} < Du.n, n >^2 < V, n > \; d\Gamma \tag{14.15}$$

$$\int_\Gamma Du..Du \; < V, n > \; d\Gamma = \int_{\partial S} < Du.n >^2 < V, n > \; d\Gamma \tag{14.16}$$

$$\int_\Gamma (divu)^2 \; < V, n > \; d\Gamma = \int_{\partial S} < Du.n, n >^2 < V, n > \; d\Gamma \tag{14.17}$$

We have $divv = 0$ in Ω. If we take v such that $v = 0$ on ∂S then

$$I_1(u)(s) = \int_{\Omega_s} < D(u \circ T_s^{-1}).v \circ T_s^{-1}, u \circ T_s^{-1} > dX$$

Thus

$$\left[\frac{\partial}{\partial s} I_1(u)(s)\right]_{s=0} =$$

$$= - \int_\Omega < D(DuV)v, u > + < DuDv.V, u > + < Duv, Du.V > dx +$$

$$+ \int_\Gamma < Du.v, u >< V, n > d\Gamma$$

Then

$$\left[\frac{\partial}{\partial s} I_1(u)(s)\right]_{s=0} =$$

$$= - \int_\Omega < Dv.V, D^*u.u > dx + \int_\Omega divv < Du.V, u > dx +$$

$$+ \int_\Gamma < Du.v, u >< V, n > d\Gamma - \int_\Gamma < Du.V, u >< v, n > d\Gamma$$

The "energy" expression for the extractor. We denote $j_V(s) = detDT_s$. By a change of variable we get

$$I_1(u)(s) = -\int_\Omega < Du(DT_s)^{-1}.v, u > j_V(s)\, dx$$

But we have $[\frac{\partial}{\partial s} j_V(s)]_{s=0} = divV$.

Thus we get

$$\left[\frac{\partial}{\partial s} I_1(u)(s)\right]_{s=0} = -\int_\Omega < DV.v, D^*u.u > + < Du.v, u > divV\, dx$$

Therefore we obtain

$$\mathcal{E}(u)(V) = [\frac{\partial}{\partial s}\{\int_\Omega \alpha Du.(DT_s)^{-1})..Du.(DT_s)^{-1} j_V(s)\, dx +$$

$$+\int_\Omega \beta(div((DT_s)^{-1}u))^2 j_V(s) dx + I_1(u)(s)\}]_{s=0}$$

So we have

$$\mathcal{E}(u)(V) = -\int_\Omega \alpha(Du.DV..Du + Du..Du.DV) + 2\beta divu\, div(DV.u)\, dx +$$

$$-\int_\Omega < DV.v, D^*u.u > +\{\alpha Du..Du + \beta(divu)^2 + < Du.v, u >\} divV\, dx$$

As $V \in V^{ad}$ $divV = 0$ then we get

$$\mathcal{E}(u)(V) = -\int_\Omega 2\beta divu\, div(DV.u) + 2\alpha Du.DV..Du + < DV.v, D^*u.u > dx \tag{14.18}$$

Finally from (11.5) and (14.18) we obtain the results of the following extractor identity:

$$\int_{\partial S} (\alpha < Du.n >^2 + \beta < Du.n, n >^2) < V, n > \, d\Gamma =$$

$$= \int_\Omega < 2(\alpha\Delta u + \beta\nabla(divu)), Du.V > dx + + \int_\Omega 2\alpha Du.DV..Du\, dx$$

$$+ \int_\Omega 2\beta divu\, div(DV.u)\, dx - \int_\Omega < Dv.V, D^*u.u > dx +$$

$$+ \int_\Omega < D^*u.u, DV.v > dx + \int_\Omega divv < Du.V, u > dx$$

From which the proposition derives.

14.2.4 The estimate.

Proposition 14.3. *Assume that $v \in H^1(D, R^3), u \in H^2(\Omega, R^3)$ with $u = 0$ on ∂S.*

$$\alpha\Delta u + \beta\nabla(divu)) \in L^2(\Omega, R^3).$$

Let $V \in V^{ad}$ with $SuppV \subset D$ such as $< V, n > \geq \Gamma > 0$. Then we have

$$\Gamma\alpha\|Du.n\|^2_{L^2(\Gamma,R^3)} + \Gamma\beta\| < Du.n, n > \|^2_{L^2(\Gamma)} \leq$$

$$\leq \int_\Omega 2 < \alpha\Delta u +$$

$$\beta\nabla(divu)), Du.V > dx \;\; + \int_\Omega 2\beta divu\; div(DV.u) - 2\alpha Du.DV..Du$$

$$+ < D^*u.u, DV.v > + divv < Du.V, u > dx$$

Proof. We set $\zeta = Du.n$. We have $\int_{\partial S}(\alpha < Du.n >^2 + \beta < Du.n, n >^2) < V, n > \, d\Gamma \geq \int_{\partial S} \alpha_\Gamma \zeta^2 + \beta_\Gamma \; < \zeta, n >^2 d\Gamma$

Hence we get

$\int_{\partial S}(\alpha < Du.n >^2 + \beta < Du.n, n >^2) < V, n > \;\; d\Gamma \geq \alpha_\Gamma \|\zeta\|^2_{L^2(\Gamma,R^3)} + \beta_\Gamma \| < \zeta, n > \|^2_{L^2(\Gamma)}$

Thus the proposition 14.2 yields these results.

14.2.5 Dense subspace in $\mathcal{H}(\Omega)$. We denote $\mathcal{H}(\Omega) = \{u \in H^1(\Omega, R^3)/ - \alpha\Delta u - \beta\nabla(divu) \in L^2(\Omega, R^3), u|_{\partial S} = 0\}$ equipped with the graph norm. Assume that $\epsilon_0 > 0$. Let $q \geq \frac{3}{2} + \epsilon_0$. We shall extend the previous estimate to non smooth element $u \in \mathcal{H}(\Omega)$. Let $v \in H^q(D)$. If $divv = 0$ in Ω then we only need $v \in H^1(D, R^3)$. We assume that the domain Ω is such as the following density property holds:

($\mathcal{D}$) $\quad S$ is such as :$\forall u \in H^1(\Omega, R^3)$ with $u = 0$ on ∂S there exists $u^l \in H^2(\Omega, R^3) \to u$ strongly in $H^1(\Omega, R^3)$ with $u^l = 0$ on ∂S and $-\alpha\Delta u^l - \beta\nabla(divu^l) \to -\alpha\Delta u - \beta\nabla(divu)$ in $L^2(\Omega, R^3)$, for $l \to \infty$.

Theorem 14.1. *Let $\Omega = D \setminus \bar{S}$ be a domain with a nonempty boundary Γ that is an $(N-1)$ Lipschitz submanifold. Let $v \in H^q(D)$ with $q > \frac{3}{2}$ or $v \in H^1(D, R^3)$ with $divv = 0$ in Ω. Assume that we have the density property ($\mathcal{D}$) then $Du.n$ is in $L^2(\Gamma, R^3)$.*

Proof. Let be

$$\mathcal{E}(u^l)(V) = \left[\frac{\partial}{\partial s}\{\int_{\Omega_s} \alpha D(u^l \circ o T_s^{-1}) + \beta(div(u^l \circ T_s^{-1}))^2 dx + I_1(u^l)(t)\}\right]_{t=0}$$

We are in the conditions of the proposition (14.3) so we obtain

$$\int_{\partial S} (\alpha_\Gamma < Du^l.n >^2 + \beta_\Gamma < Du^l.n, n >^2)\, d\Gamma \leq F(u^l)$$

where $F(u^l)$ is defined by:

$$F(u^l) = \int_\Omega < 2(\alpha \Delta u^l + \beta \nabla (divu^l)), Du^l.V > + 2\alpha Du^l.DV..Du^l\, dx + \tag{14.19}$$

$$+ \int_\Omega 2\beta divu^l\, div(DV.u^l) + < D^* u^l.u^l, DV.v > + divv < Du^l.V, u^l > dx$$

If we set $\zeta^l = Du^l.n$ then we get

$$\int_{\partial S} (\alpha_\Gamma (\zeta^l)^2 + \beta_\Gamma < \zeta^l, n >^2) < V, n > \, d\Gamma \leq F(u^l) \tag{14.20}$$

We have $u^l \to (weakly)$ u in $H^1(\Omega, R^3)$, for $l \to \infty$.
Therefore we obtain $Du^l \to Du$ in $L^2(\Omega, R^3 \times R^3)$, for $l \to \infty$.
And $divu^l \to divu$ in $L^2(\Omega)$, for $l \to \infty$

Let

$$F(u) = \int_\Omega < 2(\alpha \Delta u + \beta \nabla (divu)), Du.V > dx + \int_\Omega 2\alpha Du.DV..Du\, dx + \tag{14.21}$$

$$+ \int_\Omega 2\beta divu\, div(DV.u) + < D^* u.u, DV.v > + divv < Du.V, u > dx$$

u is in $\mathcal{H}(\Omega)$ so $F(u)$ is well defined. Hence $F(u^l) \to F(u)$ for $l \to \infty$. As we have

$$\alpha_\Gamma \|\zeta^l\|^2_{L^2(\Gamma, R^3)} + \beta_\Gamma \| < \zeta^l, n > \|^2_{L^2(\Gamma)} \leq F(u^l) \tag{14.22}$$

then we get

$$\alpha_\Gamma \|\zeta^l\|^2_{L^2(\Gamma, R^3)} + \beta_\Gamma \| < \zeta^l, n > \|^2_{L^2(\Gamma)} \leq \liminf F(u^l) = F(u) \tag{14.23}$$

Thus we obtain that there exists a constant

$$M = \left(\frac{F(u)}{\alpha \Gamma}\right)^{\frac{1}{2}} > 0$$

such that

$$\|\zeta^l\|_{L^2(\Gamma)} \leq M \tag{14.24}$$

Hence, there exists a subsequence (ζ^{l_k}) such that $\zeta^{l_k} (weakly) \to \zeta$ weakly in $L^2(\Gamma, R^3)$, for $k \to (weakly)\infty$. Then $< \zeta^{l_k}, n >$ $(weakly) \to < \zeta, n >$ weakly in $L^2(\Gamma)$, for $k \to \infty$

Let $u \in \mathcal{H}(\Omega)$. We have for all φ in $H^1(\Omega, R^3)$

$$< Du.n, \varphi|_\Gamma >_{H^{-\frac{1}{2}}(\Gamma,R^3)\times H^{\frac{1}{2}}(\Gamma,R^3)} = \qquad (14.25)$$

$$=< \Delta u, \varphi >_{H^{-1}(\Omega,R^3)\times H^1(\Omega,R^3)} + \int_\Omega Du..D\varphi\, dx. \qquad (14.26)$$

But for each l we have

$$< \zeta^l, \varphi|_\Gamma >_{H^{-\frac{1}{2}}(\Gamma,R^3)\times H^{\frac{1}{2}}(\Gamma,R^3)} =$$

$$=< \Delta u^l, \varphi >_{H^{-1}(\Omega,R^3)\times H^1(\Omega,R^3)} + \int_\Omega Du^l..D\varphi\, dx.$$

We set $\varphi|_\Gamma = 1$ then $\zeta^l(weakly) \to Du.n$ weakly in $H^{-\frac{1}{2}}(\Gamma, R^3)$, for $l \to \infty$. But $\zeta^l(weakly) \to \zeta$ weakly in $L^2(\Gamma, R^3)$, for $l \to \infty$. Hence $\zeta^l(weakly) \to \zeta$ weakly in $H^{-\frac{1}{2}}(\Gamma, R^3)$ so $\zeta^l(weakly) \to Du.n$ in $L^2(\Gamma, R^3)$, for $l \to \infty$.

14.2.6 Ω and S star-shaped domains. We consider Ω and S two star-shaped domains with respect to $0 \in \overline{S}$: for $0 < \zeta < 1$, we have

for $x \in S \bigcap SuppV$ (respectively $\Omega \bigcap SuppV$) then $(1+\zeta)x \in S$ (respectively Ω).

Let ρ_r be a mollifier. ρ_r is a smooth $C_c^\infty(R^3, R^3)$ function (or $C_c^\infty(R^3)$), such that $\rho_r \geq 0$ and

$$\int_{R^N} \rho_r dx = 1$$

Let ρ_ζ be the function $x \to \frac{1}{\zeta^3}\rho_r(\frac{x}{\zeta})$. ρ_ζ converges in the distribution sense to the Dirac distribution, for $\zeta \to 0$.

Let $u \in \mathcal{H}(\Omega), u^0 = \begin{cases} u/\,(d_{\Omega^c})^\alpha & \text{in } \Omega \\ 0 \ outside \end{cases}$ $\quad u_\zeta(x) = u^0(\,(1+\zeta)^{-1}\,x)$ and

$u^\zeta = (d_{\Omega^c})^\alpha\ \rho_\zeta * u_\zeta = (d_{\Omega^c})^\alpha \int_{R^3} \rho_\zeta(x-y)u_\zeta(y)dy$

We have $u_\zeta(x) \in H^1(R^3, R^3)$ $u^\zeta(x) \to u \in H^1(\Omega, R^3)$ and $u^\zeta \in C_c^\infty(R^3, R^3)$

Proposition 14.4. *Let $\Omega = D \setminus \bar{S}$ be a star-shaped domain with S also star-shaped. Then $u^\zeta \to u$ in $\mathcal{H}(\Omega)$.*

Proof. We give an idea of the proof by taking $\alpha = zero$.

Lemma 14.1. *$-\alpha\Delta u^\zeta - \beta\nabla(div u^\zeta) \to -\alpha\Delta u^0(x) - \beta\nabla(div u^0(x))$, strongly in $L^2(R^3)$, for $\zeta \to 0$*

We use to

Corollary 14.1.

$$div u_\varsigma = (1+\varsigma)^{-1}((div u^0)\circ(1+\varsigma)^{-1}Id)(x)$$
$$\nabla(div u_\varsigma) = (1+\varsigma)^{-2}((\nabla(div u^0))\circ(1+\varsigma)^{-1}Id)(x)$$
$$\Delta u_\varsigma = (1+\varsigma)^{-2}((\Delta u^0)\circ(1+\varsigma)^{-1}Id)(x)$$

Proof. The composed functions derivative lead to the above results.

$$div u_\varsigma = ((div u^0)\circ(1+\varsigma)^{-1}Id)(x).\nabla(1+\varsigma)^{-1}x =$$
$$= (1+\varsigma)^{-1}((div u^0)\circ(1+\varsigma)^{-1}Id)(x)$$
$$\nabla(div u_\varsigma) = \nabla((1+\varsigma)^{-1}((div u^0)\circ(1+\varsigma)^{-1}Id)(x)) =$$
$$(1+\varsigma)^{-2}((\nabla(div u^0))\circ(1+\varsigma)^{-1}Id)(x)$$

Similarly

$$\Delta u_\varsigma = (1+\varsigma)^{-1}((\Delta u^0)\circ(1+\varsigma)^{-1}Id)(x).\nabla(1+\varsigma)^{-1}x$$

Corollary 14.2. $-\alpha\Delta u^\varsigma - \beta\nabla(div u^\varsigma) = \rho_\varsigma * (-\alpha\Delta u_\varsigma - \beta\nabla(div u_\varsigma))$

Proof. We use to the change of variable $X = x - y$

$$-\alpha\Delta u^\varsigma - \beta\nabla(div u^\varsigma) =$$
$$= -\alpha\Delta_x\int_{R^3}\rho_\varsigma(X)u_\varsigma(x-X)dX - \beta\nabla_x div_x\int_{R^3}\rho_\varsigma(X)u_\varsigma(x-X)dX$$
$$= \rho_\varsigma * (-\alpha\Delta u_\varsigma - \beta\nabla(div u_\varsigma))$$

Hence deriving from the corollary 14.1 we have

$$-\alpha\Delta u^\varsigma - \beta\nabla(div u^\varsigma) =$$
$$= (1+\varsigma)^{-2}\rho_\varsigma * (-\alpha((\Delta u^0)\circ(1+\varsigma)^{-1}Id)(x) - \beta((\nabla(div u^0))\circ(1+\varsigma)^{-1}Id)(x)$$

$$= g_\varsigma(x)$$

As u is in $\mathcal{H}(\Omega)$, then $g_\varsigma(x)$ is in $L^2(\Omega, R^3)$ and from the classical result $\rho_\varsigma * g_\varsigma(x)$ converges strongly to $g_\varsigma(x)$ in $L^2(\Omega, R^3)$), for $\varsigma \to 0$

But $g_\varsigma(x) \to -\alpha\Delta u^0(x) - \beta\nabla(div u^0(x))$ strongly in $L^2(\Omega, R^3)$), for $\varsigma \to 0$ So $\rho_\varsigma * g_\varsigma(x)$ converges strongly to $-\alpha\Delta u^0(x) - \beta\nabla(div u^0(x))$ in $L^2(\Omega, R^3)$), for $\varsigma \to 0$. We set $u^l = u^\varsigma$ with $l = \frac{1}{\varsigma}$

The boundary of the support of u_ς is $\Gamma_\varsigma = \{x \in \Omega, (1+\varsigma)^{-1}x \in \partial S\}$. We have $u_\varsigma = 0$ on Γ_ς

But $Support u^\varsigma \subset Support\rho_\varsigma + Support u_\varsigma$ and $Support\rho_\varsigma \subset B(x,l)$ so $u^l = 0$ on ∂S

We are in the assumption (K) and we are in the conditions of the theorem 14.1 so we obtain the result.

14.2.7 The pressure approximation. We choose the penalization term β in the form $\beta = \nu + \frac{1}{\sigma}$. Let $f \in L^2(D, R^3)$ and $v \in H_1^{div}$.

We consider the two following problems: u_σ in $H^1(\Omega, R^3)$ solves

$$\begin{cases} -\nu\Delta u_\sigma + Du_\sigma\, v - (\nu + \frac{1}{\sigma})\nabla(divu_\sigma) = f \text{ in } \Omega \\ \sigma p_\sigma + divu_\sigma = 0 \text{ in } \Omega \\ u_\sigma = 0 \text{ on } \partial S \\ u_\sigma = u_\infty \text{ on } \partial D \end{cases} \tag{14.27}$$

(u, p) in $H^1(\Omega, R^3)\times L^2(\Omega)$ solves

$$\begin{cases} -\nu\Delta u + Du.v + \nabla p = f \text{ in } \Omega \\ divu = 0 \text{ in } \Omega \\ u = 0 \text{ on } \partial S \\ u = u_\infty \text{ on } \partial D \end{cases} \tag{14.28}$$

Proposition 14.5. *Let Ω be a bounded and Lipschitz open domain of R^3. Assume that u_σ is in $H^1(\Omega, R^3)$ and is a solution of (14.27). Assume that $(u,p) \in H^1(\Omega, R^3) \times L^2(\Omega)$ is a solution of (14.28) then we have u_σ which converges to u strongly in $H^1(\Omega, R^3)$ and we have $-\frac{1}{\sigma}divu_\sigma$ that converges to p strongly in $L^2(\Omega)$, for $\sigma \to 0$.*

Proof. The systems (14.27) and (14.28) yield the following equation

$$-\nu\Delta(u_\sigma - u) + D(u_\sigma - u)v - (\nu\sigma + 1)\nabla(\frac{1}{\sigma}divu_\sigma) = \nabla p \tag{14.29}$$

Using the equation (14.29) we obtain the following weak formulation for all φ in $H^1(\Omega, R^3)$

$$\int_\Omega \nu D(u_\sigma - u)..D\varphi\, dx + \int_\Omega (\nu\sigma + 1)(\frac{1}{\sigma}divu_\sigma)div\,\varphi\, dx + \int_\Omega D(u_\sigma - u).v\varphi\, dx$$
$$= -\int_\Omega p\, div\,\varphi\, dx + \int_\Gamma p\,\varphi.n\, d\Gamma\ +$$
$$+\int_\Gamma \nu D(u_\sigma - u).n\,\varphi\, d\Gamma + \int_\Gamma (\nu\sigma + 1)divu_\sigma\,\varphi.n\, d\Gamma$$

Thus we obtain for $\varphi = u_\sigma - u$ that

$$\nu||D(u_\sigma - u)||_{L^2}^2 + (\nu\sigma + 1)\frac{1}{\sigma}||divu_\sigma||_{L^2}^2 + \int_\Omega D(u_\sigma - u).v(u_\sigma - u)\, dx$$
$$= -\int_\Omega p\, divu_\sigma\, dx$$

From the above and from the Cauchy-Schwarz inequality, we have

$$\|D(u_\sigma - u)\|_{L^2}^2 + (\nu\sigma + 1)\frac{1}{\sigma}\|divu_\sigma\|_{L^2}^2 + \int_\Omega D(u_\sigma - u).v(u_\sigma - u)\, dx$$
$$\leq \|p\|_{L^2}\|divu_\sigma\|_{L^2}$$

But we have

$$\|p\|_{L^2}\|divu_\sigma\|_{L^2} \leq \frac{1}{2\sigma}\|divu_\sigma\|_{L^2}^2 + \frac{\sigma}{2}\|p\|_{L^2}^2 \tag{14.30}$$

and

$$\int_\Omega D(u_\sigma - u).v(u_\sigma - u)\, dx = -\frac{1}{2}\int_\Omega divv(u_\sigma - u)^2 dx \geq 0$$

So we get

$$\nu\|D(u_\sigma - u)\|_{L^2}^2 + (\nu\sigma + \frac{1}{2})\frac{1}{\sigma}\|divu_\sigma\|_{L^2}^2 + \int_\Omega D(u_\sigma - u).v(u - u_\sigma)\, dx$$
$$\leq \frac{\sigma}{2}\|p\|_{L^2}^2$$

Finally we obtain that u_σ strongly converges to u in $H^1(\Omega, R^3)$, when $\sigma \to 0$.

Using the equation (14.29) and from the above, we have $\Delta(u_\sigma - u) \to 0$ strongly in $H^{-1}(\Omega, R^3)$, for $\sigma \to 0$. Then we deduce that $-\frac{1}{\sigma}\nabla(divu_\sigma) \to \nabla p$ in $H^{-1}(\Omega, R^3)$, for $\sigma \to 0$.

Ω is a bounded and Lipschtiz open domain; hence there is a cônstant C=C(Ω) depending only on Ω, such as

$$\|p + \frac{1}{\sigma}divu_\sigma\|_{L^2} \leq C\left\{\int_\Omega p + \frac{1}{\sigma}divu_\sigma dx + \|\nabla(p + \frac{1}{\sigma}divu_\sigma)\|_{H^{-1}}\right\} \tag{14.31}$$

$\int_\Omega p + \frac{1}{\sigma}divu_\sigma\ dx = 0$ so from (14.31) we have $-\frac{1}{\sigma}divu_\sigma \to p$ strongly in $L^2(\Omega)$

Lemma 14.2. *$divu_\sigma \to 0$ strongly in $L^2(\Omega)$, for $\sigma \to 0$.*

Proof. The proposition 14.5 yields that there exists a sequence $p_\sigma = -\frac{1}{\sigma}divu_\sigma \to p$, strongly in $L^2(\Omega)$ for $\sigma \to 0$ so $\sigma p_\sigma \to \sigma p = 0$, strongly in $L^2(\Omega)$ for $\sigma \to 0$.

14.2.8 The regularity problem.

Theorem 14.2. *Let $\Omega = D \setminus \bar{S}$ be a domain with a nonempty boundary Γ that is an (N-1) Lipschitz submanifold. Let $v \in H_1^{div}$ Assume that $u \in H^1(\Omega, R^3)$ is a solution of (14.28) then $Du.n$ is in $L^2(\Gamma, R^3)$.*

Proof. We take α such as $\alpha = \nu$, β such as $\beta = \nu + \frac{1}{\sigma}$. Let $v \in H_1^{div}$ and assume that $u' \in H^2(\Omega, R^3) \to u_\sigma$ in $\mathcal{H}(\Omega)$, which is a solution of (14.27). Hence we have $-\nu \Delta u_\sigma - (\nu + \frac{1}{\sigma})\nabla(divu_\sigma) = f - Du_\sigma.v$ If $< V, n > \geq \Gamma > 0$, then the proposition 14.3 leads to

$$\Gamma\nu \int_{\partial S} (< Du_\sigma.n >)^2 \, d\Gamma \leq \int_\Omega -2f.Du_\sigma.V \, dx +$$

$$+ \int_\Omega 2\nu Du_\sigma.DV..Du_\sigma u_\sigma \, dx + \int_\Omega 2(\nu + \frac{1}{\sigma})divu_\sigma \; div(DV.u_\sigma) \, dx +$$

$$+ \int_\Omega < D^* u_\sigma.u_\sigma, DV.v > + < 2Du_\sigma.v, Du_\sigma.V > dx$$

So we can deduce that there is a constant $M_0 > 0$ such as

$$\int_{\partial S} < Du_\sigma.n >^2 \; d\Gamma \leq M_0 \tag{14.32}$$

So we can extract a subsequence $Du_\sigma.n$ that converges weakly to ζ in $L^2(\Gamma, R^3)$ when $\sigma \to 0$. Therefore $Du_\sigma.n(weakly) \to \zeta$ weakly in $H^{-\frac{1}{2}}(\Gamma, R^3)$, for $\sigma \to 0$.

We have $u_\sigma \to u$, strongly in $H^1(\Omega, R^3)$ for $\sigma \to 0$.

We notice that the term $Du.n$ is defined in the weak form, we have for all φ in $H^1(\Omega, R^3)$ $< Du.n, \varphi|_\Gamma > = < Du.n, \varphi|_\Gamma >_{H^{-\frac{1}{2}}(\Gamma,R^3) \times H^{\frac{1}{2}}(\Gamma,R^3)}$ and

$$< Du.n, \varphi|_\Gamma > = \int_\Omega Du..D\varphi \, dx + \frac{1}{\nu} \int_\Omega < \nabla p^{k+1} + Du.u^k - f, \varphi > \; dx \tag{14.33}$$

But for each σ we have

$$< Du_\sigma.n, \varphi|_\Gamma > = < Du_\sigma.n, \varphi|_\Gamma >_{H^{-\frac{1}{2}}(\Gamma,R^3) \times H^{\frac{1}{2}}(\Gamma,R^3)} = \int_\Omega Du_\sigma..D\varphi \, dx +$$

$$+ \int_\Omega < \frac{1}{\nu} Du_\sigma.u^k - (1 + \frac{1}{\nu\sigma})\nabla(divu_\sigma) + \frac{f}{\nu}, \varphi > dx$$

Then from (14.33) we obtain

$$\nu < D(u_\sigma - u).n, \varphi|_\Gamma > = nu \int_\Omega D(u_\sigma - u)..D\varphi \, dx + \int_\Omega D(u_\sigma - u).u^k \; \varphi \, dx +$$

$$+\int_{\Omega} < (1+\frac{1}{\sigma})\nabla(divu_\sigma) + \nabla p, \varphi > \, dx$$

Given $\varphi = 1$ we obtain that $Du_\sigma.n(weakly) \to Du.n$ weakly in $H^{-\frac{1}{2}}(\Gamma, R^3)$, for $\sigma \to 0$.

So $Du_\sigma.n(weakly) \to Du.n$ weakly in $L^2(\Gamma, R^3)$, for $\sigma \to 0$.

Using the regularity of the solution u we are going to show that the shape gradient of the fineness is well defined.

14.3 Shape optimization problem

This section deals with shape optimizations problem governed by the Oseen equations. The objective is to improve the shape of a body (sail or flat plate, ...) that leads to a minimum for the viscous drag force and a maximum for the lift force. Hence, we make a sensitivity analysis of this shape control problem. We look for the relationship between available control parameters and responses of the state variables and cost functional (fineness) to change in the parameters, so we compute the derivatives of the state variables and cost functional . We use the material derivative method and the adjoint equation technique to simplify the calculation of the shape gradient.

The mean curvature H of Γ is defined as the trace of the matrix D^2b

$$H = Tr(D^2b) = \Delta b$$

b is C^k in the neighborhood of the boundary if and only the domain Ω is C^k itself.

14.3.1 Shape gradient with linear splitted flow. The objective is to improve the shape of the body in order to minimize the force in a given direction e_1 and to maximize it in another given direction e_2. Assume that $E_i \in H^2(D, R^3), i = 1..2$, is solution of

$$\begin{cases} \Delta^2 E_i + \nabla q = 0 \text{ in } \Omega \setminus \overline{B} \\ E_i = e_i \text{ in } \overline{B} \\ \frac{\partial E}{\partial n} = 0 \text{ on } \partial B \\ E_i = 0 \text{ on } \partial D \end{cases} \tag{14.34}$$

If the viscosity is small enough, we can miss the viscous force and we focus on the minimization of the cost functional $J(\Omega)$ that is

Definition 14.2.

$$J(\Omega) = \frac{F_{e_2}(\Omega)}{F_{e_1}(\Omega)} = \frac{\int_{\partial S} \beta divu \; e_2.n \; d\Gamma}{\int_{\partial S} \beta divu \; e_1.n \; d\Gamma} = \frac{\int_{\Gamma} \beta divu \; E_2.n \; d\Gamma}{\int_{\Gamma} \beta divu \; E_1.n \; d\Gamma}$$

where u is solution of (11.1).

So we consider this following extremal problem:

$$\min_{\Omega \in \mathcal{D}_{ad}} J(\Omega) \tag{14.35}$$

Let $v \in H_1^-(D)$. We define $\lambda_i \in H^1(\Omega, R^3), i = 1..2$, to be the adjoint state of the vector u associated to $F_{e_i}(\Omega)$ and solution of

$$\begin{cases} -\alpha\Delta\lambda_i - D\lambda_i.v - divv\lambda_i - \beta\nabla(div\lambda_i) = 0 \text{ in } \Omega \\ (\lambda_i)_\Gamma = 0 \text{ on } \Gamma \\ \lambda_i.n = \frac{\beta}{\alpha+\beta} E_i.n \text{ on } \Gamma \end{cases} \tag{14.36}$$

We introduce

$$\lambda_g = \frac{F_{e_1}(\Omega)\lambda_2 - F_{e_2}(\Omega)\lambda_1}{(F_{e_1}(\Omega))^2},$$

and

$$E_g = \frac{F_{e_1}(\Omega)E_2 - F_{e_2}(\Omega)E_1}{(F_{e_1}(\Omega))^2}$$

$\lambda_g \in H^1(\Omega, R^3)$ is solution of

$$\begin{cases} -\alpha\Delta\lambda_g - D\lambda_g.v - divv\lambda_g - \beta\nabla(div\lambda_g) = 0 \text{ in } \Omega \\ (\lambda_g)_\Gamma = 0 \text{ on } \Gamma \\ \lambda_g.n = \frac{\beta}{\alpha+\beta} E_g.n \text{ on } \Gamma \end{cases} \tag{14.37}$$

We have in Ω $(\lambda_{gs}^c \circ T_s)' =< \frac{\beta}{\alpha+\beta} E_g, (\mathcal{N}_s \circ T_s)' > \mathcal{N} + < \frac{\beta}{\alpha+\beta} E_g, \mathcal{N} > (\mathcal{N}_s \circ T_s)'$

with $(\mathcal{N}_s \circ T_s)' =< D^*V\mathcal{N}, \mathcal{N} > \mathcal{N} - D^*V.\mathcal{N}$

The formulation of the gradient , $dJ(\Omega, V)$, is given by

Theorem 14.3. *We denote e_g by e, λ_g by λ. Assume that $\partial\Omega$ is lipschitzian and that $v \in H_1(D)$ then the functional $J(\Omega)$ is shape differentiable, and we have $\forall\, V \in V^{ad}$,*

$$dJ(\Omega, V) = \int_\Omega \theta(\Omega, V)\, dx$$

with

$$\theta(\Omega, V) = -2\alpha(Du.\epsilon(V))..D\lambda - < Du.DV.v, \lambda > +$$

$$+\alpha Du..D(\lambda_{gs}^c \circ T_s)' + Du.v.(\lambda_{gs}^c \circ T_s)' + \beta divu\, div(\lambda_s^c \circ T_s)' - f(\lambda_s^c \circ T_s)' +$$

$$- \beta(divuD\lambda..D^*V + div\lambda Du..D^*V)\, dx$$

and then as θ is linear and continuous on V we get

$$dJ(\Omega, V) =< G(\Omega), V >$$

where G is a distribution supported by the boundary, $G = \Gamma^*(g\, n)$.

Moreover, If v is more regular, $v \in H_1^{div}(D)$ or $v \in H^q(D)$ with $q > \frac{3}{2}$ then the density $g(\Gamma) \in L^1(\Gamma)$, hence

$$dJ(\Omega, V) = \int_{\partial S} g(\Gamma) < V, n > d\Gamma$$

and we have the two following explicit expressions of g:

$$g_m^e(\Gamma) = < \beta\nabla(divu), e > -(\alpha+\beta) < D\lambda.n, n >< Du.n, n > +$$

$$- < \alpha(D\lambda.n)_\Gamma, Du.n > + \frac{\beta}{\alpha+\beta} < \alpha(D^2b - Hnn^*)\, e - \beta Hnn^* e, Du.n >$$

and

$$g_c^e(\Gamma) = \frac{\alpha\beta}{\alpha+\beta} < e, n > div_\Gamma(Du.n) + \beta < D_\Gamma(Du.n)e_\Gamma, n > +$$

$$+ \alpha < Du.n, (D^*\lambda.n)_\Gamma - (D\lambda.n)_\Gamma > -(\alpha+\beta) < D\lambda.n, n >< Du.n, n > +$$

$$+ \beta < (D^2b - Hnn^*)e, Du.n > + \frac{\alpha\beta}{\alpha+\beta}(< (D^2b - Hnn^*)e, Du.n > +$$

$$- < e, n >< \frac{f}{\alpha}, n >)$$

The regularity of $(u, \lambda), u \in H^1(\Omega, R^3), \lambda \in H_0^1(D, R^3)$ and of $V \in W^{1,\infty}$ yields the existence of the third distributed expression of $dF_e(\Omega, V)$

Consequently the two previous boundary expressions of $dF_e(\Omega, V)$ are well defined since they represent the same distribution G.

Proof. We consider u solution of the problem (11.1) and R of (14.3). Let $\beta = \nu + \frac{1}{\sigma}$ and $\alpha = \nu$. Let $u = \begin{cases} U + R \text{ in } \Omega \setminus \overline{B} \\ U \text{ in } \overline{B} \end{cases}$

U is a solution of the problem (14.4). Let $\lambda_D \in H_0^1(D, R^3)$ such as $\lambda_D = \lambda$ in Ω. Let E be defined as in (14.34). Assume that v is defined as in (11.1) and $\lambda \in H^1(\Omega, R^3)$ is the adjoint state of U. λ is solution of

$$\begin{cases} -\alpha\Delta\lambda - D\lambda.v - divv\lambda - \beta\nabla(div\lambda) = 0 \text{ in } \Omega \\ \lambda_\Gamma = 0 \text{ on } \Gamma \\ \lambda.n = \frac{\beta}{\alpha+\beta} < E, n > \text{ on } \Gamma \end{cases} \quad (14.38)$$

Proposition 14.6. *There exists one unique solution $\lambda \in H^1(\Omega, R^3)$ of (14.38).*

Proof. Let $R_\lambda \in H^2(\Omega, R^3)$ such that $\begin{cases} (R_\lambda)_\Gamma = 0 \text{ on } \Gamma \\ R_\lambda.n = \frac{\beta}{\alpha+\beta} < E, n > \text{ on } \Gamma \end{cases}$

Let $w = \lambda - R_\lambda$ and $F_\lambda = \alpha \Delta R_\lambda + DR_\lambda.v + divvR_\lambda + \beta\nabla(divR_\lambda)$. w is solution of

$$\begin{cases} -\alpha\Delta w - Dw.v - divvw - \beta\nabla(divw) = F_\lambda \text{ in } \Omega \\ w = 0 \text{ on } \Gamma \end{cases} \tag{14.39}$$

We consider the continuous and bilinear form on $H_0^1(\Omega, R^3)$ defined by

$$a(w,\varphi) = \int_\Omega \alpha Dw..D\varphi \, dx - \int_\Omega Dw.v\varphi \, dx -$$

$$+ \int_\Omega divv \, w\varphi \, dx + \int_\Omega \beta \, divwdiv\varphi \, dx$$

For $\varphi = w$ we get

$$a(w,w) = \int_\Omega \alpha Dw..Dw \, dx - \int_\Omega Dwvw \, dx - \int_\Omega divvw^2 dx + \int_\Omega \beta \, (divw)^2 \, dx$$

But as $divv \leq 0$ in Ω we have $\int_\Omega Dwvw \, dx = -\frac{1}{2}\int_\Omega divv \, w^2 dx \geq 0$ so

$$a(w,w) = \int_\Omega \alpha Dw..Dw \, dx + \int_\Omega \beta \, (divw)^2 \, dx - \frac{1}{2}\int_\Omega divv \, w^2 dx$$

¿From the Poincare inequality and as α and β are strictly positive we have

$$a(w,w) \geq \frac{\alpha}{C^2(\Omega)+1} \|w\|^2_{H^1(\Omega,R^3)}$$

From the Lax-Milgram theorem we get an unique solution $w \in H_0^1(\Omega, R^3)$ such as

$$-\alpha\Delta w - \beta\nabla(divw) = F_\lambda + Dw.v + divvw$$

We compute the derivative of the cost functional $J(\Omega)$ with Ω

$$dJ(\Omega, V) = \frac{dF_{e_2}(\Omega,V)F_{e_1}(\Omega) - dF_{e_1}(\Omega,V)F_{e_2}(\Omega)}{(F_{e_1}(\Omega))^2}$$

$$= < \frac{G_2 F_{e_1}(\Omega) - G_1 F_{e_2}(\Omega)}{(F_{e_1}(\Omega))^2}, V >$$

$$= \int_\Gamma \frac{g_2 F_{e_1}(\Omega) - g_1 F_{e_2}(\Omega)}{(F_{e_1}(\Omega))^2} < V, n > d\Gamma$$

where G_i $(i = 1..2)$ is the gradient of $F_{e_i}(\Omega)$ at Ω with $suppG_i \in \Gamma$. G_i is given by $G_i = \Gamma_\Gamma^*(g_i n)$ with Γ_Γ is the trace operator.

Thus we obtain that

$$dJ(\Omega, V) =< G(u, \lambda_g), V >$$

$$= \int_\Gamma g(u, \lambda_g) < V, n > d\Gamma$$

where G is the gradient of $J(\Omega)$ at Ω with $suppG \in \Gamma$. G is given by $G = \Gamma_\Gamma^*(g\,n)$.

Let

$$F_e(\Omega) = f_e(\Omega, u_v) = \int_{\partial S} \beta divu \ < e, n > d\Gamma =$$

$$= \int_\Gamma \beta divu \ < E, n > d\Gamma = \int_\Gamma \beta divU \ < E, n > d\Gamma$$

where we set $e = e_1$ or $e = e_2$, $E = E_1$ or $E = E_2$.

We introduce the following convex set $K(\Omega)$

Definition 14.3. $K(\Omega) = \{\lambda \in H^1(\Omega, R^3)/\lambda_\Gamma = 0 \ on \ \Gamma, \lambda.n = \frac{\beta}{\alpha+\beta} < E, n > on \ \Gamma\}$

$\lambda = \lambda_D|_\Omega$, adjoint state associated to the solution u belongs to $K(\Omega)$

For convenience, we shall write λ instead of λ_D. We compute the shape gradient of $F_e(\Omega)$ with a min max formulation and the adjoint technique. Given a speed field $V \in V^{ad}$ we take the velocity method and we use to the three following different ways: first, we compute the gradient using the derivative of integrals on moving domains. Then, we calculate the gradient by the introduction of the convex set $K(\Omega)$ where the adjoint λ lies. Finally, we give the expression of the gradient derived by change of variable in the distributed sense.

The expression of the gradient derived using derivative of integrals on moving domains. We consider the Lagrangian formulation to derive the functional and we use itto the derivation of the $min - max$ of Cuer-Zolesio. We have an unique saddle point of the problem, hence the expression of the gradient is written in the solution of the direct state u and adjoint state λ equations.

Let the functional be

$$\mathcal{L}_0(\Omega, v, u, \lambda) = F_e(\Omega) + \int_\Omega \alpha Du..D\lambda + \beta divu \ div\lambda + < Du.v - f, \lambda > dx \ +$$

$$- \int_\Gamma < \alpha Du.n + \beta divu \ n, \lambda > d\Gamma$$

We maximize $\mathcal{L}_0(\Omega, v, u, \lambda)$ with $\lambda \in H_0^1(D, R^3)$. We have

$$F_e(\Omega) = \min_{u \in H(\Omega)} \max_{\lambda \in H_0^1(D)} \mathcal{L}_0(\Omega, v, u, \lambda)$$

The formulation of the gradient $dF_e(\Omega, V)$ is equal to

Proposition 14.7. *Assume that $(u, \lambda) \in H(\Omega) \times H_0^1(D, R^3)$ is the unique saddle point. We set*

$$g^1(\Gamma) = -(\alpha+\beta) < D\lambda.n, n >< Du.n, n > - < \alpha(D\lambda.n)_\Gamma, Du.n >$$

and

$$g^2(\Gamma) = \beta\nabla(divu)e$$

and

$$g^{cur}(\Gamma) = \frac{\alpha\beta}{\alpha+\beta} < (D^2b - Hnn^*)\, e, Du.n > - < \frac{\beta^2}{\alpha+\beta} Hnn^* e, Du.n >$$

then

$$dF_e(\Omega, V) = \int_{\partial S} \{g^1(\Gamma) + g^2(\Gamma) + g^{cur}(\Gamma)\} < V, n > \, d\Gamma$$

Proof. We rewrite the boundary integrals in the distributed sense hence we get

$$F_e(\Omega) = \int_\Omega \beta div(divu\; E)\, dx$$

$$\int_\Gamma \alpha < Du.n, \lambda > d\Gamma = \int_\Omega \alpha div(D^* u\lambda)\, dx$$

$$\int_\Gamma \beta divu\lambda.n \, d\Gamma = \int_\Omega \beta div(divu\lambda)\, dx$$

Therefore the Lagrangian functional becomes

$$\mathcal{L}_0(\Omega, v, u, \lambda) = \int_\Omega \beta div(divu\; E) + \alpha Du..D\lambda + Du.v.\lambda + \beta divu\; div\lambda - f\lambda +$$

$$- \alpha div(D^* u\lambda) - \beta div(divu\lambda)\, dx$$

Let $U = u - R$ be in Ω, thus $U = \begin{cases} u - R \text{ in } \Omega \setminus \overline{B} \\ u \text{ in } \overline{B} \end{cases}$ where R is defined as in (14.3). We change of variable thus

$$\mathcal{L}_0(\Omega, v, u, \lambda) = \mathcal{L}_0(\Omega, v, U, \lambda) =$$

$$= \int_\Omega \beta div(divU\; E) + \alpha DU..D\lambda + < DU.v, \lambda > + \beta divU\; div\lambda -$$

$$+ f\lambda \; - \; \alpha div(D^* U\lambda) - \beta div(divU\lambda)\, dx - \int_{\Omega\setminus B} F(R, \lambda)\, dx$$

with

$$F(R, \lambda) = -\beta div(divR\; E) - \alpha D\lambda..DR - < DR.v, \lambda > - \beta divR div\lambda +$$

$$+ \alpha div(D^* R\lambda) + \beta div(divR\lambda)$$

We maximize $\mathcal{L}_0(\Omega, v, u, \lambda)$ with $\lambda \in H_0^1(D, R^3)$. We have

$$F_e(\Omega) = \min_{u\in H(\Omega)} \max_{\lambda\in H_0^1(D,R^3)} \mathcal{L}_0(\Omega, v, u, \lambda) =$$

$$\min_{U\in H_0^1(\Omega,R^3)} \max_{\lambda\in H_0^1(D,R^3} \mathcal{L}_0(\Omega, v, U, \lambda)$$

(U_v, λ_v) is the unique solution of respectively (14.4) and (14.38) so this problem has an unique saddle point (U_v, λ_v) for each Ω and v then from the derivation of a min max of Cuer-Zolesio we have

$$dF_e(\Omega, V) = \frac{\partial \mathcal{L}_0}{\partial s}(\Omega_s, v, U_v, \lambda_v)|_{s=0}$$

We denote (U_v, λ_v) by (U, λ).

The expression of the shape gradient is obtained by computing the derivative of an integral over a moving domain:

$$\mathcal{L}_0(\Omega_s, v, U, \lambda) =$$

$$\int_{\Omega_s} \alpha D(U\circ T_s^{-1})..D\lambda + D(U\circ T_s^{-1}).v.\lambda + \beta div(U\circ T_s^{-1})\ div\lambda +$$

$$- < f, \lambda > + \beta div(div(U\circ T_s^{-1})E) - \alpha div(D^*(U\circ T_s^{-1})\lambda) +$$

$$\beta div(div(U\circ T_s^{-1})\lambda)\ dx \ - \int_{\Omega_s} F(R\circ T_s^{-1}, \lambda)\ dx$$

So its derivative gives

$$\frac{\partial \mathcal{L}_0}{\partial s}(\Omega_s, v, U, \lambda)|_{s=0} = \int_\Omega \alpha D(-DU.V)..D\lambda + D(-DU.V)v\lambda +$$

$$+\beta div(-DU.V)\ div\lambda + \beta div(div(-DU.V)E) + \alpha div(D^*(DU.V)\lambda) +$$

$$+\beta div(div(DU.V)\lambda)\ dx \ + \int_\Gamma \{\alpha DU..D\lambda + \beta divU\ div\lambda +$$

$$+\beta divU\ div\lambda + < DU.v - f, \lambda > + \beta div(divU\ E) - \alpha div(D^*U\lambda) +$$

$$-\beta div(divU\lambda)\} < V, n > \ d\Gamma - \int_\Omega F(-DR.V, \lambda)\ dx \ - \int_\Gamma F(R, \lambda)V.n\ d\Gamma$$

Hence we have

$$\frac{\partial \mathcal{L}_0}{\partial s}(\Omega_s, v, U, \lambda)|_{s=0} = \int_\Omega \alpha D(-DU.V)..D\lambda + D(-DU.V)v\lambda +$$

$$+\beta div(-DU.V)\ div\lambda + \beta di\dot{v}(div(-DU.V)E) + \alpha div(D^*(DU.V)\lambda) +$$

$$+\beta div(div(DU.V)\lambda)\ dx \ + \int_{\partial S} \{\alpha DU..D\lambda + \beta divU\ div\lambda + < DU.v - f, \lambda > +$$

$$+\beta div(divU\ E) - \alpha div(D^*U\lambda) \ + -\beta div(DivU\lambda)\} < V, n > \ d\Gamma +$$

$$- \int_\Omega F(-DR.V, \lambda)\ dx \ - \int_\Gamma F(R, \lambda)V.n\ d\Gamma$$

Let us return in the variable u. We set $\eta = Du.V$. We have $SuppV \subset D$. By the Green's theorem we rewrite the following distributed integrals

$$-\int_\Omega \alpha D\eta..D\lambda dx = \int_\Omega \alpha \Delta\lambda\eta\, dx - \int_{\partial S} \alpha D\lambda.n\eta\, d\Gamma$$

$$-\int_\Omega D\eta v\lambda dx = \int_\Omega < D\lambda.v + divv\lambda, \eta > dx - \int_{\partial S} v.n\lambda\eta\, d\Gamma$$

$$-\int_\Omega \beta div\eta\, div\lambda\, dx = \int_\Omega \beta\nabla(div\lambda)\eta\, dx - \int_{\partial S} \beta div\lambda\eta.n\, d\Gamma$$

Consequently the derivative leads to

$$\frac{\partial \mathcal{L}_0}{\partial s}(\Omega_s, v, u, \lambda)|_{s=0} = \int_\Omega < \alpha\Delta\lambda + D\lambda.v + divv\lambda + \beta\nabla(div\lambda), \eta > dx +$$

$$\int_{\partial S} \{\alpha Du..D\lambda + \beta divu\, div\lambda + < Du.v - f, \lambda > + \beta div(divu\, E) +$$

$$-\alpha div(D^* u\lambda) - \beta div(divu\lambda)\} < V, n > \; - v.n\lambda - \beta div\lambda\, n - \alpha D\lambda.n, \eta > \; d\Gamma +$$

$$+ \int_\Gamma -\beta div\eta < E, n > + \alpha D^*\eta\lambda.n + \beta div\eta\lambda.n\, d\Gamma$$

We shall use the following lemma to simplify the expression of the derivative.

Lemma 14.3. *Let A and B be two vectors and l a scalar. Then we have*
$div(DAB) = \nabla(divA) + DA..D^*B$
$div(D^*A\, B) = \Delta A.B + DA..DB$
$div(divA\, B) = \nabla(divA).B + divAdivB$
$div(l\, A) = \nabla l.A + ldivA$

Proof. $div(DAB) = \partial_i(DAB)_i = \partial_i((DA)_{ik}B_k) = \partial_i(\partial_k A_i\, B_k) = \partial^2_{ik} A_i\, B_k + \partial_k A_i\, \partial_i B_k$ $div(D^*AB) = \partial_i(D^*AB)_i = \partial_i((D^*A)_{ik}B_k) = \partial_i(\partial_i A_k\, B_k) = \partial^2_{ii} A_k\, B_k + \partial_i A_k\, \partial_i B_k$
$div(divA\, B) = \partial_i(\partial_j A_j B_i) = \partial^2_{ij} A_j B_i + \partial_i B_i \partial_j A_j$ $div(l\, A) = \partial_i(lA_i) = \partial_i l A_i + l\, \partial_i A_i$

Using the lemma 14.3 $div(D^*u\lambda) = \Delta u.\lambda + Du..D\lambda$

And λ is solution of (14.38) so

$$\frac{\partial \mathcal{L}_0}{\partial s}(\Omega_s, v, u, \lambda)|_{s=0} = \int_{\partial S} \{- < v.n\lambda + \beta div\lambda.n + \alpha D\lambda.n, \eta > +$$

$$+ \beta div(divu\, E) - \alpha\Delta u\lambda - \beta\nabla(divu)\lambda - f\lambda\} < V, n > \, d\Gamma +$$

$$+ \int_\Gamma \beta div\eta(\lambda.n - < E, n >) + \alpha D^*\eta\lambda.n\, d\Gamma$$

With the help of lemma 14.4 we can simplify on Γ the divergence $divu$ and the matrix Du.

Lemma 14.4. *We have on Γ $divu =< Du.n, n >$ and $Du = Du.n.n^*$*

Proof. As on Γ $divu = div_\Gamma u + < Du.n, n >= div_\Gamma u + < \epsilon(u).n, n >$, the matrix $Du = D_\Gamma u + Du.n.n^*, u$ is constant on Γ then we have $div_\Gamma u = 0$ and $D_\Gamma u = 0$ Hence we get on Γ $divu =< Du.n, n >$ and $Du = Du.n.n^*$.

Using the previous lemma we get $\eta|_\Gamma = Du.n < V, n >$ so

$$\frac{\partial \mathcal{L}_0}{\partial s}(\Omega_s, v, u, \lambda)|_{s=0} =$$

$$\int_{\partial S} \{\beta divu\, divE - \alpha \Delta u \lambda + \beta \nabla(divu)(\lambda - E) + < v.n\, Du.n - f, \lambda >\} < V, n > +$$
$$- < v.n\, Du.n - f, \lambda >\} < V, n > - < v.n\lambda + \beta div\lambda.n + \alpha > D\lambda.n, \eta > \, d\Gamma +$$
$$+ \int_\Gamma \beta div\eta(\lambda.n - < E, n >) + \alpha < D\eta.n, \lambda > \, d\Gamma$$

But on Γ we can write $div\eta =< D\eta.n, n > + div_\Gamma \eta$ thus we have

$$\frac{\partial \mathcal{L}_0}{\partial s}|_{s=0}(\Omega_s, v, u, \lambda) =$$
$$= \int_{\partial S} \{\beta \nabla(divu)(E - \lambda) + \beta divu divE + < v.n\, Du.n - \alpha \Delta u - f, \lambda > +$$
$$- < v.n\lambda + \beta div\lambda.n + D\lambda.n, Du.n >\} < V, n > d\Gamma +$$
$$+ \int_\Gamma \beta div_\Gamma \eta(\lambda.n - < E, n >) + (\alpha + \beta) < D\eta.n, n > \lambda.n$$
$$- \beta < D\eta.n, n >< E, n > \, d\Gamma$$

And with the boundaries conditions of λ we obtain

$$\frac{\partial \mathcal{L}_0}{\partial s}(\Omega_s, v, u, \lambda)|_{s=0} =$$
$$= - \int_\Gamma \frac{\alpha\beta}{\alpha + \beta} div_\Gamma \eta < E, n > d\Gamma + \int_{\partial S} \{< \beta div\lambda.n + D\lambda.n, Du.n > +$$
$$- \beta \nabla(divu)E - \beta divu divE\} < V, n > \, d\Gamma$$

But by parts integration on Γ it follows that

$$\int_\Gamma -div_\Gamma \eta < E, n > \, d\Gamma = \int_\Gamma \nabla_\Gamma(< E, n >)\eta - H < E, n >< \eta, n > D\Gamma$$

But we have

$$\int_\Gamma -div_\Gamma \eta < E, n > \, d\Gamma = \int_\Gamma < (D^2 b - Hnn^*)E, \eta > D\Gamma$$

Thus we obtain

$$\frac{\partial \mathcal{L}_0}{\partial s}(\Omega_s, v, u, \lambda)|_{s=0} = \int_{\partial S} \{\beta\nabla(divu)e + \frac{\alpha\beta}{\alpha+\beta} < (D^2b - Hnn^*)\, e, Du.n > +$$

$$-(\beta div\lambda + \alpha < D\lambda.n, n >) < Du.n, n > -$$

$$+\alpha < (D\lambda.n)_\Gamma, Du.n >\} < V, n > \, d\Gamma$$

But on ∂S

$$div\lambda = div_\Gamma \lambda_\Gamma + div_\Gamma(< \lambda, n > n) + < D\lambda.n, n >=$$

$$= \frac{\beta}{\alpha+\beta} H < e, n > + < D\lambda.n, n >$$

So we get

$$\frac{\partial \mathcal{L}_0}{\partial s}(\Omega_s, v, u, \lambda)|_{s=0} =$$

$$\int_{\partial S} \{\beta\nabla(divu)e + \frac{\alpha\beta}{\alpha+\beta} < (D^2b - Hnn^*)\, e, Du.n > +$$

$$-(\beta+\alpha) < D\lambda.n, n >< Du.n, n > +$$

$$- < \frac{\beta^2}{\alpha+\beta} Hnn^*e + \alpha(D\lambda.n)_\Gamma, Du.n >\} < V, n > \, d\Gamma$$

We notice that $\nabla(divu)|_\Gamma$ is a second order derivative at the boundary. We can see the geometrical terms deriving from the oriented distance $b(x)$ on Γ, $D^2b - Hnn^*$. We recall that the normal vector $n = \nabla b$, the curvatures' matrix $D^2b = D(\nabla b)$ and the mean curvature $H = \Delta b = Tr(D^2b)$ in the neighborhood of the boundary.

Expression of the gradient by introduction of the convex set K where the adjoint λ lies. We use to the Lagrangian functional defined in the previous section but we maximize $\mathcal{L}_0(\Omega, v, u, \lambda))$ with the adjoint state $\lambda \in K(\Omega)$. In the above section, we take $\lambda \in H_0^1(D, R^3)$ Hence we happen to have an extremal problem on a convex set:

$$F_e(\Omega) = \min_{u \in H(\Omega)} \max_{\lambda \in K(\Omega)} \mathcal{L}_0(\Omega, v, u, \lambda)$$

We also obtain an unique saddle point $(u, \lambda) \in H(\Omega) \times K(\Omega)$. And we use to the derivation of the $min - max$ of Cuer-Zolesio.

Proposition 14.8. *There exists an unique solution $\lambda \in K(\Omega)$ to the following problem in Ω*

$$-\alpha\Delta\lambda - D\lambda.v - divv\lambda - \beta\nabla(div\lambda) = 0$$

Proof. Assume that $\varphi \in K(\Omega)$

Let the bilinear form be

$$a(\lambda,\varphi) = \int_\Omega \alpha D\lambda..D\varphi - D\lambda.v\varphi - divv\lambda\varphi + \beta div\lambda div\varphi \, dx +$$

$$- \int_\Gamma \alpha D\lambda.n\varphi + \beta div\lambda\varphi.n \, d\Gamma$$

From Stampacchia's theorem, there exists an unique $\lambda \in K(\Omega)$ such that for all $\varphi \in K(\Omega)$ $a(\lambda, \varphi - \lambda) \geq 0$

We set $\varphi = \lambda - \phi$, for all $\phi \in \mathcal{D}(\Omega, R^3)$. $\lambda \in K(\Omega)$ hence $\varphi \in K(\Omega)$

So $a(\lambda, \varphi - \lambda) = a(\lambda, \phi) \geq 0$ But if we set $\phi = -\phi$ then $a(\lambda, \varphi - \lambda) = -a(\lambda, \phi) \geq 0$. As a consequence we get $a(\lambda, \phi) = 0$ for all $\phi \in \mathcal{D}(\Omega, R^3)$.

$$a(\lambda,\phi) = \int_\Omega \alpha D\lambda..D\phi - D\lambda.v\phi - divv\lambda\phi + \beta div\lambda \; div\phi \, dx$$

Therefore

$$a(\lambda,\phi) = \int_\Omega < -\alpha\Delta\lambda - D\lambda.v - divv\lambda - \beta\nabla(div\lambda), \phi > dx = 0$$

for all $\phi \in \mathcal{D}(\Omega, R^3)$. Finally there exists an unique $\lambda \in K(\Omega)$ such as $-\alpha\Delta\lambda - D\lambda.v - divv\lambda - \beta\nabla(div\lambda) = 0$ in Ω.

The expression of the gradient $dF_e(\Omega, V)$ is equal to

Proposition 14.9. *Assume that* $(u, \lambda) \in H^1(\Omega, R^3) \times K(\Omega)$ *is the unique saddle point. We set*

$$g^1(\Gamma) = \alpha < Du.n, (D^*\lambda.n)_\Gamma - 2(D\lambda.n)_\Gamma > -$$

$$+(\alpha+\beta) < D\lambda.n, n >< Du.n, n >$$

and

$$g^2(\Gamma) = \beta < D_\Gamma(Du.n)e_\Gamma, n > + \frac{\alpha\beta}{\alpha+\beta}(< e, n > div_\Gamma(Du.n)$$

and

$$g^{cur} = \frac{\beta^2 + 2\alpha\beta}{\alpha+\beta} < (D^2b - Hnn^*)e, Du.n > - \frac{\beta}{\alpha+\beta} < e, n >< f, n >$$

then

$$dF_e(\Omega, V) = \int_{\partial S} \{g^1(\Gamma) + g^2(\Gamma) + g^{cur}(\Gamma)\} < V, n > \, d\Gamma$$

Let the functional be $\mathcal{L}(\Omega, v, u, \lambda)$ defined by

$$\mathcal{L}(\Omega, v, u, \lambda) = \int_\Omega \alpha Du..D\lambda + < Du.v - f, \lambda > + \beta divu \, div\lambda \, dx$$

then we have the following lemma

Lemma 14.5.

$$F_e(\Omega) = \min_{u \in H(\Omega)} \max_{\lambda \in K(\Omega)} \mathcal{L}(\Omega, v, u, \lambda)$$

Proof. We know that $divu|_\Gamma =< Du.n, n >$ and as $\lambda \in K(\Omega)$ then

$\int_\Gamma \alpha < Du.n, \lambda > + \beta divu \lambda.n \, d\Gamma = \int_\Gamma (\alpha + \beta)\lambda.n < Du.n, n > d\Gamma = \int_\Gamma \beta < E, n >< Du.n, n > d\Gamma = F_e(\Omega)$

Therefore

$$\min_{u \in H(\Omega)} \max_{\lambda \in K(\Omega)} \mathcal{L}_0(\Omega, v, u, \lambda) = \min_{u \in H(\Omega)} \max_{\lambda \in K(\Omega)} \mathcal{L}(\Omega, v, u, \lambda)$$

But the convex set $K(\Omega)$ is also defined by

Definition 14.4. $K(\Omega) = H_0^1(\Omega, R^3) \oplus < \frac{\beta}{\alpha+\beta} E, n > n$

Consequently $\lambda \in K(\Omega)$ leads to

$$\lambda = \lambda^0 + \lambda^1$$

where $\lambda^0 \in H_0^1(\Omega, R^3)$ and $\lambda^1 = \frac{\beta}{\alpha+\beta} < E, \mathcal{N} > \mathcal{N}$ in Ω with $\mathcal{N}$ which is an extension of the normal vector field n in the neighborhood of Γ We assume that $\mathcal{N} = \frac{1}{\|\mathcal{N}\|}\mathcal{N}$ in the neighborhood of Γ.

We denote in Ω_s

$$\lambda_s^c = \lambda^0 \circ T_s^{-1} + \lambda_s^1 \tag{14.40}$$

Consequently $\lambda_s^c|_{s=0} = \lambda$ and we have the following proposition

Proposition 14.10.

$$\lambda_s^c \in K(\Omega_s) \iff \phi_s = \lambda_s^c - \frac{\beta}{\alpha + \beta} < E, \mathcal{N}_s > \mathcal{N}_s \in H_0^1(\Omega_s, R^3)$$

$$\phi_s \in H_0^1(\Omega_s, R^3) \iff \phi_s \circ T_s \in H_0^1(\Omega, R^3)$$

Proof. We have $X = T_s(V)(x)$, $\Omega_s = T_s(V)(\Omega)$ and $\Gamma_s = T_s(V)(\Gamma)$

If $\lambda_s^c \in K(\Omega_s)$ then $< \lambda_s^c, n_s > = < \frac{\beta}{\alpha+\beta} E, n_s >$ on Γ_s and $(\lambda_s^c)_{\Gamma_s} = 0$
We know that $\lambda^0 \in H_0^1(\Omega, R^3)$ consequently $\lambda^0 \circ T_s^{-1} \in H_0^1(\Omega_s, R^3)$ with the transport lemma $\phi_s = \lambda_s^c - \frac{\beta}{\alpha+\beta} < E, \mathcal{N}_s > \mathcal{N}_s = \lambda^0 \circ T_s^{-1} + \lambda_s^1 - \frac{\beta}{\alpha+\beta} < E, \mathcal{N}_s > \mathcal{N}_f = \lambda^0 \circ T_s^{-1}$. Thus $\phi_s \in H_0^1(\Omega_s, R^3)$

If $\phi_s \in H_0^1(\Omega_s, R^3)$ then $\lambda_s^c = \phi_s + \frac{\beta}{\alpha+\beta} < E, \mathcal{N}_s > \mathcal{N}_s \in K(\Omega_s)$.
Indeed $(T_s, T_s^{-1}, \mathcal{N}_s)$ is smooth enough to get $\lambda_s^c \in H^1(\Omega_s, R^3))$ and $< \lambda_s^c, n_s > = < \frac{\beta}{\alpha+\beta} E, n_s > n_s$ and $(\lambda_s^c)_{\Gamma_s} = 0$

Using the transport lemma if $\phi_s \in H_0^1(\Omega_s, R^3)$ then $\phi_s \circ T_s \in H_0^1(\Omega, R^3)$ and it's also right in the another way. So we obtain the result above.

The expression of the derivative of λ_s^c is given by

Lemma 14.6. *In Ω*

$$(\lambda_s^c)' = -D\lambda^0.V + < \frac{\beta}{\alpha+\beta} E, \mathcal{N} > (\mathcal{N}_s)' + < \frac{\beta}{\alpha+\beta} E, (\mathcal{N}_s)' > \mathcal{N}$$

On Γ

$$(\lambda_s^c)' = -D\lambda^0.V + < \frac{\beta}{\alpha+\beta} E, n > (\mathcal{N}_s)' + < \frac{\beta}{\alpha+\beta} E, (\mathcal{N}_s)' > n$$

with $(\mathcal{N}_s)' = (n_s)' = \frac{1}{\|\mathcal{N}\|}\left(\frac{1}{\|\mathcal{N}\|^2}(< D^*V\mathcal{N}, \mathcal{N} > \mathcal{N} + < D\mathcal{N}.V, \mathcal{N} > \mathcal{N}) - D^*V.\mathcal{N} - D\mathcal{N}.V\right) = -((D^*Vn)_\Gamma + D^2bV_\Gamma)$ *in the neighborhood of Γ.*

We have $\mathcal{N}_s = (\|(A_s^*)^{-1}\mathcal{N}\|^{-1}(A_s^*)^{-1}\mathcal{N}) \circ T_s^{-1}$ in the neighborhood of Γ_s with $\mathcal{N} = \nabla b = n$ on Γ.

We use to the following corollary to obtain the derivative formulation of $\mathcal{N}_s$ and λ_s^c.

Corollary 14.3.

$$\frac{\partial}{\partial s}(T_s^{-1}) = -(DT_s)^{-1} \circ T_s^{-1} V$$

$$(DT_s \circ T_s^{-1})' = DV$$

$$(D^*T_s^{-1} \circ T_s^{-1})' = -D^*V$$

$$\frac{\partial \|D^*T_s^{-1}.\mathcal{N}\|_{R^3}^{-1} \circ T_s^{-1}}{\partial s}|_{s=0} = \frac{1}{\|\mathcal{N}\|^3}(< D^*V\mathcal{N}.\mathcal{N} > + < D\mathcal{N}.V, \mathcal{N} >)$$

Proof. We have $\frac{\partial}{\partial s}(T_s \circ T_s^{-1}) = 0$ and $\frac{\partial}{\partial s}(T_s \circ T_s^{-1}) = (V(s,X) \circ T_s) \circ T_s^{-1} + (DT_s)^{-1} \circ T_s^{-1} \frac{\partial}{\partial s}(T_s^{-1})$ thus

$$\frac{\partial}{\partial s}(T_s^{-1}) = -(DT_s)^{-1} \circ T_s^{-1} V(s,X) \tag{14.41}$$

We have $D(T_s \circ T_s^{-1}) = Id = (DT_s) \circ T_s^{-1}.D(T_s^{-1}) = D(T_s^{-1}) \circ T_s.DT_s$ so we deduce that

$$(DT_s) \circ T_s^{-1} = (D(T_s^{-1}))^{-1} \tag{14.42}$$

$$(D^*T_s)^{-1} \circ T_s^{-1} = D^*(T_s^{-1}) \tag{14.43}$$

And we know that $B_s B_s^{-1} = Id$ hence $\frac{\partial}{\partial s}(B_s B_s^{-1}) = \frac{\partial}{\partial s}(B_s)B_s^{-1} + B_s \frac{\partial}{\partial s}(B_s^{-1}) = 0$. It follows that

$$\frac{\partial}{\partial s}(B_s^{-1}) = -B_s^{-1}\frac{\partial}{\partial s}(B_s)B_s^{-1} \tag{14.44}$$

So $(D(T_s^{-1}))^{-1})' = -Id.(D(T_s^{-1}))'.Id = DV$ and $(D^*(T_s^{-1}))' = -D^*V$
And we also have

$$\frac{\partial \|D^*T_s^{-1}.\mathcal{N}\|_{R^3}^2 \circ T_s^{-1}}{\partial s}|_{s=0} = -2 < D^*V\mathcal{N}, \mathcal{N} > -2 < D\mathcal{N}.V, \mathcal{N} >$$

and

$$\frac{\partial \|D^*T_s^{-1}.\mathcal{N}\|_{R^3}^2 \circ T_s^{-1}}{\partial s}|_{s=0} = 2\|\mathcal{N}\|_{R^3}\frac{\partial \|D^*T_s^{-1}.\mathcal{N}\|_{R^3} \circ T_s^{-1}}{\partial s}|_{s=0}$$

So we obtain that

$$\frac{\partial \|D^*T_s^{-1}.\mathcal{N}\|_{R^3} \circ T_s^{-1}}{\partial s}|_{s=0} = -\frac{1}{\|\mathcal{N}\|}(< D^*V\mathcal{N}, \mathcal{N} > + < D\mathcal{N}.V, \mathcal{N} >)$$

If we denote$l = \|D^*T_s^{-1}.\mathcal{N}\|_{R^3} \circ T_s^{-1}$ then $(l^{-1})' = -\frac{l'}{l^2}$.

Hence we get $(\mathcal{N}_s)' = \frac{1}{\|\mathcal{N}\|}(\frac{1}{\|\|^2}(< D^*V\mathcal{N}, \mathcal{N} > \mathcal{N} + < D\mathcal{N}.V, \mathcal{N} > \mathcal{N}) - D^*V.\mathcal{N} - D\mathcal{N}.V)$ in the neighborhood of Γ. Thus we obtain the formulation of the derivative $(\mathcal{N}_s)'|_\Gamma$.

$$(\mathcal{N}_s)'|_\Gamma = (n_s)' = -((D^*Vn)_\Gamma + D^2bV_\Gamma) = -\nabla_\Gamma(< V, n >) \tag{14.45}$$

Let $U = u - R$ be in Ω, thus $U = \begin{cases} u - R \text{ in } \Omega \setminus \overline{B} \\ u \text{ in } \overline{B} \end{cases}$ where R is defined as in (14.3). Set

$$F_1(R, \lambda) = \alpha D\lambda..DR - < DR.v, \lambda > -\beta divR\, div\lambda$$

By changing variables we have

$$\mathcal{L}(\Omega_s, v, u, \lambda) = \mathcal{L}(\Omega_s, v, U, \lambda) =$$

$$\int_{\Omega_s} \alpha D(U \circ T_s^{-1})..D\lambda_s^c + < D(U \circ T_s^{-1}).v - f, \lambda_s^c > +$$

$$+ \beta div(U \circ T_s^{-1}) div\lambda_s^c - F_1(R \circ T_s^{-1}, \lambda_s^c)\, dx$$

$(U_v, \lambda_v) \in H_0^1 \times K(\Omega)$ are the unique solution of respectively (14.4) and (14.38) so this problem has a unique saddle point (U_v, λ_v) for each Ω and v then from the derivation of a min max of Cuer-Zolesio we have

$$dF_e(\Omega, V) = \frac{\partial \mathcal{L}}{\partial s}(\Omega_s, v, U_v, \lambda_v)|_{s=0}$$

The expression of the shape gradient is obtained by computing the derivative of an integral over a moving domain: We denote $\lambda_s^c|_{s=0}$ by λ^c, (U_v, λ_v) by (U, λ) hence

$$\frac{\partial \mathcal{L}_0}{\partial s}(\Omega_s, v, U, \lambda)|_{s=0} =$$

$$\int_\Omega \alpha D(-DU.V)..D\lambda^c + D(-DU.V)v\lambda^c + \beta div(-DU.V)\, div\lambda^c\, dx\ +$$

$$+ \int_\Omega \alpha DU..D(\lambda_s^c)' + < DU.v - f, (\lambda_s^c)' > + \beta divU div(\lambda_s^c)'\, dx\ +$$

$$+ \int_\Gamma \{\alpha DU..D\lambda^c + DU.v\lambda^c + \beta divU\ div\lambda^c - f\lambda^c\} < V, n > d\Gamma$$

We return in the variable u. We set $\eta = Du.V$. And by the Green's theorem we get

$$\frac{\partial \mathcal{L}_0}{\partial s}(\Omega_s, v, u, \lambda)|_{s=0} =$$

$$\int_\Omega < \alpha\Delta\lambda^c + D\lambda^c.v + divv\lambda^c + \beta\nabla(div\lambda^c), \eta > dx\ +$$

$$+ \int_\Omega < -\alpha\Delta u + Du.v - \beta\nabla(divu) - f, (\lambda_s^c)' > dx\ +$$

$$+ \int_\Gamma \{\alpha Du..D\lambda^c + \beta divu\ div\lambda^c + < Du.v - f\lambda^c >\} < V, n > +$$

$$< -v.n\lambda^c - \beta div\lambda^c\, n - \alpha D\lambda^c.n, \eta > + < \alpha Du.n + Du.v + \beta divu\, n, (\lambda_s^c)' > d\Gamma$$

But $\lambda^c = \lambda$ and λ is solution of (14.38) and u of (11.1), thus we get

$$\frac{\partial \mathcal{L}_0}{\partial s}(\Omega_s, v, u_v, \lambda_v)|_{s=0} =$$

$$\int_\Gamma \{\alpha Du..D\lambda + \beta divu\ div\lambda + < Du.v - f\lambda\} < V, n > +$$

$$< -v.n\lambda - \beta div\lambda\ n - \alpha D\lambda.n, \eta > +$$

$$< \alpha Du.n + Du.v + \beta divu\ n, (\lambda_s^c)' > d\Gamma$$

We use the following corollary and lemma in order to simplify the expression of the gradient: we reduce the expression of differential matrix of vectors and the gradient of a scalar term.

Corollary 14.4. *Let A, B be two vectors and l be a scalar then*

$$D(l\ A) = A.\nabla^* l + l\ DA$$

$$\nabla < A, B >= D^* A.B + D^* B.A$$

$$D(DB.A) = D(DB).A + DB.DA$$

$$D(D^* B.A) = D(D^* B).A + D^* B.DA$$

$$D^*(DB.A) = D^*(DB).A + D^* A.D^* B$$

where we denote the third order tensor

$$D(DB)_{ijk} = \partial^2_{jk} B_i,\ D(D^* B)_{ijk} = \partial^2_{ji} B_k$$

and $D^*(DB)_{ijk} = \partial^2_{ik} B_j$

Proof. We have

$$D(l\ A)_{ij} = \partial_j(lA_i) = \partial_j l A_i + l\ \partial_j A_i$$

$$(\nabla < A, B >)_i = \partial_i(A_j B_j) = B_j \partial_i A_j + A_j \partial_i B_j$$

$$D(DB.A)_{ijk} = \partial_j(\partial_k B_i A_k) = \partial^2_{jk} B_i A_k + \partial_k B_i \partial_j A_k$$

$$D(D^* B.A)_{ijk} = \partial_j(\partial_i B_k A_k) =$$

$$\partial^2_{ji} B_k A_k = +\partial_i B_k \partial_j A_k D^*(DB.A)_{ijk} = \partial_i(\partial_k B_j A_k) =$$

$$\partial^2_{ik} B_j A_k = +\partial_k B_j \partial_i A_k$$

We simplify the expression of the following differential matrix on ∂S.

Lemma 14.7. *We have on ∂S*

$$D(< e, n > n) = (n\ e^* + < e, n > Id) D^2 b$$

$$D_\Gamma(< e, n > n) = (n\ e^* + < e, n > Id) D^2 b$$

$$D_\Gamma(e_\Gamma) = -(n\ e^* + < e, n > Id) D^2 b$$

Proof. Using the previous corollary we obtain the results above so

$$D(< e,n > n) = n.\nabla^*(< e,n >)+ < e,n > D^2b = n\, e^* D^2b+ < e,n > D^2b$$

$$D_\Gamma(< e,n > n) = D(< e,n > n) - D(< e,n > n)nn^* =$$
$$= n\, e^* D^2b+ < e,n > D^2b - (n\, e^* D^2b+ < e,n > D^2b)nn^* =$$
$$= n\, e^* D^2b+ < e,n > D^2b$$
$$D_\Gamma(e_\Gamma) = D_\Gamma e - D_\Gamma(< e,n > n) = -n\, e^* D^2b- < e,n > D^2b$$

We have on ∂S

$$D\lambda.V = D_\Gamma\lambda^0.V + D\lambda^0.nn^*.V =< V,n > D\lambda^0.n =$$
$$=< V,n > D\lambda.n- < V,n > D\lambda^1.n = D\lambda.n < V,n >$$

But the derivative of λ_s^c on Γ is given by

$$(\lambda_s^c)'|_\Gamma = -D\lambda.V- < \frac{\beta}{\alpha+\beta}E,n > \nabla_\Gamma(< V,n >)-$$
$$< \frac{\beta}{\alpha+\beta}E_\Gamma, \nabla_\Gamma(< V,n >) > n$$

It follows that on ∂S

$$(\lambda_s^c)' = - < V,n > D\lambda.n- < \frac{\beta}{\alpha+\beta}e,n > \nabla_\Gamma(< V,n >)-$$
$$< \frac{\beta}{\alpha+\beta}e_\Gamma, \nabla_\Gamma(< V,n >) > n \qquad (14.46)$$

But $div u|_\Gamma =< Du.n,n >$, also $Du|_\Gamma = Du.nn^*$ and using the previous expression of $(\lambda_s^c)'$ then

$$\frac{\partial \mathcal{L}_0}{\partial s}(\Omega_s,v,u,\lambda)|_{s=0} =$$
$$= \int_\Gamma \{\alpha < Du.n, D^*\lambda.n - 2D\lambda.n > -\beta < D\lambda.n,n >< Du.n,n > +$$
$$- f\lambda\} < V,n > - < \alpha Du.n, \frac{\beta}{\alpha+\beta} < e,n > \nabla_\Gamma(< V,n >) > +$$
$$-div u < \beta e, \nabla_\Gamma(< V,n >) > d\Gamma.$$

Integrating by parts on Γ the boundary integrals become

$$\int_\Gamma div u < e_\Gamma, -\nabla_\Gamma(< V,n >) > d\Gamma =$$

$$\int_\Gamma div_\Gamma(div u\; e_\Gamma) < V,n > -H div u\; e_\Gamma.n < V,n > \; d\Gamma =$$

$$= \int_\Gamma \{ < \nabla_\Gamma(divu), e_\Gamma > - H < e, n > divu \} < V, n > d\Gamma =$$

$$= \int_\Gamma \{ < D^2 b Du.n, e_\Gamma > + < D_\Gamma(Du.n) e_\Gamma, n > +$$
$$- H < e, n > < Du.n, n > \} < V, n > d\Gamma =$$

$$= \int_\Gamma \{ < (D^2 b - Hnn^*) e, Du.n > + < D_\Gamma(Du.n) e_\Gamma, n > \} < V, n > d\Gamma$$

We also have

$$\int_\Gamma < e, n >< Du.n, -\nabla_\Gamma(< V, n >) > d\Gamma =$$

$$\int_\Gamma div_\Gamma(< e, n > Du.n) < V, n > +$$
$$- H < Du.n, n > < e, n > < V, n > d\Gamma$$

$$= \int_\Gamma \{ < e, n > div_\Gamma(Du.n) + < (D^2 b - Hnn^*) e, Du.n > \} < V, n > d\Gamma$$

Finally with the new expressions of the boundaries integrals we obtain

$$dF_e(\Omega, V) =$$

$$\int_{\partial S} \{ \alpha < Du.n, (D^* \lambda.n)_\Gamma - 2(D\lambda.n)_\Gamma > +$$
$$-(\alpha + \beta) < D\lambda.n, n >< Du.n, n > + \beta < (D^2 b - Hnn^*) e, Du.n > +$$
$$\beta < D_\Gamma(Du.n) e_\Gamma, n > + \frac{\alpha\beta}{\alpha + \beta} (< e, n > div_\Gamma(Du.n) +$$
$$+ < (D^2 b - Hnn^*) e, Du.n > - < e, n >< \frac{f}{\alpha}, n >) \} < V, n > d\Gamma$$

We notice that we can write on Γ $div_\Gamma(Du.n) = div_\Gamma((Du.n)_\Gamma) + \nabla_\Gamma(< Du.n, n >).n + H < Du.n, n >= div_\Gamma((Du.n)_\Gamma) + H < Du.n, n > D_\Gamma(Du.n)$ and $div_\Gamma(Du.n)$ give us tangential second order derivatives of u.

The totally distributed expression of the gradient derived by change of variable. We maximize $\mathcal{L}_0(\Omega, v, u, \lambda)$ with $\lambda \in K(\Omega)$ as in the previous section but we calculate the gradient in the distributed sense. First we do a change of function and then a change of variable $X = T_s(V)(x)$ in order to obtain a fixed domain. We denote $j = detDT_s$.

Let $U = u - R$ be in Ω, thus $U = \begin{cases} u - R & \text{in } \Omega \setminus \overline{B} \\ u & \text{in } \overline{B} \end{cases}$ where R is defined as in (14.3). F_1 is given in the previous section. First we rewrite the formulation with U thus

$$\mathcal{L}(\Omega_s, v, u, \lambda) = \mathcal{L}(\Omega_s, v, U, \lambda) =$$

$$\int_{\Omega_s} \alpha D(U \circ T_s^{-1})..D\lambda_s^c + D(U \circ T_s^{-1}).v.\lambda_s^c + \beta div(U \circ T_s^{-1}) div\lambda_s^c +$$

$$-F_1(R \circ T_s^{-1}, \lambda_s^c)\, dx$$

Using the expression of $(\lambda_{gs}^c \circ T_s)'$ given in the following lemma we compute the derivative of $\mathcal{L}(\Omega_s)$

Lemma 14.8. *We have in Ω*

$$(\lambda_s^c \circ T_s)' = < \frac{\beta}{\alpha+\beta} E, (\mathcal{N}_s \circ T_s)' > \mathcal{N} + < \frac{\beta}{\alpha+\beta} E, \mathcal{N} > (\mathcal{N}_s \circ T_s)'$$

with

$$(\mathcal{N}_s \circ T_s)' = < D^*V\mathcal{N}, \mathcal{N} > \mathcal{N} - D^*V.\mathcal{N}$$

Proof. We have in Ω

$$\lambda_s^c \circ T_s = \lambda^0 + \lambda^1 \circ T_s$$

Hence

$$(\lambda_s^c \circ T_s)' = (\lambda^1 \circ T_s)' = < \frac{\beta}{\alpha+\beta} E, (\mathcal{N}_s \circ T_s)' > \mathcal{N} + < \frac{\beta}{\alpha+\beta} E, \mathcal{N} > (\mathcal{N}_s \circ T_s)'$$

From the lemma 14.6 and the corollary 14.3 we get $(\mathcal{N}_s \circ T_s)' = < D^*V\mathcal{N}, \mathcal{N} > \mathcal{N} - D^*V.\mathcal{N}$ And we obtain the result above

Proposition 14.11. *Assume that $(u, \lambda) \in H(\Omega) \times K(\Omega)$ is the unique saddle point then*

$$dF_e(\Omega, V) = \int_\Omega -2\alpha(Du.\epsilon(V))..D\lambda - < Du.DV.v, \lambda > +$$

$$+\alpha Du..D(\lambda_s^c \circ T_s)' + Du.v.(\lambda_s^c \circ T_s)' + \beta divu\; div(\lambda^c \circ T_s)' - f(\lambda_s^c \circ T_s)' +$$

$$-\beta(divuD\lambda..D^*V + div\lambda Du..D^*V)\, dx$$

Proof. We get $D(U \circ T_s^{-1}) = DU \circ T_s^{-1}.D(T_s^{-1})$ from the derivation of composed functions. Since $D(T_s^{-1}) \circ T_s = (DT_s)^{-1}$ we have

$$\mathcal{L}(\Omega_s) = \int_{\Omega_s} \alpha DU \circ T_s^{-1} D(T_s^{-1})..D\lambda_s^c + DU \circ T_s^{-1} D(T_s^{-1}).v_s.\lambda_s^c +$$

$$+\beta(j)^{-2} div(j(DT_s)^{-1}U)\circ T_s^{-1} div(j(DT_s)^{-1}\lambda_s^c \circ T_s)\circ T_s^{-1} - F_1(R\circ T_s^{-1}, \lambda_s^c)\, dx$$

By change of variable (we apply the transport T_s) we have

$$\mathcal{L}(\Omega_s, v, U, \lambda) = \int_{\Omega} \{\alpha DU D(T_s^{-1})\circ T_s..(D\lambda_s^c)\circ T_s + DU D(T_s^{-1})\circ T_s.v_s.\lambda_s^c \circ T_s +$$

$$+\beta(j)^{-2} div(j(DT_s)^{-1}U) div(j(DT_s)^{-1}\lambda_s^c \circ T_s) - F_1(R, \lambda_s^c \circ T_s)\ \} j\, dx$$

So

$$\mathcal{L}(\Omega_s, v, U, \lambda) =$$

$$\int_{\Omega} \{\alpha DU(DT_s)^{-1}..D(\lambda_s^c \circ T_s)(DT_s)^{-1} + DU(DT_s)^{-1}.v_s.\lambda_s^c \circ T_s +$$

$$+\beta(j)^{-2} div(j(DT_s)^{-1}U) div(j(DT_s)^{-1}\lambda_s^c \circ T_s) - F_1(R, \lambda_s^c \circ T_s)\ \} j\ \ dx$$

And $\lambda_s^c \circ T_s = \lambda^0 + < \frac{\beta}{\alpha+\beta} E, \mathcal{N}_s \circ T_s > \mathcal{N}_s \circ T_s$ in Ω

$$\mathcal{L}(\Omega_s, v, U, \lambda) =$$

$$= \int_{\Omega} \{\alpha DU(DT_s)^{-1}..D(\lambda_s^c \circ T_s)(DT_s)^{-1} + DU.(DT_s)^{-1} v_s.\lambda_s^c \circ T_s - f\lambda_s^c \circ T_s$$

$$+DR.(DT_s)^{-1} v_s.DT_s\lambda_s^c \circ T_s + \alpha DR(DT_s)^{-1}..D(\lambda_s^c \circ T_s)(DT_s)^{-1}\ \} j +$$

$$+(j)^{-1}\{\beta div(j(DT_s)^{-1}R) div(jDT_s^{-1}.\lambda^c \circ T_s) +$$

$$+\beta div(j(DT_s)^{-1}U) div(j(DT_s)^{-1}\lambda_s^c \circ T_s)\}\, dx$$

Thus we get

$$\frac{\partial \mathcal{L}}{\partial s}(\Omega_s, v, U, \lambda)|_{s=0} =$$

$$= \int_{\Omega} \alpha D(U+R).(-DV)..D\lambda^c + \alpha D(U+R)..D\lambda^c(-DV) +$$

$$+\alpha D(U+R)..D((\lambda_s^c \circ T_s)') + D(U+R).v.(\lambda_s^c \circ T_s)' + \beta div(U+R) div((\lambda^c \circ T_s)') +$$

$$-f(\lambda_s^c \circ T_s)'\, dx + \int_{\Omega} D(U+R).(-DV)v\lambda^c + \beta div(U+R) div(-DV\lambda^c) +$$

$$+\beta div(-DV(U+R)) div\lambda^c\, dx + < D(U+R).v - f, \lambda^c >\} divV +$$

$$\int_{\Omega} \{\alpha D(U+R)..D\lambda^c + +\beta div(divV(U+R)) div\lambda^c +$$

$$+\beta div(U+R)(div(divV\lambda^c) - divV div\lambda^c)\, dx$$

We return with the variable u hence

$$\frac{\partial \mathcal{L}}{\partial s}(\Omega_s, v, u, \lambda)|_{s=0} =$$

$$\int_\Omega \alpha Du.(-DV)..D\lambda^c + \alpha Du..D\lambda^c(-DV) + Du.(-DV)v\lambda^c +$$

$$+\alpha Du..D(\lambda_s^c \circ T_s)' + Du.v.(\lambda_s^c \circ T_s)' + \beta divu\, div(\lambda^c \circ T_s)' - f(\lambda_s^c \circ T_s)'\, dx +$$

$$+ \int_\Omega \beta div(-DV.u) div\lambda^c + \beta divu\, div(-DV\lambda^c) - \beta divu\, divV\, div\lambda^c\, dx +$$

$$+ \int_\Omega \{\alpha Du..D\lambda^c + < Du.v - f, \lambda^c >\} divV + \beta div(divVu) div\lambda^c +$$

$$+\beta divu\, div(divV\lambda^c)\, dx$$

The lemma 14.3 yields $div(DV.u) = \nabla(divV).u + Du..D^*V$ and $div(DV.\lambda^c) = \nabla(divV).\lambda^c + D\lambda^c..D^*V$

It follows that

$$\frac{\partial \mathcal{L}}{\partial s}(\Omega_s, v, u, \lambda)|_{s=0} =$$

$$= \int_\Omega -2\alpha(Du.\epsilon(V))..D\lambda^c - < Du.DV.v, \lambda^c > +$$

$$+\alpha Du..D(\lambda_s^c \circ T_s)' + Du.v.(\lambda_s^c \circ T_s)' + \beta divu\, div(\lambda^c \circ T_s)' - f(\lambda_s^c \circ T_s)' +$$

$$+ \{\alpha Du..D\lambda^c + < Du.v - f, \lambda^c > + \beta divu\, div\lambda^c\} divV +$$

$$-\beta(divuD\lambda^c..D^*V + div\lambda^c Du..D^*V)\, dx$$

But $\lambda^c = \lambda$, consequently we obtain the formulation of $dF_e(\Omega, V)$

$$dF_e(\Omega, V) = \int_\Omega -2\alpha(Du.\epsilon(V))..D\lambda - < Du.DV.v, \lambda > +$$

$$+\alpha Du..D(\lambda_s^c \circ T_s)' + Du.v.(\lambda_s^c \circ T_s)' + \beta divu\, div(\lambda_s^c \circ T_s)' - f(\lambda_s^c \circ T_s)' +$$

$$+ \int_\Omega \{\alpha Du..D\lambda + < Du.v - f, \lambda > + \beta divu\, div\lambda\} divV +$$

$$-\beta(divuD\lambda..D^*V + div\lambda Du..D^*V)\, dx$$

$V \in V^{ad}$ thus $divV = 0$ in Ω finally we have

$$dF_e(\Omega, V) = \int_\Omega -2\alpha(Du.\epsilon(V))..D\lambda - < Du.DV.v, \lambda > +$$

$$+\alpha Du..D(\lambda_s^c \circ T_s)' + Du.v.(\lambda_s^c \circ T_s)' + \beta divu\, div(\lambda_s^c \circ T_s)' - f(\lambda_s^c \circ T_s)' +$$

$$- \beta(divuD\lambda..D^*V + div\lambda Du..D^*V)\, dx$$

v is given in $H^q(D)$. The regularity of $(u, \lambda), u \in H^1(\Omega, R^3), \lambda \in H_0^1(D, R^3)$ and of $V \in W^{1,\infty}$ yields the distributed formulation existence of $dF_e(\Omega, V)$.

Hence, the two previous boundary expressions of $dF_e(\Omega, V)$ are well defined since they represent the same distribution G.

We assume that $\mathcal{N} = \frac{1}{\|\mathcal{N}\|}\mathcal{N}$ in a neighborhood of Γ. In order to compute the shape gradient in the distributed sense we would like to write a specific formulation of $dF_e(\Omega, V)$ with a given extension $\mathcal{N}$. Let h be a strictly real positive. We define $\mathcal{N} = 0$ in D except in the neighborhood $\mathcal{U}$ of $\Gamma, \mathcal{U} = \{x \in D / 0 \le b(x) < h\}$ where

$$\mathcal{N} = (\rho_h \circ b)\nabla b$$

with b which is the oriented function and

$$\rho_h : R \to R$$

$$z \mapsto 1 - \frac{1}{h}z$$

$\forall x \in \mathcal{U}$ we have $(\rho_h \circ b)(x) = 1 - \frac{1}{h}b(x) >= 0$

We have $\|\mathcal{N}\| = |(\rho_h \circ b)| \, \|\nabla b\| = |(\rho_h \circ b)| = 1 - \frac{1}{h}b$. Therefore, $\frac{1}{\|\mathcal{N}\|}\mathcal{N} = \frac{1-\frac{1}{h}b}{1-\frac{1}{h}b}\nabla b = \nabla b$.

So we can see that $\frac{1}{\|\mathcal{N}\|}\mathcal{N}$ is independent of the choice of the function ρ_h and it follows that $dF_e(\Omega, V)$ is also independent of the choice of the function ρ_h.

14.3.2 The second design performance functional. If we take the viscous force into account, we consider this functional:

Definition 14.5. *Assume that e_i and $E_i, i = 1..2$, are defined as in (14.34).*

$$J(\Omega) = \frac{F_{e_2}(\Omega)}{F_{e_1}(\Omega)} = \frac{\int_{\partial S} < 2\alpha\epsilon(u).n + \beta divu \, n, e_2 > d\Gamma}{\int_{\partial S} < 2\alpha\epsilon(u).n + \beta divu \, n, e_1 > d\Gamma} =$$

$$= \frac{\int_{\Gamma} < 2\alpha\epsilon(u).n + \beta divu \, n, E_2 > d\Gamma}{\int_{\Gamma} < 2\alpha\epsilon(u).n + \beta divu \, n, E_1 > d\Gamma}$$

where u is solution of (11.1).

It is the same functional in the previous section but we add the viscosity term $< 2\alpha\epsilon(u).n, E_i >$. Consider this following extremal problem:

$$\min_{\Omega \in \mathcal{D}_{ad}} J(\Omega) \tag{14.47}$$

We consider the new convex set

Definition 14.6. $K(\Omega) = \{\lambda \in H^1(\Omega, R^3) / \lambda_\Gamma = E_\Gamma$ *on* $\Gamma, \lambda.n = \frac{2\alpha+\beta}{\alpha+\beta} < e, n >$ *on* $\Gamma\}$

$K(\Omega)$ is also defined by

Definition 14.7. $K(\Omega) = H_0^1(\Omega, R^3) \oplus (E- < E, n > n) \oplus < \frac{2\alpha+\beta}{\alpha+\beta}E, n > n$

Consequently $\lambda \in K(\Omega)$ leads to

$$\lambda = \lambda^0 + \lambda^1 + \lambda^2$$

where $\lambda^0 \in H_0^1(\Omega, R^3)$ and $\lambda^1 = E- < E,\mathcal{N} > \mathcal{N}$ and $\lambda^2 = \frac{2\alpha+\beta}{\alpha+\beta} < E,\mathcal{N} >$ $\mathcal{N}$ in Ω

We assume that $\mathcal{N} = \frac{1}{\|\mathcal{N}\|}\mathcal{N} = n$ in a neighborhood of Γ.

In the second way, we maximize $\mathcal{L}_0(\Omega)$ with $\lambda \in K(\Omega)$. Hence we happen to have an problem on a convex set:

$$F_e(\Omega) = \min_{u \in H(\Omega)} \max_{\lambda \in K(\Omega)} \mathcal{L}_0(\Omega)$$

We define as in the previous section $\lambda_s^c \in K(\Omega_s), \lambda_s^c = \lambda^0 \circ T_s^{-1} + \lambda_s^1 + \lambda_s^2$.

Let $v \in H_q(\Omega)$. We introduce $\lambda_i \in H^1(\Omega, R^3), i = 1..2$, the adjoint state of the vector u associated to $F_{e_i}(\Omega)$ and a solution of

$$\begin{cases} -\alpha\Delta\lambda_i - D\lambda_i.v - divv\lambda_i - \beta\nabla(div\lambda_i) = 0 \text{ in } \Omega \\ (\lambda_i)_\Gamma = (E_i)_\Gamma \text{ on } \Gamma \\ \lambda_i.n = \frac{2\alpha+\beta}{\alpha+\beta} E_i.n \text{ on } \Gamma \end{cases} \tag{14.48}$$

We define

$$\lambda_g = \frac{F_{e_1}(\Omega)\lambda_2 - F_{e_2}(\Omega)\lambda_1}{(F_{e_1}(\Omega))^2}$$

and

$$E_g = \frac{F_{e_1}(\Omega)E_2 - F_{e_2}(\Omega)E_1}{(F_{e_1}(\Omega))^2}$$

$\lambda_g \in H^1(\Omega, R^3)$ is solution of

$$\begin{cases} -\alpha\Delta\lambda_g - D\lambda_g.v - divv\lambda_g - \beta\nabla(div\lambda_g) = 0 \text{ in } \Omega \\ (\lambda_g)_\Gamma = E_\Gamma \text{ on } \Gamma \\ \lambda_g.n = \frac{2\alpha+\beta}{\alpha+\beta} E_g.n \text{ on } \Gamma \end{cases} \tag{14.49}$$

And

$$\lambda_g^c = \lambda_g^0 + (E_g- < E - g, \mathcal{N} > \mathcal{N}) + < \frac{2\alpha+\beta}{\alpha+\beta} E_g, \mathcal{N} > \mathcal{N} \tag{14.50}$$

$$(\lambda_{gs}^c \circ T_s)' =< E, (\mathcal{N} \circ T_s)' > \mathcal{N} + < E, \mathcal{N} > (\mathcal{N} \circ T_s)' +$$
$$+ < \frac{2\alpha+\beta}{\alpha+\beta} E_g, (\mathcal{N} \circ T_s)' > \mathcal{N} + \ < \frac{2\alpha+\beta}{\alpha+\beta} E_g > (\mathcal{N} \circ T_s)'$$

with $(\mathcal{N}_s \circ T_s)' =< D^*V\mathcal{N}, \mathcal{N} > \mathcal{N} - D^*V.\mathcal{N}$

The formulation of the gradient, $dJ(\Omega, V)$, is given by

Theorem 14.4. *We denote e_g by e, λ_g by λ.*

Assume that $\partial\Omega$ is lipschitzian and that $v \in H_1(D)$ then the functional $J(\Omega)$ is shape differentiable, and we have $\forall\, V \in V^{ad}$,

$$dJ(\Omega,V) = \int_\Omega \theta(\Omega,V)\, dx$$

with

$$\theta(\Omega,V) = -2\alpha(Du.\epsilon(V))..D\lambda - < Du.DV.v,\lambda > + \alpha Du..D(\lambda^c_{gs}\circ T_s)' +$$
$$+ Du.v.(\lambda^c_{gs}\circ T_s)' + \beta divu\; div(\lambda^c_s\circ T_s)' - f(\lambda^c_s\circ T_s)' +$$
$$-\beta(divuD\lambda..D^*V + div\lambda Du..D^*V)$$

and then, as θ is linear and continuous on V, we get

$$dJ(\Omega,V) =< G(\Omega),V >$$

where G is a distribution supported by the boundary, $G = \Gamma^*(g\, n)$.

Moreover, when v is more regular, $v \in H^{div}_1(D)$ or $v \in H^q(D)$ with $q > \frac{3}{2}$, then the density $g(\Gamma) \in L^1(\Gamma)$, hence

$$dJ(\Omega,V) = \int_{\partial S} g(\Gamma) < V,n > \, d\Gamma$$

and we have the two following explicit expressions of g:

$$g^e_m(\Gamma) =< (\alpha+\beta)\nabla(divu),e > +\alpha < D^2u.n,e > +$$
$$-(\alpha+\beta) < D\lambda.n,n >< Du.n,n > - \alpha < D\lambda.n,(Du.n)_\Gamma > +$$
$$+\frac{\alpha^2}{\alpha+\beta} < D^2be, Du.n >$$

and

$$g^e_c(\Gamma) = \frac{\alpha^2}{\alpha+\beta} < e,n > \; div_\Gamma((Du.n)_\Gamma) + \alpha < D_\Gamma(Du.n)e_\Gamma,n > +$$
$$+\alpha < Du.n,(D^*\lambda.n)_\Gamma - 2(D\lambda.n)_\Gamma > -(\alpha+\beta) < D\lambda.n,n >< Du.n,n > +$$
$$+\alpha < (D^2b - Hnn^*)e, Du.n > -\frac{2\alpha+\beta}{\alpha+\beta} < e,n >< f,n > - < f,e_\Gamma >$$

Proof. It is nearly the same functional in the previous section. Indeed we only add the viscous term $< 2\alpha\epsilon(u).n, E_i >$. Similarly we compute the shape gradient as in the three previous ways in the proof of the theorem 14.3.

And we also use to these equations:

$$\int_\Gamma 2\alpha < \epsilon(u).n, E > d\Gamma = \int_\Omega 2\alpha\epsilon(u)..\epsilon(E) + \alpha < \Delta u + \nabla(divu), E > dx \tag{14.51}$$

Using the equality $< D^*u.n, E >=< Du.E, n >$ and $Du = Du\, nn^*$ on Γ we obtain that $< D^*u.n, E >=< Dunn^* E_\Gamma, n > + < Du.n, n >< E, n >=< Du.n, n >< E_\Gamma, n > + < Du.n, n >< E, n >=< Du.n, n >< E, n >$ thus

$$\int_\Gamma 2\alpha < \epsilon(u).n, E > d\Gamma =$$

$$= \int_\Gamma 2\alpha < Du.n, n >< E, n > +\alpha < (Du.n)_\Gamma, E_\Gamma > \; d\Gamma \tag{14.52}$$

Integrating by parts , the equation 14.51 yields

$$\int_\Gamma 2\alpha < \epsilon(u).n, E > d\Gamma =$$

$$= \int_\Gamma 2\alpha < \epsilon(E).n, u > +\alpha(< Du.n, E > - < D < E, n >, u > +$$

$$+divu < E, n > -divEu.n)\; d\Gamma$$

Thus we obtain the same formulation of the shape gradient in the second way (and third way) as in the theorem 14.3. The adjoint term also verifies the same equation, only the boundary conditions are changed.

14.4 Monotone approximations

14.4.1 Definition and existence result. Let C_0, and Γ be two real strictly positive numbers. Assume that $J_\Omega(.)$ is a lower bounded functional verifying :

$$\forall\Omega,\ \forall v \in H_1^{div}(\Omega), \quad J_\Omega(v) \geq -C_0$$

We consider the following extremal problem:

$$(\mathcal{P}_\Gamma^{min}) \qquad \mathcal{J}_\Gamma^{min}(\Omega) = \min_{v \in H_1^{div}(\Omega)} \left[\quad J_\Omega(v) + \Gamma \|u_v - v\|^2_{H^1(\Omega, R^3)} \right]$$

where u_v is the solution of (11.1). We take β large enough, then, from lemma 14.2, $divu_v$ is small enough. The shape functional $\mathcal{J}_\Gamma^{min}$ has to be compare with the shape functional governed by the non linear viscous flow

$$\mathcal{J}^{min}(\Omega) \; = \; J_\Omega(u_\Omega)$$

where u_Ω is the solution of the non linear penalized Navier Stokes problem in the domain Ω. Taking $v = u(\Omega)$ in $(\mathcal{P}_\Gamma^{min})$, we get $u_v = u_\Omega = v$ and then

$$\mathcal{J}_\Gamma^{min}(\Omega) \leq \mathcal{J}^{min}(\Omega)$$

and

$$-C_0 + \Gamma \|u_v - v\|^2_{H^1(\Omega,R^3)} \leq \mathcal{J}^{min}(\Omega)$$

Assuming the functional J_Ω having a upper bound, say C_1, then for a sequence Γ_n converging to ∞ we get , for any minimizer v_n,

$$\|u_{v_n} - v_n\|_{H^1} \leq C\,(\Gamma_n)^{-\frac{1}{2}}, \qquad \|\frac{u_{v_n} - v_n}{(\Gamma_n)^{-\frac{1}{2}}}\|_{H^1} \leq C$$

Then we can subtract a sequence, still denoted Γ_n and an element $\theta \in H^1$ such that

$$(u_{v_n} - v_n)(\Gamma_n)^{\frac{1}{2}}$$

weakly converges in H^1 to θ.

Proposition 14.12. *For β small enough, as $\Gamma_n \to \infty$ the element $u_n = u_{v_n}$, (where v_n is any minimizer in the problem $\mathcal{P}_{\Gamma_n}^{min}$) are bounded in H_0^1 and all the weak cluster points are solutions to the non linear β-penalized Navier Stokes problem in the domain Ω : any weakly converging subsequence u_{n_k} in $H_0^1(\Omega, R^3)$ converges to a solution u_Ω to the non linear penalized Navier Stokes flow in Ω.*

Proof. From the weak formulation of the linear problem whose u_n is solution we derive the boundedness of u_n . Let $U_n = u_n - R$ be in Ω, thus $U_n = \begin{cases} u_n - R \text{ in } \Omega \setminus \overline{B} \\ u_n \text{ in } \overline{B} \end{cases}$ where R is defined as in (14.3). $\forall \phi \in H_0^1(\Omega, R^3)$

$$\int_\Omega \alpha DU_n..D\phi + < DU_n.v_n + DR.v_n, \phi > + \beta divU_n\, div\phi\, dx =$$

$$= \int_\Omega < f + \alpha\Delta R, \phi > dx$$

We denote $f + \alpha\Delta R$ by F and choosing $\phi = U_n$ we get :

$$(\frac{\alpha}{C^2+1} - \frac{C_1}{2}\|divv_n\|_{L^2})\|U_n\|^2_{H^1} \leq$$

$$\leq (\|F\|_{L^2} + \|R\|_{L^4}\|divv_n\|_{L^2} + \|R\|_{L^4}\|v_n\|_{L^4})\|U_n\|_{H^1}$$

for any minimizer v_n we have :

$$\|divv_n\|_{L^2} \leq M/\beta$$

Then using the continuous inclusion of H^1 in L^4 we derive the boundedness as soon as β is large enough.

Corollary 14.5. *Let Γ_n be a sequence going to ∞ and v_n a minimizer in the problem $\mathcal{P}_{\Gamma_n}^{min}$, then v_n is bounded in H_0^1 and any weak cluster point of that sequence is a solution to the non linear β-penalized Navier Stokes problem in the domain Ω.*

Proof. we have

$$||u_{v_n} - v_n||_{H^1} \leq (\Gamma_n)^{-1} (\, C_0 + J_\Omega(u_\Omega)\,) \longrightarrow 0, \quad n \to 0$$

for any solution u_Ω to the Navier Stokes problem. As $u_n = u_{v_n}$ weakly converges to a u_Ω then v_n weakly converges to the same solution u_Ω.

For each Γ let us consider the shape functional

$$\mathcal{J}_\Gamma^{min}(\Omega) = Min\{J_\Omega(v) + \Gamma||u_v - v||_{H^1} \quad | \; v \in H_0^1 \;\}$$

this is an increasing family of functional (with Γ) :

$$\Gamma_1 \leq \Gamma_2 \;\text{ implies }\; \mathcal{J}_{\Gamma_1}^{min}(\Omega) \;\leq\; \mathcal{J}_{\Gamma_2}^{min}(\Omega)$$

We consider the functional $\mathcal{J}_\infty^{min} = sup\{\; \mathcal{J}_\Gamma \;|\; \Gamma > 0 \;\}$ defined over the family of admissible domains, that is :

$$\mathcal{J}_\infty^{min}(\Omega) = lim_{\Gamma\to\infty} \left[\;\; Min_{v\in H_0^1}\{ \;\; j_{\Gamma,\Omega}(v) + \Gamma\,||u_v - v||_{H_0^1} \;\; \} \;\right]$$

Then we have

Corollary 14.6.

$\mathcal{J}_\infty^{min}(\Omega) = Min\;\{\; j_\Omega(u_\Omega) \;|\; u_\Omega$ *solution to the Navier Stokes problem* $\}$

If the functional J_Ω is weakly lower semi continuous in the limit we get :

$$J_\Omega(u_\Omega) + ||\theta||_{H^1} \;\leq\; J_\Omega(u_\Omega)$$

Then

$$\theta = 0,$$

that is $(\Gamma_n)^{\frac{1}{2}}\;(u_{v_n} - v_n)$ weakly converges in H_0^1 to 0.
We denote by $J_{\Gamma,\Omega}$ the augmented functional.

$$J_{\Gamma,\Omega} : L^{3+\epsilon}(D) \to (weakly) R$$

$$v \mapsto J(v) + \Gamma||u_v - v||^2_{H^1(\Omega,R^3)}$$

Proposition 14.13. *The application*

$$L^{3+\epsilon}(D) \to H(\Omega)$$

$$v \mapsto u_v$$

is continuous.

Proof. Let (v_k) be a convergent sequence in H_1^{div} and $,v$ its limit. Then for any φ in $H^1(\Omega, R^3)$ we have

$$\int_\Omega \alpha Du_{v_k}..D\varphi + Du_{v_k}.v_k\varphi + \beta divu_{v_k}\varphi \, dx =$$

$$= \int_\Omega f\varphi \, dx + \int_\Gamma \alpha Du_{v_k}\varphi + \beta divu_{v_k}\varphi \, d\Gamma$$

$$\int_\Omega \alpha Du_v..D\varphi + Du_v.v\varphi + \beta divu_v\varphi \, dx = \int_\Omega f\varphi \, dx + \int_\Gamma \alpha Du_v\varphi + \beta divu_v\varphi \, d\Gamma$$

Let $w = u_{v_k} - u_v \in H_0^1(\Omega, R^3)$. We subtract the two previous equations and we get

$$\int_\Omega \alpha Dw..D\varphi + < Dw.v_k + Du_v(v_k - v), \varphi > + \beta divw\varphi \, dx =$$

$$= \int_\Gamma \alpha Dw\varphi + \beta divw\varphi \, d\Gamma$$

For $\varphi = w$ we obtain

$$\alpha\|Dw\|_{L^2}^2 + \beta\|divw\|_{L^2}^2 + \int_\Omega < Dw.v_k + Du_v(v_k - v), w > dx = 0$$

But with the Green's theorem we have

$$\int_\Omega < Dw.v_k, w > dx = -\frac{1}{2}\int_\Omega divv_k w^2 \, dx \geq 0$$

and using the Holder inequality we get

$$|\int_\Omega < Du_v(v_k - v), w > dx| \leq \|u_v\|_{L^2}\|v_k - v\|_{L^r}\|w\|_{L^{6-\epsilon}}$$

Where $\frac{1}{r} = \frac{1}{2} - \frac{1}{6-\epsilon}$ that is $r = 3+\epsilon_1$ Let $C = C_(\Omega)$ be the Poincare constant. As H^1 is a subspace of $L^{6-\epsilon}$, there exist a constant $C_2 = C_2(\Omega)$ such that

$$\frac{\alpha}{C^2+1}\|w\|_{H^1}^2 \leq \alpha\|Dw\|_{L^2}^2 \leq C_2\|u_v\|_{H^1}\|v_k - v\|_{L^r}\|w\|_{H^1}$$

Therefore we get

$$\frac{\alpha}{C^2+1}\|w\|_{H^1} \leq C_2\,\|u_v\|_{H^1}\|v_k - v\|_{L^r}$$

Therefore we obtain the continuity of the application.

Proposition 14.14. *There exists $v_\Gamma \in H_1^{div}$ (at least one) such as*

$$J_{\Gamma,\Omega}(v_\Gamma) = \min_{v\in H_1^{div}} J_{\Gamma,\Omega}(v)$$

Proof. We consider a minimizing sequence (v_k) in $H^v(D)$. We have, for any k,

$$J_{\Gamma,\Omega}(v_{k+1}) \le J_{\Gamma,\Omega}(v_k)$$

and

$$J_{\Gamma,\Omega}(v^k) \to \min_{v\in H^v(D)} J_{\Gamma,\Omega}(v)$$

As $J(v) \ge -C_0$, $J_{\Gamma,\Omega}(v)$ is coercive and so $||(v_k)||_{H^1}$ is bounded thus we can extract a subsequence (v_{k_i}) such that $v_{k_i}(weakly) \to v^*$, weakly in $H^1(\Omega, R^3)$, for $i \to \infty$. But $J_{\Gamma,\Omega}(v)$ is l.s.c hence we get

$$J_{\Gamma,\Omega}(v^*) \le \lim inf J_{\Gamma,\Omega}(v_{k_i})$$

therefore

$$J_{\Gamma,\Omega}(v) \le \lim inf[J(v_{k_i})] + \lim inf[\frac{\Gamma_1}{2}||u_{v_{k_i}} - v_{k_i}||^2_{H^1(\Omega,R^3)}]$$

And with the proposition 14.13 $u_{v_{k_i}} \to u_{v^*}$ strongly in $H^1(\Omega, R^3)$. Thus we obtain that

$$J_{\Gamma,\Omega}(v^*) \le J_{\Gamma,\Omega}(u_{v^*})$$

Finally we have

$$J_{\Gamma,\Omega}(v_\Gamma) = \min_{v\in H^v(D)} J_{\Gamma,\Omega}(v) = J_{\Gamma,\Omega}(u_{v^*})$$

14.4.2 For large viscosity solutions of the relaxed problem are quasi unique. By the choice of $v = u$ in he relaxed minimization problem we derive that it exists $M > 0$ such as $||u_v - v||_{H^1(\Omega,R^3)} \le \frac{M}{\Gamma_1}$. For Γ_1 large enough , $||u_v - v||_{H^1(\Omega,R^3)}$ is small enough. We have

$$\alpha||D(u_v - u)||^2_{L^2(\Omega,R^3\times R^3)} + \beta||div(u_v - v)||^2_{L^2(\Omega}+$$

$$+\int_\Omega Du_v.u_v(u_v-u)-Du.u(u_v-u)\ dx \le \int_\Omega |Du_v(v-u_v)(u_v-u)|\ dx$$

The third term in the left hand side can be written as :

$$\int_\Omega Du_v.u_v(u_v - u) - Du.u(u_v - u)\ dx =$$

$$= \int_\Omega Du_v.(u_v - u)(u_v - u) + D(u_v - u).u(u_v - u)\ dx$$

the first non linear term in the right hand side can be estimate via Holder inequality :

$$|\int_{\Omega} Du_v.(u_v - u)(u_v - u)\, dx| \leq \|Du_v\|_{L^2(\Omega,R^3\times R^3)}\|u_v - u\|^2_{L^4(\Omega,R^3)}$$

For the second one, we have :

$$\int_{\Omega} D(u_v - u).u(u_v - u)\, dx = -1/2 \int_{\Omega} divu < u_v - u, u_v - u > dx$$

and again from Holder's inequality

$$|\int_{\Omega} D(u_v - u).u(u_v - u)\, dx| \leq 1/2\, \|\, divu\, \|_{L^2(\Omega)}\, \|u_v - u\|^2_{L^4(\Omega,R^3)}$$

Now we have the continuous injection $H^1 \subset L^4(\Omega)$, we denote by $C_1 = C_1(\Omega)$ its norm and then from the two previous estimates we derive:

$$|\int_{\Omega} Du_v.u_v(u_v - u) - Du.u(u_v - u)\, dx\,| \leq$$

$$\leq C_1^2\ [\ 1/2\, \|\, div(u)\, \|_{L^2(\Omega)} + \|Du_v\|_{L^2(\Omega,R^3\times R^3)}\ ]\ \ \|u_v - u\|^2_{H^1(\Omega,R^3)}.$$

$$\alpha\|Du_v - Du\|^2_{L^2(\Omega,R^3)} \leq$$

$$\leq C_1^2\ \|Du_v\|_{L^2(\Omega,R^3\times R^3)}\|v - u_v\|_{H^1(\Omega,R^3)}\|u_v - u\|_{H^1(\Omega,R^3)}+$$

$$+C_1^2\ [\ 1/2\, \|\, div(u)\, \|_{L^2(\Omega)} + \|Du_v\|_{L^2(\Omega,R^3\times R^3)}\,]\,|u_v - u\|^2_{H^1(\Omega,R^3)}$$

We denote by $(\lambda_1(\Omega))^{-\frac{1}{2}})$ the Poincare constant, so that:

$$\|u_v - u\|_{H^1} \leq [1 + (\lambda_1(\Omega))^{-\frac{1}{2}})]\ \|Du_v - Du\|_{L^2(\Omega,R^3)}$$

Hence we obtain

$$\{\frac{\alpha}{([1 + (\lambda_1(\Omega))^{-\frac{1}{2}})])^2} -$$

$$C_1^2[\ 1/2\, \|\, div(u)\, \|_{L^2(\Omega)} = + \|Du_v\|_{L^2(\Omega,R^3\times R^3)}\,]\,\}\, \|u_v - u\|_{H^1(\Omega,R^3)}$$

$$\leq C_1\ \|Du_v\|_{L^2(\Omega,R^3\times R^3)}\|v - u_v\|_{H^1(\Omega,R^3)}$$

It follows that If $\alpha > (C^2 + 1)(\epsilon_1 + C_1\|Du_v\|_{L^2(\Omega,R^3\times R^3)})$ then

If $u_v \to v$, strongly in $H^1(\Omega, R^3)$ then $u_v \to u$, strongly in $H^1(\Omega, R^3)$. But we have

$$\leq \|v - u\|_{H^1(D,R^3)} = \|v - u_v + u_v - u\|_{H^1(\Omega,R^3)} \leq$$

$$\|v - u_v\|_{H^1(\Omega,R^3)} + \|u_v - u\|_{H^1(\Omega,R^3)}$$

Thus it leads that $v \to u$, strongly in $H^1(\Omega, R^3)$ in the same conditions.

14.4.3 Upper approximation of the functional. We assume here that the functional J_Ω is weakly upper semi continuous on H^1 and is upper bounded. We introduce the functional

$$\mathcal{J}_\Gamma^{max}(\Omega) = Max\{\ J_\Omega(v)\ - \Gamma\, ||u_v - v||^2_{H^1} \quad |\ v \in H^1\ \}$$

which is decaying to the limit functional

$$\mathcal{J}^{max}(\Omega) =$$

$$= Max\{\ J_\Omega(u_\Omega)\ |\ u_\Omega \text{ solution to the Navier Stokes problem in } \Omega\ \}$$

We have , for any $\Gamma_1 \leq \Gamma_2$:

$$\mathcal{J}_{\Gamma_1}^{min} \leq \mathcal{J}_{\Gamma_2}^{min} \leq \mathcal{J}^{min} \leq \mathcal{J}^{max} \leq \mathcal{J}_{\Gamma_2}^{max} \leq \mathcal{J}_{\Gamma_1}^{max}$$

When the viscosity α is large enough, the solution u_Ω is unique , then the functionals $\mathcal{J}^{max}$ and $\mathcal{J}^{min}$ are equals to $\mathcal{J}$ and we get monotone approximations for that shape functional. These shape functional approximations are governed by the "linear splitted" problem for which the complete shape analysis is available. In the realistic situation where the viscosity α is very small the solution is non unique and the problem

$$\mathcal{P}^{min} \qquad Max\{\ \mathcal{J}^{max}(\Omega)\ \ |\ \Omega \in \mathcal{O}_{ad}\ \ \}$$

is a reasonable modeling of the shape optimization problems as in the optimization we never know which solution u_Ω the system "would choose".

To deal with a tractable problem we propose to replace that problem by the following , for Γ large enough :

$$\mathcal{P}_\Gamma^{min} \qquad Max\{\ \mathcal{J}_\Gamma^{max}(\Omega)\ \ |\ \Omega \in \mathcal{O}_{ad}\ \ \}$$

Another advantage of the problems $\mathcal{P}_\Gamma^{min}$ and $\mathcal{P}_\Gamma^{max}$ is that the Navier Stokes problem could be easily replaced by the so call steady $k - \epsilon$ problem as far as only the splitted linear version of the equation is concerned.

14.4.4 Flow with non unique solution. We consider the three-dimensional incompressible flow of a viscous fluid and the following cost functional governed by the linear splitted formulation.

$$\mathcal{J}_\Gamma^{max}(\Omega) = Max\{\ \ J_\Gamma(\Omega; v)\ \ |\ \ v \in H^1(D, R^3)\ \}$$

where

$$J_\Gamma(\Omega; v) = J_\Omega(v)\ - \frac{\Gamma}{2}\, ||u_v - v||^2_{H^1(\Omega, R^3)}$$

and

$$J_\Omega(v) = \int_\Gamma < \beta divv\ n + 2\alpha\epsilon(v).n, E > d\Gamma =$$

$$= \int_\Omega div(< \beta divv + 2\alpha\epsilon(v), E >)\ dx$$

In order to simplify the computation of the shape gradient we use the adjoint technique. R is defined as in (14.3). We introduce the Lagrangian $\mathcal{L}$ defined by

$$\mathcal{L}(\Omega, v, w, \lambda) =$$

$$= J_\Omega(v) - \frac{\Gamma}{2}\|w+R-v\|^2_{H^1} + \int_\Omega \alpha Dw..D\lambda + Dw.v.\lambda + \beta div w \; div\lambda - F\lambda \, dx \; +$$

$$- \int_\Gamma \alpha < Dw.n, \lambda > + \beta div w \lambda.n \, d\Gamma$$

and we redefine

$$J_\Gamma(\Omega; v) = Min_{w \in H^1_0(\Omega, R^3)} Max_{\lambda \in H^1_0(D, R^3)}[\; \mathcal{L}(\Omega, v, w, \lambda) \;]$$

Let $H(\Omega) = \{u_v \in H^1(\Omega, R^3)/u_v = 0 \text{ on } \partial S, u_v = u_\infty \text{ on } \partial D\}$

For each Ω and v there is an unique saddle point (U_v, λ_v). Let $u_v = U_v + R$ in Ω, thus $u_v = \begin{cases} U_v + R \text{ in } \Omega \setminus \overline{B} \\ U_v \text{ in } \overline{B} \end{cases}$

where $u_v \in H(\Omega)$ is a solution of the problem (11.1). U_v is the solution of the problem (14.4) and $\lambda_v|_\Omega = \lambda_\Omega \in H^1_0(\Omega, R^3)$ of

$$\begin{cases} -\alpha\Delta\lambda_\Omega - D\lambda_\Omega.v - div v \lambda_\Omega - \beta\nabla(div\lambda_\Omega) = \\ \qquad = -\Gamma(\Delta(u_v - v) - u_v - v) \text{ in } \Omega \\ \lambda_\Omega = 0 \text{ on } \Gamma \end{cases} \tag{14.53}$$

Proposition 14.15. *There exists one unique solution $\lambda_\Omega \in H^1_0(\Omega, R^3)$ of (14.53).*

Proof. From the Lax-Milgram theorem as in the first section it exists an unique solution in H^1_0 of the problem (14.53)

Thus we have the uniqueness of the saddle point (U_v, λ_v) in $H^1_0(\Omega, R^3 \times H^1_0(D, R^3)$ hence $\mathcal{L}$ is differentiable with s.

Let $E(\Omega) = \{v \in H^1(D, R^3)/v$ extremal solution for $J_\Gamma(\Omega; v)\}$

For $v^* \in E(\Omega)$ we have $J_\Gamma(\Omega; v^*) \geq J_\Gamma(\Omega; v)$ for all $v \in H^1(D, R^3)$

We denote v^* by v. Using the characterization of the parameter derivative of a minimum we get

$$d\mathcal{J}^{max}_\Gamma(\Omega, V) = Max_{v \in E(\Omega)} dJ_\Gamma(\Omega; v; V)$$

Moreover

$$J_\Gamma(\Omega_s; v) = Min_{w \in H^1_0(\Omega, R^3)} Max_{\lambda \in H^1_0(D, R^3)} \mathcal{L}(\Omega_s, v, w \circ T_s^{-1}(V), \lambda)$$

This problem has a unique saddle point (U_v, λ_v) for each Ω and v then from the derivation of a min max of Cuer-Zolesio

$$d\mathcal{J}_\Gamma(\Omega, v; V) = \frac{\partial}{\partial s}[\mathcal{L}(\Omega_s, v, u_v \circ T_s^{-1}(V), \lambda_v)]_{s=0}$$

The formulation of the previous gradient is given by the

Theorem 14.5. *Let the following terms*

$$g^{\Gamma}(\Gamma, v) = \Gamma < D(u_v - v).n, Du_v.n > -\frac{\Gamma}{2}((u_v - v)^2 + D(u_v - v)..D(u_v - v))$$

and

$$g^2(\Gamma, v) =< (\beta + \alpha)\nabla(divv) + \alpha D^2 v.n, e >$$

and

$$g^1(\Gamma, v) = -(\alpha + \beta) < D\lambda_v.n, n >< Du_v.n, n > -\alpha < (D\lambda_v.n)_{\Gamma}, Du_v.n >$$

and

$$g^{cur}(\Gamma, v) = \alpha H < Dv.n, e >$$

where λ_v is solution of (14.53) then

$$d\mathcal{J}_{\Gamma}(\Omega, v; V) =$$

$$= Max_{v\in E(\Omega)} \int_{\partial S} \{g^{\Gamma}(\Gamma, v) + g^2(\Gamma, v) + g^1(\Gamma, v) + g^{cur}(\Gamma, v)\} < V, n > d\Gamma$$

Proof. Assume that in Ω_s

$$F(R\circ T_s^{-1}, \lambda_v) = f - \alpha D(R\circ T_s^{-1})..D\lambda_v - D(R\circ T_s^{-1}).v\lambda_v - \beta div(R\circ T_s^{-1}) div\lambda_v$$

But we rewrite the following boundary integrals in a distributed sense

$$\int_{\Gamma} \alpha < DU_v.n, \lambda_v > d\Gamma = \int_{\Omega} \alpha div(D^* U_v \lambda_v)\, dx$$

and

$$\int_{\Gamma} \beta div U_v \lambda_v.n\, d\Gamma = \int_{\Omega} \beta div(div U_v \lambda_v)\, dx$$

The expression of the shape gradient is obtained by computing the derivative of an integral over a moving domain:

$$\mathcal{L}(\Omega_s, v, u_v \circ T_s^{-1}(V), \lambda_v) =$$

$$= \int_{\Omega_s} \alpha D(U_v \circ T_s^{-1})..D\lambda_v + D(U_v \circ T_s^{-1}).v_s.\lambda_v \ + \ \beta div(U_v \circ T_s^{-1})\ div\lambda_v +$$

$$-F(R\circ T_s^{-1}, \lambda_v) - div((\alpha D^*(U_v \circ T_s^{-1}) + \beta div(U_v \circ T_s^{-1})\ Id)\lambda_v)\, dx +$$

$$+ \int_{\Omega_s} div(< \beta divv + 2\alpha\epsilon(v), E >)\, dx - \frac{\Gamma}{2} \|u_v \circ T_s^{-1} - v\|^2_{H^1(\Omega_s, R^3)}$$

So its derivative becomes

$$\frac{\partial \mathcal{L}}{\partial s}(\Omega_s, v, u_v \circ T_s^{-1}(V), \lambda_v)|_{s=0} =$$

$$= \int_{\Omega} \alpha D(-DU_v.V)..D\lambda_v + D(-DU_v.V)v\lambda_v \ + \beta div(-DU_v.V)\ div\lambda_v +$$

$$+div((\alpha D^*(DU_v.V)+\beta div(DU_v.V)Id)\lambda_v) \;\; - F(-DR.V,\lambda_v)\, dx \;+$$
$$+\Gamma\int_\Omega <Du_v.V, u_v - v> + D(Du_v.V)..D(u_v - v)\, dx +$$
$$+\int_\Gamma \{\alpha DU_v..D\lambda_v + \beta div U_v\; div\lambda_v + <DU_v.v,\lambda_v> - F(R,\lambda - v) +$$
$$+div(<\beta divv + 2\alpha\epsilon(v), E>) - div((\alpha D^* U_v + \beta div(U_v)\; Id)\lambda_v) - \frac{\Gamma}{2}((u_v - v)^2 +$$
$$+D(u_v - v)..D(u_v - v)\;\} <V,n>\; d\Gamma$$

We set $\eta = D(U_v + R).V = Du_v.V$. We have $\eta|_\Gamma = Du_v.n <V,n>$ We have $SuppV \subset D$. By Green's theorem we have

$$\frac{\partial\mathcal{L}}{\partial s}(\Omega_s, v, u_v\circ T_s^{-1}(V), \lambda_v)|_{s=0} =$$
$$= \int_\Omega \alpha < \alpha\Delta\lambda_v + D\lambda_v.v + divv\lambda_v + \beta\nabla(div\lambda_v) \;\; - \; \Gamma\Delta(u_v - v) +$$
$$+\Gamma(u_v - v), \eta > dx + \int_\Gamma \{\Gamma D(u_v - v).nDu_v.n + \alpha Du_v..D\lambda_v + <Du_v.v,\lambda_v> +$$
$$-f + \; div(<\beta divv + 2\alpha\epsilon(v), E>) - \alpha\; div(D^* u_v\lambda_v) \; + - \; \beta\nabla(divu_v)\lambda_v +$$
$$-\frac{\Gamma}{2}((u_v - v)^2 + D(u_v - v)..D(u_v - v)\} <V,n>\; d\Gamma + + \int_\Gamma \alpha < D\eta.n, \lambda_v >$$
$$+\beta\; div\eta\lambda_v.n - \alpha D\lambda_v.n\eta - v.n <\lambda_v, \eta> - \beta div\lambda_v <\eta, n> d\Gamma$$

By lemma 14.3 $div(D^* u_v\lambda_v) = \Delta u_v.\lambda_v + Du_v..D\lambda_v$

But λ_v is solution of (14.53), thus we obtain

$$\frac{\partial\mathcal{L}}{\partial s}(\Omega_s, v, u_v\circ T_s^{-1}(V), \lambda_v)|_{s=0} =$$
$$= \int_{\partial S}\{\Gamma < D(u_v - v).n, Du_v.n > + div(\beta divv + 2\alpha\epsilon(v))e \; +$$
$$-\frac{\Gamma}{2}((u_v - v)^2 + D(u_v - v)..D(u_v - v)) - \alpha < D\lambda_v.n, Du_v.n > +$$
$$-\beta\; div\lambda_v \; < Du_v.n, n >\} < V, n > \; d\Gamma$$

But $div\lambda_v = < D\lambda_v.n, n >$ on Γ and

$$div(\beta divv + 2\alpha\epsilon(v))e = (\beta + \alpha) < \nabla(divv), e > + \alpha < \Delta v, e > =$$
$$= (\beta + \alpha) < \nabla(divv), e > + \alpha < HDv.n + D^2 v.n, e >$$

Finally we obtain

$$dJ_\Gamma^{max}(\Omega, V) = Max_{v\in E(\Omega)} \int_{\partial S} g(v, u_v, \lambda_v) < V, n > \; d\Gamma$$

with

$$g(v, u_v, \lambda_v) = \Gamma < D(u_v - v).n, Du_v.n > - \frac{\Gamma}{2}((u_v - v)^2 +$$
$$+D(u_v - v)..D(u_v - v)) + < (\beta + \alpha)\nabla(divv) + \alpha D^2 v.n, e > +$$
$$-(\alpha + \beta) < D\lambda_v.n, n >< Du_v.n, n > - \alpha < (D\lambda_v)_\Gamma, Du_v.n > +$$
$$+ \; \alpha H < Dv.n, e >$$

We have seen that when $\Gamma \to \infty$ the diameter of the set $E(\Omega)$ tends to zero weakly in $H^1(D, R^3)$; moreover $u_v \to u$ also $v \to u$, so $u_v \to v$, weakly in $H^1(\Omega, R^3)$.

Optimal Shape Design by Local Boundary Variations

Olivier Pironneau

University of Paris 6 - email: Olivier.Pironneau @ann.jussieu.fr

1. Introduction

Hadamard (1910) may have been the first applied mathematician to derive a formula for the sensitivity of a Partial Differential Equation (PDE) with respect to the shape of its domain. This opened the field of Optimal Shape Design (OSD). But the field as we know it now, really began with Cea et al (1973) as an offspring of optimal control theory (Lions (1968)) and the calcul of variation. So OSD has borrowed the vocabulary of control theory: the design is done by minimizing a *cost* function, which depends upon a *state* variable, i.e. the solution of the PDE, itself function of a *control*, the shape.

Among others, Pironneau (1973), Murat-Simon (1976), Cea (in Haug et al (1978) gave methods to derive optimality conditions for the continuous problems and Begis et al (1976) Morice (1976) and Marrocco et al (1978), in the same school, for the discretized problems.

Theoretical results on existence of solutions were obtained by Chenais (1975), Sverak (1992) Bucur et al. (1995) and Liu et al (1999); a counter example to existence was produced by Tartar (1975) in a key paper which linked optimal shape design with homogenization theory in what is now known as "topological optimization".

Most design engineers do their optimization by hand, intuitively. But it is generally believed that intuitive optimization is not possible beyond a handfold of degrees of freedom. When the design parameters are few, say less than a hundred, sensitivity with respect to shape can be obtained by finite difference approximation (take two $\epsilon-$close shapes and approximate the derivative by the difference of the values of the cost function divided by ϵ) and essentially no additional programming is needed beyond the state equation solver. But the precision may not be sufficient and stiff problens cannot be solved this way.

There are also commercial packages which find the minimum of a functional with respect to parameters and require from the user only a subroutine to evaluate the cost function for a given design. These packages are usally based on local variation methods (Powell(1970)), involving polynomial fits of the functional from point evaluations. They are expensive here because they require $O(P^2)$ solutions of the flow solver where P is the number of design variables.

But for 3D wings for example, there are hundreds of design parameters so that shape optimization requires a complete numerical treatment with a robust differentiable optimization package and a precise sensitivity analysis with respect to the shape of the wing.

A numerical fluid solver can be vewed as a C function with an input and an output , the design variables which define the wing shape and the drag for instance. Sensitivity analysis finds the gradient of the cost function with respect to the design variables. It is difficult when the fluid is compressible. An alternative is to let the computer do it for you by using a software for "Automatic Differentiation of programs" such as ADOL-C. This approach is extremely convenient and we shall give here a brief presentation. But to understand it fully it is better to know the analytical approach as well; this is the object of the paragraph on sensistivity analysis. More details can be found in Pironneau (1983), Neittanmaki (1991), and Banichuk (1990).

2. Examples

Before going to industrial examples let us present two laboratory examples which will serve to illustrate the method of solution chosen here.

2.1 Two Laboratory Test Cases: Nozzle Optimization

For clarity we will consider an optimization problem for incompressible irrotational inviscid flows

$$\min_{\partial\Omega}\{\int_D |\nabla\varphi - u_d|^2 : \quad -\Delta\varphi = 0 \text{ in } \Omega, \ \partial_n\varphi|_{\partial\Omega} = g\}$$

or with a stream function in 2D

$$\min_{\partial\Omega}\{\int_D |\nabla\psi - v_d|^2 : \quad -\Delta\psi = 0 \text{ in } \Omega, \ \psi|_{\partial\Omega} = \psi_\Gamma\}$$

In both problems one seeks for a shape which produces the closest velocity to u_d in the region D of Ω. In the second formulation the velocity of the flow is given by

$(\partial_2\psi, -\partial_1\psi)^T$ so $v_d = (u_{d2}, -u_{d1})^T$.

An application to wind tunnel or nozzle design for potential flow is obvious but it is laboratory because these are usually used with compressible flows.

2.2 Minimum weight of structures

In 2D linear elasticity, for a structure clamped in a part Γ_1 of its boundary $\Gamma = \partial\Omega$ and subject to volume forces F and surface shear g, the displacement $u = (u_1, u_2)$ is found by solving for u:

$$u \in V_0 = \{u \in H^1(\Omega)^2 : \quad u|_{\Gamma_1} = 0\}$$

$$\int_\omega [\partial_{tt}u \cdot v + \mu\epsilon_{ij}(u)\epsilon_{ij}(v) + \lambda\epsilon_{ii}(u)\epsilon_{jj}(v)] = \int_{\partial\Omega} g.v + \int_\Omega F.v \quad \forall v \in V_0$$

$$\text{where} \quad \epsilon_{ij} = \frac{1}{2}(\partial_i u_j + \partial_j u_i),$$

Many important problems of design arise when one wants to find the structure with minimum weight yet satisfying some inequality constraints for the stress such as in the design of light weight beams for strengthening of airplane floors, or for crank shaft optimization...

For all these problems the criteria for optimisation is the weight

$$J(\Omega) = \int_\Omega \rho,$$

where ρ is the density of the material.

But there are constraints on the maximum stress (itself a linear tensor function of the displacement tensor ϵ)

$$\tau(x) \cdot d < \tau_{dmax}$$

at some points x and for some directions d.

Indeed, a wing for instance, will behave differently under spanwise and chordwise load. Moreover, due to coupling between physical phenomena, the surface stresses come in part from fluid forces acting on the wing. This implies many additional constraints on the aerodynamical (drag, lift, moment) and structural (Lamé coefficients) characteristics of the wing. Therefore, the Lamé equations of the structure must be coupled with the equations for the fluid (fluid structure interactions). This is why most optimization problems nowadays require the solution of several state equations ("multiphysics").

2.3 Wing design

An important industrial problem is the optimization of the shape of a wing to reduce the drag. The drag is the reaction of the flow on the wing, its component in the direction of flight is the drag proper and the rest is the lift. A few percents of drag optimization means a great saving on commercial planes.

For viscous drag the Navier-Stokes equations must be used. For wave drag the Euler system is sufficient.

For a wing S moving at constant speed u_∞ the force acting on the wing is in a cartesian frame

$$F = (F_x, F_y, F_z)^T = \int_S [\mu(\nabla u + \nabla u^T) - \frac{2\mu}{3}\nabla.u]n - \int_S pn$$

The first integral is a viscous force, the so called viscous drag and the second is called the wave drag. In a frame attached to the wing, and with uniform flow at infinity, the drag is the component of F parallel to the velocity at infinity (i.e. $F.u_\infty$). The viscosity of the fluid is μ and p is its pressure.

The Navier-Stokes equations govern u the fluid velocity, θ the temperature, ρ the density and E the energy:

$$\partial_t\rho + \nabla.(\rho u) = 0$$

$$\partial_t(\rho u) + \nabla.(\rho u \otimes u) + \nabla p - \mu\Delta u - \frac{1}{3}\mu\nabla(\nabla.u) = 0,$$

$$\partial_t[rhoE] + \nabla\cdot[u\rho E] + \nabla\cdot(pu) = \nabla\cdot\{\kappa\nabla\theta + [\mu(\nabla u + \nabla u^T) - \frac{2}{3}\mu)\mathrm{I}\nabla\cdot u]u\}$$

$$\text{where } E = \frac{u^2}{2} + \theta \quad p = (\gamma - 1)\rho\theta$$

The problem is to minimize

$$J(S) = F.u_\infty$$

with respect to the shape of S.

There are several *constraints*:

- A geometrical constraint: the volume of S greater than a given value, else the solution will be a point.
- An aerodynamic constraint: the lift must be greater than a given value or the wing will not fly.

The problem is difficult because it involves the compressible Navier-Stokes equations at high Reynolds number. It can be simplified by considering only the wave drag i.e. the pressure term only in the definition of F (Jameson (1987)). When the viscous terms are dropped in the Navier-Stokes equations ($\mu = \kappa = 0$). Euler's equations remain. The problem is

$$\min_S \int_S pn \cdot u_\infty \qquad \text{subject to}$$

$$\partial_t\rho + \nabla.(\rho u) = 0$$

$$\partial_t(\rho u) + \nabla.(\rho u \otimes u) + \nabla p = 0,$$

$$\partial_t[rhoE] + \nabla\cdot[u\rho E] + \nabla\cdot(pu) = 0$$

$$\text{with } E = \frac{u^2}{2} + \theta \quad p = (\gamma - 1)\rho\theta$$

However, it is now well known that viscous effects have an important impact on the final shape (Mohammadi (1997)). Indeed, in transonic flows for instance the shock position is 30 percents chord upstream due to viscous effects.

Assuming irrotational flow an even greater simplication replaces the Euler equations by the compressible potential equation ($\gamma = 1.4$ for air):

$$u = \nabla\varphi, \quad \rho = (1 - |\nabla\varphi|^2)^{1/(\gamma-1)}, \quad p = \rho^\gamma, \quad \nabla.\rho u = 0.$$

Or even, if at low Mach number, by the incompressible potential flow equation:

$$u = \nabla\varphi, \quad -\Delta\varphi = 0.$$

Constraints on admissible shapes are numerous:

- Minimal thickness, given length.
- Minimum admissible curvature
- Minimal angle at the trailing edge...

Another problem arises due to instability of optimal shapes with respect to data. It will be seen that the leading edge of the solution is a wedge. Thus if the incidence angle of u_∞ is changed the solution becomes bad. A multi-point functional must be used in the optimization, for some weighting factors β_i

$$J(S) = \sum \beta_i u^i_\infty F^i \quad \text{or} \quad J(S) = \max_i \{u^i_\infty F^i\}$$

at given lift $F^i \times u_\infty$ where the F^i are computed from Navier-Stokes equations with boundary conditions $u = u^i_\infty$.

2.4 Stealth Wings

2.4.1 Maxwell equations. The optimization of the far-field energy of a radar wave reflected by an airplane in flight requires the solution of Maxwell's equations for the electric field E and the magnetic field H:

$$\epsilon\partial_t E + \nabla \times H = 0 \quad \nabla.E = 0, \qquad \mu\partial_t H - \nabla \times E = 0 \quad \nabla.H = 0.$$

The dielectric and magnetic coefficient ϵ, μ are constant in air but not so in an absorbing medium. One variable, H for instance, can be eliminated by differentiating in t the first equation:

$$\epsilon\partial_{tt} E + \nabla \times (\frac{1}{\mu}\nabla \times E) = 0,$$

from which it is easy to see that $\nabla.E = 0$ is always zero if it is zero at initial time.

2.4.2 Helmholtz equation. Now if the geometry is cylindrical with axis z and if $E = (0, 0, E_z)^T$ then the equation becomes a scalar wave equation for E_z. Furthermore if the boundary conditions are periodic in time at infinity, $E_z = \mathcal{R}_e v_\infty e^{i\omega t}$ and compatible with the initial conditions then the solution has the form $E_z = \mathcal{R}_e v(x) e^{i\omega t}$ where v, the amplitude of the wave E_z of frequency ω , is solution of:

$$\nabla(\cdot \frac{1}{\mu} \nabla v) + \omega^2 \epsilon v = 0$$

Notice the wrong sign for ellipticity in the "Helmholtz" equation.

Remark

1. This equation arises naturally in accoustics. So the technics of this paragraph applies also there.

2. In vacuum $\mu\epsilon = c^2$,c the speed of light, so for numerical purposes it is a good idea to rescale the equation. The critical parameter is then the number of waves on the object, i.e. $\omega c/L$ where L is the size of the object.

2.4.3 Boundary conditions. The reflected signal on solid boundaries Γ satifies

$$v = 0 \quad \text{or} \quad \partial_n v = 0 \quad \text{on } \Gamma$$

depending on the type of waves (Transverse Magnetic polarization requires Dirichlet condition).

When there is no object this Helmholtz equation has a simple sinusoidal set of solutions which we call v_∞:

$$v_\infty(x) = \alpha \sin(k \cdot x) + \beta \cos(k \cdot x), \qquad \text{i.e. } E_z = \mathcal{R}_e(Ae^{i(k \cdot x + \omega t)})$$

where k is any vector of modulus $|k| = \omega c$. Radar waves are more complex but by Fourier decomposition, they can be viewed as a linear combination of such simple unidirectional waves.

Now if such a wave is sent on a object, it is reflected by it and the signal at infinity is the sum of the original wave with the reflected wave. So it is better to set an equation for the amplitude of the reflected wave only $u = v - v_\infty$.

A good boundary condition for u is difficult to set; one possibility is

$$\partial_n u + iau = 0.$$

Indeed when $u = e^{id \cdot x}$, $\partial_n u + iau = i(d \cdot n + a)u$, so that this boundary condition is "transparent" to waves of direction d when $a = -d \cdot n$. If we want this boundary condition to let all *outgoing* waves pass the boundary best when it is normal to it, we will set a=1.

To sumarize, we set for u the system in the complex plane:

$$\nabla \cdot (\frac{1}{\mu} \nabla u) + \omega^2 u = 0, \quad \text{in } \Omega,$$
$$\partial_n u + iu = 0 \quad \text{on } \Gamma_\infty$$
$$u = g \equiv -e^{ik\cdot x} \quad \text{on } \Gamma.$$

where $\partial\Omega = \Gamma \cup \Gamma_\infty$. It can be shown that the solution exists and is unique. Notice that the variables have been rescaled, ω is ωc, μ is μ/μ_{vacuum}.

Usually the criteria for optimization is a minimum amplitude for the reflected signal in a region of space D at infinity (hence D is an angular sector). For instance one can consider

$$\min_{S\in O}\{\int_{\Gamma_\infty \cap D} |\nabla u|^2 : \quad \omega^2 u + \nabla \cdot (\frac{1}{\mu}\nabla u) = 0, \quad u_{|\Gamma} = g, \quad iu + \partial_n u|_{\Gamma_\infty} = 0\}$$

where μ is different from one only in a region very near Γ and schematizes an absorbing paint.

But constraints are aerodynamical as well, (lift above a given lower limit for instance) and thus requires the solution of the fluid part as well. The design variables are:

- The shape of the wing
- The thickness of the paint
- The material characteristics (ϵ, μ) of the absorbing paint.

Here again, the theoretical complexity of the problem can be appreciated from the following question:

Would ribblets of the size of the radar wave improve the design?

Actually homogenization can answser the question as in Achdou (1991) (see also Artola (1991) and Achdou et al (1991)) It shows that indeed ribblets improve the design and in practice absorbing paints on the wing surface work in the same manner.

Homogenization shows that periodic surfacic irregularities are equivalent to new "effective" boundary conditions

$$u = 0 \quad \text{replaced by} \quad au + \partial_n u = 0$$

and so the optimization can be done with respect to a also. Hence the connections between OSD and topological optimization.

2.5 Optimal brake water

As a first approximation, the amplitude of sea waves satisfies Helmholtz' equation

$$\nabla(\mu \cdot \nabla u) + \epsilon u = 0$$

where μ is a function of the water depth and ϵ is proportional to the wave speed.

With approximate reflection and damping whenever the waves collide on a brake water S which is surrounded by rocks we have

$$\partial_n u + au = 0 \quad \text{on} \quad S.$$

At infinity a non reflecting boundary condition can be used

$$\partial_n(u - u_\infty) + ia(u - u_\infty) = 0$$

The problem is to find the best S with given length so that the waves have minimum amplitudes in a given harbour D:

$$\min_S \int_D u^2.$$

2.6 Ribblets

Consider a flat plate with groves dug on the surface parallel to the mean flow. It has been shown that such configurations have less drag per unit surface area than the flat plate (Figure 5).

The phenomenon is turbulent in its principle (Moin (1993)) because these groves or ribblets trap the large vortices and retard the formation of horse shoe vortices. It is beyond the limit of present computers to hope to solve such problems by optimal design methods. However even the laminar case leads to an optimization and it is not true that the flat plate is the best surface for drag per unit surface area for a Poiseuille flow.

Consider ribblets which are well within the logarithmic layer and near the viscous sublayer. Apply the Couette flow approximation. Then the problem is:

$$\min_\Sigma u_\infty \cdot \int_\Sigma [\nu(\nabla u + \nabla u^T) - pn]$$

with (u, p) solution of

$$\mathbf{u} = \begin{pmatrix} 0 \\ 0 \\ u(x,y) \end{pmatrix} \quad \text{and} \quad p = p(z)$$

$$-\nu\Delta u + \nabla p = \begin{pmatrix} 0 \\ 0 \\ \frac{\partial p}{\partial z} - \nu\Delta_{x,y}u \end{pmatrix} = 0$$

A solution with $p = kz$ is found and u solves

$$-\nu\Delta u + k = 0$$

The domain is 2D and with a periodic distribution of ribblets, the domain is one cell containing one ribblet Σ with $u|_\Sigma = 0$ and a Neumann condition on the upper artificial boundary which simulates the matching with the boundary layer S and periodic conditions on the lateral boundaries of the cell. The problem becomes:

$$\min_{\Sigma}(-\int_{\Sigma}\frac{\partial u}{\partial n})$$

subject to (u, k) solution of:

$$-\nu\Delta u + k = 0 \text{ in } \Omega$$
$$u = 0 \text{ on } \Sigma \quad \frac{\partial u}{\partial n} = 0 \text{ on } S$$
$$u = x - \text{periodic} \quad \int_{\Omega} u = d.$$

The last constraint on the flux has been added to fix k:

2.7 Sonic boom reduction

Some supersonic carrier are considered too noisy. An optimization of the shock wave jump and of the jet noise can be performed with respect to the far field noise. Again the full problem involves the Navier-Stokes equations but simpler approximations like Lighthill's turbulent noise source approximation can be used and in the far field it is the wave equation which is solved.

3. Existence of Solutions

3.1 Generalities

Assume that $\psi(\Omega)$ is the solution of

$$-\Delta\psi = f \text{ in } \Omega, \quad \psi|_{\partial\Omega} = 0$$

and that

$$u_d \in L^2(\Omega), \quad f \in H^{-1}(\Omega)$$

For simplicity we have translated the nonhomogeneous boundary conditions of the laboratory examples above into a right hand side in the PDE ($f = \Delta\psi_\Gamma$).

Let $O \supset D$ be two given closed bounded sets in $R^d, d = 2, 3$ and consider

$$\min_{\Omega\in\mathcal{O}} J(\Omega) = \int_D |\nabla\psi(\Omega) - u_d|^2$$

with

$$\mathcal{O} = \{\Omega \subset R^d \ : \ O \supset \Omega \supset D, \ |\Omega| = 1\}.$$

where $|\Omega|$ denotes the area in 2D and the volume in 3D.

Chenais (1975) showed that there exists a solution provided that the class $\mathcal{O}$ is restricted to Ω which are:

1. locally on one side of their boundaries,
2. verifying the *Cone Property.*

Let $D_\epsilon(x,d)$ be the intersection with the sphere of radius ϵ and center x of the cone of vertex x direction d and angle ϵ.

Cone Property: *There exists ϵ such that for every $x \in \partial\Omega$ there exists d such that $\Omega \supset D_\epsilon(x,d)$.*

These two conditions imply that the boundary cannot oscillate too much.

Denote by $\mathcal{O}_\epsilon$ this set of admissible shapes.

Theorem:
The problem

$$\min_{\Omega\in\mathcal{O}_\epsilon} J(\Omega)$$

has at least one solution

Proof
The proof is done by considering a minimizing sequence Ω^n. The cone property implies that there exists Ω such that $\Omega^n \to \Omega$ in a sense sufficiently strong so that

$$\psi(\Omega^n)|_D \to \psi(\Omega)|_D, \quad \text{in } H^1(D)$$

$$\int_\Omega \nabla\psi(\Omega)\nabla w = \int_\Omega fw \quad \forall w \in H^1(\Omega).$$

Hence $J(\Omega^n) \to J(\Omega)$ and Ω is a solution.

In 2D an important result has been obtained by Sverak (1992):

Theorem .
If $\mathcal{O} = \mathcal{O}_N$ is the set of open sets containing D (possibly with a constraint on the area such as area ≥ 1) and whose number of connected component is bounded by N then

$$\min_{\mathcal{O}_N} J(\Omega) = \int_D |\nabla\psi(\Omega) - v_d|^2 \ : \ -\Delta\psi(\Omega) = f \ \ in \ \Omega, \ \ \psi(\Omega)|_{\partial\Omega} = 0\}$$

has a solution.

In other words, two things can happen to minimizing sequences:

- Either accumulation points are solutions
- Or the number of holes in the domain tends to infinity (and their size to zero).

This result is false in 3D as it is possible to make shapes with spikes such that a 2D cut will look like a surface with holes and yet the 3D surface remains singly connected. Bucur-Zolezio (1995) obtained an extension to 3D of the same idea by using capitance (see also Liu et al. (1999) for a result using equi-continuity for boundaries having the segment property (a segment of fixed size must fit in and out of the domain with one end on the boundary, at each boundary point) for the Neumann problem).

A corollary of their result can be summarized as:

If the boundary of the domain has the flat cone insertion property (each boundary point is the vertex of a fixed size 2D truncated cone which fits inside the domain) then the problem has at least one solution.

The proof of Sverak's theorem is sketched in Appendix A for the reader to see the kind of tools which are used in such studies.

3.2 Sketch of the proof of Sverak's Theorem

The proof relies on a compactness result for the Hausdorff topology and on a result of potential theory (capacitance).
The Hausdorff distance between 2 closed sets A, B is

$$\delta(A,B) = \max\{d(B,A), d(A,B)\} \quad \text{where } d(A,B) = \sup_{x\in A} d(x,B).$$

For this distance we have

Proposition

If F_n is a uniformly bounded sequence , then there is a closed bounded set F and a subsequence converging in the sense of Hausdorff to F.

Equivalently let Ω_n be a sequence of open sets in R^d with $\Omega_n \subset O$. Then one can extract a subsequence, also denoted by Ω_n converging in the sense of Hausdorff to a Ω, that is, verifying:

$$\forall C \subset \Omega, \exists m : C \subset \Omega_n \forall n \geq m \text{ and } \forall x \in O - \Omega, \exists x_n \in O - \Omega_n : x_n \to x.$$

So a minimizing subsequence for (2) will have the following properties

$$J(\Omega_n) \to \inf J(\Omega)$$

$$-\Delta\psi_n = f \ \ in\ \Omega_n, \quad \psi \in H^1(\Omega_n) \text{ and } \psi_n \to \psi \quad \text{in } H^1(O) \text{ weakly with },$$

$$-\Delta\psi = f \ \ in\ \Omega, \quad \inf J(\Omega) = \int_D |\nabla\psi - v_d|^2.$$

But we do not know how to show that

$$\psi = 0 \ \ in \ \ O - \Omega$$

For this an information on the characteristic function χ_n of $O - \Omega_n$ is needed because

$$0 = \chi_n \psi_n \to \chi\psi, \ \Rightarrow \ \psi(x) = 0 \ pp \ si \ \ \chi(x) \neq 0.$$

Sverak uses another argument. First he shows that it is sufficient to study the case $f = 1$. If Ω^n denotes the solution in $H_0^1(\Omega^n)$ of $-\Delta\Omega^n = 1$ then the convergence of Ω^n towards its weak limit is almost uniform (this is the difficult point) when the number of connected components is finite.

This result from the theory of sub-harmonic functions is true in 3D also with an hypothesis of capacitance. Hence a generalization can be found in Bucur et al (1995) where by existence is shown under the only restriction that one can fit a flat cone (a 2D cone as in Chesnais but for a 3D surface, so it is much more general) at each point of the boundary.

Corollary

Given N and the 2D-Navier-Stokes equations for incompressible flows there exists an optimal wing profile with given area in 2D in the class of uniformely bounded domains with less than N connected components

Proof

Let Ω^n be a minimizing sequence. Let u^n be the corresponding solution of the Navier-Stokes equations :

$$-\nu\Delta u^n + \nabla.(u^n \otimes u^n) + \nabla p^n = 0, \ \ \nabla \cdot u^n = 0 \ \ in \ \ \Omega^n, \ \ u^n|_S = 0, \ \ u|_{\Gamma_\infty} = u_\infty$$

By hypothesis Ω^n is bounded by O. From the Navier-Stokes equations it is easy to see that u^n extended by 0 in O is bounded in $H_0^1(O)^2$, so there exists a subsequence which converges weakly; let u be the limit. Now

$$f^n \equiv \nabla \cdot (u^n \otimes u^n) \to f \equiv \nabla \cdot (u \otimes u) \ \text{ in } W^{-1,p}(O), \ \forall p,$$

But now if

$$-\Delta u^n + \nabla p^n = -f^n \quad \nabla \cdot u^n = 0$$
$$f^n \to f \ \text{ in } W^{-1,p}(O),$$

then u is solution of the same Stokes problem with f instead of f^n. It remains to show that $u|_{\Omega - S} = 0$ but that is done for the Stokes problem exactly as for the Laplace equation since Stokes equation is a Laplacian in the space of solenoidal fields.

4. Solution By Optimization Methods

4.1 Gradient Methods

At the basis of gradient methods is the Taylor expansion of

$$J: \ V \to \mathcal{R}$$

where, if V is a Hilbert space,

$$J(v + \lambda w) = J(v) + \lambda < \text{Grad}_v J, w > + o(\lambda\|w\|), \quad \forall v, w \in V, \quad \forall \lambda \in \mathcal{R}.$$

where V is a Hilbert space with scalar product $< \cdot, \cdot >$ and $\text{Grad}_v J$ is the element of V given by Ritz' theorem and defined by

$$< \text{Grad}_v J, w >= J'_v w, \quad \forall w \in V.$$

By taking $w = -\rho \text{Grad}_v J(v)$, with $0 < \rho << 1$ we find :

$$J(v + w) - J(v) = -\rho\|\text{Grad}_v J(v)\|^2 + o(\rho\|\text{Grad}_v J(v)\|)$$

Hence if ρ is small enough the first term on the right hand side will dominate the remainder and the sum will be negative:

$$\rho\|\text{Grad}_v J(v)\|^2 > o(\rho\|\text{Grad}_v J(v)\|) \quad \Rightarrow \quad J(v + w) < J(v)$$

Thus the sequence defined by :

$$v^{n+1} = v^n - \rho \text{Grad}_v J(v), \quad n = 0, 1, 2, \ldots$$

makes $J(v^n)$ monotone decreasing. We have the following result:

Theorem: *If J is continuous, bounded from below, and $+\infty$ at infinity, then all accumulation points v^* of v^n satisfy*

$$\text{Grad}_v J(v^*) = 0.$$

This is the so called *optimality condition* of the order 1 of the problem. If J is convex then it implies that v^* is a minimum; if J is strictly convex the minimum is unique.

By taking the best ρ in the *direction of descent* $w^n = -\text{Grad}_v J(v^n)$,

$$\rho^n = arg \min_\rho J(v^n + \rho w^n) \text{ (meaning that } J(v^n + \rho^n w^n) = \min_\rho J(v^n + \rho w^n))$$

we obtain the so called *method of steepest descent with optimal step size*

We have to remark however, that minimizing a one parameter function is not all that simple. The exact minimum cannot be found in general, except for polynomial functions J. So in the general case, several evaluations of J are required for an approximate minimum only.

A closer look at the convergence proof of the method shows that it is enough to find ρ^n with the following property (Armijo rule):

Given $0 < \alpha < \beta < 1$, find ρ such that

$$-\rho\beta||\text{Grad}_v J(v^n)||^2 < J(v^n - \rho\text{Grad}_v J(v^n)) - J(v^n) < -\rho\alpha||\text{Grad}_v J(v^n)||^2$$

It can be found by relating β to α, in the following fashion:

Choose two numbers $0 < \rho_0 < 1$, $\omega \in (0,1)$ and find $\rho = \rho_0^k$ where k is the first integer such that

$$\begin{aligned} J(v^n - \rho_0^{k+1}\text{Grad}_v J(v^n)) - J(v^n) &< -\rho_0^{k+1}\omega||\text{Grad}_v J(v^n)||^2 \\ -\rho_0^k\omega||\text{Grad}_v J(v^n)||^2 &< J(v^n - \rho_0^k\text{Grad}_v J(v^n)) - J(v^n) \end{aligned}$$

4.2 Newton Methods

Newton's method with optimal step size applied to the minization of J is

$$\begin{aligned} \text{Compute } w \text{ solution of} \quad J"_{vv}w &= -\text{Grad}_v J(v^n), \\ \text{Set} \quad v^{n+1} &= v^n + \rho w \\ \text{with } \rho = arg \min_\rho J(v^n &+ \rho w) \end{aligned}$$

Near to the solution it can be shown that $\rho^n \to 1$ so that it is also the root finding Newton method applied to the optimality condition

$$\text{Grad}_v J(v) = 0$$

It is quadratically convergent but it is expensive and usually $J"$ is difficult to compute, so a quasi-Newton, where an approximation of $J"$ is used, is prefered. For instance, a directional approximation can be found by:

$$\begin{aligned} &\text{Choose } 0 < \epsilon << 1, \quad w \text{ approximate solution of} \\ &\frac{1}{\varepsilon}(\text{Grad}_v J(v^n + \epsilon w) - \text{Grad}_v J(v^n)) = J"_{vv}(v^n).w, \end{aligned}$$

4.3 Constraints

In constrained optimization, we can have equality or inequality constraints on the optimization parameters or the state variables. When using gradient methods, equality constraints are usually taken into account by penalization in J while inequality constraints are treated by projection when they concern the optimization parameters directly. If they concern the state variables, usually they are transformed to equality constraint and then penalized.

Consider the following minimization problem under equality and inequality constraint on the parameters and state:

$$\min_x J(x, u(x)), \quad A(x, u(x)) = 0,$$

subject to

$$B(x, u(x)) \leq B_0, \quad C(x, u(x)) = C_0, \quad x_{\text{min}} \leq x \leq x_{\text{max}},$$

here A, B, C involve the parameters x and the state variable u (state constraints) while the last constraints is a box constraint on the parameters only. The problem can be approximated by "penalty"

$$\min_x J(x, u(x)) + \beta|(B - B_0)^+|^2 + \gamma|C - C_0|^2,$$

subject to

$$A(x, u(x)) = 0, \quad x_{\text{min}} < x < x_{\text{max}}.$$

β and γ are penalization parameters. They are usually difficult to choose.

At each iteration of the gradient method, the new prediction is kept inside this box x_min, x_max by projecting the gradient. To improve the treatment of constraints interior point algorithms can be used.

5. Sensitivity Analysis

Gradient and Newton methods require gradients of the cost function J and for this we need to identify an underlying Hilbert structure for the parameters of J, the shape. Two ways have been proposed:

- Assume that all admissible shapes are obtained by mapping a reference domain $\hat{\Omega}$: $\Omega = T(\hat{\Omega})$. Then the parameter of J is $T : R^d \to R^d$. A possible Hilbert space for T is the Sobolev space of order m and it seems that $m = 2$ is a good choice.
- What is important is a Hilbert structure for the tangent plane of the parameter space, meaning by this that the Hilbert structure is needed only for small variations of Ω, so that one works with local variations defined around a reference boundary Σ by

$$\Gamma(\alpha) = \{x + \alpha(x) n_\Sigma(x) \; : \; x \in \Sigma\}$$

where n_Σ is the normal to Σ at x and Ω is the domain which is on the left side of the oriented boundary $\Gamma(\alpha)$. Then the Hilbert structure is placed on α, for instance $H^m(\Sigma)$.

Comments It is generally believed that PDE-parameter optimization (here T) is more difficult than shape optimization numerically.

Before proceeding we need the following preliminary result. In most cases only one part of the boundary Γ is optimized, we call this part S.

Proposition
Consider a small perturbation S' of S given by

$$S' = \{x + \lambda \alpha n : x \in S\}$$

where α is a function of x via the curvilinear abscissa of x on S and λ is a positive number destined to tend to zero. Denote $\Omega' = \Omega(S')$. Then for any $f \in H^1(C), C \supset \Omega \cup \Omega'$

$$\int_{\Omega(S')} f - \int_{\Omega(S)} f = \int_{\Omega(S')-\Omega(S)\cap\Omega(S')} f - \int_{\Omega(S)-\Omega(S)\cap\Omega(S')} f$$
$$= \lambda \int_S \alpha f + o(\lambda\|\alpha\|)$$

$$\text{and so} \quad \lim_{\lambda\to 0} \frac{1}{\lambda}[\int_{\Omega(S')} f - \int_{\Omega(S)} f] = \int_S \alpha f$$

Remark If S has an angle not all variations S' can defined by local variation on S but it can be shown that it is a sufficient class of variations.

Similarly the following can be proved (Pironneau(1983), p87).

Proposition
If $g \in H^1(S)$ and if R denotes the mean radius of curvature of S in any local basis ($1/R = 1/R_1 + 1/R_2$ in 3D) then

$$\lim_{\lambda\to 0} \frac{1}{\lambda}[\int_{S'} g - \int_S g] = \int_S \alpha(\partial_n g - \frac{g}{R})$$

5.1 Sensitivity Analysis for the nozzle problem

Consider

$$\min_{\partial\Omega\in\mathcal{O}} \int_D |\nabla\phi - u_d|^2$$
$$\text{subject to :}$$
$$-\Delta\phi = 0 \text{ in } \Omega, \quad a\phi + \partial_n\phi = g \text{ on } \partial\Omega,$$

the class of admissible shapes $\mathcal{O}$ being the set bounded domains with Lipschitz continuous boundaries containing D; but we will not worry about this constraint set for the time being and assume all constraints are verified by all variations encountered. In practice however we may have even additional constraints such as $\mathcal{O} \subset C$.

If $a = 0$ it is the potential flow formulation and if $a \to \infty, g = af$ it becomes the stream function formulation.

Assume that some part of $\Gamma = \partial\Omega$ is fixed, the unkown part being called S.

The variational formulation of the Laplace equation with Fourier boundary condition is

$$\text{Find } \phi \in H^1(\Omega) \text{ such that}$$
$$\int_\Omega \nabla\phi \cdot \nabla w + \int_\Gamma a\phi w = \int_\Gamma gw, \quad \forall w \in H^1(\Omega).$$

The Lagrangian of the problem is

$$L(\phi, w, S) = \int_D |\nabla\phi - u_d|^2 + \int_{\Omega(S)} \nabla\phi \cdot \nabla w + \int_\Gamma (a\phi w - gw)$$

and the minimization of J is equivalent to the min-max problem

$$\min_{S,\phi} \max_v L(\phi, v, S).$$

Recall that

$$J'_S(S, \phi) = L'_S(\phi, v, S) \text{ at the solution } \phi, v \text{ of the min-max}$$

Let us write that the solution is a saddle point of L. As L is linear in w and quadratic in ϕ, stationarity in these variables is simply

$$\begin{aligned}
\partial_\lambda L(\phi + \lambda\hat{\phi}, v, S) &= 2\int_D (\nabla\phi - u_d) \cdot \nabla\hat{\phi} + \int_{\Omega(S)} \nabla\hat{\phi} \cdot \nabla v \\
&+ \int_\Gamma a\hat{\phi}v = 0, \quad \forall\hat{\phi} \\
\partial_\lambda L(\phi, v + \lambda w, S) &= \int_{\Omega(S)} \nabla\phi \cdot \nabla w \\
&+ \int_\Gamma (a\phi w - gw) = 0 \quad \forall w
\end{aligned}$$

Acording to the 2 propositions above, stationarity with respect to S is

$$\begin{aligned}
\lim_{\lambda\to 0} \frac{1}{\lambda}[L(\phi, w, S') - L(\phi, w, S)] &= \\
\int_S \alpha[\nabla\phi \cdot \nabla w \quad + \quad \partial_n(a\phi w - gw) &- \frac{1}{R}(a\phi w - gw)] = 0
\end{aligned}$$

and so we have shown that

Theorem : *The variation of J with respect to the shape deformation $S' = \{x + \alpha(x)n_S(x) \ : \ x \in S\}$ is*

$$\delta J \equiv J(S', \phi(S')) - J(S, \phi(S)) = \int_S \alpha \nabla\phi \cdot \nabla v$$

$$+ \int_S \alpha[\partial_n(a\phi v - gv) - \frac{1}{R}(a\phi v - gv)] + o(\|\alpha\|)$$

where $v \in H^1(\Omega(S))$ is the solution of

$$\int_{\Omega(S)} \nabla\hat{\phi} \cdot \nabla v + \int_\Gamma a\hat{\phi}v = 0, \quad \forall\hat{\phi} \in H^1(\Omega(S))$$

Notice that the boundary conditions for ϕ and v being

$$\partial_n\phi + a\phi = g, \qquad \partial_n v + av = 0$$

we can eliminate a from the optimality conditions and find

$$\delta J = \int_S \alpha[\partial_s\phi \cdot \partial_s v - \partial_n\phi \cdot \partial_n v - \partial_n(gv) + \frac{1}{R}v\partial_n\phi].$$

where ∂_s denotes the derivative with respect to the curvilinear coordinate of S.

Corollary : *With homogeneous Neumann conditions ($a = g = 0$)*

$$\delta J = \int_S \alpha\partial_s\phi \cdot \partial_s v + o(\|\alpha\|)$$

and with homogeneous Dirichlet conditions on S ($a \to \infty, g = 0$)

$$\delta J = -\int_S \alpha\partial_n\phi \cdot \partial_n v + o(\|\alpha\|)$$

5.2 Discretization with Triangular Elements

For discretization let us use the simplest, a *Finite Element Method* of degree 1 on triangles. Unstructured meshes are better for OSD because they are easier to deform and adapt for a general shape deformation.

More precisely, Ω is approximated by $\Omega_h = \cup_{k=1}^{n_T} T_k$ where the T_k are triangles such that

- The vertices of $\partial\Omega_h$ are on $\partial\Omega$ and the corners of $\partial\Omega$ are vertices of $\partial\Omega_h$.
- $T_k \cap T_l$, ($k \neq l$) is either a vertex or an entire edge or empty.
- Triangulations are indexed on the longest edge, of size h, and as $h \to 0$ no angle should go to zero or π.

The Sobolev space $H^1(\Omega)$ is approximated by

$$H_h = \{w_h \in C^o(\bar{\Omega}_h) : w_h|_{T_k} \in P^1 \quad \forall k\}$$

where $P^1 = P^1(T_k)$ is the space of linear polynomials. The discrete problem in variational form is

$$\min_{\{(q^i)_{i=1}^{n_v} \in Q\}} J(q^1, ..., q^{n_v}) = \int_D \|\nabla\phi_h - (u_h)_d\|^2$$

$$\text{subject to } \phi_h \in H_h \text{ solution of :}$$

$$\int_{\Omega_h} \nabla\phi_h \cdot \nabla w^j + \int_\Gamma a\phi w^j = \int_\Gamma g w^j, \forall j \in [1, ..., n_v]$$

The dimension of H_h equals n_v the number of vertices q^i of the triangulation and every function ϕ_h belonging to H_h is completely determined by its values on the vertices $\phi_h(q^i)$.

The canonical basis of H_h is the set of so-called *hat functions* defined by

$$w^i \in H_h, \quad w^i(q^j) = \delta_{ij}$$

Denoting by ϕ_i the coefficient of ϕ_h on that basis,

$$\phi_h(x) = \sum_1^{n_v} \phi_i w^i(x),$$

the PDE

$$\int_{\Omega_h} \nabla\phi_h \cdot \nabla w^j + \int_\Gamma a\phi w^j = \int_\Gamma g w^j, \forall j \in [1, ..., n_v]$$

yields a linear system for $\Phi = (\phi_i)$

$$A\Phi = F, \quad A_{ij} = \int_{\Omega_h} \nabla w^i \nabla w^j + \int_{\Gamma_h} a w^i w^j, \quad F_j = \int_{\Gamma_h} g w^j.$$

Hence in matrix form the problem is to find $Q = (q^i)$ solution of

$$\min_{Q \in \mathcal{Q}} \{J(Q) = \Phi^T B \Phi - 2U \cdot \Phi \ : \ A(Q)\Phi = F\}$$

$$\text{with } B_{ij} = \int_D \nabla w^i \nabla w^j, \quad U_j = \int_D u_d \cdot \nabla w^j.$$

where B and U are independent of Q if the triangulation is fixed within D. For simplicity we shall assume that F does not depend on Q, i.e. that $g = 0$ on S.

Remark

The method applies also to Dirichlet conditions treated by penality, as explained before. However, in practice it is necessary for numerical quality to use a lumped quadrature in the integral of the Fourier term, or equivalently to apply $\phi = \phi_\Gamma$ at all points of Γ by

$$A_{ij} = \int_{\Omega_h} \nabla w^i \nabla w^j + p\delta_{ij}\delta(q^i \in \Gamma_h), \quad F_j = p\phi_\Gamma(q^j)\delta(q^j \in \Gamma_h).$$

where p is a large number.

We present below a computation of discrete gradients for a Neumann problem but the method applies also to Dirichlet problems with this modification.

5.3 Discrete Gradients

A straightforward calculus of variation gives

$$\delta J = 2(B\Phi - U) \cdot \delta\Phi \quad \text{with } A\delta\Phi = -(\delta A)\Phi$$

Introducing Ψ solution of $A^T\Psi = 2(B\Phi - U)$ leads to

$$\delta J = (A^T\Psi) \cdot \delta\Phi = \Psi^T A\delta\Phi = -\Psi \cdot ((\delta A)\Phi)$$

To evaluate δA we need 3 lemmas. If δQ is a variation of vertex positions (i.e. each vertex q^i moves by δq^i), we define

$$\delta q_h(x) = \sum_1^{n_v} \delta q^i w^i(x), \quad \forall x \in \Omega_h$$

and denote by Ω'_h the new domain.

Lemma 1 (see Figure 4)

$$\delta w^j = -\nabla w^j \cdot \delta q_h + o(\|\delta q_h\|)$$

Lemma 2

$$\int_{\delta\Omega_h} f \equiv \int_{\Omega'_h} f - \int_{\Omega_h} f = \int_{\Omega_h} \nabla \cdot (f\delta q_h) + o(\|\delta q_h\|)$$

Lemma 3

$$\int_{\delta\Gamma_h} g \equiv \int_{\Gamma'_h} g - \int_{\Gamma_h} g = \int_{\Gamma_h} gt \cdot \partial_s \delta q_h + \int_{\Gamma_h} \delta q_h \nabla g + o(\|\delta q_h\|)$$

where ∂_s denotes the derivative with respect to the curvilinear abscissa and t the oriented tangent vector of Γ_h.

In these the integrals are sums of integrals on triangles or edges and so f and g can be piecewise discontinuous across elements or edges.

Proofs

Proofs for Lemma 1 & 2 are in Pironneau (1983), so only the proof of Lemma 3 is given here.
Consider an edge $e_l = q^j - q^i$ and an integral on that edge

$$I_l = \|q^j - q^i\| \int_0^1 g(q^i + \lambda(q^j - q^i))d\lambda.$$

Then

$$\delta I_l = (\delta q^j - \delta q^i) \cdot (q^j - q^i)\|q^j - q^i\|^{-2} I_l$$
$$+\|q^j - q^i\| \int_0^1 ((1-\lambda)\delta q^i + \lambda \delta q^j)\nabla g(q^i + \lambda(q^j - q^i))d\lambda + o(\delta q_h)$$
$$= \int_{\Gamma_h} gt \cdot \partial_s \delta q_h + \int_{\Gamma_h} \delta q_h \nabla g + o(\delta q_h)$$

Now putting the pieces together (we omit to write the remainders $o(\ )$),

$$\delta \int_{\Omega_h} \nabla w^i \cdot \nabla w^j = \int_{\delta\Omega_h} \nabla w^i \cdot \nabla w^j + \int_{\Omega_h} [\nabla \delta w^i \cdot \nabla w^j + \nabla w^i \cdot \nabla \delta w^j]$$
$$= \int_{\Omega_h} [\nabla \cdot (\delta q_h \nabla w^i \cdot \nabla w^j) - \nabla(\nabla w^i \cdot \delta q_h) \cdot \nabla w^j - \nabla w^i \cdot \nabla(\nabla w^j \cdot \delta q_h)]$$

$$\delta \int_{\Gamma_h} w^i \cdot w^j = \int_{\delta\Gamma_h} w^i \cdot w^j + \int_{\Gamma_h} [\delta w^i \cdot w^j + w^i \cdot \delta w^j]$$
$$= \int_{\Gamma_h} w^i \cdot w^j t \cdot \partial_s \delta q_h + \int_{\Gamma_h} \delta q_h \nabla(w^i \cdot w^j)$$
$$- \int_{\Gamma_h} [(\nabla w^i \cdot \delta q_h) \cdot w^j + (\nabla w^j \cdot \delta q_h) \cdot w^i]$$

giving

Proposition

$$\delta J = \int_{\Omega_h} \nabla \psi_h^T (\nabla \cdot \delta q_h - \nabla \delta q_h - \nabla \delta q_h^T) \nabla \Phi_h + a \int_{\Gamma_h} \psi_h \cdot \Phi_h t^T \nabla \delta q_h t$$
$$+ o(\|\delta q_h\|)$$

where t is the tangent vector to Γ_h, $\psi_h = \sum \Psi_i w^i$ and Ψ is solution of $A^T \Psi = 2(B\Phi - U)$.

Consequently an iterative process like the method of steepest descent to compute the optimal shape will move each vertex of the triangulation in the direction opposite to the partial derivative of J with respect to the vertex coordinates (E^k is the kth unit vector of R^d):

$$q_k^i := q_k^i - \rho[\int_{\Omega_h} \nabla\psi_h^T(\nabla\cdot(E^k w^i) - \nabla(E^k w^i) - \nabla(E^k w^i)^T)\nabla\phi_h$$
$$+a\int_{\Gamma_h} \psi_h \cdot \phi_h t^T \nabla(E^k w^i) t]$$

5.4 Implementation problems

Computation of discrete derivatives of cost functional is, as we have seen a crafty work, only reasonnable for simple problems.

Another difficulty is that for practical applications the optimization problem is changed all the time by the designer until a feasible situation is reached. A first cost function and constraint sets are set, the solution is found to violate certain unforeseen constraints so the constraint set is changed... Finally multipoint optimization is desired so the cost function and equations are changed... and each time the discrete gradients must be computed. *Automatic differentiation* is the cure but as we shall see it has its own difficulties.

Mesh distortion is also a big problem. After a few iterations the mesh is no longer feasible. A remeshing will induce interpolation errors which may cause divergence in the optimization process if done too often. *Automatic mesh adaption* and motion is the cure, it will also be explained in a coming chapter.

Finally boundary oscillation is also a frequent curse usually due to a wrong choice of scalar product in the optimization algorithm. We will give some elements of answer below.

5.5 Optimal shape design with Stokes flow

The drag and lift are the only forces at work in the absence of gravity. If the body is symmetric and its axis is aligned with the velocity at infinity then there is no lift and therefore we can equally well minimize the energy of the system, which for Stokes flow gives the following problem:

$$\min_{\Omega\in\mathcal{O}} J(\Omega) = \nu\int_\Omega |\nabla u|^2$$
$$\text{subject to :}$$
$$-\nu\Delta u + \nabla p = 0, \quad \nabla\cdot u = 0 \text{ in } \Omega$$
$$u|_S = 0, \quad u|_{\Gamma_\infty} = u_\infty$$

An example of $\mathcal{O}$ is :

$$\mathcal{O} = \{\Omega,\ \partial\Omega = S \cup \Gamma_\infty, |\tilde{S}| = 1\},$$

where $\tilde{S}$ is the domain inside the closed boundary S and $|\tilde{S}|$ is its volume or area in 2D.

Sensitivity analysis is as before; let $\Omega' \in \mathcal{O}$ be a domain "near" Ω defined by its boundary $\Gamma' = \partial\Omega'$, with

$$\Gamma' = \{x + \alpha(x)n(x), \text{ with } \alpha = \text{ regular, small, } \forall x \in \Gamma = \partial\Omega\}$$

Define also

$$\delta u = u(\Omega') - u(\Omega) \equiv u' - u$$

while extending u by zero in $\tilde{S}$. Then

$$\delta J = \nu\delta(\int_\Omega |\nabla u|^2) = \nu \int_{\delta\Omega} |\nabla u|^2 + 2\nu \int_\Omega \nabla\delta u : \nabla u + o(\delta\Omega, \delta u).$$

When ∇u is smooth, then

$$\nu \int_{\delta\Omega} |\nabla u|^2 = \nu \int_\Gamma \alpha|\nabla u|^2 + o(\|\alpha\|_{C^2}) = \nu \int_\Gamma \alpha|\partial_n u|^2 + o((\|\alpha\|_{C^2}).$$

Now $\delta u, \delta p$ satisfy

$$-\nu\Delta\delta u + \nabla\delta p = 0, \quad \nabla\cdot\delta u = 0 \text{ in } \Omega$$
$$\delta u|_{\Gamma_\infty} = 0, \quad \delta u|_S = -\alpha\partial_n u$$

Indeed the only non-obvious relation is the boundary condition on S. Now by a Taylor expansion

$$u'(x + \alpha n) = u'(x) + \alpha\partial_n u'|_{S'} + o(|\alpha|) = 0 \text{ since } u'|_{S'} = 0.$$

Now $u\ |_S = 0$ so,

$$\delta u|_S = -\alpha\partial_n u|_S.$$

Consequently ($A : B$ means $\sum_{ij} A_{ij}B_{ij}$)

$$\nu \int_\Omega \nabla\delta u : \nabla u = \nu \int_\Omega (-\Delta u)\cdot\delta u + \nu \int_\Gamma \partial_n u \cdot \delta u =$$
$$\int_\Omega p\nabla\cdot\delta u - \int_\Gamma p\delta u \cdot n + \nu \int_\Gamma \partial_n u \cdot \delta u =$$
$$\int_\Gamma (\nu\partial_n u - pn)\cdot\delta u = -\int_S \nu\alpha|\partial_n u|^2,$$

because, if s denotes the tangent component,

$$n\cdot\partial_n u = -s\cdot\partial_s u = 0 \quad \text{on } \Gamma.$$

We have proved the

Proposition 3 : *The variation of J with respect to Ω is :*

$$\delta J = -\nu \int_S \alpha|\partial_n u|^2 + o(\alpha)$$

Consequences

- If $\mathcal{O} = \{S : S \supset C\}$, as $|\partial_n u|^2 > 0$, then C is the solution (no fairing around C will decrease the drag in Stokes flow).
- If $\mathcal{O} = \{S : Vol\ \tilde{S} = 1\}$, then the object with minimum drag saisfies $\partial_n u \cdot s$ constant on S. Lighthill (cf. Pironneau (1973)) showed that near the leading and the trailing edge the only possible axisymmetric flow which can achieve this condition must have conical tips of half angle equal to 60°.
- The method of steepest descent gave a shape near to the optimal shape after one iteration (cf. Pironneau (1973)), and it was confirmed in Bourot (1976) by a Newton method.

A similar analysis can be done for the Navier-Stokes equation for incompressible flows.

The Optimal shape undeer the constraint that the volume is fixed and that the shape be axisymmetric, is given on Figure 1.

5.6 OSD for laminar flow

Consider the minimum drag/energy problem with the Navier-Stokes equations.

$$\min_{S\in\mathcal{O}} J(\Omega) = \nu \int_\Omega |\nabla u|^2 \quad \text{subject to}$$
$$-\nu\Delta u + \nabla p + u\nabla u = 0, \quad \nabla\cdot u = 0, \text{ in } \Omega$$
$$u|_S = 0, \quad u|_{\Gamma_\infty} = u_\infty$$

and with $\mathcal{O} = \{S \ : \ |\tilde{S}| = 1\}$, $\Gamma = \partial\Omega = S_\infty \cup \Gamma$.

Let us express the variation of $J(\Omega)$ in terms of the variation α of Ω.

As for the Stokes problem,

$$\delta J = J(\Omega') - J(\Omega) = \nu \int_{\delta\Omega} |\nabla u|^2 + 2\nu \int_\Omega \nabla u \nabla \delta u + o(\delta u, \alpha).$$

but now the equation of δu is no longer self adjoint

$$-\nu\Delta\delta u + \nabla\delta p + u\nabla\delta u + \delta u\nabla u = o(\delta u),$$
$$\nabla\cdot\delta u = 0$$
$$\delta u|_{\Gamma_\infty} = 0, \quad \delta u|_S = -\alpha\partial_n u$$

So an adjoint equation is introduced with an adjoint state (P, q) :

$$-\nu\Delta P + \nabla q - u\nabla P - (\nabla P)u = -2\nu\Delta u, \quad \nabla\cdot P = 0 \text{ in } \Omega$$
$$P|_\Gamma = 0.$$

Proposition

The variation of J with respect to Ω is :

$$\delta J = \nu \int_S \alpha(\frac{\partial P}{\partial n} - \partial_n u) \cdot \partial_n u + o(\alpha)$$

For the chosen admissible set $\mathcal{O}$ we have that $\delta J \geq 0$ for every α with $\int_\Gamma \alpha = 0$. So, the optimality condition for this problem is :

$$\partial_n u \cdot (\partial_n P - \partial_n u) = \text{ constant on } S.$$

Proof

Multiply the equation for (P, q) by δu and integrate by parts

$$\int_\Omega \nu \nabla P : \nabla \delta u - \int_S \partial_n P \cdot \delta u - q \nabla \cdot \delta u + P \nabla \cdot (u \otimes \delta u + \delta u \otimes u)$$
$$= 2 \int_\Omega \nu \nabla u : \nabla \delta u - 2 \int_\Gamma \nu \partial_n u \cdot \delta u.$$

Then use the equation of δu multiplied by P and integrated on Ω

$$\int_\Omega \nu \nabla P : \nabla \delta u + P \nabla \cdot (u \otimes \delta u + \delta u \otimes u) = 0$$

So

$$\delta J = \nu \int_{\delta\Omega} |\nabla u|^2 + \int_S \alpha(\partial_n P - 2\partial_n u) \cdot \partial_n u$$
$$= \nu \int_S \alpha(|\partial_n u|^2 + (\partial_n P - 2\partial_n u) \cdot \partial_n u)$$

6. Alternative ways

An alternative method to obtain the discrete optimality conditions is to see that

$$A\Phi = F \quad \text{with} \quad A_{ij} = \int_\Omega \nabla w^i \cdot \nabla w^j$$

Therefore

$$A\delta\Phi = -(\delta A)\Phi + \delta F$$

with

$$\delta A = \int_{\delta\Omega} \nabla w^i \cdot \nabla w^j + \int_\Omega \nabla \delta w^i \cdot \nabla w^j + \int_\Omega \nabla w^i \cdot \nabla \delta w^j$$

Next use Lemma 4 for the first term and lemma 3 for the two others

$$\delta A = \int_\Omega \nabla w^i \cdot \nabla w^j \nabla \cdot \delta q - \int_\Omega (\nabla \delta q \nabla w^i) \cdot \nabla w^j - \int_\Omega (\nabla \delta q \nabla w^j) \cdot \nabla w^i$$

with the convention that the function of x, $\delta q = \sum \delta q^i w^i$.

7. Problems Connected With The Numerical Implementation

7.1 Independence from J

Note that the adjoint state p depends on the criterion t. On the other hand if the software is to be provided as a black box to the industry it must be such that it is easy to :

- change the design criterion
- add geometrical contraints.

Suppose that we minimize a functional of the general form :

$$J(\phi,\Omega)=\int_D f(\phi)dx, \quad \phi=\{\phi^j\}, j=1,\dots,r.$$

Since the second member of the adjoint state equation is δE, we must be able to compute $\frac{\partial E}{\partial \phi_j}$ independently of $J(\phi,\Omega)$.

This computation can be done by finite differences because :

$$\frac{\partial E}{\partial \phi_j}\simeq\frac{J(\phi_h+\delta\phi_h,\Omega_h)-J(\phi_h,\Omega_h)}{\delta\phi_j}$$

This computation is not expensive. The number of elementary computations is of order N. Indeed, if N is the number of the mesh nodes, the calculation cost is of the order N, which is the same cost as the solution of a laplacian (cf. Arumugam(1989)).

7.1.1 Add geometrical constraints. To add geometrical constraints is easy if we give a parametrized description of the domain and its triangulation.

If the boundary to optimize is described by r parameters α_j, we can define it by a curve (ex. spline) defined by α_j and then generate the triangulation with vertices $\{q^i\}, i=1,\dots,N$ on the curve.

Since in this case only the parameters α_j move independently, we must compute the variation of E with respect to α_j. But

$$\frac{\partial E}{\partial \alpha_j}=\sum_{k,i}\frac{\partial E}{\partial q_i^k}\cdot\frac{\partial q_i^k}{\partial \alpha_j}, i=1,\dots,N, k=1,2.$$

Therefore, we must be able to compute $\frac{\partial q_i^k}{\partial \alpha_j}$ and this is done also by finite differences :

$$\frac{\partial q_i^k}{\partial \alpha_j}\simeq\frac{q_i^k(\alpha_j+\delta\alpha_j)-q_i^k(\alpha_j)}{\delta\alpha_j}$$

which is not computationaly expensive.

Remark : One could think that we can compute everything by finite differences, even

$$\frac{\partial E}{\partial q_i^k} \simeq \frac{J(q_i^k + \delta q_i^k) - J(q_i^k)}{\delta q_i^k}$$

but this is far too expensive, since we have to solve the state equation every time we compute $J(q_i^k)$. So, the computational cost is $2N * O(N) \simeq O(N^2)$ which is the cost of solution of N partial differential equations.

7.1.2 Other discretization methods. We have shown above that the finite element method is well suited to Optimal Shape Design because the same principles can be used on the discrete system. In Brackman (1987) and Makinen (1990) an extension to Iso-parametric elements can be found. Chenais (1993) shows also that with Cea's artificial domain velocity it is possible to have the discrete derivatives equal to the continuous derivatives discretized. Finally Finite Volume methods computations of derivatives can be found in Dervieux (1993).

7.1.3 Automatic Differentiation of Programs.. Usually the computer program for the PDE solver is written before hand and the optimal shape design analysis comes after.

The idea is to say that the PDE is known from a long sequence of equalities each of which is easy to differentiate. If each program line is thus differentiated a linearized solver is found. Then an adjoint equation is easier to found.

A review article on these methods can be found in (Gilbert et al (1991) for example).

Example

Consider the problem

$$\min_{u_1,u_2} \{J(u_1,u_2) = x_2^2 \ : \ x_1 = u_1; x_2 = ax_1^2 + u_2^2\}$$

Direct Method

The automatic differentier ADOLC of Griewank works as follows. The problem above is represented by the following program

$$\begin{aligned} x_1 &= u_1 \\ x_2 &= ax_1^2 + u_2^2 \\ J &= x_2^2 \end{aligned}$$

After each line one inserts the differentiated line and obtain

$$\begin{aligned} x_1 &= u_1 \\ dx_1 &= du_1 \\ x_2 &= ax_1^2 + u_2^2 \\ dx_2 &= 2ax_1dx_1 + 2u_2du_2 \\ J &= x_2^2 \\ dJ &= 2x_2dx_2 \end{aligned}$$

This is not too hard to do automatically because each line involves only usual functions whose derivative can be computed by symbolic computation. The resulting program gives, for prescribed du_i, the directional derivative

$$dJ = J'_{u_1}du_1 + J'_{u_2}du_2$$

Inverse method

From this it is possible to compute the partial derivatives J'_{u_i} by choosing $du_j = \delta_{ij}$, but the computing cost is prohibitive when the control space dimension is large. Then another strategy is possible.

Construct the Lagrangian by multiplying each line of the computer program by a lagrangian multiplier p_i:

$$L = J + p_1(x_1 - u_1) + p_2(x_2 - ax_1^2 + u_2^2)$$

Then as in control theory write that L has a saddle point at the solution:

$$\begin{aligned} L'_{x_1} &= p_1 - 2ax_1p_2 = 0 \\ L'_{x_2} &= 2x_2 + p_2 = 0 \\ L'_{u_1} &= -p_1 \\ L'_{u_2} &= -2u_2p_2 \end{aligned}$$

At the solution $J'_{u_i} = L'_{u_i}$, so the last two lines gives us the anwser. Notice that the first lines define the adjoint of the problem and they must be computed from down up (hence the name reverse method). It is not easy to set up this strategy automatically. The program *Odyssee* implements the method for FORTRAN programs with some restrictions (no GOTO...).

Handling DO loops

Consider the equation

$$-\frac{d^2u}{dx^2}^2 + \sin u = 1, \quad \forall x \in]0,1[, \quad u(0) = u(1) = 0,$$

discretized by a finite difference method and a Gauss-Seidel solution of the linear system:

```
do  i=0..N
      u_{i}=0
do  k=1..M
   do  i=1..N-1
      v_{i}=sin  u_{i}
      u_{i} =(u_{i+1}  +u_{i-1}) - (v_{i}-1)/N^2)/2
end_do
end_do.
```

As is often the case, while programming, the intermediate variable v_i is introduced.

A DO loop being in fact identical to a long sequence of program statement let us introduced a lagrange multiplyier for each line and construct the Lagrangian:

$$L = \sum_0^N p_i^0 u_i + \sum_{k=1}^{M} \sum_1^{N-1} p_i^k (v_i - \sin u_i) + p_{N+i}^k (N^2(2u_i - u_{i+1} - u_{i-1}) + v_i - 1)$$

This Lagrangian contains only simple function so it can be differentiated with respect to u and v by any formal computation program (Maple, Mathematica...) Thus the adjoint program is obtained:

$$\frac{\partial L}{\partial u_0} = p_0^0 - \sum_{k=1}^{M} p_{N+1}^k N^2$$

$$\frac{\partial L}{\partial u_i} = p_i^0 + \sum_{k=1}^{M} [-p_i^k \cos u_i + N^2(2p_{N+i}^k - p_{N+i-1}^k - p_{N+i+1}^k)$$

$$\frac{\partial L}{\partial u_N} = p_N^0 - \sum_{k=1}^{M} p_{2N-1}^k N^2$$

$$\frac{\partial L}{\partial v_i} = \sum_{k=1}^{M} (p_i^k + p_{N+i}^k)$$

While there seems to be no conceptual difficulties, there is a dramatic increase of lagrangian variables due to DO loops.

The limit of the method is the memory of the computer.

Notice however that if we set $p_i' = \sum_{k=1}^{M} p_i^k$, the usual discrete adjoint equations are obtained, as if the linear system was solved in one go. This important remark can save us from trouble when handling iterative methods for systems.

Handling IF statements

Branching instructions are no problem. consider the case where $\sin u$ is replace by $\sin |u|$ and programmed as

$$\begin{aligned} &if\ u_i > 0\ then\ v_i = \sin u_i \\ &else\ \ v_i = \sin(-u_i) \end{aligned}$$

The idea is to consider that we have 2 programs, one for each result of the if statement. Then there will be two lagrangian and after differentiation one puts them back into the if structure and obtain

$$\begin{aligned} &if\ u_i > 0\ then\ \frac{\partial L}{\partial u_i} = -p_{i+1}\cos u_i + N^2(2p_{N+i} - p_{N+i-1} - p_{N+i+1}) \\ &else\ \frac{\partial L}{\partial u_i} = p_{i+1}\cos(-u_i) + N^2(2p_{N+i} - p_{N+i-1} - p_{N+i+1}) \end{aligned}$$

8. Regularity Problems

Consider an optimal shape design problem

$$\min_{S\in\mathcal{S}} J(S)$$

with admissible shapes defined locally around Σ fixed and smooth

$$\mathcal{S} = \{\mathbf{x} + \alpha(x)\mathbf{n}(x)\ :\ x \in \Sigma, \quad \alpha \in H_0^2(\Sigma)\ \}$$

Suppose we know a $\chi \in L^2(\Sigma)$ such that

$$J(S(\alpha + \delta\alpha)) = J(S(\alpha)) + \int_\Sigma \chi(s)\delta\alpha(s)ds + o(\|\delta\alpha\|_2)$$

It is not a good idea to apply a gradient method in L^2 like

$$\alpha^{m+1} = \alpha^m - \rho\chi^m$$

because $\alpha^m \in H^2(\Sigma)$ does not imply $\alpha^{m+1} \in H^2(\Sigma)$ as one usually cannot expect $\chi \in H_0^2(\Sigma)$.

So let us define $\xi \in H_0^2(\Sigma)$ by

$$\frac{d^4\xi}{ds^4} = \chi, \quad \text{on } \Sigma, \quad \xi = \frac{d\xi}{ds} = 0 \text{ on } \partial\Sigma.$$

Then

$$\begin{aligned} J(S(\alpha + \delta\alpha)) &= J(S(\alpha)) + \int_\Sigma \chi(s)\delta\alpha(s)ds + o(\|\delta\alpha\|_2) \\ &= J(S(\alpha)) + \int_\Sigma \frac{d^2\xi}{ds^2}\frac{d^2\alpha}{ds^2}\delta\alpha(s)ds + o(\|\delta\alpha\|_2) \end{aligned}$$

Now we can do

$$\alpha^{m+1} = \alpha^m - \rho\xi^m$$

8.1 Application

Consider again the laboratory problem of Figure 2 and its discretization by the FInite Element Method (Figure 3).

$$J(S) = \int_D |u - u_d|^2 \quad : \quad -\Delta u = f, \quad u \in H_0^1(\Omega), \quad S \subset \partial\Omega$$

for which we know that for some $y(s) = x(s) + \theta\delta\alpha(s)$, $\theta \in]0,1[$ we have

$$\delta J = \int_\Sigma \frac{\partial u}{\partial n}\frac{\partial p}{\partial n}\delta\alpha - \frac{1}{2}\int_\Sigma \frac{\partial p}{\partial n}(x(s))\frac{\partial^2 u}{\partial n^2}(y(s))\delta\alpha(s)^2 ds$$

with p solution of

$$p \in H_0^1(\Omega), \quad -\Delta p = 2(u - u_d)I_D$$

With the regularity $S \in H_0^2(\Sigma)$ we have $u, p \in H^2(\Omega)$, $\chi \in W^{\frac{1}{2},1}(\Sigma)$. Setting ξ by

$$\frac{d^4\xi}{ds^4} = \chi, \text{ on } \Sigma, \quad \xi = \frac{d\xi}{ds} = 0 \text{ on } \partial\Sigma.$$

might be difficult in practice, especially in 3D because PDEs on surfaces may be tricky to solve. Instead consider $\xi = v|_\Sigma$ with

$$-\Delta w = 0, \quad \frac{\partial w}{\partial n}|_\Sigma = \chi, \qquad -\Delta v = w, \quad \frac{\partial v}{\partial n}|_\Sigma = 0.$$

Lemma 1
The operator $\chi \to \xi \equiv A\chi$ is positive definite and $\chi \in L^2(\Sigma) \Rightarrow \xi \in H^2(\Sigma)$

Proof Let

$$-\Delta w' = 0, \quad \frac{\partial w'}{\partial n}|_\Sigma = \chi', \qquad -\Delta v' = w', \quad \frac{\partial v'}{\partial n}|_\Sigma = 0.$$

and note that

$$<\chi', A\chi> = \int_\Sigma \frac{\partial w'}{\partial n}v = \int_\Omega \nabla w'\nabla v = \int_\Omega w'w$$

Proposition
The following algorithm preserves the regularity of the variables:

$$\alpha^{m+1} = \alpha^m - \rho\xi^m$$

Remark
Oden et al suggested to use $\xi = \mathbf{u}\cdot\mathbf{n}|_\Sigma$ where

$$-\lambda\Delta\mathbf{u} + \mu\nabla(\nabla\cdot\mathbf{u}) = 0, \quad \sigma_{\mathbf{nn}}(\mathbf{u})|_\Sigma = \chi, \quad \mathbf{u}\cdot\mathbf{s} = 0$$

because

$$\int_\Sigma \xi\chi' = \int_\Sigma \sigma_{nn}(\mathbf{u}')\mathbf{u}\cdot\mathbf{n} = \int_\Omega (\lambda\nabla u\nabla v + \mu\nabla\cdot u\nabla\cdot v)$$

Although it is also a smoothing process it is less regualar than the one above.

8.2 Discretization

Recall that a finite element discretization of the problem is

$$J = \int_D (u - u_d)^2, \quad \int_\Omega \nabla u \nabla w^j + \frac{1}{\epsilon} u_j I_{q^j \in \Gamma} = \int_D f w^j$$

with $u = \sum_1^N u_j w^j$, $\Omega = \cup T_k$, $T_k = \{q^{j_1}, q^{j_2}, q^{j_3}\}$.

Recall that

$$\delta w^j = -\nabla w^j \cdot \delta q_h, \quad \delta q_h = \sum_j \delta q^j w^j$$

and that

$$\int_{\Omega'} f - \int_\Omega f = \sum_k \int_{T_k} \nabla \cdot (f \delta q_h)$$

so that we have the

Proposition

$$\delta J = \int_\Omega \nabla u^T (\nabla \delta q_h + \nabla \delta q_h^T - I \nabla \cdot \delta q_h) \nabla p + o(|\delta q_h|)$$

Proof

$$\delta u = \sum_1^N \delta u_j w^j + u_j \delta w^j = \delta u_h - \nabla u_h \cdot \delta q_h$$

$$\int_\Omega \nabla(\delta u_h - \nabla u_h \cdot \delta q_h) \nabla w^j - \int_\Omega \nabla u \nabla w^j \nabla \delta q_h$$

$$+ \int_\Omega \nabla \cdot (\nabla u \nabla w^j \delta q_h) + \frac{1}{\epsilon} \delta u_j I_{q^j \in \Gamma} = 0$$

$$\delta J = \int_D 2 \delta u (u - u_d) = \int_\Omega \nabla p \nabla \delta u_h + \sum_{q^j \in \Gamma} p_j \delta u_j$$

Proposition

$$\delta J = \int_S \nabla u \cdot \nabla p \, \delta q_h \cdot n + \int_{Edges} [\nabla u \cdot \nabla p] \delta q_h \cdot n + o(|\delta q_h|)$$

This is because

$$\delta(\nabla Q_h \nabla Q_h^T \det Q_h^{-1}) = \nabla \delta q_h + \nabla \delta q_h^T - I \nabla \cdot \delta q_h + o(|\delta q_h|)$$

8.3 Consequence

- It is clear the the discrete optimization process tends to the continuous one. But... do we have the necessary regularity?
- It is not necessary to account for the motion of the inner mesh points if $h << 1$
- One should not use the gradient with respect to the inner points to move them because it is an order of magnitude smaller: $W^j = (w^j, w^j)^T$

$$q^j \leftarrow q^j - \rho \int_\Omega \nabla u^T (\nabla W^j + \nabla W^{jT} - I \nabla \cdot W^j) \nabla p$$

Use the smoothers, so the complete algorithm is

1. Solve
$$\int_\Omega \nabla u \nabla w^j + \frac{1}{\epsilon} u_j I_{q^j \in \Gamma} = \int_D f w^j$$
2. Solve
$$\int_\Omega \nabla p \nabla w^j + \frac{1}{\epsilon} p_j I_{q^j \in \Gamma} = 2 \int_D (u - u_d) w^j$$
3. Solve
$$\int_\Omega \lambda \nabla U : \nabla W^j + \mu \nabla \cdot U \nabla \cdot W^j = \int_\Omega \nabla u^T (\nabla W^j + \nabla W^{jT} - \nabla \cdot W^j) \nabla p$$
4. Move the points of the mesh by
$$q^j \leftarrow q^j - \rho U^j$$

9. Consistent Approximations

OSD is expensive; there is a great economical advantage to combine the optimization algorithm with mesh refinement so as to obtain a speed up similar to multigrid.

For standard optimization, E. Polak (1998) developped a tool which he calls the theory of consistent approximation" which we apply here. The following is a summary of results obtained jointly with N. Dicesare and E. Polak. For more details see Dicesare et al (1998).

9.1 Algorithm

The problem is to minimize $J(z)$ in $\mathcal{O}$. The discrete problem is indexed by a discretization parameter h: minimize $J_h(z_h)$ in $\mathcal{O}_h$.

Assume that $\mathcal{O}$ is a Hilbert space. Let

$$\theta(x) = -\|Grad_z J(z)\| \quad \text{and} \quad \theta_h(z) = \|Grad_z J_h(z)\|$$

Algorithm 1

1. Choose a converging sequence of discretization spaces $\{\mathcal{O}_{h_n}\}$ with $\mathcal{O}_{h_n} \subset \mathcal{O}_{h_{n+1}}$ for all n. Choose $z^0, \epsilon^0, \beta \in]0,1[$.
2. Set $n = 0, \epsilon = \epsilon^0, h = h_0$
3. Compute z^n_m by performing m iterations of a descent algorithm on $\mathcal{P}_h$ from starting point z^n so as to achieve

$$\theta_h(z^n_m) > -\epsilon$$

4. Set $\epsilon = \beta\epsilon, h = h_{n+1}, z^{n+1} = z^n_m, n = n+1$ and go to Step 3.

The mathematical result is that if $\mathcal{P}_h$ epi-converge to $\mathcal{P}$ then any accumulation point z^* of $\{z^n\}$ generated by Algorithm 1 satisfies $\theta(z^*) = 0$.

9.2 Problem Statement

Consider a simple model problem where the shape is to be found that brings u, solution of a PDE, nearest to u_d in a subregion D of the entire domain Ω.

The unknown shape Γ is a portion of the entire boundary $\partial\Omega$: it is parametrized by its distance α to a reference smooth boundary Σ. To prevent an excess of oscillation the problem is regularized.

More concretely with the following notations ($\epsilon << 1$),

$$D \subset \Omega,\ u_d \in H^1(D),\ g \in H^1(\Omega),\ I \subset K \subset \mathcal{R},\quad \Sigma = \{x(s)\ :\ s \in K\}$$

we consider

$$\min_{\alpha \in H^2_0(I)} J(\alpha) = \int_D (u - u_d)^2 + \epsilon \int_\Sigma |\frac{d^2\alpha}{ds^2}|^2$$

$$\text{subject to } u - \Delta u = 0 \text{ in } \Omega(\alpha), \quad \frac{\partial u}{\partial n}|_{\Gamma(\alpha)} = g|_{\Gamma(\alpha)},$$

$$\text{where } \Gamma(\alpha) = \partial\Omega(\alpha) = \{x(s) + \tilde{\alpha}(s)n(x(s))\ :\ s \in K\}$$

where $\tilde{\alpha}$ is the extension by zero in K of α which is only defined on I.

Recall that

$$H^2_0(I) = \{\alpha \in L^2(I)\ :\ \alpha', \alpha'' \in L^2(I),\quad \alpha(a) = \alpha'(a) = 0\ \ \forall a \in \partial I\}$$

and that $\|\alpha''\|_0 = \|d^2\alpha/ds^2\|_0$ is a norm in that space.

Let us denote the unknown part of the boundary by

$$S(\alpha) = \{x(s) + \alpha(s)n(x(s)) \ : \ s \in I\}$$

For simplicity let us assume that g is always zero on S.

9.3 Discretization

The discrete problem is

$$\min_{\alpha \in L_h \subset H_0^2(I)} J(\alpha) = \int_D (u - u_d)^2 + \epsilon \int_\Sigma |\frac{d^2\alpha}{ds^2}|^2$$

$$\text{subject to} \quad \int_{\Omega(\alpha)} (uv + \nabla u \nabla v) = \int_{\Gamma(\alpha)} gv \ \forall v \in V_h, \quad u \in V_h$$

where V_h is the usual Lagrange Finite Element space of degree 1 on triangles except that the boundary triangles have a curved side because $S(\alpha)$ is a cubic spline.

The space L_h is the finite dimensional subspace of $H_0^2(I)$ defined as the set of cubic splines which passes through the vertices which would we would have used otherwise to define a feasible polygonal approximation of the boundary. This means that the discretization of Ω is done as follows

1. Give a set of n_f boundary vertices $q^{i_1}, ..., q^{i_{n_f}}$, construct a polygonal boundary near Σ
2. Construct a triangulation of the domain inside this boundary with an automatic mesh generator, i.e. Mathematically the inner nodes are theoretically linked to the outer ones by a map
$$q^j = Q^j(q^{i_1}, ..., q^{i_{n_f}}), \quad nf < j < nv$$
3. Construct $\Gamma(\alpha)$, the cubic splines from the $q^{i_1}, ..., q^{i_{n_f}}$, set α to be the normal distance from Σ to $\Gamma(\alpha)$.
4. Construct V_h by using triangular finite elements and overparametric curved triangular elements on the boundary.

This may seem complex but it is a handy construction because the discrete cost function J_h coincide with the continuous J and because L_h is a finite subspace of the (infinite) set of admissible parameters H_0^2.

We proceed and verify the hypothesis of the theorem to apply Algorithm 1.

9.4 Optimality Conditions: the continuous case

As before, by calculus of variations

$$\delta J = 2\int_D (u-u_d)\delta u + 2\epsilon\int_\Sigma \frac{d^2\alpha}{ds^2}\frac{d^2\delta\alpha}{ds^2}$$

with $\delta u \in H^1(\Omega(\alpha))$ and

$$\int_{\Omega(\alpha)} (\delta u v + \nabla\delta u\nabla v) + \int_\Sigma \delta\alpha(uv + \nabla u\nabla v) = 0 \ \forall v \in H^1(\Omega(\alpha)).$$

Introduce an adjoint $p \in H^1(\Omega(\alpha))$

$$\int_{\Omega(\alpha)} (pq + \nabla p\nabla q) = 2\int_D (u-u_d)q, \quad \forall q \in H^1$$

i.e.

$$p - \Delta p = I_D u, \quad \frac{\partial p}{\partial n} = 0$$

Then

$$\delta J = -\int_\Sigma \delta\alpha(up + \nabla u\nabla p - 2\epsilon\frac{d^4\alpha}{ds^4})$$

9.4.1 Definition of θ. So we should take

$$\theta = -\|up + \nabla u\nabla p - 2\epsilon\frac{d^4\alpha}{ds^4}\|_{-2}$$

i.e. solve

$$\frac{d^4\theta}{ds^4} = up + \nabla u\nabla p - 2\epsilon\frac{d^4\alpha}{ds^4} \quad on\ I, \quad \theta = \frac{d\theta}{ds} = 0 \quad on\ \partial I$$

9.5 Optimality Conditions: the discrete case

Let w^j be the hat function attached to vertex q^j. If some vertices q^j vary by δq_j we define

$$\delta q_h(x) = \sum_j \delta q_j w^j(x)$$

and we know that (Pironneau[1983])

$$\delta w^k = -\nabla w^k \cdot \delta q_h$$

$$\int_{\delta\Omega} f = \int_\Omega \nabla\cdot(f\delta q_h) + o(|\delta q_h|)$$

Hence

$$J(\alpha+\delta\alpha) = 2\int_D (u_h - u_{dh})\delta u_h + 2\epsilon\int_\Sigma \frac{d^2\alpha}{ds^2}\frac{d^2\delta\alpha}{ds^2},$$

Furthermore and by definition of δu_h

$$\delta \sum_i u_i w^i = \sum_i (\delta u_i w^i + u_i \delta w^i) = \delta u_h + \delta q_h \cdot \nabla u_h$$

the partial variation δu_h is found by

$$\delta \int_{\Omega(\alpha)} (u_h w^j + \nabla u_h \nabla w^j) = \int_\Omega (\nabla \cdot (u w^j \delta q_h) + \delta u_h w^j + \nabla \delta u_h \nabla w^j)$$

$$+ \int_\Omega (u_h \delta q_h \cdot \nabla w^j + \nabla u_h \nabla \delta q_h \nabla w^j + u_h \delta w^j + \nabla u_h \nabla \delta w^j) = 0$$

Hence

$$\int_\Omega (\delta u_h w^j + \nabla \delta u_h \nabla w^j =$$

$$\int_\Omega (\nabla u_h (\nabla \delta q_h + \nabla \delta q_h^T) \nabla w^j - (u_h w^j + \nabla u_h \cdot \nabla w^j) \nabla \cdot \delta q_h)$$

So introduce an adjoint $p_h \in V_h$

$$\int_\Omega (p_h w^j + \nabla p_h \nabla w^j) = 2 \int_D (u_h - u_{dh}) w^j \quad \forall j$$

And finally

$$\delta J_h = \int_\Omega (\nabla u_h (\nabla \delta q_h + \nabla \delta q_h^T) \nabla p_h - (u_h p_h + \nabla u_h \cdot \nabla p_h) \nabla \cdot \delta q_h) + 2\epsilon \int_\Sigma \frac{d^2 \alpha}{ds^2} \frac{d^2 \delta\alpha}{ds^2}$$

9.6 Definition of θ_h

Let $e^1 = (1,0)^T$, $e^2 = (0,1)^T$ be the coordinate vectors of R^2, let χ^j be the vector of R^2 of components

$$\chi_k^j = \int_\Omega (\nabla u_h (\nabla w^j e^k + (\nabla w^j e^k)^T) \nabla p_h - (u_h p_u + \nabla u_h \cdot \nabla p_h) \nabla \cdot w^j e^k.$$

Because the inner vertices are linked to the boundary ones by the maps Q^j, let us introduce

$$\xi_k^j = \chi_k^j + \sum_{q^i \notin \Gamma} \chi^i \partial_{q_k^j} Q^i.$$

Then obviously

$$\delta J = \sum_1^{nf} \xi^j \cdot \delta q^j + 2\epsilon \int_\Sigma \frac{d^2 \alpha}{ds^2} \frac{d^2 \delta\alpha}{ds^2}$$

It is possible to find a β so as to express the first discrete sum as an integral on Σ of $\frac{d^2\beta}{ds^2}\frac{d^2\delta\alpha}{ds^2}$; it is some sort of variational problem in L_h:

$$\int_\Sigma \frac{d^2\beta}{ds^2}\frac{d^2\delta\lambda^j}{ds^2} = \xi^j \cdot n_\Sigma, \quad j = 1,...,nf; \quad \beta \in L_h$$

where λ^j is the cubic spline obtained by a unit normal variation of the boundary vertex q^j only.
Then the "derivative" of J_h is the function $s \in I \to \beta(s) + 2\epsilon\alpha(s)$ and the function θ_h is

$$\theta_h = -\|\beta\|_{H_0^2(I)}$$

Remark This may be unnecessarily complicated in practice. A pragmatic summary of the above is that β is solution of a fourth order problem, so why not set a discrete fourth order problem on the normal component of the vertex themselves. In the case $\epsilon = 0$ this would be

$$\frac{1}{h^4}[q'^{j+2}_n - 4q'^{j+1}_n + 6q'^j_n - 4q'^{j-1}_n + q'^{j-2}_n] = \xi^j,$$
$$q'^0_n = q'^1_n = q'^{nf-1}_n = q'^{nf}_n = 0$$

and then the norm of the second derivative of the result for θ_h

$$\theta_h \approx -(\sum_j \frac{1}{h^2}[q_n^{j+1} - 2q_n^j + q_n^{j-1}]^2)^{\frac{1}{2}}$$

9.7 Hypothesis of the Theorem

The following is shown in Dicesera et al (1999)

- *Inclusion* $h' < h \Rightarrow O_h \subset O_h$.
- it Continuity The cost functions are continuous in z

$$\alpha^n \overset{H_0^2(I)}{\longrightarrow} \alpha, \quad \Rightarrow J^n \to J.$$

 Similarly in the discrete case, the spline is continuous with respect to the vertex position so

$$q^{i\ n} \to q^i \Rightarrow \alpha_h^n \overset{H_0^2(I)}{\longrightarrow} \alpha_h, \quad \Rightarrow J_h^n \to J_h.$$

- *Consistency* $\forall\alpha, \exists\alpha_h \to \alpha$ with $J_h \to J$.
 if the following is observed:
  - Corners of the continuous curve are vertices of the discrete curves
  - the distance between boundary vertices converges uniformly to zero.
- *Continuity of θ Conjecture* : There exists ε such that $\alpha \in H_0^2 \quad \Rightarrow u \in H^{3/2+\epsilon}(\Omega)$.
 Arguments: We know that $\alpha \in C^{0,1} \quad \Rightarrow u \in H^{3/2}$(Jerisson and Kenig [???]) and $\alpha \in C^{1,1} \quad \Rightarrow u \in H^2$(Grisvard [???]).
 This technical point of functional analysis is need for the continuity of θ.

$$\alpha^n \overset{H^2}{\longrightarrow} \alpha, \quad \Rightarrow \nabla u^n \nabla p^n|_\Sigma \overset{L^2}{\longrightarrow} \nabla u \nabla p|_\Sigma$$

- *Continuity of $\theta_h(\alpha_h)$*
  Recall that a variation $\delta\alpha_h$ (i.e. a boundary vertex variation $\delta q^j, j \in \Sigma$) implies variations of all inner vertices $\delta\alpha, \delta q^k, \forall k$
  The problem is that θ is a boundary integral on Σ and θ_h is a volume integral! We must explain why

$$\delta J_h = \int_\Omega (\nabla u_h(\nabla\delta q_h + \nabla\delta q_h^T)\nabla p_h - \nabla u_h \cdot \nabla p_h \nabla \cdot q_h$$

$$+2\epsilon \int_\Sigma \frac{d^2\alpha}{ds^2}\frac{d^2\delta\alpha}{ds^2} \xrightarrow{?} \delta J = -\int_\Sigma \delta\alpha(up + \nabla u \nabla p + 2\epsilon\frac{d^4\alpha}{ds^4})$$

This is due to the fact that if $\nabla X = I + \nabla Q$, the jacobian matrix of the mapping $x \to X = x + Q(x)$ of $\mathcal{R}^2 \to \mathcal{R}^2$ is the linearization of the operator which appears in the change of variable $x \to X(x)$:

$$\delta(\nabla X^T \nabla X \det X^{-1}) = \nabla Q + \nabla Q^T - I\nabla \cdot Q + o(\|Q\|).$$

So δJ_h is almost a surface integral:

$$\delta J_h = -\int_\Sigma (\delta q_h \cdot n_\Sigma(u_h p_h + \nabla u_h \nabla p_h) + 2\epsilon\frac{d^4\alpha_h}{ds^4})$$

$$-\int_E [\delta q_h \cdot n_E(u_h p_h + \nabla u_h \nabla p_h)] + o(\delta q_h) + o(h)$$

where E is the set of edges of the triangulation, [.] the jump across the edges and n_E the normal to the edge E (the sign of this expression depends on the choice of the normal n_E).

9.8 Algorithm 3

An adaptation of Algorithm 1 to this case is

1. Choose an initial set of boundary vertices.
2. Construct a finite element mesh, construct the spline of the boundary.
3. Solve the discrete PDE and the discrete adjoint PDE.
4. Compute θ_h (or its approximation (cf. remark above))
5. if $\theta_h > -\epsilon$ add points to the boundary mesh, update the parameters and go back to Step 2.

There are still several hypothesis to verify to make sure that Algorithm 3 converges. We proceed in a loose fashion and give only the general idea of the proof.

9.9 Convergence

It comes from the theory of Finite Element Error Analysis (Ciarlet[1975]):

Lemma

$$|\int_{\Sigma} \nabla u_h \nabla p_h - \nabla u \nabla p \,| \leq C h^{1/2}(\|p\|_2 + \|u\|_2)$$

and the following triangular inequalities

- $$|a_h b_h - ab| = (a_h - a)(b_h - b) + b(a_h - a) + (b_h - b)a$$
  $$\leq |b||a_h - a| + |a||b_h - b| + |a_h - a|^2 + |b_h - b|^2$$
- $$|\nabla u_h - \nabla u|_{0,\Sigma} \leq |\nabla(u_h - \Pi_h u)|_{0,\Sigma} + |\nabla(\Pi_h u - u)|_{0,\Sigma}$$

plus an inverse inequality for the first term and an interpolation for the second.

10. Numerical Results

Numerical results with the local boundary variation method just described have been obtained my PhD students. For details we send the reader to their thesis, mostly at the Université Paris 6:

- F. Angrand for a wing optimization with the transonic equation
- G. Arumugam for the optimization of ribblets in laminar flow
- A. Vossinis for the choice of a numerical algorithm, Newtown, GMRES or Conjugate Gradient.
- F. Baron for the stealth wing problem and the harbour optimization

But very impressive results have been obained by Marrocco for the design of an electromagnet and by Mohammadi for the design of 3D aircrafts and wings by using automatic differentiation of programs.

Thanks to this last piece of work the method is now mature and efficient.

1. Y. Achdou: Effect of a metallized coating on the reflection of an electromagnetic wave. Note CRAS (1992).
2. G. Anagnostou, E. Ronquist, A. Patera: A computational procedure for part design. Comp. Methods Applied Mech and Eng. July (1992).
3. F. Angrand: Numerical methods for optimal shape design in aerodynamics, Thesis in French, Univ. Paris 6, (1980).
4. M. Artola, M. Cessenat: Propagation des ondes electromagnetiques dans un milieu composite. note CRAS 311(1):77-82, (1990).

5. G. Arumugam: Optimum design et applications a la mecanique des fluides. Universite Paris 6 These. 1989.
6. G. Arumugam, O. Pironneau : On the problems of riblets as a drag reduction device, Optimal Control Applications & Methods, Vol. 10, (1989).
7. N.V. Banichuk: *Introduction to optimization of structures*. Springer (1990).
8. D. Begis, R. Glowinski: Application of FEM to approximation of an optimum design problem. Appl. Math. Optim 2(2) (1975).
9. F. Beux, A. Dervieux: Exact-gradient shape optimization of a 2D Euler flow. INRIA report 1540 (1991).
10. D. Bucur, J.P. Zolezio: -Dimensional Shape Optimization under Capacitance Constraint. J. Diff. Eqs. Vol 123, No 2, pp504-522 (1995).
11. M. Crouzeix: Variational approach of a magnetic shaping problem. Eur. J. Mech B/fluids 10, 5:527-536 (1991).
12. J. Cea: Conception optimale ou identification delta forme: calcul rapide delta la dérivée directionelle delta la fonction cout. Modélisation Math Anal, AFCET, Dunod (1986).
13. J. Cea, A.Gioan, J. Michel: Some results on domain identification. Calcolo 3/4 (1973).
14. D. Chenais: Shape otptimization in shell theory. Eng. Opt. 11:289-303 (1987).
15. J.C. Gilbert, G. Le Vey, J. Masse: La differentiation automatique des fonctions représentéés par des programmes. INRIA report 1557 (1991).
16. J. Haslinger, P. Neittaanmäki: *Finite element approximations for optimal shape design*. Wiley 1989.
17. E.J. Haug, J. Cea: *Optimization of distributed parameter structures* vol I and II , Sijthoff and Noordhoff (1981).
18. A. Jameson: Automatic design of transonic airfoils to reduce the shock induced pressure drag. Proc. 31st Israel Annual conf on aviation and aeronautics. Feb 1990.
19. W.G. Litvinov: The problem of the optimal shape of an hydrofoil. J Optimization, theory and appl. (to appear)
20. W.B. Liu, P. Neittaanmäki, D. Tiba: Existence for Shape Optimization Problems in Arbitrary Dimension. Universität Jyväskyla, Mathematisches Institut preprint 208, April 1999.
21. J. Hadamard: Lecon sur le calcul des variation. Gauthier-Villards (1910)
22. R. Mäkinen: Finite Element design sensitivity analysis for non linear potential problems. Comm Applied Numer Methods. 6:343-350 (1990).
23. A. Marrocco, O. Pironneau: Optimum design with Lagrangian Finite Element. Comp. Meth. Appl. Mech . Eng. 15-3 (1978).
24. M. Masmoudi: Conception delta circuit passif delta très haute fidélité. Matapli no 31 (1990).
25. F. Moens: Réalisation d'une méthode d'optimisation numérique pour la définition delta profils hypersustentés. Rapport ONERA 43/1736, 1991.
26. F. Murat, J. Simon: Etude delta problèmes d'optimum design. Proc. 7th IFIP conf. Lecture notes in Computer sciences, 41, 54-62, 1976.
27. P. Neittaanmäki, A. Stachurski: Solving some optimal control problems using the Barrier Penalty function method. Appl Math Optim 25:127-149 (1992)
28. P. Neittaanmäki: Computer aided optimal structural design Surv. Math. Ind. 1:173-215, 1991.
29. O. Pironneau: On optimal shapes for Stokes flow. J. Fluid Mech, (1973).
30. O. Pironneau:*"Optimal shape design for elliptic systems"*, Springer-Verlag, (1984).

31. B. Rousselet: Shape design sensitivity of a membrane. J Optimization Theory and
appl. 40, 4:595-623 (1983).
32. J. Sokolowski, J.P. Zolezio: Introduction to Shape Optimization. Springer Series in Computational Mathematics (1991).
33. A. Sverak: On existence of solution for a class of optimal shape design problems. Note C.R.A.S. 1992.
34. S. Ta'asan, G. Kuruvila: Aerodynamic Design and optimization in one shot. AIAA paper 92-0025
35. L. Tartar: Control problems in the coefficients of PDE. In Lecture notes in Economics and Math systems. A. Bensoussan ed. Springer, (1974).
36. A. Vossinis : Optimization Algorithms for Optimum Shape design problems (to appear).
37. G. Volpe: Geometric and surface pressure restrictions in airfoil design. AGARD report 780 (1990).
38. D. Young, R. Melvin, F. Johnson, J. Bussoletti, L. Wigton, S. Samant: Application of sparse matrix solvers as effective preconditionners. SIAM J. Sci. Stat. Comput. 10-6:1118-1199, (1989).
39. J.P. Zolesio: Les dérivées par rapport au noeuds des triangulations et leurs utilisations en identification de domaines. Ann, Sc. Math, Quebec 8, 97-120, (1984).

LIST OF C.I.M.E. SEMINARS

1954	1.	Analisi funzionale	C.I.M.E
	2.	Quadratura delle superficie e questioni connesse	"
	3.	Equazioni differenziali non lineari	"
1955	4.	Teorema di Riemann-Roch e questioni connesse	"
	5.	Teoria dei numeri	"
	6.	Topologia	"
	7.	Teorie non linearizzate in elasticità, idrodinamica, aerodinamic	"
	8.	Geometria proiettivo-differenziale	
			"
1956	9.	Equazioni alle derivate parziali a caratteristiche reali	"
	10.	Propagazione delle onde elettromagnetiche	"
	11.	Teoria della funzioni di più variabili complesse e delle funzioni automorfe	"
1957	12.	Geometria aritmetica e algebrica (2 vol.)	"
	13.	Integrali singolari e questioni connesse	"
	14.	Teoria della turbolenza (2 vol.)	"
1958	15.	Vedute e problemi attuali in relatività generale	"
	16.	Problemi di geometria differenziale in grande	"
	17.	Il principio di minimo e le sue applicazioni alle equazioni funzionali	"
1959	18.	Induzione e statistica	"
	19.	Teoria algebrica dei meccanismi automatici (2 vol.)	"
	20.	Gruppi, anelli di Lie e teoria della coomologia	"
1960	21.	Sistemi dinamici e teoremi ergodici	"
	22.	Forme differenziali e loro integrali	"
1961	23.	Geometria del calcolo delle variazioni (2 vol.)	"
	24.	Teoria delle distribuzioni	"
	25.	Onde superficiali	"
1962	26.	Topologia differenziale	"
	27.	Autovalori e autosoluzioni	"
	28.	Magnetofluidodinamica	"
1963	29.	Equazioni differenziali astratte	"
	30.	Funzioni e varietà complesse	"
	31.	Proprietà di media e teoremi di confronto in Fisica Matematica	"
1964	32.	Relatività generale	"
	33.	Dinamica dei gas rarefatti	"
	34.	Alcune questioni di analisi numerica	"
	35.	Equazioni differenziali non lineari	"
1965	36.	Non-linear continuum theories	"
	37.	Some aspects of ring theory	"
	38.	Mathematical optimization in economics	"
1966	39.	Calculus of variations	Ed. Cremonese, Firenze
	40.	Economia matematica	"
			"

	41. Classi caratteristiche e questioni connesse	"
	42. Some aspects of diffusion theory	
1967	43. Modern questions of celestial mechanics	"
	44. Numerical analysis of partial differential equations	"
	45. Geometry of homogeneous bounded domains	"
1968	46. Controllability and observability	"
	47. Pseudo-differential operators	"
	48. Aspects of mathematical logic	"
1969	49. Potential theory	"
	50. Non-linear continuum theories in mechanics and physics and their applications	"
	51. Questions of algebraic varieties	"
1970	52. Relativistic fluid dynamics	"
	53. Theory of group representations and Fourier analysis	"
	54. Functional equations and inequalities	"
	55. Problems in non-linear analysis	"
1971	56. Stereodynamics	"
	57. Constructive aspects of functional analysis (2 vol.)	"
	58. Categories and commutative algebra	"
1972	59. Non-linear mechanics	"
	60. Finite geometric structures and their applications	"
	61. Geometric measure theory and minimal surfaces	"
1973	62. Complex analysis	"
	63. New variational techniques in mathematical physics	"
	64. Spectral analysis	"
1974	65. Stability problems	"
	66. Singularities of analytic spaces	"
	67. Eigenvalues of non linear problems	"
1975	68. Theoretical computer sciences	"
	69. Model theory and applications	"
	70. Differential operators and manifolds	"
1976	71. Statistical Mechanics	Ed. Liguori, Napoli
	72. Hyperbolicity	"
	73. Differential topology	"
1977	74. Materials with memory	"
	75. Pseudodifferential operators with applications	"
	76. Algebraic surfaces	"
1978	77. Stochastic differential equations	"
	78. Dynamical systems	Ed. Liguori, Napoli & Birkhäuser
1979	79. Recursion theory and computational complexity	"
	80. Mathematics of biology	"
1980	81. Wave propagation	Ed. Liguori, Napoli & Birkhäuser
	82. Harmonic analysis and group representations	"
		"

	83. Matroid theory and its applications		
1981	84. Kinetic Theories and the Boltzmann Equation	(LNM 1048)	Springer-Verlag
	85. Algebraic Threefolds	(LNM 947)	"
	86. Nonlinear Filtering and Stochastic Control	(LNM 972)	"
1982	87. Invariant Theory	(LNM 996)	"
	88. Thermodynamics and Constitutive Equations	(LN Physics 228)	"
	89. Fluid Dynamics	(LNM 1047)	"
1983	90. Complete Intersections	(LNM 1092)	"
	91. Bifurcation Theory and Applications	(LNM 1057)	"
	92. Numerical Methods in Fluid Dynamics	(LNM 1127)	"
1984	93. Harmonic Mappings and Minimal Immersions	(LNM 1161)	"
	94. Schrödinger Operators	(LNM 1159)	"
	95. Buildings and the Geometry of Diagrams	(LNM 1181)	"
1985	96. Probability and Analysis	(LNM 1206)	"
	97. Some Problems in Nonlinear Diffusion	(LNM 1224)	"
	98. Theory of Moduli	(LNM 1337)	"
1986	99. Inverse Problems	(LNM 1225)	"
	100. Mathematical Economics	(LNM 1330)	"
	101. Combinatorial Optimization	(LNM 1403)	"
1987	102. Relativistic Fluid Dynamics	(LNM 1385)	"
	103. Topics in Calculus of Variations	(LNM 1365)	"
1988	104. Logic and Computer Science	(LNM 1429)	"
	105. Global Geometry and Mathematical Physics	(LNM 1451)	"
1989	106. Methods of nonconvex analysis	(LNM 1446)	"
	107. Microlocal Analysis and Applications	(LNM 1495)	"
1990	108. Geometric Topology: Recent Developments	(LNM 1504)	"
	109. H_∞ Control Theory	(LNM 1496)	"
	110. Mathematical Modelling of Industrial Processes	(LNM 1521)	"
1991	111. Topological Methods for Ordinary Differential Equations	(LNM 1537)	"
	112. Arithmetic Algebraic Geometry	(LNM 1553)	"
	113. Transition to Chaos in Classical and Quantum Mechanics	(LNM 1589)	"
1992	114. Dirichlet Forms	(LNM 1563)	"
	115. D-Modules, Representation Theory, and Quantum Groups	(LNM 1565)	"
	116. Nonequilibrium Problems in Many-Particle Systems	(LNM 1551)	"
1993	117. Integrable Systems and Quantum Groups	(LNM 1620)	"
	118. Algebraic Cycles and Hodge Theory	(LNM 1594)	"
	119. Phase Transitions and Hysteresis	(LNM 1584)	"

1994	120. Recent Mathematical Methods in Nonlinear Wave Propagation	(LNM 1640)	Springer-Verlag
	121. Dynamical Systems	(LNM 1609)	"
	122. Transcendental Methods in Algebraic Geometry	(LNM 1646)	"
1995	123. Probabilistic Models for Nonlinear PDE's	(LNM 1627)	"
	124. Viscosity Solutions and Applications	(LNM 1660)	"
	125. Vector Bundles on Curves. New Directions	(LNM 1649)	"
1996	126. Integral Geometry, Radon Transforms and Complex Analysis	(LNM 1684)	"
	127. Calculus of Variations and Geometric Evolution Problems	(LNM 1713)	"
	128. Financial Mathematics	(LNM 1656)	"
1997	129. Mathematics Inspired by Biology	(LNM 1714)	"
	130. Advanced Numerical Approximation of Nonlinear Hyperbolic Equations	(LNM 1697)	"
	131. Arithmetic Theory of Elliptic Curves	(LNM 1716)	"
	132. Quantum Cohomology	(LNM)	"
1998	133. Optimal Shape Design	(LNM 1740)	"
	134. Dynamical Systems and Small Divisors	to appear	"
	135. Mathematical Problems in Semiconductor Physics	to appear	"
	136. Stochastic PDE's and Kolmogorov Equations in Infinite Dimension	(LNM 1715)	"
	137. Filtration in Porous Media and Industrial Applications	(LNM 1734)	"
1999	138. Computational Mathematics driven by Industrial Applications	(LNM 1739)	"
	139. Iwahori-Hecke Algebras and Representation Theory	to appear	"
	140. Theory and Applications of Hamiltonian Dynamics	to appear	"
	141. Global Theory of Minimal Surfaces in Flat Spaces	to appear	"
	142. Direct and Inverse Methods in Solving Nonlinear Evolution Equations	to appear	"

4. Lecture Notes are printed by photo-offset from the master-copy delivered in camera-ready form by the authors. Springer-Verlag provides technical instructions for the preparation of manuscripts. Macro packages in TEX, LATEX2e, LATEX2.09 are available from Springer's web-pages at

http://www.springer.de/math/authors/b-tex.html.

Careful preparation of the manuscripts will help keep production time short and ensure satisfactory appearance of the finished book.

The actual production of a Lecture Notes volume takes approximately 12 weeks.

5. Authors receive a total of 50 free copies of their volume, but no royalties. They are entitled to a discount of 33.3 % on the price of Springer books purchase for their personal use, if ordering directly from Springer-Verlag.

Commitment to publish is made by letter of intent rather than by signing a formal contract. Springer-Verlag secures the copyright for each volume. Authors are free to reuse material contained in their LNM volumes in later publications: A brief written (or e-mail) request for formal permission is sufficient.

Addresses:

Professor F. Takens, Mathematisch Instituut,
Rijksuniversiteit Groningen, Postbus 800,
9700 AV Groningen, The Netherlands
E-mail: F.Takens@math.rug.nl

Professor B. Teissier
Université Paris 7
UFR de Mathématiques
Equipe Géométrie et Dynamique
Case 7012
2 place Jussieu
75251 Paris Cedex 05
E-mail: Teissier@math.jussieu.fr

Springer-Verlag, Mathematics Editorial, Tiergartenstr. 17,
D-69121 Heidelberg, Germany,
Tel.: *49 (6221) 487-701
Fax: *49 (6221) 487-355
E-mail: lnm@Springer.de